# Radiation, Ionization, and Detection in Nuclear Medicine

Tapan K. Gupta

# Radiation, Ionization, and Detection in Nuclear Medicine

Tapan K. Gupta
Radiation Monitoring Devices Research
Nuclear Medicine
Watertown
Massachusetts, USA

ISBN 978-3-642-34075-8    ISBN 978-3-642-34076-5 (eBook)
DOI 10.1007/978-3-642-34076-5
Springer Heidelberg New York Dordrecht London

Library of Congress Control Number: 2013933011

© Springer-Verlag Berlin Heidelberg 2013
This work is subject to copyright. All rights are reserved by the Publisher, whether the whole or part of
the material is concerned, specifically the rights of translation, reprinting, reuse of illustrations,
recitation, broadcasting, reproduction on microfilms or in any other physical way, and transmission or
information storage and retrieval, electronic adaptation, computer software, or by similar or dissimilar
methodology now known or hereafter developed. Exempted from this legal reservation are brief excerpts
in connection with reviews or scholarly analysis or material supplied specifically for the purpose of being
entered and executed on a computer system, for exclusive use by the purchaser of the work. Duplication
of this publication or parts thereof is permitted only under the provisions of the Copyright Law of the
Publisher's location, in its current version, and permission for use must always be obtained from
Springer. Permissions for use may be obtained through RightsLink at the Copyright Clearance Center.
Violations are liable to prosecution under the respective Copyright Law.
The use of general descriptive names, registered names, trademarks, service marks, etc. in this
publication does not imply, even in the absence of a specific statement, that such names are exempt
from the relevant protective laws and regulations and therefore free for general use.
While the advice and information in this book are believed to be true and accurate at the date of
publication, neither the authors nor the editors nor the publisher can accept any legal responsibility for
any errors or omissions that may be made. The publisher makes no warranty, express or implied, with
respect to the material contained herein.

Printed on acid-free paper

Springer is part of Springer Science+Business Media (www.springer.com)

*The book is dedicated to the memory of my father Dr. Gopeswar Gupta and my mother Dr. Kanaklata Gupta.*

# Preface

Cancer is the second leading cause of death in the United States, surpassed only by breast disease. The good news about the disease is that the recent technological developments, especially, in the field of detection technology had brought much hope to combat the disease more effectively and efficiently. As for example, the most recent development of dual modality PET (positron emission tomography)/CT (computed tomography) scanners prompts the comparative study of *numerous scintillators* to select the best one, which could be used simultaneously in PET detectors working in the pulse mode and in the CT detectors working in the current mode. These *numerous scintillators* (*radiation detectors*) have been fabricated from numerous materials (*organic and inorganic*) like anthracene, stilbene, BGO (bismuth germanate, $B_4Ge_3O_{12}$), GSO (gadolinium sulfoxylate, $Gd_2O_2S$), LGSO (lanthanum gadolinium sulfoxylate), LSO (lutetium oxyorthosilicate, $Lu_2SiO_5$), LYSO (lutetium yttrium oxyorthosilicate ($Lu_{1-y} Y_y)_{2(1-x)} SiO_5$), LaCl$_3$ (lanthanum chloride), La Br$_3$ (lanthanum bromide), NaI: Tl (thallium-doped sodium iodide), CsI: Tl (thallium-doped cesium iodide), PbI$_2$ (lead iodide), HgI$_2$ (mercuric iodide), and CWO (cadmium tungstate, $CdWO_4$) and many other scintillators have been explored in diagnosing early stage of cancer. There are also some other scintillating devices made out of new scintillating materials, and these devices are in the process of developments.

It has been found that due to the presence of the *cancerous cell*, the body of the tumor might be grossly necrotic, hemorrhagic, or otherwise. Thanks to research and development of the recent detectors and instruments that have enabled the surgeons to identify the cancerous cells easily and to remove or destroy the affected cells by surgery or by radiation therapy. However, the margins of the tumor and fingers of the tumor extending into normal tissue are very difficult to identify. In many cases, numerous tissues are biopsied to check if the cancerous cells are present in the tumor. The process is painstaking and typically undersamples the tissue bed, and the procedure is extraordinarily time-consuming.

Therefore, to ease the painstaking method of surgery and to pinpoint the location of the tumor and its extension to the healthy cells, the development of the intraoperative imaging probes has been a top priority. The problem is pervasive throughout oncology, and thus the development of a radiation detector and the design of a modular system that will accept probes with specific designs optimized for various regional tumor extensions are sensible. It is hoped that with these radiation-detecting materials, devices could be made that would be able to provide a better image to pinpoint the position of the cancerous cell.

PET/CT dual modality imaging provides the physicians a powerful tool for improved diagnostics of cancer at an early stage by combining the functional information of PET with anatomical information of X-ray CT. The state of the art of PET/CT scanners combines a CT and PET as two separate units but sharing the same patient bed. It allows the CT and PET scanning to be performed without a patient repositioning but sequentially with axial bed movement.

According to the *experts from National Institutes of Health*, the continuing advances in the development of the sophisticated detection materials and technology and the therapies like monoclonal antibodies and tumor vaccines, cancer detection and treatment might consume nearly 200 billion dollars by the year 2015. Therefore, it will not be unrealistic to think that as cost soars and reimbursement wanes, many patients are unlikely to reap the benefits of some of the most promising results of *nuclear radiation* and *associated instruments* that are being used as diagnostic and therapeutic tools for cancer treatment.

Modern 3-D neuroimaging techniques provide an estimate of the gross 3-D localization of the tumor; however, both MRI and X-ray CT have difficulty in differentiating tumor from the surrounding edema due to altered blood–brain barrier. Even if it were possible to identify all of the tumors and set up a stereotactical surgery system, the changes in the geometry of the brain that often occur intraoperatively due to edema, ventricular drainage, and maneuvers to dehydrate the brain (use of hyperventilation and osmotic agents) can invalidate the preoperative measurements. Even intraoperative ultrasound will frequently miss infiltrative cells at the margin of the tumor. Therefore, one can say that though these procedures have importance in the preoperative staging of the tumor, the more sensitive and more effective detection devices are not less important. Moreover, there are no imaging tools available to the surgeons for inoperative localization of the subclinical tumoral mass remaining after the surgeon extricated the obvious tumor tissue.

The area of X-ray and gamma ($\gamma$)-ray imaging has diverse applications including medical diagnostic imaging systems, such as those used in radiography, X-ray computed tomography (CT), single photon emission computed tomography (SPECT), and positron emission tomography (PET). In addition to their uses in nuclear medicine, X-ray and $\gamma$-ray imaging systems are also used in high-energy physics, solar flare imaging, X-ray sky surveys, and surveys of various galactic regions from satellite-borne instruments.

*Nuclear radiation* associated with nuclear energy is referred to as *ionizing radiation* that arises from hundreds of different kinds of unstable atoms, and many of them exist in nature. Occupational *exposures* to *nuclear radiation* come from industries including medical, educational, and research establishments. *Ionizing radiation* has many uses. As for example, an X-ray is used in medicine to *kill cancerous cells*, and radiography is performed by means of gamma ($\gamma$) rays. In biology, radiation is mainly used for sterilization and enhancing mutations. It is also used in sterilizing medical hardware or food.

# Preface

In Chap. 1 some of the aspects of nuclear radiation, the dose and the energy of the incoming radiation ($\alpha$-,$\beta$-,$\gamma$-, and X-rays), and their origins have been discussed. In addition to that the shielding materials to protect the unwanted radiation and the attenuation coefficient which is measure of the efficiency of the shielding materials have also been discussed.

*Nuclear energy* is a promising energy source for mankind, and it is argued that it is the safest, economical, and clean energy source. It gives certainty that the *nuclear energy* will never be halted; however, over the decades, public perception of the risk with nuclear energy can be classified as (a) environmental and physical aspects, (b) psychological aspects, and (c) sociopolitical implications. Chapter 2 has been devoted to the effects of ionization radiation on consumer products together with its biological effects (deterministic and probabilistic).

*Mathematical modeling* has been used routinely in the design and analysis of semiconductor radiation detectors because it saves development timing in the initial stages and saves the manufacturing cost as a whole. Moreover, one can easily change parameters in a computer simulation, such as *trapping and de-trapping times* ($\tau_t$, $\tau_D$), *electric field strengths* ($\varepsilon$), *electron hole mobility* ($\mu_n$, *and* $\mu_h$), and *electrode designs*, and ultimately can minimize *polarization* and overall the detection efficiency ($\eta$) of the detector without spending a lot of time in the laboratory. Thus, in the model these parameters can be changed partially or fully until the model spectrum matches the real spectrum. Chapter 3 has been devoted to the theory and mathematical modeling to help and understand the inner meanings of nuclear radiation and its effects and to save development timing in the initial stages and the manufacturing cost as a whole.

Chapter 4 is a subdivision of nuclear medicine which deals with medical imaging—the techniques and processes used to create images of the human body or animals (or parts thereof) for clinical purposes or medical therapy. As a discipline in its widest sense or as a subdiscipline, *medical imaging* is the perturbation of the *cellular molecules* to identify the true identity of the cells or tissues of a living organism.

The detection of ionizing radiation by *scintillation light* produced by certain materials is one of the oldest and useful methods adopted in nuclear medicine even today. Chapter 5 has been totally devoted to different types of radiation detectors, organic and inorganic, their classifications, and their working principles.

*Threat reduction, nuclear nonproliferation*, and *medical imaging* activities require improved radiation detectors. The performance of these detectors can be significantly enhanced if the materials currently entrusted with the detection of radiation placed with optimized materials. In order to gain unique expertise knowledge about the requisite optimization of the radiation-detecting materials, one has to study the *theory and nucleation of the growth* of the materials either in the form of *a crystal, a polycrystal*, or in the form of *a thin* or *a thick film*.

In Chaps. 6 and 7, I have tried to accommodate some of the basic principles and experimental techniques to deposit scintillating materials either in the form of single crystal or in the form of thin and/or thick films.

Silicon (Si) or germanium (Ge) pixel detectors can collect on the order of 5,000–10,000 electrons at an electrode. The spatial resolution in these pixelated devices is governed by pixel granularity, which can vary from a few to about 10 μm with pixel dimension of 50–127 μm. For mammography, a pixel dimension of 60 μm or less is recommended. However, for some autoradiography applications, submicron resolution is required, and unfortunately for these applications, proper pixelated devices (TFT arrays) are not yet available in the market.

In order to improve the performance of the pixel detectors, several ideas and developments are being pursued such as capacitive coupling between pixels, MCM-D module technology, and active edge 3-D technology.

For space applications, these devices have some advantages over charge-coupled devices (CCDs) in that pixels are directly addressed so that there are no proton-induced charge transfer losses. For space applications, these hybrid arrays are radiation hardened. However, the pixelated devices suffer from increased dark current, dark current nonuniformity, cross talk, and fixed pattern noise. On the other hand, the drawback with the cryogenic CMOS over room temperature CMOS is dose effect.

In Chap. 8, the optical and electrical behaviors of different scintillators have been discussed elaborately starting from HPGe (high-purity germanium), which is the only radiation detection technology, which can offer sufficient information to accurately and reliably identify radionuclides from their passive γ-ray emission to CZT (cadmium zinc telluride) detectors.

Radiopharmaceuticals introduced inside the patient's body mostly have two components: a *radionuclide*, an excited state of atom, which emits energy so that the atom can convert to a stable form, and a *carrier molecule* that travels through the body until it interacts with its target cell tissue or organ system. No single isotope dominates in radioisotope therapy as $^{99m}$Tc does in nuclear imaging. The design of a successful therapeutic radiopharmaceutical requires: (a) selection of the targeting molecule to deliver the radioisotope to the diseased site properly, (b) accurate calculations of the amount of the dose to destroy the affected cells, and (c) development of a method for destroying the bad cells without adversely affecting the good cells. The widespread availability of the radioisotope therapy depends upon the availability of: the therapeutic doses, appropriate legend chemistry techniques, emitters at reasonable costs, and long-term therapy without complications such as bone marrow toxicity and renal damage.

In Chap. 9, the instruments that are very frequently used for *therapeutic examinations* and medical imaging have been discussed, for example, *linear accelerator (LINC)*, *Gamma Knife*, stereotactic *radio surgery (SRS)*, *angiography*, *computed tomography (CT)*, *positron emission tomography (PET)*, *CT/PET*, *interventional radiology (IR)*, *magnetic resonance imaging (MRI)*, *mammography*, *nuclear medicine (NM)* (*general*, *cardiac*, pediatric, and *PET*), *ultrasound (sonography)*, and *X-ray radiography*.

Winchester, MA, USA                                                                 Tapan K. Gupta

# Acknowledgments

The book *Radiation, Ionization, and Detection in Nuclear Medicine* has been developed from my research work at Radiation Monitoring Devices, Inc., (RMD), Watertown, Massachusetts, since January 1996. The president of the company Dr. Gerald Entine has kindly given me the permission to include some of my research works that I performed at RMD with my different colleagues, like Dr. M. Squillante, Mr. Kanai Shah, Mr. Paul Bennett, and Dr. V. Nagarkar. I express my sincerest gratitude for all the benefits I have gotten from the association of these scientists including Dr. Gerald Entine.

The research in the field of nuclear medicine at RMD is funded by different federal agencies like National Institutes of Health (NIH), National Institute of Standards and Technology (NIST), National Aeronautics and Space Administration (NASA), and National Science Foundation (NSF). During my research at RMD, I have the opportunity to work with different scientists and medical professionals who are well-known figures in their respective fields, like Prof. Daniel Kopans, Professor, Harvard Medical School, Harvard, MA; Prof. Larry Antonuk, Professor, Oncology Department, Michigan University, Michigan; Dr. W. W. Moses, Lawrence Berkeley Laboratory, USA; Prof. C. L. Melcher, University of Tennessee, Knoxville, USA; Dr. S. E. Derenzo, Lawrence Berkeley Laboratory; Prof. Simon R. Cherry, Professor, Bio-Medical Engineering, University of California, Davis, USA; Drs. Larry Partain, George Zentai, and Mike Green, from Varian Medical System, Palo Alto, CA; and Dr. Gerald R. Castelluci, NIST, Gaithersburg, MD, USA.

A book of such diversity could not have been possible to write without direct and indirect help of many researchers, book authors, and publishers. I want to acknowledge my sincerest thanks and gratitude to the authors and the publishers of those journals and books. I want to express my sincerest thanks to the editorial board of the publishing company for their interest and encouragement in the project and my thankful gratitude to the anonymous reviewers for their painstaking reading and valuable time.

Lastly, I express my sincerest thanks to my wife Mrs. Arundhati Gupta and my daughter Ms. Atreyee Gupta for their love and understanding during the tiresome work.

Winchester, MA, USA                                          Tapan K. Gupta

# Contents

**1 Nuclear Radiation, Ionization, and Radioactivity** ............... 1
- 1.1 Introduction ........................................ 2
- 1.2 Ionizing Radiation and Consequences ................. 3
- 1.3 Visual Demonstration of Radiation .................... 4
- 1.4 Definition .......................................... 5
- 1.5 Sources of Nuclear Radiation ........................ 15
- 1.6 Attenuation Coefficient of $\gamma$-Rays ..................... 21
- 1.7 Half-Value Layer .................................... 22
- 1.8 Neutrons ........................................... 23
- 1.9 Neutron Scattering .................................. 25
- 1.10 The Cross-Section Concept .......................... 27
- 1.11 Thermal Neutrons ................................... 27
- 1.12 Neutron Sources .................................... 28
- 1.13 Neutron Shielding .................................. 28
- 1.14 X-Rays ............................................ 29
- 1.15 Interactions of X-Rays .............................. 37
- 1.16 Photoelectric Effect ................................ 40
- 1.17 Pair Production ..................................... 41
- 1.18 Use of Natural Forces for the Material Improvement
  of Mankind ........................................ 43
- 1.19 Radionuclides ...................................... 44
- 1.20 Production of Radionuclides ......................... 44
- 1.21 Developments and Uses of Radionuclides .............. 47
- 1.22 Summary .......................................... 51
- References ............................................. 52

**2 Radiation Exposure: Consequences, Detection,
and Measurements** .................................... 59
- 2.1 Introduction ........................................ 60
- 2.2 Sources of Radiation Exposure ....................... 60
- 2.3 Biological and Related Effects of Radiation ............ 61
- 2.4 Effects of Radiation on Consumable Products .......... 67
- 2.5 Effects of Radiation ................................. 68
- 2.6 Detection of Radiation ............................... 73

|  |  |  |
|---|---|---|
| 2.7 | Neutron Detection | 75 |
| 2.8 | Boron Reaction | 77 |
| 2.9 | Lithium Reaction | 78 |
| 2.10 | Instrumentation | 78 |
| 2.11 | Photomultiplier | 87 |
| 2.12 | Modes of Detector Operation | 90 |
| 2.13 | Recording and Measurement Techniques | 95 |
| 2.14 | Statistical Fluctuations in Nuclear Process | 115 |
| 2.15 | Chi-Square Distribution | 123 |
| 2.16 | Pros and Cons of Radiation Energy | 125 |
| 2.17 | Summary | 126 |
| | References | 126 |

**3 Mathematical Modeling of Radiation** .......................... 135

|  |  |  |
|---|---|---|
| 3.1 | Introduction | 135 |
| 3.2 | Trapping and De-trapping | 137 |
| 3.3 | Polarization | 141 |
| 3.4 | Electrode Design | 143 |
| 3.5 | Frisch Grid Design | 146 |
| 3.6 | Coplanar Design | 150 |
| 3.7 | Pixelated Design | 153 |
| 3.8 | Digital Radiation Detector | 157 |
| 3.9 | Direct Conversion Efficiency | 162 |
| 3.10 | Measurement of Alpha, Beta, and Gamma Radiation | 166 |
| 3.11 | Noise in a Radiation Detector | 168 |
| 3.12 | Noise and Its Effect on Medical Imaging | 175 |
| 3.13 | Dead Time | 177 |
| | References | 180 |

**4 Medical Imaging** ........................................... 187

|  |  |  |
|---|---|---|
| 4.1 | Introduction | 188 |
| 4.2 | Radiation and Carcinogen | 189 |
| 4.3 | Molecular (Medical) Imaging | 191 |
| 4.4 | More Advanced Technology for Medical Imaging | 195 |
| 4.5 | Advanced Tools (Instruments) for Medical (Molecular) Imaging | 197 |
| 4.6 | X-Ray Computed Tomography (CT) | 209 |
| 4.7 | Nuclear Medicine Imaging | 213 |
| 4.8 | Image Acquisition | 215 |
| 4.9 | Imaging Technology | 227 |
| 4.10 | Dependency of the Quality of the Medical Imaging System | 236 |
| 4.11 | Energy Resolution | 242 |
| 4.12 | Digital Image Acquisition System | 244 |
| 4.13 | Summary | 246 |
| | References | 247 |

# Contents

xv

**5 Basic Principles of Radiation Detectors** ..................... 251
5.1 Introduction ...................................... 251
5.2 Working Principle of the Detectors Used in Nuclear Medicine ................................... 253
5.3 Organic Scintillators ............................... 255
5.4 Light Output in an Organic Scintillator .................. 257
5.5 Kinetics of Quenching in Organic Scintillators ........... 259
5.6 Scintillation Efficiency of an Organic Scintillator ......... 261
5.7 Structural and Electronic Properties of Scintillators ........ 262
5.8 Detector Counting Efficiency ($\eta_C$) .................... 265
5.9 Time Resolution of an Inorganic Scintillator ............. 266
5.10 Interaction of Ionizing Radiation with Scintillators ........ 266
5.11 Ionization Losses ................................. 269
5.12 Inorganic Scintillators ............................. 272
5.13 Defect Formation by Ionizing Radiation ................ 277
5.14 Solid-State Detector ............................... 278
References ............................................ 282

**6 Theoretical Approach of Crystal and Film Growths of Materials Used in Medical Imaging System** ....................... 287
6.1 Introduction ...................................... 287
6.2 Theory of Crystal Growth ........................... 289
6.3 Theoretical Modeling of Growing Single Crystal/Polycrystal Used in Radiation Detection and Medical Imaging ......... 293
6.4 Growth of a Crystal on a Seed ....................... 295
6.5 Physical Vapor Transport (PVT) and Bridgman–Stockbarger (BS) Processes ................................... 299
6.6 Traveling Heater Method (THM) ..................... 304
6.7 Metal Solution Growth ............................. 305
6.8 Purification of Crystal ............................. 307
6.9 Thin- and Thick-Film Technology ..................... 308
6.10 Solgel Coating (SGC) .............................. 310
References ............................................ 311

**7 Device Fabrication (Scintillators/Radiation Detectors)** .......... 315
7.1 Introduction ...................................... 316
7.2 Compound Halides ................................. 316
7.3 Halides of Heavy Metals ............................ 321
7.4 Lanthanide Halides ................................ 333
7.5 Complex Oxides with High Atomic Number .............. 342
7.6 Compound Semiconductor ........................... 351
7.7 Cadmium Zinc Telluride (CZT) ....................... 353
7.8 Elemental Semiconductor ........................... 354
References ............................................ 360

**8 Characterization of Radiation Detectors (Scintillators) Used in Nuclear Medicine** ... 367

8.1 Introduction ... 368

8.2 Compound Halides ... 368

8.3 Alkali Halides ... 370

8.4 Halides of Heavy Metals ... 384

8.5 Lanthanide (Ln) Halides ... 404

8.6 Cerium-Activated Lutetium Oxyorthosilicate ($LSO$, $Lu_2SiO_5$) ... 417

8.7 Complex Oxides with High Atomic Number ... 420

8.8 Cadmium Telluride (CdTe) Crystals ... 422

8.9 Cadmium Zinc Telluride (CZT) ... 426

8.10 Elemental Semiconductor ... 429

References ... 442

**9 Instrumentation and Its Applications in Nuclear Medicine** ... 451

9.1 Introduction ... 451

9.2 Administration of the Radionuclides ... 454

9.3 Preparation of Radionuclides ... 455

9.4 Radiation Dose ... 459

9.5 Radiation Therapy ... 460

9.6 Intensity-Modulated Radiation Therapy (IMRT) ... 461

9.7 Scintigraphy ... 464

9.8 Fusion or Hybrid Technology ... 485

9.9 Final Remarks ... 487

References ... 490

**Glossary** ... 495

**Index** ... 501

# Nuclear Radiation, Ionization, and Radioactivity

# 1

## Contents

| | | |
|---|---|---|
| 1.1 | Introduction | 2 |
| 1.2 | Ionizing Radiation and Consequences | 3 |
| 1.3 | Visual Demonstration of Radiation | 4 |
| 1.4 | Definition | 5 |
| | 1.4.1 Roentgen | 5 |
| | 1.4.2 Fluence | 7 |
| | 1.4.3 Quality Factor | 8 |
| | 1.4.4 Radiation Dose | 8 |
| | 1.4.5 Rem | 9 |
| | 1.4.6 Energy | 9 |
| | 1.4.7 Relationship Between Kerma, Exposure, and Absorbed Dose | 10 |
| | 1.4.8 Radioactivity | 11 |
| | 1.4.9 Radioactive Decay | 11 |
| | 1.4.10 Positron | 13 |
| 1.5 | Sources of Nuclear Radiation | 15 |
| | 1.5.1 Alpha ($\alpha$) Particle | 15 |
| | 1.5.2 Beta ($\beta$) Particles | 17 |
| | 1.5.3 Gamma ($\gamma$) Rays | 20 |
| 1.6 | Attenuation Coefficient of $\gamma$-Rays | 21 |
| 1.7 | Half-Value Layer | 22 |
| 1.8 | Neutrons | 23 |
| 1.9 | Neutron Scattering | 25 |
| | 1.9.1 Elastic Scattering | 25 |
| | 1.9.2 Energy Distribution of Elastically Scattered Neutrons | 25 |
| | 1.9.3 Inelastic Scattering | 26 |
| 1.10 | The Cross-Section Concept | 27 |
| 1.11 | Thermal Neutrons | 27 |
| 1.12 | Neutron Sources | 28 |
| 1.13 | Neutron Shielding | 28 |
| 1.14 | X-Rays | 29 |
| | 1.14.1 Production and Properties of X-Ray | 31 |
| | 1.14.2 Beam Quality | 34 |
| 1.15 | Interactions of X-Rays | 37 |
| 1.16 | Photoelectric Effect | 40 |
| 1.17 | Pair Production | 41 |

T.K. Gupta, *Radiation, Ionization, and Detection in Nuclear Medicine*,
DOI 10.1007/978-3-642-34076-5_1, © Springer-Verlag Berlin Heidelberg 2013

| 1.18 | Use of Natural Forces for the Material Improvement of Mankind | 43 |
| 1.19 | Radionuclides | 44 |
| 1.20 | Production of Radionuclides | 44 |
| | 1.20.1 Cyclotron | 45 |
| | 1.20.2 Reactor | 46 |
| 1.21 | Developments and Uses of Radionuclides | 47 |
| 1.22 | Summary | 51 |
| References | | 52 |

## 1.1 Introduction

The history of *nuclear medicine*—a branch of *medicine* dealing with *radioactive materials* and *chemicals/die*—is rich with the contributions from gifted scientists across different disciplines: physics, chemistry, engineering, and medicine [1–5]. The *multidisciplinary* nature of *nuclear medicine* makes it difficult for medical historians to determine the birth date of *nuclear medicine*. Many historians, however, consider the discovery of artificially produced radioisotopes, used for nuclear radiation therapy, which is a part of *nuclear radiation physics* as the most significant milestone in *nuclear medicine*. Indeed, *nuclear radiation physics* made a dramatic entry into *medicine* with the discovery of *X-rays* and *natural radioactivity* more than a century ago, and the potential for *medical imaging* and *therapy* based on these discoveries was very quickly recognized [6]. *Nuclear radiation* associated with nuclear energy is referred to as *ionizing radiation*. Figure 1.1 shows the schematic of nonionizing radiation and ionizing radiation and their uses in different fields.

All *nuclear radiations* are *ionizing radiations*, but the reverse is not true. As for example, *X-ray* is included among *ionizing radiations* even though it does not originate as a result of the disintegration of the atomic nuclei. On the other hand, during various nuclear processes, electromagnetic radiation is emitted from atomic nucleus [7]. Ionizing radiation, even though very small in energy compared to nonionizing radiation, is very damaging to life. As for example, both *gamma rays* and *X-rays* are ionizing radiation, and they can do great damage to cells in living organisms, disrupting cell function. Microwave, on the other hand, cannot cause ionization and cannot break atoms or molecular bonds typical of the molecules found in living things, but can excite the movements of atoms and molecules which is equivalent to heating the samples. Theoretical calculation shows that 1.5 million joules of nonionizing radiation will be required to kill a man. On the other hand, ionizing radiation like gamma rays and X-rays will be able to kill a man with a dose of only 300 J.

Everything on Earth is exposed to some sort of radiation either natural or man-made. 85.5 % of the total radiation (natural and artificial) is made up of 71 % telluric radiation and about 14.5 % cosmic radiation. Radiation particularly associated with nuclear medicine and the use of nuclear energy, along with X-rays, is ionizing. However, all the radiation sources do not have sufficient energy

## 1.2 Ionizing Radiation and Consequences

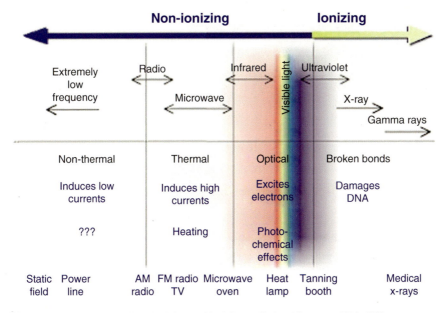

Fig. 1.1 The schematic of nonionizing and ionizing radiation (Courtesy: EPA, US)

Table 1.1 Different radiation sources with respective frequencies and energies

| Radiation | Typical frequency (s$^{-1}$) | Typical energy (kJ/mol) |
|---|---|---|
| Alpha (α) particles |  | $4.1 \times 10^8$ |
| Electromagnetic radiation |  |  |
| Cosmic rays | $6 \times 10^{21}$ s$^{-1}$ | $2.4 \times 10^9$ ionizing radiation |
| Gamma (γ) rays | $3 \times 10^{20}$ s$^{-1}$ | $1.2 \times 10^8$ ionizing radiation |
| X-rays | $3 \times 10^{17}$ s$^{-1}$ | $1.2 \times 10^5$ |
| Ultraviolet (UV) | $3 \times 10^{15}$ s$^{-1}$ | 1,200 nonionizing radiation |
| Visible | $5 \times 10^{14}$ s$^{-1}$ | 200 nonionizing radiation |
| Infrared (IR) | $3 \times 10^{13}$ s$^{-1}$ | 12 nonionizing radiation |
| Microwaves | $3 \times 10^9$ s$^{-1}$ | $1.2 \times 10^{-3}$ nonionizing radiation |
| Radio waves | $3 \times 10^7$ s$^{-1}$ | $1.2 \times 10^{-5}$ nonionizing radiation |

to interact with matter, especially the human body, and produce ions. Table 1.1 shows some of the rays that human beings encountered.

## 1.2 Ionizing Radiation and Consequences

The *ionizing radiation,* which has sufficient energy to interact with matter, especially in the human body is capable of producing ions, that is, it can eject an electron from an atom of the DNA (*deoxyribonucleic acid*) [8–11]. As a result, the bonding

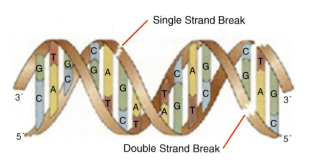

**Fig. 1.2** The damaged molecules of DNA that regulates vital cell process (Courtesy: Canadian Nuclear Association Canada)

properties of the DNA atom change, causing a physical change of the DNA. Figure 1.2 shows the damaged molecule of DNA that regulates vital cell process.

Thus the activity or the *radioactivity* of an ionizing radiation is very damaging to life. Therefore, the biological impact of terrestrial *cosmic rays* has been studied primarily to evaluate the damage to human body in air and space travel [12, 13]. Indeed, the *detection* of the *ionizing radiation* in *nuclear medicine* had been the primary goal of the research community since the discovery of artificial radioactivity in 1934 and the production of radionuclides in 1946.

## 1.3 Visual Demonstration of Radiation

1. *Spark Chamber*: An ideal instrument for visual demonstration of the *electrically charged particles* from ionizing radiation is a *spark chamber*. The device is used mostly in particle physics for detecting electrically charged particles. It consists of metal plates placed in a sealed box filled with helium or neon or the mixture of the two. As charged particle moves through detector, it ionizes the gas between the plates. However, ionization remains invisible until a high voltage is applied between the plates. The high voltage creates a spark along the trajectory taken by the cosmic ray. Figure 1.3 shows a proton–antiproton interaction at 540 GeV. The picture shows particle tracks inside a spark chamber.
2. *Bubble Chamber*: Bubble chamber, invented by Donald Glaser, is a device to demonstrate the interactions of the charged particles received from an invisible radiation source. It is basically a vessel filled with superheated liquid. The trajectories of the energetic particles passing through the superheated liquid can be made visible and photographed [14]. Figure 1.4 shows the schematic of a simple bubble chamber conception.
3. *Cloud Chamber*: Another early invention to demonstrate the *electrically charged particle track* through a supersaturated gas of water or alcohol is the *cloud chamber* or the C.T.R. Wilson chamber. Energetic particles like alpha and

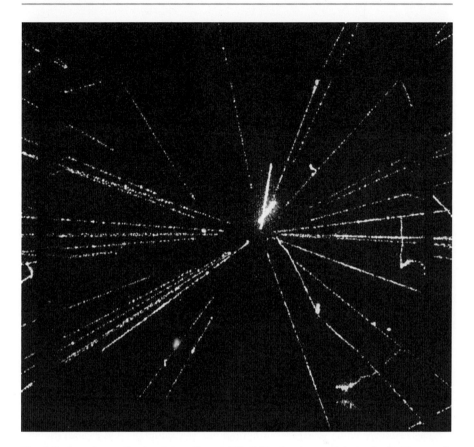

**Fig. 1.3** A proton–antiproton interaction at 540 GeV, showing particle tracks inside a spark chamber (Photo courtesy: Wikimedia Commons)

beta when interact with supersaturated gas to form ions. A small electrostatic gradient is provided to sweep out ions between gas expansions. The apparatus in the modified form makes tracks of visible nuclear bullets in the form of cloud; hence, it is called a *cloud chamber* (shown in Fig. 1.5).

## 1.4 Definition

### 1.4.1 Roentgen

*Roentgen* (R) is defined as the energy dissipated by ionizing radiation in air. It is equivalent to the dissipation of 87.6 ergs per gram of dry air, which results in the generation of one electrostatic unit of charge per 0.001293 g (1 cm$^3$ STP) of air.

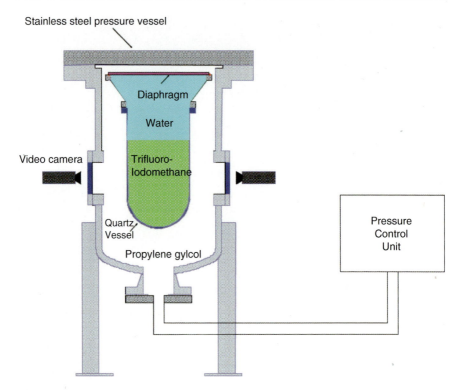

**Fig. 1.4** The schematic of a simple bubble chamber conception (Courtesy: Fermi Lab. Chicago, IL.)

The *exposure* value is expressed in coulomb per kilogram (C/kg), and the unit roentgen (R) is related to exposure as

$$1 \, \text{R} = 2.58 \times 10^{-4} \, \text{C/kg} \tag{1.1}$$

Therefore, we can say that the exposure is the effect of a given flux of gamma ($\gamma$) rays on a test volume of air. It is a function of the intensity of the source, geometry between the source and the test volume, and attenuation of the gamma rays that may take place between the two. The exposure of X-rays and $\gamma$-rays is measured on the assumptions that (a) the source is very small, (b) there is no attenuation in air between the source and the measured point, and (c) there is negligible scattering. When all of these conditions are satisfied, we can write the rate (dQ/dt) as

$$\left(\frac{dQ}{dt}\right) = \Gamma_\delta \left(\frac{\alpha}{d^2}\right) \tag{1.2}$$

where $\Gamma_\delta$ is defined as the *exposure rate constant* for the specific radioisotope of interest, $\alpha$ is the activity of the source and $d$ is the distance in centimeter between

## 1.4 Definition

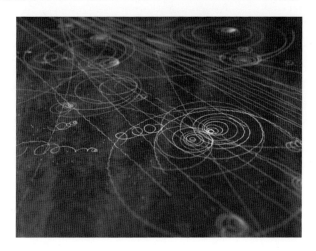

**Fig. 1.5** The cloud-chamber photograph of very fast electron colliding with a stationary one (Courtesy: WIKI)

**Table 1.2** Exposure rate constant for some radioisotope gamma-ray sources

| Nuclide | $\Gamma_0$ |
|---|---|
| Antimony (Sb$^{124}$) | 9.8 |
| Cesium (Cs$^{137}$) | 3.3 |
| Cobalt (Co$^{57}$) | 0.9 |
| Cobalt (Co$^{60}$) | 13.2 |
| Iodine (I$^{125}$) | ~0.7 |
| Iodine (I$^{131}$) | 2.2 |
| Manganese (Mn$^{54}$) | 4.7 |
| Radium (Ra$^{226}$) | 8.25 |
| Sodium (Na$^{22}$) | 12.0 |
| Sodium (Na$^{24}$) | 18.4 |
| Technetium (Te$^{99m}$) | 1.2 |
| Zinc (Zn$^{65}$) | 2.7 |

Source: Ref. [16]

the source and the measuring point, and the subscript $\delta$ applies to the energy of either that of the X-rays or $\gamma$-rays emitted by the source which has sufficient penetration energy [15]. The unit of exposure rate constant is expressed in (R·cm$^2$)/(h·mCi). The value of $\Gamma_\delta$ for some common isotope gamma-ray sources for $\delta = 0$ is given in Table 1.2.

### 1.4.2 Fluence

Next we want to establish a relation between *fluence* and *dose* [17, 18]. The effective dose equivalent ($H_{eq}$) can be written as

$$H_{eq} = h_{eq}\varphi \tag{1.3}$$

where $\varphi = N/4\pi d^2$. $N$ is the number of photons or neutrons emitted by the source, $d$ is the distance from the source, and $h_{eq}$ is the *fluence* to dose equivalent factor. However, the International Commission on Radiation Protection (ICRP) in its report 60 has introduced the definition of equivalent dose as $H_{T,R}$ which relates to the absorbed dose ($D_{T,R}$) averaged over a tissue or organ T due to radiation R multiplied by a radiation weighting factor $W_R$ and can be expressed as

$$H_{T,R} = W_R \times D_{T,R} \tag{1.4}$$

If there is a mixed radiation, then the total equivalent dose $H_T$ is given by

$$H_T = \Sigma_R H_{T,R} = \Sigma_R W_R \times D_{T,R} \tag{1.5}$$

And the effective dose can be expressed in terms of equivalent dose $H_T$ and weighting factor as

$$E_{eff} = \Sigma W_T \times H_T \tag{1.6}$$

### 1.4.3 Quality Factor

*Quality factor* (QF) is defined as the relative damage per each radiation. Mathematically, it can be expressed as

$$QF == \{\text{Sievert (Sv)}\}/\{\text{Gray (Gy)}\} \tag{1.7}$$

The SI unit of *rad* is gray (Gy), which is equivalent to 100 rads, and 1 Sv is equal to 100 rem. In nuclear medicine, 1 rad is equivalent to 100 ergs of energy per absorbed per gram of tissue. Different types of radiation do different amounts of biological damage, and we do need to know the effectiveness of an absorbed dose. The values for the Q-factors recommended by the *International Commission on Radiological Protection* (ICRP) are given in Table II [19].

### 1.4.4 Radiation Dose

It is a measure of how much energy is deposited in a material from a source of radiation. The unit of dose is called rad. However, meaning of dose in nuclear medicine is different from the dose we mean during radiation, because in the former case, it records radiation emitting from within the body rather than radiation that is generated by external source.

1.4 Definition 9

**Table 1.3** For the Q-factors

| Types of radiation | Quality factor | Adsorbed dose equal to a unit dose equivalent |
|---|---|---|
| X-ray or gamma rays | 1 | 1 |
| Beta particles | 1 | |
| Neutrons and protons of unknown energy | 10 | 0.1 |
| Singly charged particles of unknown energy with rest mass greater than 1 amu | 10 | |
| Alpha particle | 20 | 0.05 |
| Particles of multiple or unknown charge of unknown energy | 20 | |

## 1.4.5 Rem

In order to account for the differences in biological damage and risk among different types of radiation, a measurement unit called *rem* is established. *Rem* is defined as the amount of biological damage caused by ionization in human body. One *rem* of alpha ($\alpha$) exposure is qualitatively same as 1 rem of beta ($\beta$) or gamma ($\gamma$) radiation. A *committed dose equivalent* (in *rem*) considers all of the dose equivalents caused by a radionuclide during 50 years of following intake. A committed dose equivalent is defined as a dose, which has occurred and will continue to occur because of radioactive material that has been taken into the body [20].

The other unit of ionizing radiation measurement is called *curie* (Ci), which is a measure of the number of atomic disintegration ($3.7 \times 10^{10}$) in a unit time. In other way, it can be defined as the activity of 1 g of pure radium ($^{226}$Ra). However, the SI unit *becquerel* (Bq) is a more convenient practical unit used in the literature instead of *curie* (1 Bq $= 2.073 \times 10^{-11}$ Ci).

## 1.4.6 Energy

The radiation energy is measured in electron volt (eV). It is defined as the kinetic energy gained by an electron by its acceleration through a potential difference of 1 V. The SI unit of energy is expressed in joule (J).

One joule is equivalent to $6.241 \times 10^{18}$ eV. The energy of radiation ($E$) can be related to wavelength ($\lambda$) of the photon as

$$E = (1.240 \times 10^{-6}/\lambda) \tag{1.8}$$

where $\lambda$ is in meters and $E$ is in eV. Table 1.3 presents the quality factors of different radiation and adsorbed dose.

Table 1.4 presents the SI units of ionizing radiation, their conversion units, and the equivalence.

**Table 1.4** Conversion unit

| SI unit | Conversional unit | Equivalence |
| --- | --- | --- |
| Becquerel (Bq) | Curie (Ci) | 1 Bq $=2.703 \times 10^{-11}$ |
| Gray (Gy) | Rad | 1 Gy $=100$ rad |
| Coulombs per kilogram (C/kg) | Roentgen (R) | 1 C/kg $= 3.9 \times 10^3$ R |
| Sievert (Sv) | Rem | 1 Sv $= 100$ rem |

### 1.4.7 Relationship Between Kerma, Exposure, and Absorbed Dose

*Kerma* in medical science is defined as the kinetic energy which is released in the medium [21]. Or in other words, it is the ratio of the sum of the initial kinetic energy (KE) $dE$ of all the charged ionizing particles (electrons and positrons) to the liberated by uncharged particles (photons) in a material of mass $m$. Mathematically, we can express *kerma* as

$$Kerma\ (k) = (dE/dm) \tag{1.9}$$

The *kerma* $(k)$ is again related to the *fluence* $(\varphi)$ of the photons and is expressed mathematically as

$$K = \varphi\ (\bar{\mu}/\rho) \tag{1.10}$$

where $(\bar{\mu}/\rho)$ is the mass energy transfer coefficient for the medium averaged over the energy *fluence* spectrum of photons.

The *exposure* $(X_\mathrm{r})$ [22] is defined as the ionization equivalent of the collision *kerma* $(K)$ in air, and these two quantities are related as

$$X_\mathrm{r} = (K)_\mathrm{air}(e/\bar{E}) \tag{1.11}$$

where $e$ is the electronic charge $(1.602 \times 10^{-19}$ J) and $\bar{E}$ is the mean energy required to produce an ion pair in dry air. The SI unit of $X_\mathrm{r}$ is coulomb (C) per kilogram (kg) $(X_\mathrm{r} = $ C/kg) and the special unit is roentgen (R), where 1 R $= 2.58 \times 10^{-4}$ (C/kg).

Next, we want to find out a relation between the absorbed dose $D$ and the fluence and hence to *kerma*. The dose $D$ is closely approximated by

$$D = \int_0^{E_\mathrm{n}} \Delta\psi_\mathrm{E}(L/\rho)_\mathrm{col.\Delta}dE \tag{1.12}$$

where $\psi_\mathrm{E}$ is the differential distribution of *fluence* $(\varphi)$ and is equal to $(d\psi_\mathrm{E}/dE)$ and $(L/\rho)_\mathrm{col.\Delta}$ is the restricted mass collision stopping power of a material [23, 24].

## 1.4.8 Radioactivity

*Nuclear radiation* can occur when a *radioactive* substance with unstable nucleus disintegrates. The *nucleus* is supposed to contain *proton A* and *(A–Z) electrons and neutrons*. *Protons* are positively charged, while the *electrons* are negatively charged. However, the nucleus cannot be composed of protons alone, since $A$ is always equal to Z. Therefore, there must be some new particles, which together with protons constitute the nucleus. These particles are called *neutrons,* and they are neutral, having masses equal to $2.5m_e$, where $m_e$ is the mass of an electron. *Neutrons* are heavier than protons and $1.5m_e$ heavier than the sum of the masses of proton and electron.

The discovery of neutrons and the investigation of its interaction with matter led to one of the most significant achievements in nuclear physics. In nuclear medicine, the *ionizing radiation,* which can damage the living tissues, is cleverly applied to *diagnose* (medical imaging) and *damage* (nuclear therapy) the cancerous cells without damaging the healthy tissues [25, 26].

All *nuclei* contain two types of elementary particles or *nucleons,* namely, *protons* and *neutrons*. A *nuclide* is a *nucleus,* which contains specified numbers of *neutrons* and *protons*. Therefore, *nuclides* are composite particles of nucleons. *Nuclides* that have the same number of protons but different numbers of neutrons are called *isotopes*. Certain combinations of neutrons and protons are unstable and break up spontaneously, a process referred to as *radioactive decay*. Such nuclides are called *radionuclides* or *radioisotopes* [27].

## 1.4.9 Radioactive Decay

All unstable nuclei, as well as nuclei in excited states, undergo spontaneous transformations leading to a change in their composition and/or internal energy of the nucleus. Such spontaneous nuclear processes are called radioactive processes since the *radioactive decay* [28] is governed by a fundamental law and is expressed mathematically as

$$(dN/dt) = -(\lambda N) \tag{1.13}$$

where $(dN/dt)$ is differential equation of the number of radioactive nuclei $N$ with time $t$ in second and $\lambda$ is the decay constant. The decay of a given isotope may lead to a daughter product. The average lifetime $\tau$ of a radioactive nucleus is the reciprocal of the disintegration constant $\lambda$ [29].

Another unit called as *specific activity* is also used to define the activity of a radioactive source per unit mass. It is expressed as

$$Specific\ activity = \{(\lambda N)(A_v)/(NM)\} = (\lambda A_v)/(M) \tag{1.14}$$

**Fig. 1.6** The arrangements of protons, neutrons, and electrons (outside the orbit) in a helium atom ($^4$He$_2$)

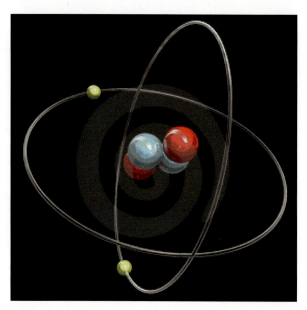

where $M$ is molecular weight of the sample, $A_v$ is Avogadro's number ($=6.02 \times 10^{23}$ nuclei/mol), and $\lambda$ is radioactive decay constant ($=(\ln 2)/(\text{half-life})$).

Figure 1.6 shows the arrangements of protons (positively charged, red), neutrons (neutral, blue), and electrons (negatively charged, outside circle) in a helium atom. Helium atom ($^4$He$_2$) has 4 nucleon numbers (protons and neutrons together) and 2 nuclear charge number (proton number).

*The radioactive nucleus or the radioisotope (or radionuclide, a specific nucleus with given mass number A and electric charge Z, is sometimes called nuclide)* normally decays by spontaneous emission of a particle from the nucleus. The emission may be alpha ($\alpha$) particle, in which case we call it $\alpha$-decay or the emission may emit an electron and we refer the case as $\beta$-decay. All these decays are statistical in character (Fig. 1.7).

In the beginning of the nineteenth century, the *radioisotopes* utilized were naturally occurring ones, such as *radium-226* ($^{226}$Ra), *radium-224* ($^{224}$Ra), *radon-222* ($^{222}$Rn), *polonium-210* ($^{210}$Po), *tritium* (hydrogen-3), and *carbon-14* ($^{14}$C). Joseph Hamilton was the first to use radioactive tracers to study circulatory physiology in the year 1937, and then in 1939 iodine-131 ($^{131}$I) was used as *radioisotope* in nuclear medicine followed by technetium-99m ($^{99m}$Tc). In 1940, the first particle accelerator was built, and a number of radioisotopes were available.

All atomic nuclei can be broadly divided into two groups as *stable* and *unstable* or *radioactive* nuclei. The excess of neutrons in heavy nuclei ($^{208}_{82}$Pb, $^{226}_{88}$Ra, $^{238}_{92}$U) can be explained by Coulomb repulsion of protons. For $\beta$-stable nuclei, the relation between Z and A is described by the empirical formula

## 1.4 Definition

**Fig. 1.7** The schematic of radioactive decay

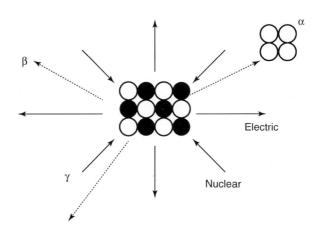

$$Z = \{(A)/(1.98 + 0.015\, A^{2/3})\} \qquad (1.15)$$

where $Z$ is the charge number and $A$ is the atomic number of the element. Nucleus in which this relation is violated is called a *β-radioactive element*. A *β-radioactive element* in course of time can decay a beta particle (an electron, $\beta^-$, or a positron, $\beta^+$). Neutron-rich nuclei, on the other hand, can emit an electron while proton rich nuclei can emit a *positron* [30].

### 1.4.10 Positron

*Positrons* are antiparticles of electrons with identical mass and charge. It can also be identified as positively charged electron ($\beta^+$). An unstable nucleus of some *radioisotopes* can emit positron due to excess number of protons and a positive charge, and then it can stabilize itself through the conversion of a proton to a neutron. Figure 1.8 shows a schematic of a positron-emitting radioisotope.

When a positron comes in contact with an electron, the two particles annihilate and the mass of the two particles turned into two 511-keV gamma rays ($\gamma$-rays), and they are emitted on the two sides of the event (Fig. 1.9). One of the attractive aspects of $e^+$–$e^-$ annihilation is the relative simplicity of the interaction. The two-body system decays into two back-to-back photons, each carrying an energy $m_e c^2$. This feature has provided the basis of medical imaging techniques called *positron emission tomography* (PET). In PET, a positron-emitting radioisotope is introduced usually by injection, which promptly combines with a nearby electron resulting in the simultaneous emission of two identifiable gamma ($\gamma$) rays in opposite directions. These two gamma rays are detected in a PET camera.

According to quantum theory, the annihilation of positrons with electrons can produce high-energy photons—gamma quanta. Annihilation can take place directly or via the formation of *positronium* I, a state in which an electron and a positron

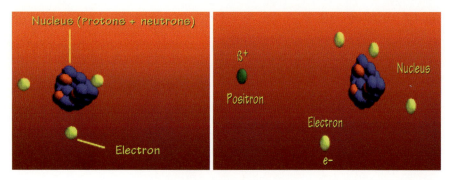

**Fig. 1.8** A schematic of a positron-emitting radioisotope (Photo courtesy: Tutorial UCLA, CA)

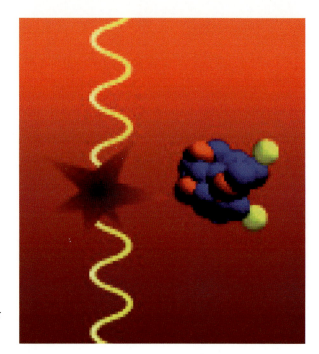

**Fig. 1.9** The schematic of the 511-keV gamma rays emitted at 180° to each other (Photo courtesy Tutorial UCLA, CA)

form a *light hydrogen atom* bound by Coulomb attraction [31, 32]. In condensed matter, the interactions between positronium and the substrate can provide information about the details of the substrate structure via the effects on the momentum distribution at the instant of annihilation [33].

Figure 1.9 shows the conversion of two light energy photons in the form of $\gamma$-rays. The $\gamma$-rays on two opposite sides can easily escape from the human body and can be recorded by external *detector*/s [34]. Compact gamma ($\gamma$)-ray imager based on a silicon drift detector (SDD) coupled to a single scintillator crystal is very much in use in *medical imaging* system [35].

## 1.5  Sources of Nuclear Radiation

In *medical imaging system*, radiation detectors are used to diagnose cancerous cell. The benefit of a *medical imaging* examination in terms of its ability to yield an accurate diagnosis depends upon the quality of both the image acquisition and the image interpretation [36]. The special branch of the science dealing with medical imaging [37] and the diagnosis of the cancerous cells is called *radiology*, and the treatment for curative or adjuvant cancer treatment with radiation is called *radiotherapy* [38]. During the past century, both *radiology* and *radiotherapy* have grown tremendously due to the advances in image detector systems and computer technology [39].

## 1.5  Sources of Nuclear Radiation

*Nuclear radiation* arises from hundreds of different kinds of unstable atoms, and many of them exist in nature. As a matter of fact, in major cases under certain conditions, *nuclear radiation* is observed during nuclear reactions. Indeed, *ionizing radiation*, which damages living tissues, is emitted during the spontaneous *decay* of different kinds of atoms [40]. The principal kinds of ionizing radiation are (1) *alpha* ($\alpha$) particles, (2) *beta* ($\beta$) particles, (3) *gamma* ($\gamma$) rays, and (4) X-*rays*.

### 1.5.1  Alpha ($\alpha$) Particle

Alpha particles consist of two protons and two neutrons bound together into a helium particle identical to a helium nucleus, which is produced in the process of alpha decay. It has a very large mass (more than 700 times the mass of beta particle), charge, a very short range (less than a tenth of a millimeter inside the body), and a great destructive power of damaging fast-growing membrane and living cells [41]. Alpha decay is a radioactive process, which occurs in very heavy elements such as uranium (U), thorium (Th), and radium (Ra). The nuclei of these atoms are very neutron rich that make the emission of the alpha particle possible. When an atom emits an alpha particle during alpha decay, the mass number of the atom is decreased by four due to loss of the four nucleons in the alpha particle. The atomic number of the atom goes down exactly by two, as a result of the loss of two protons, and the atom becomes a new element. In contrast to beta decay, the fundamental interactions responsible for alpha decay are a balance between the electromagnetic force and nuclear force. Alpha decay results from Coulomb repulsion between alpha particle and the rest of the nucleus. Figure 1.10 shows alpha decay with two protons and two neutrons from a parent nucleus.

Heavy nuclei are energetically unstable against spontaneous emission of an alpha ($\alpha$) particle (or $^{4}$He nucleus). The decay process proceeds as

$$^{A}X_{Z} \rightarrow {}^{A-4}Y_{Z-2} + {}^{4}\alpha_{2} \tag{1.16}$$

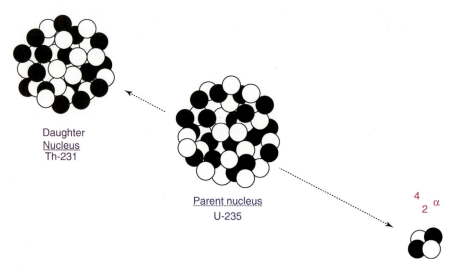

**Fig. 1.10** Radiation of alpha particle

where X and Y are the initial and final nuclear species; A, the atomic number; and Z, the atomic weight. For each distinct transition between initial and final nucleus, a fixed energy difference ($E_b$–$E_e$, where $E_b$ is binding energy of the incident particle and $E_e$ is the binding energy of escaping—particles relative to the compound nucleus) or Q-value characterizes the decay.

It has been found that there is a strong correlation between alpha (α) particle energy and half-life of the parent isotope and the particle with the shortest half-life has the highest energy. The most common calibration source of α-particle is americium-241 ($^{241}$Am). Monoenergetic α-particle sources are prepared in very thin layers, because α-particle loses energy very rapidly in materials.

Alpha particles, like helium nuclei, have a net spin of zero and have a classical total energy of about 5 MeV. These particles are highly ionized and have low penetration depth. Currently alpha (α)-emitting radioisotopes are finding much interest in nuclear medicine for radiotherapy and imaging [42] (Table 1.5). Isotope emitting alpha particle causes more damage to the body cells compared to the other ionizing particles due to their high relative biological effectiveness.

The binding energy of α-particle is variable and can go from zero to as high as 28.3 MeV. This contrasts with the binding energy of helium-3 ($^3$He) which is only 8.0 MeV. Because of its high mass and energy, it interacts with the matter very heavily, and thus it has the shortest range among the three main particles (α, β, and γ) which are of importance in the field of nuclear medicine.

Figure 1.11 shows the penetrating capabilities of different radiation sources. From Fig. 1.11, it is clear that γ-particle has the highest penetrating power. In order to prevent unwanted nuclear radiation, shielding materials are very frequently used [43].

Shielding materials that are frequently used to protect from nuclear radiation are lead (Pb), concrete, and steel [44, 45]. Lead has the highest atomic number (82) and has highest density (11.32 g/cm$^3$) among the shielding materials that are frequently

## 1.5 Sources of Nuclear Radiation

**Table 1.5** Alpha-emitting radioisotopes of current interest

| Radioisotope | Energy of the emitting α-particle (MeV) | Half-life | Production method |
|---|---|---|---|
| Terbium-149 ($^{149}$Tb) | 3.9 | 4.1 h | Accelerator |
| Astatine-211 ($^{211}$At) | 6.8 | 7.2 h | Accelerator |
| Bismuth-212 ($^{212}$Bi) | 7.8 | 60.5 min | Decay of lead-212 ($^{212}$Pb) |
| Bismuth-213 ($^{213}$Bi) | 5.8 | 45 min | Decay of actimium-225 ($^{225}$Ac) |

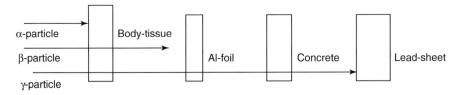

**Fig. 1.11** The penetrating capabilities of different radiation sources

used. It is mostly used in the form of rectangular blocks in the construction of gamma-ray shielding. Solid-shaped lead cast can also be used as shielding material. However, these solid cast materials sometimes contain voids and are not suitable for shielding.

Ordinary lead normally contains significant amount of natural activity due to low-level contaminants. Therefore, lead is specially refined to use as shielding material. Even when it is refined, it shows detectable radioactivity due to impurities from natural fallout. Sometimes lead mixed with plastic or epoxy compositions is used as shielding material. The photoelectric absorption cross section predominates up to gamma-ray energies as high as 0.5 MeV.

Iron or steel shielding is also used when lead shielding alone becomes expensive. In such circumstances, lead with a steel outer lining is used. However, the density and atomic number of iron is lower than lead (Pb). As a result, to make steel or iron shielding as effective as lead, the thickness of the steel shielding is made several tens of centimeters thicker than lead.

Concrete is also used as a shielding material. Because of its low cost, outer lining of concrete with inner lining of lead or steel is used. A special formulation of the concrete known as *barytes concrete* is preferably used as the material is much more effective in gamma-ray shielding. Table 1.6 shows the absorption coefficients of the radiation sources used in nuclear medicine. The absorption coefficients of the radiated sources are calculated on the basis of 511-keV photons [46].

### 1.5.2 Beta (β) Particles

β-particles are just high-energy electrons ejected from the nucleus. Because of its higher energy than the energy of alpha (α) particle, it penetrates more inside the body tissue compared to the equal mass of α-particle. Beta (β) emission is accompanied by the emission of electron antineutrino which shares the momentum

**Table 1.6** Absorption coefficients of the radiation sources used in nuclear medicine

| Shielding material | $\alpha$ (cm$^{-1}$) | $\beta$ (cm$^{-1}$) | $\gamma$ (cm$^{-1}$) |
|---|---|---|---|
| Lead | 1.7772 | −0.5228 | 0.5457 |
| Concrete | 0.1539 | −0.1161 | 2.0752 |
| Steel | 0.5704 | −0.3063 | 0.6326 |

and energy of the decay. The emission of the electron's antiparticle, the positron, is also called beta decay. The $\beta^-$-decay can be represented by

$$^{A}X_Z \rightarrow {}^{A}Y_{Z+1} + \beta^- + v^- \tag{1.17}$$

where X is the mother nucleus, Y is the final nucleus, and $v^-$ is the antineutrino [47, 48]. The neutrinos and antineutrinos are extremely small, and their interaction probability with matter is undetectable. The recoil nucleus Y passes almost zero energy, and the total energy is shared between the beta ($\beta$) particle and the invisible *neutrino* with no mass and charge. The $\beta$-particle, on the other hand, changes its energy from decay to decay, which can range from zero to the end-point energy. A fixed decay energy or Q-value characterizes each specific $\beta$-decay.

$\beta$-decay is relatively a slow process and may lead to population of the excited state in the daughter nucleus having a shorter average lifetime. The excited daughter nucleus can come down to the ground state (non-excited state) through the emission of gamma-ray ($\gamma$-ray) photon whose energy is essentially equal to the difference in energy between the initial and final nuclear states. The fact that $\beta$-decay involves emission of electrons, that is, particles constituting the atom, from the nucleus stems has been established from the following considerations:

(a) The mass and charge of $\beta^-$-particles are identical to those of atomic electrons.
(b) $\beta^+$-particles are annihilated by the atomic electrons.
(c) Atomic electrons are captured by the nuclei.
(d) Pauli's exclusion principle is valid for mixed ensemble of particles contains atomic electrons and $\beta^-$-particles.

Figure 1.12 shows the decay of cobalt-60 ($^{60}Co_{27}$) to nickel-60 ($^{60}Ni_{28}$). Another example of gamma ($\gamma$)-ray radiation is the formation of neptunium-237 ($^{237}Np$) from americium-241 ($^{241}Am$) during $\alpha$-decay. In some cases, the $\gamma$-emission spectrum is simple, for example, $^{60}Co/^{60}Ni$. But in $^{241}Am/^{237}Np$ and $^{192}Ir/^{192}Pt$, the $\gamma$-emission spectra are complex, and they reveal the existence of a series of nuclear energy levels. The fact that an $\alpha$-spectrum can have a series of different peaks with different energies reinforces the idea that several nuclear energy levels are possible. But it is completely different when $\gamma$-ray is accompanied by $\beta$-particle emission [49, 50]. It has been found that the $\beta$-spectrum never shows a sharp peak, because a $\beta$-decay is accompanied by the emission of a *neutrino*, which carries away the energy (Fig. 1.13) [51, 52]. The transmission curve for $\beta$-particles emitted by a radioisotope differs significantly because of continuous distribution of their energies.

The absorption coefficient ($\alpha$) of a specific absorber is defined as

$$(I/I_0) = \exp(-\alpha t)$$

## 1.5 Sources of Nuclear Radiation

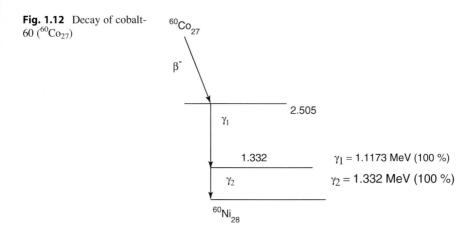

**Fig. 1.12** Decay of cobalt-60 ($^{60}\text{Co}_{27}$)

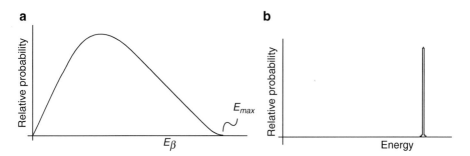

**Fig. 1.13** (a) The peak of an alpha spectrum and (b) beta spectrum without a sharp peak

or

$$(-\alpha) = (1/t) \log (I/I_0) \quad (1.18)$$

where $I$ is the initial counting rate, $I_0$ the final counting rates with absorber, and $t$ the thickness of the absorber in g/cm². It has been found that the value of $\alpha$ correlates very well with the end-point energy of the β-emitter for a specific absorber. The energy deposited within the absorber can be represented mathematically by

$$\Delta E = \{(-dE/dx)_{\text{avg}}\} t \quad (1.19)$$

where $(-dE/dx)$ is the linear stopping power averaged over the energy of the particle and t is the thickness of the absorber. If the energy loss within the absorber is small, the stopping power does not change much; otherwise, it is difficult to obtain a properly weighted $(-dE/dx)_{\text{avg}}$ value directly from the data. On the other hand, the stopping time ($t_{\text{stop}}$) required to stop the charged particles within an absorber can be written as

$$t_{stop} = (R/<v>) = (R/Kc)\,(mc^2/2E)^{1/2} \tag{1.20}$$

where $m$ is the mass of the particle, $v$ is the initial velocity of the particle, $<v>$ is the average velocity, $c$ is the velocity of light, $E$ is the kinetic energy in MeV, $R$ is the range is in meters, and $K$ is a constant and equal to $(<v>/v)$. The linear stopping power $(S)$ of a charged particle in a given absorber is defined as the differential energy loss $(-dE)$ of the charged particle within a specific differential length $(dx)$. Mathematically, the expression can be represented as

$$S = (-dE/dx) = \left[\{(4\pi\,e^4 Z^2)/(m_0 v^2)\}\,NB\right] \tag{1.21}$$

where $Z$ is the atomic number, $N$ the density of the absorber, $e$ the electronic charge, $m_0$ the rest mass, and $B = Z\{\ln(2m_0 v^2/I) - \ln(1 - v^2/c^2) - (v^2/c^2)\}$, where $I$ is the average excitation and ionization potential of the absorber. The expression for $B$ shows that it varies slowly with particle energy.

### 1.5.3 Gamma (γ) Rays

*Gamma (γ) ray is a short wavelength electromagnetic radiation* of nuclear origin, having energy varying from 10 keV to 5 MeV. It is the spontaneous emission of gamma *(γ) quanta* by the nucleus and was discovered by a French chemist and physicist, Paul Ulrich Villard, in 1900. The γ-quanta emitted by the nucleus during transition to a lower energy state may carry different angular momenta.

The γ-rays are the highest range of energetic form of electromagnetic radiation and are more energetic than alpha (α) and beta (β) radiation but less than the energy possessed by an ionizing radiation. They damage the body tissues and skin, similar to that caused by X-rays, such as *burns, cancer,* and *genetic mutation.* Gamma rays are produced by the nuclear transitions, while X-rays are produced by energy transitions due to accelerating electrons. Both of these rays are used in nuclear medicine for diagnosis and therapy of the abnormal cells.

The major interactions of *γ rays* with matter that play an important role in the measurements of radiation are *photoelectric effect, Compton effect,* and *pair production. Photoelectric effect* is due to the low-energy interactions of γ-rays with the absorber material of high atomic number $(Z)$. *The photoelectric effect* is the dominant energy transfer mechanism for X-rays and γ-rays with photon energies below 50 keV.

The Compton effect, on the other hand, is the principal absorption mechanism for γ-rays in the intermediate energy ranges between 100 keV and 10 MeV. The interaction process of Compton effect is due to the interaction between the incident gamma-ray (γ) photon and the electron of the absorbing material. Novel X-ray imaging method relying on Compton effect depends on a tracking detector like germanium (Ge).

The method employs a two-dimensional segmentation, pulse shape analysis, and tracking of Compton scattered γ-rays. The radiological imaging method is applied in single-photon emission computed tomography (SPECT) and positron emission

## 1.6 Attenuation Coefficient of γ-Rays

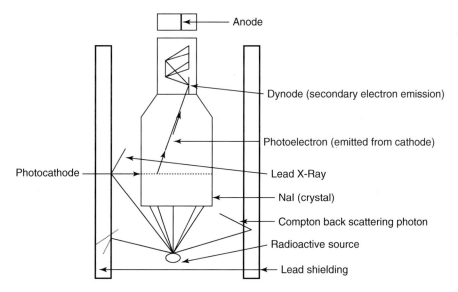

**Fig. 1.14** The interactions that occur when powerful gamma rays from a radioactive source are being focused on a sodium iodide scintillation detector

tomography (PET) processes. Another area of application is security inspection [53, 54]. It is the most predominant interaction mechanism for γ-ray energies typical of radioisotope sources.

The third interaction—the *pair production*—is energetically possible when the energy of the gamma-ray photon goes in excess of the equivalent rest mass of an electron (1.02 MeV).

Figure 1.14 shows the interactions that occur when powerful gamma rays from a radioactive source are being focused on a sodium iodide scintillation detector. The powerful nature of gamma (γ) rays is useful for sterilization of medical equipment (to kill bacteria), to treat some kind of cancers, in Gamma Knife surgery, for diagnostic purpose in nuclear medicine.

### 1.6 Attenuation Coefficient of γ-Rays

When a monoenergetic γ-ray passes through an absorber, a portion of the energy is transferred to the absorber, and thus the energy of γ-ray photons is attenuated [55, 56]. The amount of attenuation can be expressed mathematically as

$$(I/I_0) = \exp(-\mu t) = \exp(-\mu/\rho)\, \rho t \tag{1.22}$$

where $I$ is the number of transmitted photons, $I_0$ the number without an absorber, $(-\mu)$ the linear attenuation coefficient, $t$ the thickness of the absorber, and $\rho t$ the mass thickness of the absorber. The mean free path $\lambda$ of the γ-ray photon is defined as

$$\lambda = \left\{ \left( \int_0^\infty x \exp\left(-\mu x\right) dx \right) \Big/ \left\{ \int_0^\infty \exp\left(-\mu x\right) dx \right\} \right. = (1/\mu) \tag{1.23}$$

The mean free path $(1/\mu)$ of a material increases as photon energy increases [57]. It has been observed that computed tomography (CT) with higher-energy gamma ray ($\gamma$-ray) can produce a good contrast for a large-scale system [58].

## 1.7 Half-Value Layer

The *half-value layer* (HVL) is defined as the thickness of an absorber of specified composition required to attenuate the intensity of the beam to half to the original value [59]. In *medical science*, the quality of an X-ray beam is quantified in terms of *half-value layer* (HVL) and *the peak voltage* (keV) [60, 61].

Although all beams can be described in terms of HLV, the quality of gamma ($\gamma$)-ray beam is usually stated in terms of the energy of the $\gamma$-rays or its nuclide of origin which has a known emission spectrum. In case of low-energy X-ray beam (below megavoltage), it is customary to describe the quality in terms of HVL together with kVp, although HVL alone is adequate to for most clinical applications [62, 63]. On the other hand, in megavoltage X-ray range, the quality is specified by the peak energy and rarely by HVL, because the beam is heavily filtered through the transmission target and the flattening filter.

Combination of filters containing plates of aluminum (Al), copper (Cu), and tin (Sn) has been designed to increase the resulting *HVL* of *orthovoltage* beams without reducing the intensity of the beam. Such filters are called Thoraeus (Th) filters and are designated as Th I (0.2-mm Sn, 0.2-mm Cu, and 1-mm Al), Th II (0.4-mm Sn, 0.2-mm Cu, and 1-mm Al), and Th III (0.5-mm Sn, 0.2-mm Cu, and 1-mm Al).

In diagnostic and superficial X-ray energy range, primarily aluminum (Al) filters are used to harden the beam. In *orthovoltage* range, combination of filters is used to obtain HVL layers in the range of about 1- to 4-mm Cu. For cesium (Ce) and cobalt (Co) therapy, filters are not needed because of the monochromatic nature of the beam [64].

*HVL* is inversely proportional to linear attenuation coefficient ($\mu$) (HVL $= 0.693/\mu$), and it increases with increasing filter thickness as beam becomes increasingly harder, that is, contains a greater proportion of higher energy of photons.

Now we can write the linear attenuation coefficient $\mu$ in terms of the ratio of the number of photons $I$ to the number of photons without an absorber $I_0$ as

$$(I/I_0) = e^{-\mu x} \tag{1.24}$$

The above Eq. (1.24) can be rearranged as

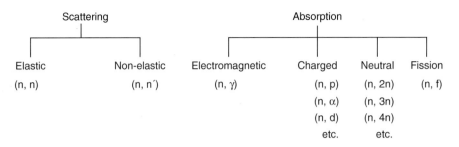

**Fig. 1.15** The interaction of neutron with matter and the consequences

$$(I/I_0) = \exp(-\mu\rho)(\rho x) \tag{1.25}$$

where $\rho$ is the density of the medium and the product $(\rho x)$ is known as the mass thickness of the absorber, which is an important parameter that determines the degree of attenuation of the absorber. The thickness of the absorber is measured in terms of mass thickness rather than physical thickness $(x)$, and it is more appropriate in radiation measurements. The mass thickness is also a useful concept when discussing the energy loss of charged particles and fast electrons.

## 1.8  Neutrons

*Neutron* is a *neutral* particle $(Z = 0)$ with a spin equal to half and a negative magnetic moment. Since neutrons have no charge, it can easily enter into a nucleus and cause reaction. Neutrons interact primarily with the nucleus of an atom, except in special case of magnetic scattering where the interaction involves the neutron spin and the atomic magnetic moment. These uncharged particles are detected indirectly via scattering against a charged particle or via nuclear reactions creating charged particles or gamma ($\gamma$) rays. Neutron *scattering* is therefore an important part of neutron detection [65].

The interaction of a neutron may be one of the two major types: (1) scattering and (2) absorption (Fig. 1.15). When a neutron is scattered by a nucleus, its speed and direction change but the nucleus is left with the same number of protons and neutrons it had before the interaction.

After the interaction with a matter, the nucleus will have some recoil velocity, and it may be left in an excited state that will lead to the eventual release of radiation. On the other hand, when a neutron is absorbed by a nucleus, a wide range of radiation can be emitted or fission can be induced.

In the Fig. 1.16, $M_1$ is the mass of the neutron with energy $E_1$ which collides with a target nucleus having mass $M_2$ and energy $(E_2 = 0)$ at room temperature, resulting in the emission of the outgoing particle (mass $M_3$ and energy $E_3$) and the residual recoiling nucleus (mass $= M_4$ and energy $E_4$). Using the conservation

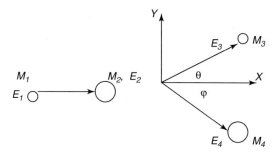

**Fig. 1.16** The collision between incident particle 1 and target particle, which is at rest, leading to the emission of third particle at an angle $\theta$ and a fourth recoiling particle

of total energy and linear momentum, and nonrelativistic kinematics, we can set up a mathematical equation as follows:

$$(E_1 + M_1 c^2) + M_2 c^2 = (E_3 + M_3 c^2) + (E_4 + M_4 c^2) \tag{1.26}$$

Now the momentum $p_1 = p_3 + p_4$ where $p_4^2 = (p_1 - p_3)^2 = p_{12} + p_{32} - 2p_1 p_3$ and $\cos\theta = 2M_4 E_4$, and we can recall the reaction equation $Q$ as [66]

$$Q = (M_1 + M_2 - M_3 - M_4)\, c^2 = E_3 + E_4 - E_1 \tag{1.27}$$

Finally we can write

$$Q = E_3\{1 + (M_3/M_4)\} - E_1\{1 - (M_1/M_4)\} \\ - \{2/M_4 (M_1 M_3 E_1 E_3)^{1/2}\cos\theta \tag{1.28}$$

The above equation is independent of the mechanism of reaction. The kinetic energies $E_1$ and $E_3$ of the product particle and the angle $\theta$ can be measured in the laboratory coordinates.

In the case of *elastic scattering*, there is no excitation of the nucleus and we can assume $Q = 0$, and we further assume that whatever energy lost by the neutron is gained by the recoiling target nucleus. Let us assume $M_1 = M_3 = m\ (M_n)$ and $M_2 = M_4 = M = Am$. Then the above $Q$ equation becomes

$$E_3\{1 + (1/A)\} - E_1\{1 - 1/A)\} - \{(2/A)\,(E_1 E_3)^{1/2}\cos\theta \tag{1.29}$$

Now for all finite $\theta$, $E_3$ has to be less than $E_1$. It is also true that the maximum energy loss by the neutron occurs at $\theta = \pi$, which corresponds to backscattering,

$$E_3 = \alpha E_1, \tag{1.30}$$

where $\alpha = \{(A - 1)/(A + 1)\}$

## 1.9 Neutron Scattering

### 1.9.1 Elastic Scattering

*Scattering* events in neutron can be subdivided into *elastic and inelastic scattering*. In *elastic* scattering [67, 68], the total kinetic energy of the neutron is unchanged by the interaction. During the interaction, a fraction of the neutron's kinetic energy is transferred to the nucleus. For elastic scattering, we can establish a relation between outgoing energy $E_3$ and the angle of scattering, $\theta$, of neutron as

$$(E_3)^{1/2} = \{1/(A+1)\}\,(E_1)^{1/2}[\cos\theta + \{A_3 - \sin^2\theta\}^{1/2}] \qquad (1.31)$$

### 1.9.2 Energy Distribution of Elastically Scattered Neutrons

From Fig. 1.16, we can see that the neutron when it strikes a target can be scattered through an angle $\theta$ and recoils through an angle $\varphi$. The probability that the scattered neutron $P(\Omega_s)$ will be distributed in the two angular variables $\varphi$ and $\theta$ when normalized can be represented by

$$\int_0^{2\pi} d\varphi_s \int_0^{\pi} \cos\theta_s \, d\theta_s \, P(\Omega_s) = 1 \qquad (1.32)$$

where $(\Omega_s)$ is a unit vector in angular space. Since there is a one-to-one relation between $\theta_s$ and $E_3$, we can transform $P(\Omega_s)\, d\Omega_s$ to obtain a probability distribution in the outgoing energy, $E_3$. Now let us define the probability of the scattering angle $\theta_s$ as $G\,(\theta_s)$, and by simply integrating $P(\Omega_s)\, d\Omega_s$, we get

$$G(\theta_s) = \int_0^{2\pi} d\varphi \, \sin\theta_s \, d\theta_s P(\Omega_s) = \frac{1}{2}\sin\theta_s \, d\theta_s \qquad (1.33)$$

Now if we set the energy before collision $E_1 = E$ and $E_3 = E'$, then the probability distribution for energy $E'$ which is the energy after collision, as $F\,(E \to E')$, we can write the previous equation (1.33) as

$$F(E \to E') = G\,(\theta_s)|(d\theta_s/dE)| \qquad (1.34)$$

Now we have $E_3 = \frac{1}{2} E_1\,[(1+\alpha) + (1-\alpha)\,\theta_s$, and we have assumed $E_1 = E$ and $E_3 = E'$, so the Jacobian transformation gives rise to

$$F(E \to E') = \{1/[E(1-\alpha)]\} \qquad \alpha E \leqslant E' \leqslant E \qquad (1.35)$$

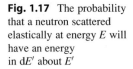
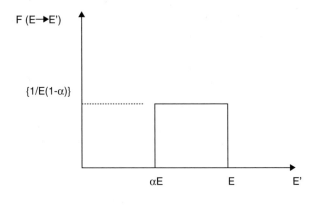

**Fig. 1.17** The probability that a neutron scattered elastically at energy $E$ will have an energy in $dE'$ about $E'$

$$= 0 \text{ otherwise} \tag{1.36}$$

The distribution is sketched in Fig. 1.17. Now since $F$ is a distribution, its dimension will be the reciprocal of its argument—an energy. Knowing the probability distribution of $F$, we can construct the energy differential cross section

$$(d\sigma_s/dE') = \sigma_s(E) F(E \to E') \tag{1.37}$$

such that

$$\int dE' \, (d\sigma_s/dE') = \sigma_T \tag{1.38}$$

where $\sigma_T$ is the total scattering cross section.

### 1.9.3 Inelastic Scattering

It is similar to elastic scattering except that the incoming neutron excites the target nucleus so it leaves the ground state and goes to the excited state at an energy $E^*$ above the ground state. The total kinetic energy of the outgoing neutron and the nucleus is then the kinetic energy of the incoming neutron. Now part of the original kinetic energy is used to place the nucleus into the excited state, and we can write the $Q$ equation as: $Q = -E^*$ ($E^* > 0$). Let us assume that the neutron mass is $m$, the target nucleus mass is $M$ (ground state), and $M^*$ is its mass at excited state, where $M^* = \{M + (E^*/c^2)\}$. The negative sign in $Q$ equation indicates that it is an endothermic process requiring energy to be supplied before the reaction can take place. On the other hand, in elastic scattering, the $Q$ equation can give

$$-E^* = -E_{th}\{(M_4 - M_1)/(M_4)\} \tag{1.39}$$

1.11 Thermal Neutrons

or,

$$E_{th} \sim E^*(\{1 + (1/A)\}) \qquad (1.40)$$

where $E_{th}$ is the minimum value of $E_1$.

## 1.10 The Cross-Section Concept

When a large number of neutrons are emerging in a thin layer of interacting material, some may pass through with no interaction; some may be absorbed and may have some interactions. The interactions of the neutrons with the material change their directions and energies. The *cross section* for the neutrons being absorbed is the probability of neutrons being absorbed divided by the real atom density. This type of *cross section* describes the probability of neutron interaction with a single nucleus. It is called the *microscopic cross section* and can be represented by a *symbol* $\sigma$. Depending on the type of interaction, the cross section may be of different types, for example, total cross section $\sigma_T$, total scattering cross section $\sigma_{SC}$, absorption or capture cross section $\sigma_{EL}$, and inelastic cross section $\sigma_i$. The cross section has the dimensions of area, and cross sections can vary with neutron energy and the target nucleus [69].

On the other hand, the concept of macroscopic cross section arises from the transmission of a parallel beam of neutrons through a thick target sample. The thick sample may be thought of a series of atomic layers, and the intensity of the uncollided neutron beam is

$$I(x) = I_0 \exp(N\sigma x) \qquad (1.41)$$

where $I_0$ is the intensity of the beam before it enters the sample, $N$ is the density of the atom, $\sigma$ is the cross section, and $x$ is the depth of the sample. The equation is analogous to the linear attenuation coefficient for gamma ($\gamma$) rays.

## 1.11 Thermal Neutrons

Thermal neutrons are said to be in thermodynamic equilibrium, that is, the level of kinetic energy (KE) possessed by the neutrons, and they are on an average similar to the KE of the molecules at room temperature [70]. At room temperature, the KE of the average energy of a thermal neutron is around 0.025 eV ($\sim 4.0 \times 10^{-21}$ J, 2.4 mJ/kg, hence the speed is approximately 2.2 km/s). *Thermal neutrons* have a much larger cross section than fast neutrons and can therefore be absorbed more easily by any atomic nuclei that they collide with, creating a heavier element similar to the unstable isotope of chemical element [71]. A comparative analysis of different kinds of neutrons and their respective energies is given below (Table 1.7):

| Neutron | Energy (eV) |
|---|---|
| Fast | $>1$ |
| Slow | $<0.4$ |
| Hot | $0.025–1$ |
| Cold | $5 \times 10^{-5}$ to $0.025$ |
| Thermal | $0.025$ |

**Table 1.7** Neutrons and their energies

## 1.12 Neutron Sources

Nuclei created with excitation energy greater than the neutron binding energy can decay by neutron emission. These highly excited states are produced as a result of any convenient radioactive decay [72]. As for example, by mixing an $\alpha$-emitting isotope with a suitable target material, it is possible to fabricate a small self-contained neutron source. Only two target nuclei $^{9}$Be and $^{2}$H are of practical importance for radioisotope phenomenon [73]. It has been found that beryllium (Be) is a good target for neutron source and the reaction of alpha ($\alpha$) particle with Be target is as follows:

$$^{1}\alpha_2 + {}^{9}Be_4 \; {}^{12}C_6 + {}^{1}n_0 \tag{1.42}$$

which has a Q-value of +5.71 MeV.

The free neutron ($n_0$) is unstable and like photon it can ionize indirectly. However, unlike photon, interactions with atomic electrons are not significant. So, only neutron–nucleus reactions need to be considered. When it is free, it can decay beta ($\beta$) particle and produces proton with an 11.7-min half-life. During *elastic scattering*, a neutron with energy of few hundred kilo-electron volt (keV) can excite a nucleus and can leave it albeit with a reduced energy. However, during *inelastic scattering*, neutrons combine with the target during capture reactions and decays by the emission of a gamma ($\gamma$)-ray or charged particle (n, 2n).

## 1.13 Neutron Shielding

The shielding of neutrons is different from shielding of gamma rays [74, 75]. The neutrons are quickly moderated to low energies where it can readily be captured in materials with high absorption cross section. The most effective moderators are elements with low atomic number. As for example, boron $^{10}$B (n, $\alpha$) with paraffin is used as a moderator. But lithium $^{6}$Li (n, $\alpha$) has been found to be a better material for neutron shielding, because lithium (Li) reaction proceeds directly to ground state and no gamma rays are emitted after reaction. Cadmium (Cd) is also used as a thermal neutron absorber as the material is opaque to thermal neutrons due to its very high cross section. But the subsequent reaction produces secondary gamma ray in the background.

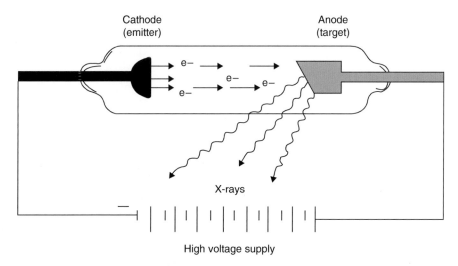

**Fig. 1.18** The generation of X-rays, due to the accelerated electrons from the hot cathode, when a high voltage is applied between the anode and the cathode

## 1.14 X-Rays

X-ray, or the *roentgen ray*, is a form of electromagnetic radiation [76]. The wavelength of X-ray is in the range of 10–0.01 nm, frequencies in the range of 30 peta hertz to 30 exahertz ($3 \times 10^{16}$ Hz to $3 \times 10^{19}$ Hz), and energies in the range of 120 eV to 120 keV. The main difference between X-rays and gamma rays is that the electrons outside the nucleus emit the X-rays, whereas the gamma rays are emitted by the nucleus itself [77]. X-rays are primarily used for diagnostic radiography, medical imaging, crystallography, and high-energy physics.

*X*-ray production typically involves bombarding a metal target inside an X-ray tube with high-speed electrons (Fig. 1.18). The bombarding electrons can also eject electrons from the inner shell of the atoms of the metal target. As a result of the ejection of electrons from the atom, vacancies will be created by the absence of electrons. These vacancies are quickly filled up by electrons dropping down from higher levels and will give rise to a sharp well-defined *X-ray*.

The energy liberated in the transition from excited states to the ground state takes the form of a characteristic X-ray photon whose energy is given by the energy difference between the initial and final states. The energy change during the transitions is accompanied by electromagnetic radiation across a broad X-ray spectrum, which is continuous due to the nature of the production [78, 79].

The X-ray exposure is proportional to the tube current (in milliampere) and also to the exposure time. In diagnostic radiography, the tube current and the exposure time are selected to produce a high-quality radiograph with correct contrast and film density. The X-ray spectrum for diagnostic purpose has an energy of 20–150 keV, and its wavelength can be calculated from the relation $\lambda = (1.24/E)$, where $E$ is the energy in keV and $\lambda$ is the wavelength in nanometer (nm).

**Fig. 1.19** The X-ray image of human teeth being taken by a dentist

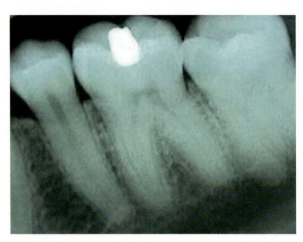

Body tissues and other substances are classified according to the degree to which they allow the passage of X-rays (radiolucency) or absorb X-rays (radiopacity). Bones and deposits of calcium salts are moderately radiopaque, whereas muscle, skin, blood, and cartilage and connective tissue have intermediate density. Bones absorb more X-rays than skin. As a result, silhouettes of the bones or teeth are left on the X-ray film when skin appears transparent. Figure 1.19 shows the X-ray picture of human teeth that show cavities, hidden dental structures (such as wisdom teeth), and bone loss that cannot be seen during visual examination. Therefore, during radiotherapy, the amount of X-ray dose is applied in a calculated way [80, 81]. X-rays are also used to identify people in forensic dentistry [82, 83].

Figure 1.20 shows the wavelength versus intensity of an X-ray spectrum produced when 35-keV electrons strike molybdenum (Mo.) target. The energy of the X-ray emitted depends on the choice of the target materials. The vacancy created in the K shell of an atom will liberate the characteristic K X-ray. On the other hand, if the electron comes from L shell, a $K_\alpha$ photon is produced. Likewise, if the filling electron comes from M shell, then $K_\beta$ photon will be produced (Fig. 1.20).

For an atom in an excited state, the ejection of an *Auger electron* is a competitive process to the emission of the characteristic of X-rays, and it is favored with low-Z (Z is atomic number) elements having small binding energies of the electrons. *Auger electrons* [84, 85] are roughly the analog of internal conversion electrons when the excitation energy originates in the atom rather than in the nucleus [86]. *Auger electron spectroscopy* is a powerful tool to obtain the chemical composition of a solid surface (ranging from 5 to 20 Å near the surface). Auger electrons have energies from 100 eV to a few keV, and they are strongly absorbed by the specimen within few angstrom unit of the surface. Therefore, Auger spectroscopy is used for studying surface of a specimen. Due to ionization, an incident high-energy electron generates a hole in the inner shell (here K shell). When the hole in the K shell is filled by an electron from an outer shell (here $L_2$), the superfluous energy is

## 1.14 X-Rays

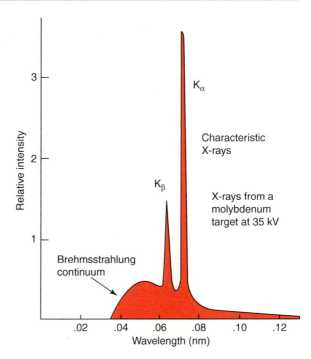

**Fig. 1.20** The X-ray spectrum produced by molybdenum target at 35 keV

transferred to another electron which is subsequently ejected (here from $L_3$ level) as *Auger electron* (Fig. 1.21).

Alpha ($\alpha$) particle excitation avoids the complication of bremsstrahlung associated with electron excitation and is therefore capable of generating a relatively clean characteristic X-ray spectrum [87]. Figure 1.22 shows the X-ray picture of Wilhelm Rontgen's (1845–1923) wife Anna's hand, taken on December 22nd, 1895. The picture is presented to Professor Ludwig Zehnder of the Physik Institut, University of Freiburg, on January 1896.

### 1.14.1 Production and Properties of X-Ray

Glass has been the primary X-ray tube envelope, but recent commercial systems also use metal ceramic X-ray tube envelopes. The metal ceramic tubes have several advantages over glass, for example, less scattered electrons, improved heat conduction, and less problem with the vaporized tungsten from the *anode disk*. The *anode disk* is made of an alloy of molybdenum (Mo), titanium (Ti), and zirconium (Zr).

A wide variety of X-ray detectors is available in the market, some count single photons, some provide only measurements of count rate of total flux, and others measure energy, position, and/or incidence time of each X-ray [88]. Table 1.8

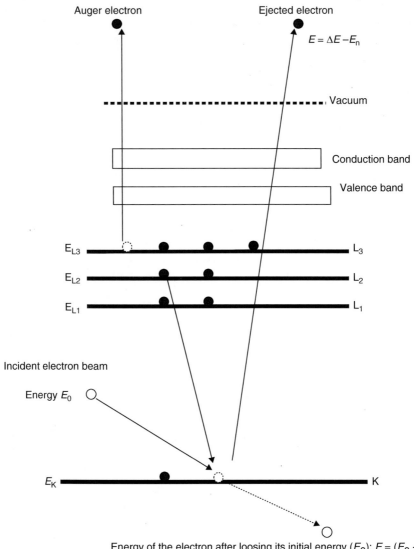

**Fig. 1.21** The sketch of emission of Auger electron due to ionization

shows typical values for useful energy range, energy resolution, dead time per event, and maximum count rate capability for X-ray detectors that are being used today.

The manufacturing of X-ray tube started in the year 1915, and the development of the sophisticated tube manufacturing with modern computer controlled system is going to be unabated. In the year 2005, the mobile C-arm and anode-grounded tube with dual support dynamic systems were developed.

## 1.14 X-Rays

**Fig. 1.22** The X-ray picture of a hand of Wilhelm Rontgen's (1845–1923) wife Anna

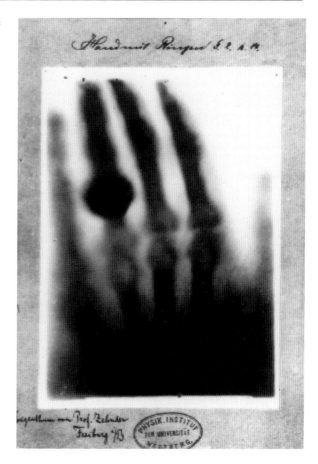

The power of a typical X-ray tube is fed from a typical generator. The alternating current (AC) is converted into a direct current (DC) by using rectifiers, and the power is fed into the X-ray tube. The voltage fed to the X-ray tube is about 150 kV with maximum current of 30 mA. The exposure timing and the switching are provided on the low-voltage circuit. Timers are mainly electronic. Current equipments available in the market do not rely on the timer mechanism for halting the exposure, and it is accomplished by feedback from an ionization chamber placed in the X-ray beam.

The principal criteria involved in the selections of X-ray are the maximum tube *voltage, current rating, focal spot size,* and the *thermal loadability* for short, medium, and long exposure time. *Loadability* is defined as the duration of time of the applied power (workload) that can be applied without rapid rise of temperature of the focal spot [89].

Ion chambers are commonly provided with automatic exposure control (AEC) for maintaining optimum film density for each kV and mA setting and are used in

**Table 1.8** Properties of common X-ray detectors

| Nomenclature | Energy range (keV) | $\Delta E/E$ @ 5.9 keV (%) | Dead time event ($\mu$s) | Maximum count rate (1/s) |
|---|---|---|---|---|
| Gas ionization (current mode) | 0.2–50 | N/A | N/A | $10^{11}$ |
| Gas proportional | 0.2–50 | 15 | 0.2 | $10^6$ |
| Multi-wire and microstrip proportional | 3–50 | 20 | 0.2 | $10^6/mm^2$ |
| Scintillator material (NaI) | 3–10,000 | 40 | 0.25 | $2 \times 10^5$ |
| Energy-resolving semiconductor | 1–10,000 | 3 | 0.5–30 | $2 \times 10^5$ |
| Surface barrier (current mode) | 0.1–20 | N/A | N/A | $10^8$ |
| Avalanche photodiode | 0.1–50 | 20 | 0.001 | $10^8$ |
| CCD | 0.1–70 | N/A | N/A | N/A |
| Superconducting | 0.1–4 | <0.5 | 100 | $5 \times 10^3$ |
| Image plate | 4–80 | N/A | N/A | N/A |

*fluoroscopy, conventional radiography,* and *digital subtraction angiography* (DSA) (Fig. 1.23) [90, 91]. On the other hand, *mammography* uses a special balanced detector system behind the cassette [92]. Recently, molybdenum (Mo) or rhodium (Rh) anodes are used to produce filtered X-ray spectra which are much more optimally tailored for high image quality with low exposure [93].

In general for radiography in *dentistry* or for small mobile applications, only a short-term loadability is recommended. On the other hand, *chest radiography* uses high power and short-time loadability. However, since the source to image distance (*SID*) is large, the focal spot is kept large for safety issue [94]. When both radiography and fluoroscopy are combined, high tube loadability is used particularly when the *SID* is long. However, for under table work where short *SIDs* are used, the anode angle becomes critical for a practical field of view. On the other hand, both *cine-cardiography* and *angiography* require series loadability. In *computer tomography* (CT) where long-term loadability is of prime importance, the X-ray tube is cooled during operation to avoid overheating and malfunctioning. However, in *mammography* anode heating is not a limiting factor [95, 96]. Figure 1.24 shows the picture of an X-ray tube assembly in a computer tomography system.

## 1.14.2 Beam Quality

The applied voltage controls the X-ray *beam quality*, and the *beam quantity* is controlled by the current in milliampere. In other words, one can say that the applied voltage actually controls the *contrast* of the *image* of the film, and the current on the other hand controls the *number of photons* and *the density* of the radiograph [97, 98]

We describe X-ray beam in terms of *photon fluence* and *energy fluence*. It is also characterized by its ability to penetrate inside different materials of known

## 1.14 X-Rays

composition. Thus the ideal way to describe the *quality of an X-ray beam* is to specify its special distribution that is the *energy fluence* in each energy interval.

Health risk to human and nonhuman biota to radiation-hardened devices to low-dose ionizing radiation remains ambiguous due to both gap junction intercellular communication and oxidative metabolism. They have shown to mediate adaptive and bi-standard effects in mammalian cells exposed in vitro to low-dose/low-fluence ionizing radiation [99]. Therefore, for various purposes (irradiation protection around high-energy accelerators, shielding calculations, aircrew dose assessment, radiation therapy, and space activity), conversion coefficients for higher energies and other kinds of radiation are needed [100–102]. Different methods have been applied to calculate the relation between dose and fluence [103, 104]. The relation between *fluence* and *effective dose* can be presented mathematically [100] as

$$f_E(\epsilon) = \{E(\epsilon)\}/\varphi(\epsilon)\} \tag{1.43}$$

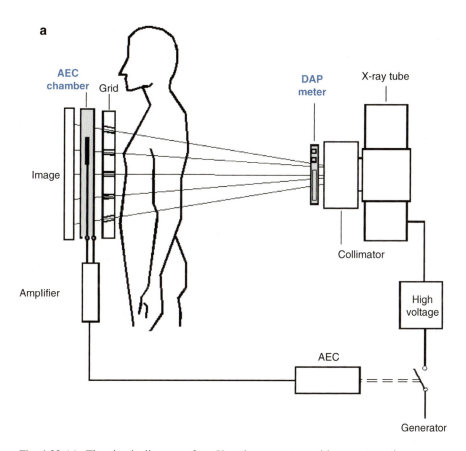

**Fig. 1.23** (a). The circuit diagram of an X-ay image system with an automatic exposure controller (AEC) (Courtesy: Vacu. Tech) and (b). The picture of an automatic exposure controlled X-ay unit (Courtesy: Dotop S&T, Beijing, China)

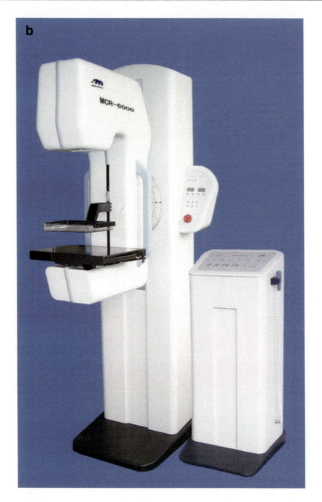

**Fig. 1.23** (continued)

where $E\ (\epsilon)$ is the *effective dose* and is a function of particle energy for various kinds of radiation and $\varphi\ (\epsilon)$ is the *fluence* of primary particle of energy ($\epsilon$). The fluence and the effective dose conversion coefficient $f_E\ (\epsilon)$ is calculated in terms of effective dose per unit fluence and is expressed in Sv cm$^2$. The scientific unit of measurement of radiation dose commonly referred to as effective dose in the millisievert (mSv).

The X-ray produced at anode passes through a variety of materials, starting from the glass envelope of the X-ray tube, before emerging from the machine as the useful beam. The glass envelope, the surrounding oil, and the exit window of the tube housing attenuate the energy of the X-ray spectrum, because of *filtering* which is called *the inherent filtration*. In most X-ray tubes, the inherent filtration is approximated to an equivalent attenuation of a 1-mm aluminum (Al) foil. In case

## 1.15 Interactions of X-Rays

**Fig. 1.24** The picture of an X-ray tube assembly commonly used in computer tomography (CT) (Courtesy: K.A. University, SA)

of a standard X-ray tube, the inherent filtration is approximately 0.5 mm Al, and in special purpose tubes, such as those for mammography, it may be as low as 0.1 mm Al. Materials purposely introduced into the incoming X-ray beam to further reduce the presence of low-energy X-rays relative to high-energy X-rays (hardening of the beam) are often referred to as *added filtration*. *Filtration*, the process of attenuating and hardening an X-ray beam, is traditionally quantified in units of mm Al or the thickness of aluminum (Al) filters that would the same effect on the beam.

## 1.15 Interactions of X-Rays

When a beam of X-rays passes through a media, a part of its energy is attenuated. Figure 1.25 (a) shows the incident beam is being attenuated by the aluminum foil before it hits a radiation detector [105]. The interaction of the beam of X-rays with the electrons of the foil will give rise to scattering of the electrons in all directions as is shown in Fig. 1.25 (b).

Let us assume:

(a) The wavelength $\lambda$ of the X-rays is small compared to the characteristic dimensions of the atom, (b) energy of a quantum of the X-rays is large compared to the binding energy of atomic electrons, and (c) the quantum energy ($h\nu$) of the X-rays is small compared to rest mass of the electron.

From the above assumptions, we can establish a relation between the incident radiation intensity $I$ with the average energy density $\bar{\rho}$ as

$$I = \bar{\rho}\,v \tag{1.44}$$

where $v$ is the volume in cm$^3$. Again, the average energy emitted in all directions per second by the electron $\bar{s}$ is proportional to $I$, and their relation can be expressed mathematically as

$$s = \sigma_T I \tag{1.45}$$

where $\sigma_T = \{(8\pi/3)\,(e^2/mc^2)\}$ is called the Thomson cross section of the electron. If the above assumptions (a) and (b) are satisfied, the problem is trivial because each of the Z electrons of the atom scatters X-rays independently, the total intensity scattered by the atom is Z times the intensity scattered by a single electron, and Eq. (1.45) will transform into

$$\bar{s} = Z\sigma_T I \tag{1.46}$$

Now $Z\sigma_T = \sigma_{sa}$ = scattering cross section per atom. Thus the differential scattering cross section per atom will give rise to a mathematical form as

$$(d\sigma_{sa}/d\Omega) = (Z\,d\sigma_T/d\Omega) \tag{1.47}$$

By integrating $(d\sigma_{sa}/d\Omega)$ over all solid angles, the predicted value of the scattering cross section per atom $\sigma_{sa}$ is obtained as an expression involving the electron density $\rho(r)$. As long as the wavelength of the X-rays is greater than $0.2 \times 10^{-8}$ cm, the results of the scattering cross section are reasonably in good agreement with the experiment. In fact, calculations of the cross section for Compton scattering have

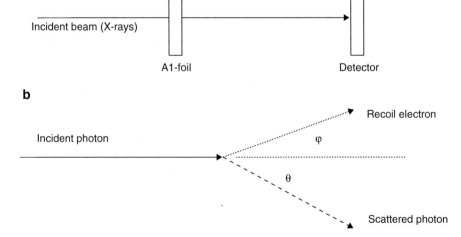

**Fig. 1.25** (a) The attenuated incident beam of X-rays being received by the detector and (b) the deflection of the incident photon

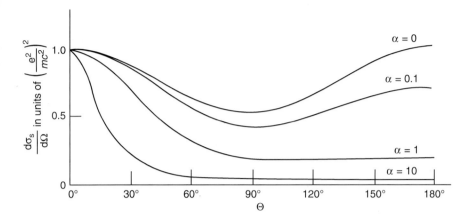

**Fig. 1.26** The differential scattering cross section for Compton scattering from a free electron having different values of α (Adopted from, Eisberg, Wiley)

been carried out in 1928 by Klein and Nishina, who used Dirac relativistic theory of quantum mechanics. The differential cross section for Compton scattering can be written (following Klein–Nishina) mathematically as [106]

$$(d\sigma_s/d\Omega) = (d\sigma_T/d\Omega)(\{1/(1+2\alpha\sin^2\theta/2)\}\{1+[(4\alpha\,2\sin^4\theta/2)/(1+\cos^2\theta)(1+2\alpha\sin^2\theta/2)]\} \tag{1.48}$$

where $\sigma_s = \sigma_T \{3 (1 + \alpha)/4\alpha\}[ \{(2 + 2\alpha)/(1 + 2\alpha)\} - \{\ln ((1 + 2\alpha)/\alpha\}]^{-3/4} [(1 + 3\alpha)/(1 + 2\alpha)^2 - \ln ((1 + 2\alpha)/2\alpha]$ and $\alpha = (h\nu)/mc^2 = $ (quantum energy/electron rest mass)

From the above equation, the differential scattering cross sections for Compton scattering from a free electron having different values of α and the scattering cross section for Compton scattering from a free electron are plotted in Figs. 1.26 and 1.27.

The scattered X-ray suffers an energy loss (ignoring binding effects) which is given by

$$(E'/E) = [1/\{1 + k(1 - \cos\theta)\}] \tag{1.49}$$

or, in terms of wavelength shift,

$$(\lambda' - \lambda) = \lambda c\,(1 - \cos\theta) \tag{1.50}$$

where $\lambda c = (h/mc) = 2.426 \times 10^{-12}$ m. The kinetic energy of the recoil electron (Fig. 1.25b) is just the energy lost by the photon in this approximation:

$$E_e = E[\{k(1 - \cos\theta)\}/\{1 + k(1 - \cos\theta)\}] \tag{1.51}$$

where $k = E/mc^2$, the photon energy measured in units of electron rest energy.

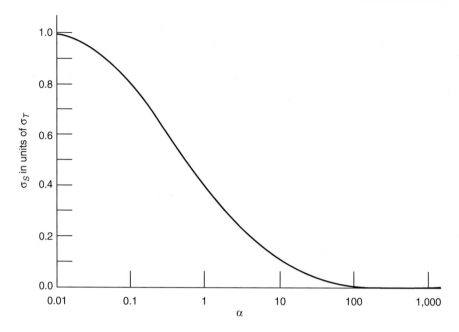

**Fig. 1.27** The scattering cross section for Compton scattering from a free electron (From Eisberg [106], p 504)

## 1.16 Photoelectric Effect

In photoelectric effect, the quantum energy of the X-rays is very small compared to the electron rest mass. As a result, the incident quantum is completely absorbed by the atom. The complete absorption of the incident quantum in the photoelectric effect does not violate energy–momentum conservation and allows the nucleus to participate because in this process the atomic electron concerned is not essentially free, but instead is bound by electrostatic forces to the massive nucleus. However, when the quantum energy is smaller than the binding energy of the electron, the probability of photoelectric absorption on that electron abruptly drops to zero. This behavior has been observed in the measured cross section of $^{82}$Pb atoms.

The effect of photoelectric absorption is the liberation of a photoelectron and is an ideal process for the measurement of energy of the original radiation. The probability for photoelectric absorption is expressed in terms of a photoelectric cross section per atom $\sigma_{PEa}$ and is defined in terms of $\bar{A}$, the average number of quanta absorbed by this process per second per atom in a beam of $\bar{I}$ quanta per second per square centimeter. Mathematically, it can be expressed as

$$\bar{A} = \bar{I}\,\sigma_{PEa} \tag{1.52}$$

The predicted cross section per atom for photoelectric absorption on a K shell electron is

$$\sigma_{PEa} = \sigma T4 \sqrt{2} \left(1/137\right)^4 Z^5 \left(mc^2/hv\right)^{7/2} \tag{1.53}$$

## 1.17 Pair Production

When the quantum energy of the incident radiation is in excess of two times the electron rest mass energy, one can observe a *photoelectric effect* where the radiation energy will be completely absorbed during interaction with matter. The process is known as *pair production*, which involves complete disappearance of a quantum and the subsequent materialization of part of its energy hv into rest mass energy of a pair production.

*Pair production* cannot take place in free space because disappearance of a quantum, followed by the appearance of an electron and positron, cannot conserve both total relativistic energy and momentum. The pair production process is complicated by the fact that the positron is not a stable particle. Once its kinetic energy (KE) becomes very low (comparable to the thermal energy of normal electrons in the interacting material), the positron will annihilate or combine with a normal electron in the absorbing medium (interacting material). The cross section per atom for the absorption of radiation energy by pair production process can be deducted by following Dirac theory and can be written mathematically as

$$\sigma_{PRa} = \sigma_T \left(3/\pi\right) \left(Z^2/137\right) f(\alpha) \tag{1.54}$$

where $\alpha \equiv (hv/mc^2)$. $Z$ is the atomic number, $Z^2$ is the dependence of the pair production cross section per atom, and $\sigma_{PRa}$ is the manifestation of the fact that the required momentum transferring interactions between the nucleus and the pair of charged particles is more readily achieved for atoms of large $Z$. The function $f(\alpha)$ has much the same form for all atoms (Fig. 1.28).

It has been noticed that in most of the cases, the effective energy of a monoenergetic X-ray shows typically one-third to one-half the peak energy. As the total filtration of an X-ray tube increases, $\mu_{eff}$ decreases, and effective energy increases. An X-ray beam with lower energy spectrum is most likely to be absorbed in a thick attenuator. As a result, the transmitted energy through the attenuator only has a beam of higher energy, which is technically termed as beam *hardening*. The *beam hardening* is dependent on the characteristics of the absorber. Higher $Z$ absorbers (e.g., bone (calcium) and iodine) cause more beam hardening per unit thickness than soft tissues, partially due to higher density (*Compton scattering*) and partially due to the $Z^3/E^3$ dependency of *photoelectric absorption*.

A single X-ray beam when positioned to a distant detector produces diverging beam with distributed point sources, each producing its own view of object. The geometry of the beam results *geometric blurring* and causes a loss of object resolution. *Geometric blurring* can be reduced with less magnification or with a smaller focal spot. In *computer tomography* (CT), magnification is the first criterion of the system geometry, and the moving detector array needs to be located at a safe

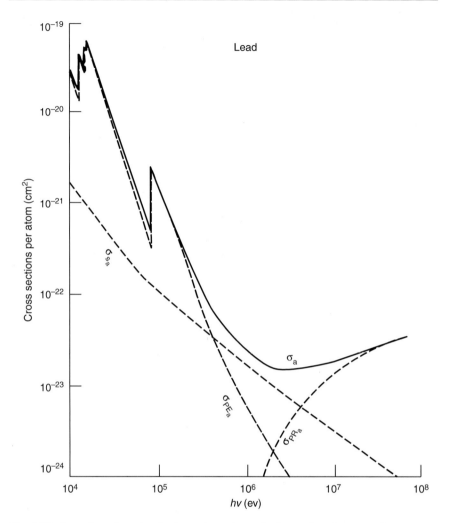

**Fig. 1.28** Scattering, photoelectric absorption, pair production, and total cross section per atom for lead ($^{82}$Pb) (Eisberg [106], p 510)

distance from the patient. Moreover, for CT, higher focal spot size is required due to the higher beam current and long duration of the CT scan.

The intensity of the X-ray beam has to be balanced during general diagnostic imaging [107], where the technologist has the flexibility to orient the X-ray tube. For example, it is prudent to orient the long axis of the X-ray tube parallel to the spine, with anode side of the field directed toward the apex of the lungs, but for diagnostic imaging of the slice thickness, the direction of the beam is not so critical [108].

*Compton scattering* rather than *photoelectric absorption* in soft tissues and bone is largely responsible for the subject contrast. The amount of scattering increases with object thickness and the scattering to primary ratio (S/P) is reduced by a factor equal to $\{1/[1 + (S/P)]\}$, which is known as *contrast reduction factor*. Typical (S/P) ratios encountered in diagnostic radiology depend upon the thickness of the object, the field size, and thickness. Scattering with conventional CT scanning is not a significant problem, because the volume of the patient irradiated at any instant is small, and the scattering is small due to light collimation (*slice thickness*). However, collimation in the slice thickness direction is increased to accommodate larger multi-row detector arrays that provide multi-slice acquisition capability. Nevertheless, scattering will become a greater problem that will affect the image quality of the computer tomography (CT) images. Scattering can be reduced either by using smaller field size or by using anti-scatter grids, or implementing air gaps between the object and the detector. Each of the methods has its pros and cons.

In $\gamma$-ray imaging system [109], scattering is generally rejected with the use of energy discrimination windows tuned to the $\gamma$-ray photon [53]. On the other hand, X-ray detectors are energy integrators and have the ability to reject scattering based on its energy due to the polyenergetic spectrum (broad X-ray spectrum) used for X-ray and CT, but mostly due to extremely high X-ray photon fluence.

## 1.18 Use of Natural Forces for the Material Improvement of Mankind

By far the most important application of nuclear radiation for the material improvement of mankind is in the field of *nuclear medicine or radiation oncology*. *Nuclear medicine or radiation oncology* is a special branch of *medical science* and *medical imaging* that uses the nuclear properties of matter to *diagnose* the abnormality of cells or tissues and treatment of the abnormal cells or tissues by *radiation therapy* [110, 111]. More specifically, *nuclear medicine* is a part of *molecular imaging* or *medical imaging*, which produces images of the biological processes at the cellular and subcellular levels [112].

*Medical imaging,* on the other hand, refers to the techniques and processes used to create images of the creatures including human body (or parts thereof) for *clinical* purposes [113]. It can be said that it was a part of biological imaging system with the concepts that incorporate *radiology, radiological sciences, endoscopy* (medical), *thermography, medical photography,* and *microscopy* [114]. A recent study shows that 95 % of radiation oncologists use advanced imaging techniques such as *fluoroscopy, computer tomography* (CT), *positron emission tomography* (PET), *single-photon emission tomography* (SPECT), *PET* and *SPECT* combined, *magnetic resonance imaging* (MRI), and *ultrasound* [115–117].

In *clinical field*, medical imaging is being used to diagnose some abnormality of a normal tissue, cells, or an organ or a part of the organ of a living animal [118]. Thus it can be said that *medical imaging* is a special subbranch of *biological engineering* or *medical physics* depending on the context [119–121]. The procedures involved in *nuclear medicine* use pharmaceuticals that have been labeled with *radionuclides*

[122]. The major *radionuclides* that are frequently used are *iodine* (I)-131, *technetium* (Tc)-99, *cobalt* (Co)-60, *iridium* (Ir)-192, *americium* (Am)-241, and *cesium* (Cs)-137. However, some of the *radionuclides* like Co-60, Cs-137, Am-241, and I-131 are particularly vulnerable for the environment.

In the field of internal radionuclide therapy, a small radiation source is administered or planted (e.g., iodine-131 is commonly used to treat thyroid cancer), which is usually a beta ($\beta$) or gamma ($\gamma$) emitter, in the target area. A new field called targeted alpha ($\alpha$) therapy (TAT) or alpha radio-immunotherapy has emerged in the field of nuclear medicine very recently. Cancer and heart diseases are the number one killers, approaching 90 % of all deaths in the United States of America. Improved *image quality* to diagnose the abnormality of the cells/tissues is essential. Indeed, shorter examination time, shift to outpatient testing, and noninvasive imaging are becoming the primary concern for the oncologists [123].

## 1.19  Radionuclides

*Radionuclide* is an atom with an unstable nucleus and is referred to by chemists and physicists as *radioactive isotope* or *radioisotope*. Radionuclides with suitable half-lives are used for diagnosis, treatment, and research in nuclear medicine. The radionuclide can be a natural one or an artificial one, produced in a laboratory.

In the field of nuclear medicine and biological research, almost 20 radionuclides are used. Among the radionuclides that are used very much in nuclear medicine are phosphorus (P)-32, technetium (Tc)-99m, thallium (Tl)-201, iodine (I)-131, iodine (I)-125, cobalt (Co)-60, iridium (Ir)-192, cesium (Cs)-137, strontium (Sr)-90, krypton (Kr)-85, tritium (Tr), and plutonium (Pu). The radiation energy from these radionuclides can change the metabolic and physicochemical properties of living beings.

Different nuclides with their half-lives and decay times used in medical therapy are presented in Table 1.9. The meanings of the different notations that are used in the Table 1.5 are as follows: (1) $\lambda$ is the decay constant, (2) EC means electron capture, (3) IC means internal conversion, (4) $n$ is neutron, (5) $f$ is fission, and (6) $d$ means decay (second column). In the second column, the metal with atomic number indicates target atom, and within parenthesis the first letter indicates bombarding particle and the second one indicates emitted particle.

## 1.20  Production of Radionuclides

Radioisotopes are used in medical imaging, diagnosis, and treatment of the abnormal tissues and cells and high-energy physics [124, 125]. Radioactive nuclides are prepared by a wide variety of particle accelerators and nuclear reactors.

## 1.20 Production of Radionuclides

**Table 1.9** Different nuclides that are used in nuclear medicine

| Nuclide | Production method target atom (bombarding particle, emitted particle) | Half-life | Decay mode | $E_{\beta avg}$ keV | γ-proton energies | | Decay product | Half-life decay product |
|---|---|---|---|---|---|---|---|---|
| | | | | | keV | abundance | | |
| $^{99m}$Tc | $^{99}$Mo–$^{99m}$Tc | 6 h | IC, λ | – | 140 | 0.88 | $^{99}$Tc | $2.1 \times 10^5$ year |
| $^{67}$Ga | $^{68}$Zn (d, n) $^{67}$Ga | 78 h | EC, λ | – | 93 | 0.38 | $^{67}$Zn | Stable |
| | | | | | 185 | 0.20 | | |
| | | | | | 300 | 0.17 | | |
| | | | | | 393 | 0.05 | | |
| $^{111}$In | $^{109}$Ag (α, 2n) | 67 h | EC, λ | – | 172 | 0.90 | $^{11}$Cd | Stable |
| | | | | | 245 | 0.94 | | |
| $^{131}$I | $^{235}$U (n, f) $^{131}$I | 8 day | β⁻, λ | 284 | 364 | 0.83 | $^{131}$Xe | Stable |
| $^{123}$I | $^{121}$Sb (α, 2n) $^{123}$I | 13 h | EC, λ | – | 159 | 0.83 | $^{123}$Te | $1.2 \times 10^{13}$ year |
| $^{133}$Xe | $^{235}$U (n, f) $^{133}$Xe | 5 day | (β⁻, λ) | 110 | 81 | 0.37 | $^{133}$Cs | Stable |
| $^{201}$Tl | $^{203}$Tl (d, 2n) $^{201}$Pb | 72 h | EC, λ | – | X-ray;70–80 | 0.93 | $^{200}$Hg | Stable |
| $^{57}$Co | $^{56}$Fe (d,n) $^{57}$Co | 270 day | EC, λ | – | 122 | 0.86 | $^{57}$Fe | Stable |
| $^{89}$Sr | $^{88}$Sr (d, p) $^{89}$Sr | 51 day | β⁻, λ | 583 | 909 | 100 | $^{89}$Y | Stable |
| $^{32}$P | $^{32}$S (n, p) $^{32}$P | 14 day | β⁻ | 735 | | | $^{32}$S | Stable |
| $^{18}$F | $^{18}$O (p, n) | 109 min | β⁺ | 242 | | | $^{18}$O | Stable |
| $^{15}$O | $^{14}$N (d, n) $^{15}$O | 124 s | β⁺ | 735 | | | $^{15}$N | Stable |
| $^{13}$N | $^{10}$B (α, n) $^{13}$N | 10 min | β⁺ | 491 | | | $^{13}$C | Stable |
| $^{11}$C | $^{11}$B (p, n) $^{11}$C | 20 min | β⁺ | 385 | | | $^{11}$B | Stable |
| $^{82}$Rb | $^{82}$Sr-82Rb | 76 s | β⁺ | 1,409 | | | $^{82}$Kr | Stable |

### 1.20.1 Cyclotron

In *cyclotron,* charged particles such as protons, deuterons, and alpha (α) particles are bombarded to the target nuclei. Unfortunately, the bombardment with the target in cyclotron does not produce an isotropic product. As a result, after chemical separations, a product of high specific activity is obtained [126]. Figure 1.29 shows the picture of a modern cyclotron.

A *second mode* of operation in reactors is by fission process itself as the majority of the fission products are radioactive and cover a wide range of atomic numbers of varying abundance. Fission process is accompanied by the generation of 2–3 neutrons, γ-rays, and two fission fragments, and they are unstable and radioactive. The half-lives of fission fragments can vary from a fraction of a second to several years. The long-lived fission products built up during the operation of a reactor can

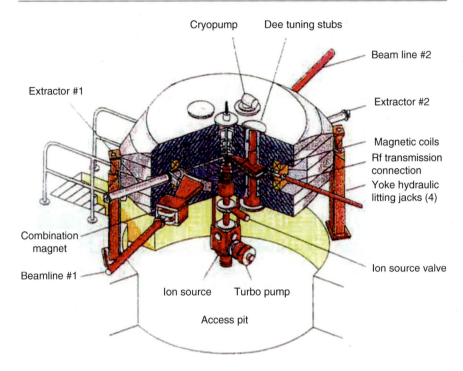

**Fig. 1.29** The picture of a modern cyclotron (Courtesy: IAEA, Vienna, Austria, Tech. Rept. 468)

have billions of curie of radioactivity per second. The radiation fields due to $\gamma$-rays and neutrons produced during fission as well as radiation emitted by the built-up fission products may reach $10^6$ to $10^8$ Gy/h inside the reactor core.

The *third mode* of production of radionuclides is as daughters of parent activities, made either by cyclotron or reactor irradiation. Among the three modes, which one will be adopted depends on the length of time for production, separation and purification processes, and the time from production to shipping.

### 1.20.2 Reactor

The reactor instruments can be divided into two groups, in-core boiling water reactor (BWR) and out-of-core pressurized water reactor (PWR). The BWR has three systems called source, intermediate, and power range monitors. The source system typically consists of four in-core fission chambers operating in pulse mode, which provides good discrimination against gamma ($\gamma$) rays. Figure 1.30 shows the schematic of a boiling water reactor (BWR).

In PWR three sets of sensors with overlapping operating ranges are used to cover the entire power range of the reactor. At intermediate power level pulse mode operation is possible because of the excessive neutron exchange rate. But when the instrument is operated at its full operating range, the neutron flux is very large and

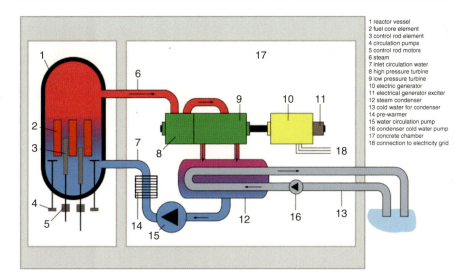

**Fig. 1.30** The schematic of a boiling water reactor (BWR) (Courtesy: WIKI)

gamma-induced currents within ion chambers are no longer significant, and the uncompensated ion chambers are commonly used as principal neutron sensor. The majority of the neutron sensors are of *gas-filled* type because of its inherent gamma-ray discrimination properties. The gas-filled chambers, whether they are ionization or proportional counters, can be operated mainly in two modes (either pulse mode or current mode). When flux levels become high enough so that pulse mode operation is no longer possible, neutron detectors are often operated in current mode.

## 1.21 Developments and Uses of Radionuclides

The ongoing research in the development of medical radioisotopes, their production and processing, the development of new radionuclide generator systems, the design and evaluation of new pharmaceuticals for applications in nuclear medicine, oncology, and interventional cardiology has been focused as number one priority throughout the world for further preclinical testing and clinical evaluation of agents. The development of high flux isotope reactor (HFIR) in the USA provides a maximum steady-state thermal neutron flux of about $2.3 \times 10^{15}$ neutrons per square centimeter per second. The instrument is a key resource for production of radioisotope used in nuclear medicine [127, 128]. A key example of HFIR production is rhenium-188 ($^{188}$Re) from the parent nucleus tungsten-188 ($^{188}$W).

$$_{186}W \xrightarrow{n,\gamma} {}_{187}W \xrightarrow{n,\gamma} {}^{188}W \xrightarrow{\beta} {}^{188}Re \quad (1.55)$$

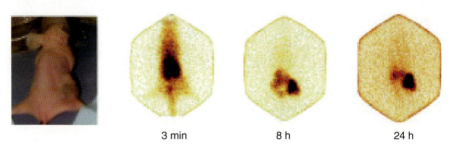

**Fig. 1.31** Scintigraphic biodistribution of $^{188}$Re as a function of time (Courtesy: F.F. (Russ), Knap Jr.)

$^{188}$Re is used in clinical radioisotope for bone pain palliation (hydroxy ethylidene, HEDP), restenosis therapy (after percutaneous transluminal coronary angioplasty, PTCA), bone marrow ablation, and therapies for liver cancer, lung tumor, and radiation synovectomy/joint diseases. The $^{188}$Re-labeled α-melanotropin peptide analog Re-CCMSH (cytochrome C-type biogenesis protein CCMH precursor) shows promise for melanoma therapy [129].

Figure 1.31 shows scintigraphic biodistribution of $^{188}$Re (P2045 peptide) as a function of time in a melanoma-bearing mouse. Recent research also includes production of alpha (α)-emitting radioisotope actinium-225 ($^{225}$Ac), which is used in radiotherapy of cancer. Actinium is routinely extracted from 229-thorium ($^{229}$Th) [130].

$$^{226}\text{Ra}\,(3n, \gamma) \rightarrow {}^{229}\text{Ra}\,(\beta^-) \rightarrow {}^{299}\text{Th} \rightarrow {}^{225}\text{Ac} \qquad (1.56)$$

Actinium can be produced by any one of the following processes:
1. Extraction from uranium-233 ($^{233}$U)
2. Reactor production from radium-226 ($^{226}$Ra)
3. Direct production from radium-226 ($^{226}$Ra)
4. Decay method of thorium-229 ($^{229}$Th)

Figure 1.32 shows the production of actinium and thorium by α- and β-decays from the mother element uranium. It also shows the half-life value of the individuals. Figure 1.33, on the other hand, shows the range of electrons in soft tissue.

The following Table 1.10 gives some of the radioactive materials (radioisotopes) that are being used in our daily lives.

Table 1.11 gives the comparisons of effective X-ray dose with background radiation exposure for several radiological procedures:

## 1.21 Developments and Uses of Radionuclides

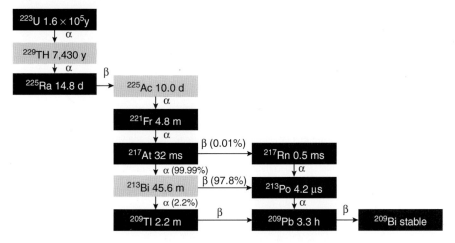

**Fig. 1.32** The production of actinium and thorium by α- and β-decays and the half-life of the individuals (Courtesy, Dr. F.F. (Russ) Knap, Jr.)

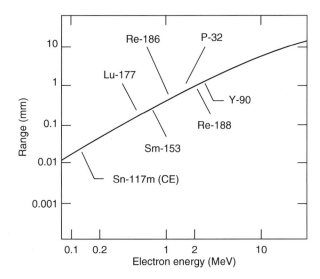

**Fig. 1.33** The range of electrons in soft tissue (maximum β-energy) (Courtesy, Dr. F.F. (Russ) Knap, Jr.)

**Table 1.10** Radioactive materials and their frequent use in daily life*

| Radioactive materials | Uses |
| --- | --- |
| Americium-241 | Used in smoke detectors, to measure toxicity of lead paint, and source to calibrate radiation |
| Cadmium-109 | To analyze metal alloys |
| Calcium-47 | To study cell function and bone formation |
| Californium-252 | To inspect explosives and moisture content in soils |
| Carbon-14 | To ensure that potential new drugs are metabolized without forming harmful products and radiocarbon dating |
| Cesium-137 | To treat cancer, to measure dose of pharmaceuticals, to measure oil flow in a pipe line, and calibration source |
| Chromium-51 | To study red blood cells' survival |
| Cobalt-51 | Standard radioactive source (SRS), as radionuclides, radio-imaging to diagnosis diseases (e.g., anemia) |
| Cobalt-60 | As SRS, to sterilize surgical instruments, to improve safety and reliability of fuel oil burners, and to preserve foods |
| Copper-60 | Injected with monoclonal antibodies into a cancer patient |
| Curium-244 | Used in mining to analyze drilled materials |
| Iodine-123 | To diagnosis thyroid disorders |
| Iodine-129 | Used to check some radioactivity counters |
| Iodine-131 | To diagnose thyroid disorders |
| Iridium-192 | To test the integrity of pipeline welds, boilers, and aircraft parts |
| Iron-55 | To analyze electroplating solutions |
| Krypton-85 | Used in indicator lights, to gauge the thickness of metal and plastics, and to measure dust and pollutant levels |
| Nickel-63 | Used to detect explosives and as voltage regulators and current surge protectors |
| Phosphorus-32 | Used in molecular biology and genetic research |
| Plutonium-147 | Fissionable isotope and to power air space craft |
| Polonium-210 | To reduce static charge in photographic films and phonograph records |
| Promethium-147 | Used in electric blanket, to gauge thickness of thin sheets of plastics, rubber, textiles, etc. |
| Radium-226 | Makes lightning rods more effective, to standardize the radioactive source (curie) |
| Selenium-75 | Used in protein studies in life science research |
| Sodium-24 | Used to study and locate leaks in pipelines |
| Strontium-85 | Used to study bone formation and metabolism |
| Technetium-99m | The mostly used radioisotope for diagnostic studies in nuclear medicine |
| Thallium-204 | To measure dust pollutant and thickness of thin sheets |
| Thallium-201 | To diagnostic studies of coronary artery disease and location of lymphomas |
| Thorium-229 | Helps fluorescent lights to last longer |
| Thorium-230 | Provides coloring and fluorescence |
| Tritium | Use for life science and drug metabolism studies |
| Uranium-234 | Used in dental fixture |
| Uranium-235 | Used in fission and chain reaction, power plants, and naval propulsion system |
| Xenon-133 | Used in nuclear medicine for lung ventilation and blood flow studies |

*adopted from different sources

## 1.22 Summary

**Table 1.11** Comparative analysis between the effective radiation dose and background radiation for different radiological procedures

| Procedure | Radiation dose | Natural background radiation |
|---|---|---|
| Computer tomography (CT)—abdomen | 10 mSv | 3 years |
| Computed tomography (CT)—body | 10 mSv | 3 years |
| Computed tomography (CT)—colonography | 5 mSv | 20 months |
| Intervenous pyelogram (IVP) | 1.6 mSv | 6 months |
| Radiography—lower GI Tract | 4 mSv | 16 months |
| Radiography—upper GI Tract | 2 mSv | 8 months |
| Bone: | | |
| Radiography—spine | 1.5 mSv | 6 months |
| Radiography—extremity | 0.001 mSv | Less than 1 day |
| General nervous system: | | |
| Computed tomography (CT)—head | 2 mSv | 8 months |
| Computed tomography—spine | 10 mSv | 3 years |
| Myelography | 4 mSv | 16 months |
| Chest: | | |
| Computed tomography (CT)—chest | 8 mSv | 3 years |
| Radiography—chest | 0.1 mSv | 10 days |
| Children's imaging: | | |
| Voiding cystourethrogram | 5–10 years old: 6 mSv | 6 months |
| | Infant: 0.8 mSv | 3 months |
| Face and neck: | | |
| Computed tomography (CT)—sinuses | 0.6 | 2 months |
| Heart: | | |
| Cardiac CT for calcium scoring | 2 mSv | 8 months |
| Men's imaging: | | |
| Bone densitometry (DEXA) | 0.01 mSv | 1 day |
| Women's imaging | | |
| Bone densitometry (DEXA) | 0.01 mSv | 1 day |
| Galactography | 0.7 mSv | 3 months |
| Hysterosalpingography | 1 mSv | 4 months |
| Mammography | 0.7 mSv | 3 months |

## 1.22 Summary

All materials on Earth are constantly exposed to radiation. The radiation can be nonionized or ionized according to their energy. The classification of radiation as ionizing is essentially a statement that has enough quantum energy to eject an electron. The ionizing radiation can produce a number of physiological effects, such as those associated with risk of mutation or cancer, which nonionizing radiation cannot directly produce at that intensity. The mechanism of interactions of the ionizing radiation in the form of X-rays and gamma rays includes the *photoelectric effect, Compton scattering,* and at high enough energies *electron–positron pair production.*

*Occupational exposures* to nuclear radiation also come from industries, medical, educational, and research establishments. Several factors combine to heighten the public anxiety about both short-range and long-range effects of radiation. Because living tissue contains 70–90 % water by weight, the dividing line between radiation and that excites electrons and radiation that forms ion is often assumed to be equal to the ionization of water. However, the good side of the radiation is the use of nuclear radiation (mostly neutrons, the X-ray, and the gamma rays) in *radiation therapy* and *medical imaging* to diagnose the cancerous cells or the abnormal tissues.

In *radiation therapy*, different radioisotopes are used. As a result, the research in the development of medical isotopes, their production and processing, and the development of new nuclide generator systems has become part and parcel of medical research. On the other hand, in the field of preclinical testing, clinical evaluation of the agents, and evaluation of new pharmaceuticals for applications in nuclear medicine, oncology and interventional cardiology have been focused as number one priority.

Utmost care is being taken during medical examination and treatment so that the normal tissues and the cells are not unnecessarily exposed to radiation dose. In order to prevent the unnecessary exposure during diagnosis and radiation therapy, different kinds of shielding are used.

Both gamma ($\gamma$) rays and X-rays are used for medical purposes. Both of these penetrating rays not only exhibit *wave*like properties but they also exhibit *discrete particle* characteristics. However, unlike discrete energies of $\gamma$-rays emitted from radioactive materials, the X-ray beam consists of a spectrum of energies over an energy range determined by the peak kilovoltage (kVp), the generator waveform, and the amount of inherent and added filtration. Most of the time, measurement of the spectral distribution of X-rays and the relative penetrability or the quality of the X-rays is useful instead of the measurements of the X-ray spectrum. Transmission measurements of the intensity of X-rays through a series of attenuator (e.g., aluminum foil) are useful to characterize the beam in terms of half-value layer (HVL).

## References

1. Gupta TK, Antonuk LE et al (2012) Investigation of active matrix flat-panel imagers (AMFPIs) employing thin layers of polycrystalline $HgI_2$ photoconductor for mammographic imaging. SPIE Proceedings, San Diego, 4–9 Feb 2012 and also Gupta TK et al (2004) $LuI_3$: Ce-a new scintillator for gamma ray spectroscopy. IEEE Trans Nucl Sci NS-51 (2302–2305) and also Gupta TK et al (1999) High speed X-ray imaging camera using structured CsI (Tl) scintillator. IEEE Trans Nucl Sci 46(3):232–236

# References

2. Gupta TK et al (2002) $RbGd_2Br_7$:Ce, scintillators for gamma ray and thermal neutron detection. IEEE Trans Nucl Sci NS-49:1655 and also Gupta TK et al (2004) High resolution scintillator spectrometer. IEEE Trans Nucl Sci 51(5):2395
3. Gerardin S, Paccagnella A (2010) Present and future non-volatile memories for space. IEEE Trans Nucl Sci 57(6):3016
4. Qu W (2012) Preparation and characterization of L-[5-[11]C]-Glutamine for metabolic imaging of tumers. J Nucl Med 53(1):98 and also Guerra AS (2004) Ionizing radiation detectors for medical imaging. World Scientific, Singapore
5. Vanderhoek M, Berlman SB, Jeraj R (2012) Impact of the definition of peak standardized uptake value on quantification of treatment response. J Nucl Med 53(1):4
6. Curry H (2009) Most radiation oncologists utilizes advanced medical imaging techniques. JACR. 25 Nov 2009 and also Bushberg JT, Siebert JA, Leidholdt EM, Boone JM (2002) Essential physics of medical imaging. Lippincott Williams & Wilkins, Philadelphia
7. Mukhin KN (1987) Experimental nuclear physics, vols I and II. Mir Pub, Moscow and also Shultis JK, Faw RE (2007) Fundamentals of nuclear science and engineering, 2nd edn. CRC Press, Boaca Raton, Fl
8. Rogers EM (1960) Physics for the inquiring mind. Princeton University Press, Princeton, Chap. 43, p 682 and also Kahn J (1996) From radioisotopes to medical imaging, history of nuclear medicine written at Berkeley, 9 Sept 1996, Science Articles Archive, Berkeley
9. National Research Council (2008) Radiation source use and replacement. The National Academy Press, Washington
10. da EN, Andrade C (1956) The birth of the nuclear atom. Sci Am 195(5):93–107, and also Buddemeier B (2003) CHP, understanding radiation and its effect, U. Calif. Lawrence Livermore Nat. Lab., Under auspices of the DOE Contract # W-7405-Eng-48, June (3-6-2003), UCRL-Pres-149818-Rev-2
11. Stanev T (2010) Ultra high energy cosmic rays: origin and propagation. Mod Phys Lett A 25 (18):1467 and also Lehnerst S (2007) Biomolecular action and ionizing radiation. Institute of Physics, Lehnerst, United Kingdom
12. Shaviv NJ (2002) Cosmic ray diffusion from Galatic spiral arms. Phys Rev Lett 89(5):051102 and Fermi E (1974) Nuclear physics. University of Chicago, Chicago Chichester; (2004) Springer, Chichester
13. Ionizing radiation fact book, EPA, USA, EPA 402-F-05-061, March 2008 and also Kume T (2006) Application of radiation in agriculture. Proceedings of international workshop on biotechnology in agriculture, Nong Lam University, Ho Chi Min City, 20–21 Oct 2006
14. Ouellette J (2011) Discovery news, NASA, 1 Nov 2011, Thirty years of bubble chamber physics, CERN courier, July/Aug 2003, p 19
15. Glen GF (2000) Radiation detection and measurements, 3rd edn. Wiley, New York, Chap. 2
16. Shleien B, Terpilak MS (1984) The Health Physics and Radiological Health Handbook. Nucleon Lectern Associates, Olney
17. Iwai S et al (2005) Calculation of fluence to dose conversion coefficients in partial exposure. Appl Radiat Isot 63(5–6):639
18. Tack D et al (2007) Dose from adult and pediatric multidetector computer tomography. Springer, New York, p 140
19. ICRP (2009) Relative biological effectiveness, quality factor (QF), and radiation weighting factor, vol 3394. J. Valentin (ed) Annals of the ICRP Pub. 1992, Elsevier, pp 1–121
20. (2000) Low-dose of ionizing radiation: biological effects and regulatory control. Int. Conf. in Seville, Spain, 17–21 Nov. 1997 (IAEA TEC.DOC. 976, Pub. Vienna)
21. Vienot KG, Hertel NE (2007) Photon extremely absorbed dose and kerma conversion coefficients for calibration geometries. Health Phys 92(2):179, Lippincott Hagerstown, MD
22. Beyer HE, Kluge HJ, Sherelko VP (1997) X-ray Radiation of Highly Charged Ions, Sept. 19, Springer, Berlin
23. L'Annunziata MF et al (2003) Handbook of radioactivity. Academic, San Diego, p 1169
24. Dendy PP, Heaton B (1999) Physics for diagnostic radiology. CRC Press, Boca Raton

25. National Research Council (US) (2007) Advancing nuclear medicine through innovation. National Academic Press, Washington, DC
26. Noz ME, Maguire GQ Jr (1999) Radiation protection in health sciences. World Scientific, New York
27. Mock WG (2005) Introduction to isotope hydrology. Taylor & Francis, New York
28. Seife C (2000) Science 288(5471):1564
29. Hendee WR, Ritenour ER (2002) Medical imaging physics. Wiley, New York, p 28
30. Bartusiak M et al (1994) A positron named Priscilla. National Academic Press, Washington, DC
31. Murtagh DJ, Cook DA, Laricchia G (2009) Excited state positronium formation from helium argon and xenon. Phys Rev Lett 102:133202
32. Cassidy DB, Mills AP Jr (2007) The production of molecular positronium. Nature Lett 449:195
33. Boronski E (2006) Positron-electron annihilation rates in electron gas studied by variational Monte Carlo simulation. Euro Phys Lett 75:475
34. Gupta TK et al (2004) LuI$_3$:Ce, a new scintillator for gamma ray spectroscopy. IEEE Trans Nucl Sci 51(5):2302
35. Gola C et al (2006) Gamma-ray imager for medical imaging. IEEE Nucl Sci Symp 6:3581
36. Henlee WR, Russell E (2002) Medical imaging. Wiley, Hoboken
37. Webb S (1988) Physics of medical imaging. CRC Press, Boca Raton
38. Mayles P, Nahum A, Rosenwald JC (2000) Hand book of radiotherapy physics. Taylor & Francis, New York
39. Giger ML (2006) Computer aided diagnosis in medical imaging – a new era in image interpretation, business briefing: next generation health care. University of Chicago, Chicago, p 75
40. Turner JE (2007) Atoms, radiation and protection, 3rd edn. Wiley, Weinheim
41. Cember H (1996) Introduction to health physics. McGraw Hill, New York
42. Webb S (2001) The physics of three dimensional radiation therapy. Institute of Physics, London
43. Kaplan MF (1999) Concrete radiation shielding. Longman Scientific, New York
44. Shultis JK, Faw RE (1996) Radiation shielding. Prentice Hall, Englewood Cliff, digitized version 2007
45. Jaeger RG (2007) Engineering compendium on radiation shielding. Springer, Berlin and also Shapiro J (2002) Radiation protection, 4th edn. Harvard University Press, Cambridge
46. Martin MC (2004) On fusion imaging and multimodalities. National symposium, Kansas, Fgeb, 18–20
47. Mahn K (2009) Neutrino and antineutrino disappearance in booster neutrino beam line, APS meeting, Denver, 2–5 May 2009
48. Giunti C, Kim CW (2007) Neutrino physics and astrophysics. Oxford University Press, Oxford
49. Hanna SS, Lavier EC, Miclass C (1954) Beta- alpha correlation in the decay of lithium (Li)8. Phys Rev 95:110
50. Ueza Y, Hashimoto T (2003) New aspects of time interval analysis method in the determination of artificial alpha nuclides. J Rad Ann Nucl Chem 255(1):87
51. L'Anunziata MF (2000) Radioactivity introduction and history. Elsevier, New York, p 120
52. Kroeninge K, Pandola L, Tretyak V (2007) Feasibility study of the observation of the neutrino accompanied double beta decay of Ge-76. Ukr J Phys 52:1036
53. Ando S et al (2004) Neutron beta decay in effective field theory. Phys Lett B 595(1–4):250
54. Gerl J (2005) Gamma ray imaging exploiting the Compton effect. Nucl Phys A 752:688C
55. Kemeny AA (2004) Gamma knife radio surgery, vol 148 (4), Acta neurochiorursgica. Springer, New York
56. Cunha RM et al (2000) Two media method for gamma ray attenuation coefficient measurement of archaeological ceramic samples. Appl Radiat Isot 53(6):1011
57. Khater AEM, Ebaid YY (2008) A simplified gamma ray self attenuation correction in bulk samples. Appl Radiat Isot 66(3):407

# References

58. Tsoulfanides N (1981) Measurement and detection of radiation. Hemisphere Publishing Corporation, New York
59. (2008) Mosby's medical dictionary, 8th edn. Elsevier Health Sci. Elsevier, Amsterdam, Netherlands
60. Khan FM (2003) The physics of radiation therapy. Lippincott Williams & Wilkins, New York
61. Shapiro J (2002) Radiation protection. Editorial UPR, p 47
62. Kruger RL, McCollough CH, Zink FE (2000) Measurements of HVL in X-ray computed tomography. IEEE Proc Annu Conf Eng Med Biol 1(1):98
63. Maria AF, Caldas LVE (2006) A simple method for evaluation of half value layer variation in CT equipment. Phys Med Biol 51:1595
64. Bomford CK, Kunkler IH, Walter J, Miller H (2003) Text book on radiotherapy. Elsevier Health Sci, New York, p 155
65. Liang L, Rinaldi R, Schober H (eds) (2009) Neutron scattering applications and techniques. Springer, New York
66. Evans RD (1955) The atomic nucleus. McGraw Hill, New York, p 411
67. Ramstrim E, Goransson PA (1977) Neutron elastic scattering cross section measurements at back angle. Nucl Phys A 284(3):461
68. Bodansky D (2004) Nuclear energy principles, practice, and prospects. Springer, New York, p 123
69. Banner TW, Slattery JC (1959) Non-elastic scattering cross section for 8-20 MeV neutrons. Phys Rev 113(4):1088
70. McNaught AD, Wilkinson A (1977) The gold book, 2nd edn. Blackwell Sci. Pub., Oxford
71. Tremsin AS, Feller WB, Downing RG, Mildner DF (2005) The efficiency of thermal neutron detection and collimation with microchannel plates of square and circular geometry. IEEE Trans Nucl Sci 52(5):1739
72. Keimer B, Sackmann E, Whiters PJ (2002) The case of neutron sources. Science 298:542
73. Park H, Kim J, Choi KO (2007) Neutron calibration facility with radioactive neutron sources at KRISS. Rad Prot Dosim 126(1–4):159
74. Spratt JP, Aghara S, Fu B, Lichethan JD, Leadon R (2005) A conformal coating for shielding against naturally occurring thermal neutrons. IEEE Trans Nucl Sci 52(6):2340
75. Asano Y, Sugita T, Suzuki T, Hirose H (2005) Comparison of thermal neutron distributions within shield materials. Rad Prot Dosim 116(1–4):284
76. Sharon S (2007) X-rays, the electromagnetic spectrum. NASA retrieved on 12 Mar 2007
77. Grupen C, Cowan G, Eidelman SD, Stroh T (2005) Astroparticle physics. Heidelberg, Germany, p 109
78. Swinbourne R (2008) The X-ray tube and image intensifier for present day application. Aust Radiol 22(3):204
79. Drenth J (1999) Principles of protein X-ray crystallography, 2nd edn. Springer, Heidelberg, Chap. 2
80. Metcafe P, Butson M, Quach K, Bengua G, Hoban P (2002) 22nd annual conference. IEEE Eng Bio 4:2928
81. Butson MJ et al (2008) Measurement of radiotherapy superficial X-ray dose under eye shields with radiochromic films. Phys Med 24(1):29
82. Jain AK, Chen H (2004) Matching of dental X-ray images for human identification. Pattern Recognit 37:1519
83. Pretty IA, Sweet D (2001) A look at forensic dentistry. Br Dent J 190(7):359
84. Polok M (1995) Ion excited low energy Auger-electron emission from Ti and TiNi. J Phys Cond Matt 7:5275
85. Mariani G, Bodel L, Adelstein SJ, Kassis AI (2000) Emerging roles for radiometabolic therapy of tumors based on Auger electron emission. J Nucl Med 41(9):1519
86. Choursia AR, Chopra DK (1997) Auger spectroscopy. In: Settle F (ed) Handbook of instrument techniques for analytical chemistry. Prentice Hall, Englewood Cliffs, Chap. 2

87. Maeo S, Kramer M, Utaka T, Taniguchi K (2009) Development of microfocus spectrometer using multiple target anode monochromatic X-ray sources. J X-Ray Spectrom 38(4):333
88. Michel T et al (2009) Reconstruction of X-ray spectra with the energy sensitive photon counting detector Medipix. Nuclr Instr Meth A 598(2):510
89. Harding G (2005) Radiat Phys Chem 73(2):69
90. Houston JD, Davis M, Davis M (2001) Fundamental of fluoroscopy. W.B. Saunders Co., Philadelphia, PA
91. Cusma JT, O'Hara MD (2001) In: Balter S (ed) Interventional fluoroscopy, technology and safety. Wiley, New York
92. Robson KJ (2001) A parametric method for determining mammographic X-ray tube output and half value layer. Br J Radiol 74:335
93. Schriene-Karoussou A (2007) Review of image quality standards to control digital X-ray systems. Radiat Prot Dosim 177(1–3):23
94. Haring G, Thran A, David B (2003) Liquid metal anode X-ray tubes and their potential for high power operation. Radiat Phys Chem 67(1):7
95. Mskinley RL, Torani MP, Samei E, Bradshaw ML (2005) Initial study of past quasi-monochromatic X-ray beam performance for X-ray computed mammography. IEEE Trans Nucl Sci 52(5):1243
96. Seeram E (2001) Computed tomography: physical principles, clinical applications, and quality control. Saunders, Philadelphia
97. Ren B et al (2008) The effect of Tomo-synthesis X-ray Pulse width on Measured Beam Quality. Proceedings of 9th international workshop on digital mammographic breast density, vol 5116, Tucson
98. Graham DT, Cloke P (2001) Principles of radiological physics. Churchchill Livingstone, New York, p 281
99. Azzam EI, DeToledo SM, Pandey BN, Venkatachalam P (2007) Mechanism underlying the expression and propagation of low dose/low fluence ionizing radiation. Int J Low Rad 4(1):61
100. Pelliccianoni M (2000) Overview of fluence to effective dose and fluence to ambient dose equivalent conversion coefficients for high energy radiation calculated using FLUKA code. Rad Prot Dosim 88:297
101. Chen J (2006) Estimated fluence to absorbed dose conversion coefficients for use in radiological protection of embryo and foetus against external exposure to photons from 50 keV to 10 GeV. Rad Prot Dosim 121(4):358
102. Beck R, Latocha M, Dorman L, Pelliccioni M, Rollet S (2007) Measurements and simulations of radiation exposure to air craft workplaces due to cosmic radiation in the atmosphere. Rad Prot Dosim 126(1–4):564
103. Sihver L, Mancusi D (2008) Improved dose and fluence calculations by using tabulated cross sections in PHITS. IEEE nuclear science symposium, Dresden, 19–25 Oct 2008
104. Bozkurt A, Xu XG (2004) Fluence to dose conversion coefficients for monoenergetic photon beams based on the VIP-man anatomical model. Rad Prot Dosim 112(2):219
105. Midgley SM (2005) Measurements of X-ray linear attenuation coefficient for low atomic number, materials at energies 32–66 and 140 keV. Radiat Phys Chem 72(4):525
106. Eisberg RM (1961) Fundamentals of modern physics. Wiley, New York
107. Aichinger H, Dierker J, Barfuk SJ, Sabel M (2004) Radiation exposure and image quality in X-ray diagnostic radiology. Springer, Berlin
108. Siebert JA, Boone JM (2005) X-ray imaging physics for nuclear medicine. J Nucl Med Technol 33(1):3
109. Orito R et al (2003) A novel design of the MeV gamma ray imaging detector with microTPC. Nucl Instrum Meth A 513(1–2):408
110. Washington CM, Leaver DT, Leaver D (2003) Principles and practice of radiation therapy. Elsevier, London
111. Bentel GC (1995) Radiation therapy planning, 2nd edn. McGraw Hill, New York

# References

112. Khan FM (2004) Treatment, planning radiation oncology, 2nd edn. Lippincott Williams & Wilkins, New York
113. Bushong SC (2008) Radiologic science for technologists: physics, biology, and protection. Mosby, Inc., St. Louis
114. Graham DT, Cloke P (2003) Principles of radiological physics, 4th edn. Churchill Livingstone, Edinburgh
115. Sarkar S et al (2007) A Linogram/Sinogram cross-connection method for motion correction in planar SPECT imaging. IEEE Trans Nucl Sci 54(1):71–79
116. Beyer T et al (2000) J Nucl Med 41(8):1369
117. Nassalski A et al (2007) IEEE Trans Nucl Sci 54(1):3
118. Seeram E (2008) Computed tomography: physical principles, clinical applications and quality control. W.B. Saunders Co., Boston
119. Bushong SC (1993) X-ray interaction with matter, 5th edn. Mosby Year Book Inc., St. Louis
120. Hollins M (2001) Medical physics, 2nd edn. Nelson Thrones, New York
121. (2007) 11th mediterrian conference on medical and bioengineering, Ljubljana, Slovenia, 26–30 June 2007, Vol.16, Springer Pub. NY
122. Elgazzar AH, Silberstein EB (2004) Orthopedic nuclear medicine. Springer, New York
123. Iniewski K (2009) Medical imaging principles, detectors and electronics. Wiley, Hoboken
124. Kim J, Sung S, Kim J (2006) A study on industrial gamma ray CT with a single source detector pair. Nucl Eng Technol 38(4):383
125. Ruddy FH, Dullo AR, Siedal JG, Petrovi B (2009) Separation of alpha emitting radioisotopes actinium 225 and bismuth 213 from thorium 229 using alpha recoil method. Nucl Instrum Meth A B213:351
126. IAEA (2009) Cyclotron produced radionuclides: principle and practice, vol 465, Tech-series. IAEA Publications, Vienna, Pergamon Press, London
127. Glatstein E (2007) Radiotherapy in practice. In: Hoskid PJ (ed) Radio therapy. Oxford University Press, Oxford
128. Santesmases MJ (2004) Peace propaganda and biomedical experimentation. J Hist Biol 39 (4):765
129. Subin YN (2001) Model calculations and evaluation of nuclear data for medical radioisotopes production. Radiochimica Acta 89(4–5):317
130. Knap FF Jr (2003) New developments with unsealed sources for targeted therapy. Nuclear medicine program. 41st annual meeting, German Society of Nuclear Medicine Essen, Germany, 2–6 April 2003

# Radiation Exposure: Consequences, Detection, and Measurements

**2**

## Contents

| | | |
|---|---|---|
| 2.1 | Introduction | 60 |
| 2.2 | Sources of Radiation Exposure | 60 |
| | 2.2.1 Natural Sources | 60 |
| 2.3 | Biological and Related Effects of Radiation | 61 |
| 2.4 | Effects of Radiation on Consumable Products | 67 |
| 2.5 | Effects of Radiation | 68 |
| | 2.5.1 Plants and Vegetations | 68 |
| | 2.5.2 Electronics and Associated Active and Passive Elements | 70 |
| 2.6 | Detection of Radiation | 73 |
| | 2.6.1 Detection of Alpha ($\alpha$) and Beta ($\beta$) Particles | 74 |
| | 2.6.2 Detection of Gamma Ray | 75 |
| | 2.6.3 Detection of X-Ray | 75 |
| 2.7 | Neutron Detection | 75 |
| | 2.7.1 Thermal Neutrons | 76 |
| | 2.7.2 Slow Neutrons | 77 |
| 2.8 | Boron Reaction | 77 |
| 2.9 | Lithium Reaction | 78 |
| 2.10 | Instrumentation | 78 |
| | 2.10.1 Ion Chamber | 79 |
| | 2.10.2 Free-Air Ionization Chamber | 80 |
| | 2.10.3 Proportional Counter (PC) | 81 |
| | 2.10.4 Position-Sensitive Proportional Counter | 84 |
| | 2.10.5 Parallel-Plate Avalanche Counter | 85 |
| | 2.10.6 Geiger–Muller (GM) Counter | 86 |
| 2.11 | Photomultiplier | 87 |
| | 2.11.1 Tube | 87 |
| | 2.11.2 Solid-State Photomultiplier (SSPM) | 89 |
| 2.12 | Modes of Detector Operation | 90 |
| | 2.12.1 Current Mode | 90 |
| | 2.12.2 Mean-Square Voltage Mode (MSV) | 91 |
| | 2.12.3 Pulse Mode | 91 |
| 2.13 | Recording and Measurement Techniques | 95 |
| | 2.13.1 Spectroscopy and Spectrometry | 95 |
| | 2.13.2 Gamma-Ray Spectroscopy | 95 |
| | 2.13.3 Neutron Spectroscopy | 98 |

T.K. Gupta, *Radiation, Ionization, and Detection in Nuclear Medicine*,
DOI 10.1007/978-3-642-34076-5_2, © Springer-Verlag Berlin Heidelberg 2013

| | 2.13.4 | X-Ray Spectroscopy | 103 |
| | 2.13.5 | Characteristic Parameters of the Spectrum | 104 |
| | 2.13.6 | Dose Received During Medical Activities | 108 |
| | 2.13.7 | Measurement of Dose Received | 109 |
| | 2.13.8 | Dosimeters | 110 |
| | 2.13.9 | Commercial Dosimeters | 111 |
| | 2.13.10 | MOSFET Dosimeter | 111 |
| | 2.13.11 | Miniature-Type Dosimeters | 112 |
| 2.14 | Statistical Fluctuations in Nuclear Process | | 115 |
| | 2.14.1 | The Binomial Distribution | 118 |
| | 2.14.2 | The Poisson Distribution (PD) | 120 |
| | 2.14.3 | Gaussian Distribution (GD) | 122 |
| 2.15 | Chi-Square Distribution | | 123 |
| 2.16 | Pros and Cons of Radiation Energy | | 125 |
| 2.17 | Summary | | 126 |
| References | | | 126 |

## 2.1 Introduction

*Exposure to Radiation Energy*: The materials existing on Earth are exposed to both natural and man-made radiation. Natural background radiation comes from three primary sources: *cosmic radiation*, *external terrestrial radiation*, and *radon radiation*. In addition to these radiations, there are man-made radiations that come from medical X-rays, gamma rays, nuclear medicine, (radionuclides/isotopes), and consumer products [1]. A schematic of the sources of radiation exposure that all the living and the nonliving materials are exposed to is listed below in Fig. 2.1.

## 2.2 Sources of Radiation Exposure

### 2.2.1 Natural Sources

Among the natural radiations that concern most are *radon, cosmic radiation, galactic*, and *solar particles*. The alpha ($\alpha$) radiation emitted by *radon* is same as the $\alpha$-radiation radiated from plutonium (Pu). It is a naturally occurring radioactive gas. You cannot see it, smell it, or taste it. The primary routes of potential human exposure to radon are inhalation and ingestion. Humans, particularly in underground and work areas such as mines and buildings, are exposed to elevated concentration of radon and its decay products. The next natural radiation that affects the environment is the *galactic* and *solar particles* that have free access to spacecraft outside of the *magnetosphere* and were identified in early 1960 during the malfunction of the spacecraft electronics.

Figure 2.2 shows EGRET gamma ray surveyed by NASA, USA. Some *galactic cosmic rays* interact with the interstellar medium and produce *gamma* ($\gamma$) *rays* [2]. The terrestrial and cosmic rays induce single events and produce an annual dose

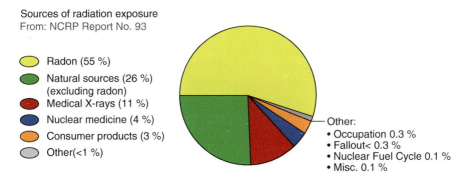

**Fig. 2.1** The total percentage of exposure of radiation energy (Courtesy of US EPA, March 2007)

equivalent to 3.1 millisievert (mSv) [3, 4]. Table 2.1 shows the list of the radiation sources that affect the environment and the surrounding.

Levels of natural or background cosmic radiation, however, can vary greatly from one location to another depending on the altitude. The *biological effects* due to natural or background cosmic radiation are so small that they may not be counted for.

## 2.3 Biological and Related Effects of Radiation

The *commercial*, *industrial*, and *medical* activities contribute about 310 mrem to our annual radiation exposure. It has been estimated that there is a continuous release of ~200-MeV energy per fission that results during the generation of electricity from the chain reactions of uranium-235 ($U^{235}$). To understand the *biological effects* of radiation, we must first understand the difference between *ionizing* and *nonionizing radiation* [5–7]. *Ionizing radiation* (shorter wavelengths) being absorbed experiences sufficient energy to disrupt and separate electrons from atoms (Fig. 2.3) and produces *ions* with a net electric charge. *Nonionizing radiation* (longer wavelengths), on the other hand, has energy to excite atoms, or electrons, but not sufficient to remove electrons from their orbits or to cause the formation of ions [8].

As the threshold for *ionization* lies somewhere in the ultraviolet region of the electromagnetic spectrum, we can consider all *X-rays and gamma ($\gamma$) rays* as ionizing radiation. At the same time, we have observed that all forms of *nuclear radiation* are *ionizing radiation* because of their extremely high energy. When the tissues absorb radiation, excitation occurs immediately, and for higher-energy radiation, we can expect removal of an electron from an atom or molecule [9, 10].

Living tissue contains about 70–90 % water by weight. When the radiation energy exceeds 1,216 kJ/mol, it can ionize the water molecule ($H_2O \rightarrow H_2O^+ + e^-$) forming free-radical ion which contains an unpaired valence shell electron [11]. The radicals formed during *ionization* are strongest oxidizing agents. At the molecular level, these oxidizing agents destroy biologically active molecules by either

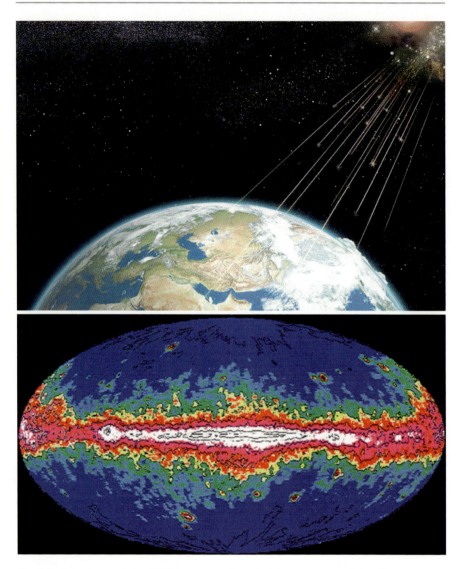

**Fig. 2.2** Galactic cosmic ray (GCR), a source of terrestrial gamma-ray radiation (Photo courtesy of NASA USA)

removing electrons or removing hydrogen atoms. This will lead to the damage of the membrane, nucleus, chromosomes, or mitochondria [12, 13].

Radiation of various energies from X-ray and $\gamma$-rays produces *alpha* ($\alpha$) particles, *beta* ($\beta$) particles, neutrons ($n$) and photons ($p$). These $\alpha$, $\beta$, $n$ and $p$ can interact with complex molecules such as proteins and nucleic acid (NA). Radiation sources having sufficient energy can fracture the molecule and prevent their proper functioning. Fast neutrons, however, collide with light nuclei present in the living cell and produce

## 2.3 Biological and Related Effects of Radiation

**Table 2.1** Annual dose from ground radiation

| Source | Type | Annual dose (mSv) |
|---|---|---|
| Terrestrial | Natural | 0.26 |
| Inhaled radionuclides | Natural | 1.0 |
| Internal radionuclides | Natural | 0.26 |
| Cosmic radiation | Natural | 0.28 |
| Cosmogenetic radionuclides | Natural | 0.01 |
| Medical diagnostic | Man-made | 0.92 |
| Atmosphere weapons testing | Man-made | 0.05 |
| Airline travel | Natural | 1.6/crew, 0.3/passenger |
| Consumer products | Man-made | 0.04 |
| Nuclear power | Man-made | <<0.01 |

Birth et al. [1]

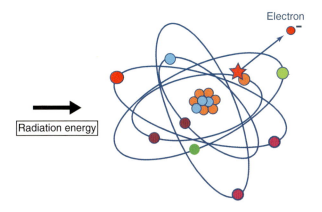

**Fig. 2.3** Energetic ionizing radiation and the detachment of an electron from an atom

*indirect ionizing radiation* [14]. This can result in loss of cell vitality, decreased enzyme activity, initiation of cancer, and genetic mutations depending on the energy of the radiated particle [15, 16]. Figure 2.4 shows the schematic of a healthy cell and the effects of radiation on the healthy cell.

Not all forms of radiation have the efficiency for damaging biological organisms. The energetic initial radiation loses energy as it passes through the tissue and causes more damage. As for example, when the energy of the radiation is more than 100 Gy (1 Gy = 100 rad = 1 J/kg), it affects the central nervous system, resulting in loss of coordination (including breathing problems), with death occurring within 1 or 2 days. When the dose ranges from 9 to 90 Gy, it can damage the gastrointestinal tract causing nausea, vomiting, and diarrhea. Even low doses (2–8 Gy) can damage bone marrow and other hematopoietic tissues [17]. It is expected that the research community, the physicians, and the manufacturers should be aware of the *competitive issue—the dose*. The exposure rate constant ($A_S$) for some common radioisotopes is given below (in Table 2.2, in general terms).

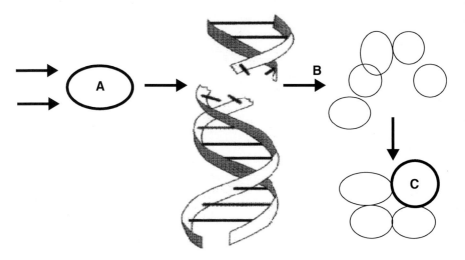

**Fig. 2.4** The ionization radiation can damage a healthy cell (*A*) and can produce a cancerous cell (*B*), which can ultimately turn into a tumor (*C*)

**Table 2.2** Exposure rate constant for some common radionuclides γ-ray sources

| Radionuclide | Exposure rate constant (R-cm$^2$/h mCi): (As) |
|---|---|
| Antimony ($^{124}$Sb)-124 | 9.8 |
| Cesium ($^{137}$Cs)-137 | 3.3 |
| Cobalt ($^{57}$Co)-57 | 0.9 |
| Cobalt ($^{60}$Co)-60 | 13.2 |
| Iodine ($^{125}$I)-125 | ~0.7 |
| Iodine ($^{131}$I)-131 | 2.2 |
| Manganese ($^{54}$Mn)-54 | 4.7 |
| Radium ($^{226}$Ra)-226 | 8.25 |
| Sodium ($^{22}$Na)-22 | 12.0 |
| Sodium ($^{24}$Na)-24 | 18.4 |
| Technetium (99mTc)-99m | ~0.7 |
| Zinc ($^{65}$Zn)-65 | 2.7 |

*Source*: Radiological Health Hand Book, US Department of Health

Different parts of the mammal's body have different mass, size, shape, and compositions. As a result, the attenuation characteristics and the distribution of the radionuclide will not be uniform throughout the whole body or different organs of the body. At the same time, the ionization density to biological material when exposed to X-rays or gamma (γ) rays is quite low compared to the exposure to neutrons, protons, or alpha (α) particles. Therefore, in order to identify the *radiation biological effectiveness* (RBE), a second unit of absorbed dose has been defined as the *roentgen equivalent man* (rem). It is defined as the absorbed dose in rad times the RBE of radiation (rem = rads × RBE) [18, 19]. Table 2.3 shows the values for the radiation biological effectiveness (RBE) of different forms of radiation.

## 2.3 Biological and Related Effects of Radiation

**Table 2.3** The radiation biological effectiveness (RBE) of various forms of radiation

| Radiation | RBE |
| --- | --- |
| X-rays and gamma rays | 1 |
| Electrons | 1 |
| Particles with energies $< 0.03$ MeV | 1 |
| Particles with energies $> 0.3$ MeV | 1.7 |
| Slow-moving thermal neutrons | 3 |
| Fast-moving neutrons or protons | 10 |
| Alpha ($\alpha$) particles or heavy ions | 20 |
| Neutrons (energy dependent) | 5–20 |

Courtesy of International Commission on Radiological Protection (ICRP)

All cells are not equally sensitive to radiation exposure. In general, the cells that produce blood are more sensitive to irradiation even at a very lower dose. An acute radiation dose (10 rad or higher to the whole body) can cause a pattern of clearly identifiable symptoms (syndromes), and the condition is called an *acute radiation syndrome*. As for example, the effect of radiation to the tumors and normal tissues is the result of reactive oxygen species (ROS) and reactive nitrogen oxide (NO) species (RNOS) generation. Hydroxyl ($-OH$) radical is the most reactive species in vivo that contributes to the peroxidation of lipids and nucleic acids following radiation therapy [20].

From the above *biological effects* of radiation on mammals, we can categorize *biological effects* in two sections: (1) deterministic and (2) probabilistic. The first one is generally the reduction of blood cells, skin reddening and blistering, induction of sterility, etc. These arise out of massive cell damage or cell killing due to exposure of the biological system to ionizing radiation.

The second category of the *biological effect* is known as *stochastic effect*, and it results from the *mutagenic* action of ionizing radiation, which can lead to the loss of control over the cell division. It eventually results in the induction of cancer as a result of *mutagenic* disturbance [21, 22]. On the other hand, if it happens to be a germ cell, the mutated information can be passed on to the progeny leading to *genetic* effects.

*Ionizing radiation* thus can cause *mutation* either by directly affecting the *deoxyribonucleic acid* (DNA) or indirectly by producing active chemical species in its vicinity, which can affect DNA—*a double-standard helical macromolecule present in chromosomes inside the nucleus* (Fig. 2.10). In simpler term, *nucleic acid* is a linear polymer of nucleotides connected by phosphodiester bonds [23].

So far, we have discussed the dark side of the radiation. As a matter of fact, radiation, like other aspects of nuclear science, can both be *destructive* and *beneficial*. Some of the intelligent uses of radiation that are beneficial to human being are treatment of cancer, medical diagnosis, and food preservations.

Figure 2.5 shows a simplified two-dimensional diagram of DNA. It has been identified that a prominent cause of DNA damage is its exposure to radiation. The ionizing radiation simulates the cell to form pyrimidine dimer, most frequently

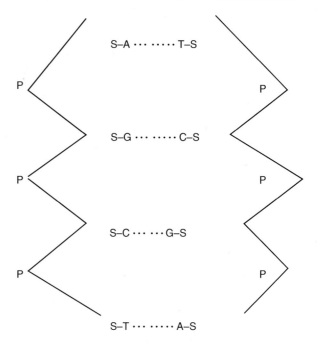

**Fig. 2.5** Schematic of a simplified two-dimensional structure of DNA. *S* sugar, *P* phosphate, *A* adenine, *G* guanine, *C* cytosine, *T* thymine

between adjacent thymine bases in the same DNA strand, and results cross-linkage of the molecule. However, it is possible to repair various kinds of DNA damage. But, if the human being is mutant in the ability to perform radiation repair, he/she can suffer from a disease known as *xeroderma pigmentosum* [24].

Radiation-induced gnomic instability (RIGI) is a delayed, long-lasting effect of ionizing radiation. What initiates RIGI and the cause that makes it to persist over time is unknown. However, it is presumed that the potential mechanism of RIGI is the oxidative stress caused by a high level of reactive oxygen species (ROS). Experimental observations reveal that increased ROS level in different radiation-induced genomically unstable cell system is the cause of RIGI [25, 26]. Figure 2.6 shows a model for the role of mitochondria in RIGI.

In the last 35 years, a lot of research has been focused on the effects of low-level radiation. But many of the findings have failed to establish the so-called *linear hypothesis*. However, some evidence suggests that there exists a threshold below which no harmful effects of radiation occur. As a matter of fact, the issue of low radiation and its effect on any increase in cancer modality has become very controversial.

For creatures, exposure to *nuclear radiation* can cause loss of hair, abnormal growth of thyroid, fatigues, and drop in the number of red blood cells. Although *radiation* may cause *cancer* at high doses and high dose rates, public data regarding low levels of exposure below about 1,000 mrem (10 msv) are harder to interpret. However, *light radiation sickness* begins at about 50–100 rad (0.5–1.0 gray (Gy), 0.5–1.0 sv, 50–100 rem, and 50,000–10,00,000 mrem) [27, 28].

## 2.4 Effects of Radiation on Consumable Products

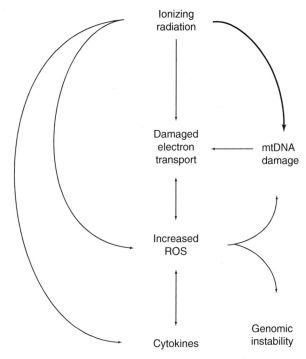

**Fig. 2.6** A model for the role of mitochondria in RIGI (Kim et al. Permission from Oxford University Press)

Figure 2.7 shows the effect of radiation on humans being exposed to a 15-megaton bomb that exploded at a test site 100 miles apart from Marshall Islands, Bikini, north of the equator in the western Pacific Ocean, on March 1, 1954.

In order to meet the criteria and the health and safety regulations, the public and the people working with radioactive materials and in the nuclear plants should be aware of the facts that proper treatments, conditioning, and final disposal of the wastes must be carried out cautiously. Radiation exposures during normal operations are to be considered in two parts: (1) exposures to occupational workers and (2) exposures to environment. Therefore, as a safety measure, the *detection* of the amount of exposure to the ionization energy is essential [29, 30].

## 2.4 Effects of Radiation on Consumable Products

People should be aware of the facts that certain foods such as bananas and nuts contain higher levels of radiation than other foods. Other consumer products that may be the source of radiation are toys and certain types of cosmetics, like hair products, contact lens solution, removing irritants, and allergens. Smoke detector is also a source of radiation, which uses americium-241 in gold matrix. During sterilization of the bandages and the surgical instruments, radiation sources are

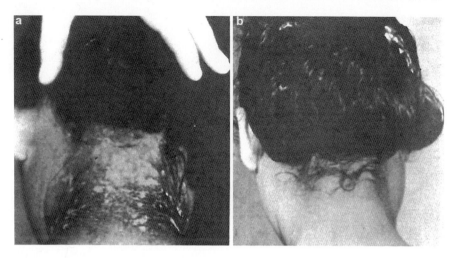

**Fig. 2.7** (a) Twenty-eight days after exposure to beta particle and (b) after recovery (Courtesy of Armed Forces Institute of Pathology)

used. The last but not least, the people should know that the brick and stone homes have higher natural radiation levels than homes made of other building materials such as wood.

## 2.5 Effects of Radiation

### 2.5.1 Plants and Vegetations

So far, we have discussed mostly about the biological effects of ionizing radiation on living animals in context of human species. But when we are talking about the effect of ionizing radiation on the environment and the surrounding, we must consider the plants and vegetations that are part and parcel to our environment. An obvious question would be: How the ionizing radiation affects the plants and the vegetations that constitute the major part of the environment? Real and his coworkers [31] have reported a comprehensive review on the effects of ionizing radiation exposure to plants, fish, and mammals.

Figure 2.8 shows radiation effects on plants in a field near Pripyat in the Ukrainian Soviet Socialist Republic. On April 26, 1986, reactor number four at Chernobyl Nuclear Power Plant near Pripyat in the Ukrainian Soviet Socialist Republic exploded. The plume drifted over extensive parts of Western Soviet Union, Eastern Europe, Western Europe, Northern Europe, and eastern North America with light nuclear rain falling as far as Ireland. After the disaster, three square miles of pine forest in the immediate vicinity of the reactor turned ginger brown and died, earning the name of the *Red Forest* [32–34].

**Fig. 2.8** Radiation effects on plants in a field near Pripyat in the Ukrainian Soviet Socialist Republic (Photo courtesy of Wikipedia)

Effects of ionizing radiation on plants and animals are considered at two levels: (1) on the individual and (2) on the population [35]. The population effect arises from the individual effects, and it is only the statistical aspect, which is important at the population level. For plants, population effects are loss of foliage, reduced growth, and loss of reproductive ability and ultimate extinction of the species. In the case of animals, it is generally considered on the basis of the health and overall survival of the animals suffering from chronic diseases.

Higher plants, which are more sensitive to radiation, are affected by ionization energy less than 10–1,000 Gy. Extensive studies on pine-birch forests have shown that trees of the same species and age have different dose sensitivities depending on external conditions such as light interception, soil fertility, and wind erosion. For example, several studies on the general threshold in the *oak-pine forest at Brookhaven* have shown that approximately 0.5 Gy/day (d) is detrimental to the trees. Most of the species are found to be affected when the radiation label reaches approximately 3 Gy/d. Several plants such as *algae, fungi*, and symbiotic *lichens* are resistant to gamma-ray radiation. However, older plants and mostly the woody species are seen to be affected with a dose of 10 Gy/d or more. The *Erigeron canadensis* are seen to show detrimental effects at 6.4 Gy/d, where most of the *Mediterranean species* are found more resistant to ionizing radiation. However, the sub-Mediterranean species are not as much resistive to ionizing radiation compared to the Mediterranean species. Deserts shrubs are seen to be affected when the ionization goes beyond 0.001–0.1 Gy/h and above 0.07 Gy/d are seen to be detrimental to all vegetation.

It has been found that gamma radiation produces enhanced *protease activities* in some seeds like *peanut* and *wheat* and induces oxidative stress, which damages the protein quantity of the seeds. *Gamma radiation* also damages the *enzyme* of the *citrus fruits* when the dose exceeds 15 Gy. Considering reproductive viability, even though the size and rate of pollen are temporarily affected with a low dose of 0.7 Gy/d, they get restored within 3 years for irradiation up to 12 Gy [36].

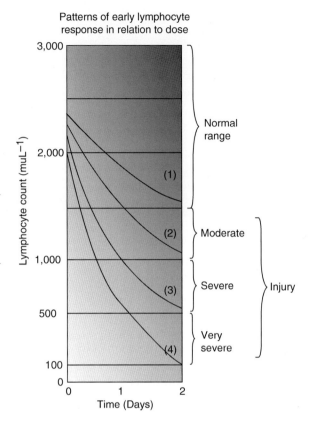

**Fig. 2.9** Classical Andrews lymphocyte depletion curves. *Curve 1–3*: 1 Gy, *curve 2–4*: 4 Gy, *curve 3–4*: 6 Gy, and *curve 4*: 7.1 Gy (Courtesy: R. Goans, ORAU)

From the studies of the radiation exposure and the consequences that are so far presented, it seems that mammals are the most sensitive to ionizing radiation, and a dose of 5–15 Gy can be lethal dose to them. At the other extreme are the bacteria and protozoa that are affected with a dose ranging from $10^3$ to $10^4$ Gy. Classical Andrews lymphocyte depletion curves (1–4) and accompanying clinical severity ranges correspond roughly the following graph (Fig. 2.9) [38]:

Table 2.4 shows the comparative radiosensitivity of groups of organisms.

### 2.5.2 Electronics and Associated Active and Passive Elements

Besides being its detrimental effects on health of the living cells and tissues of creatures, *nuclear radiation* induces defects in several *semiconductor devices* and *electronics* leading to enhanced generation/recombination currents. As a result, charge collection signal is reduced, and there is a drift in operating point.

## 2.5 Effects of Radiation

**Table 2.4** Comparative radiosensitivity of groups of organisms

| Group | Lethal dose (range) in rad |
|---|---|
| Bacteria | 100,000–1,000000 |
| Insects | 5,000–1000,000 |
| Fish | 1,000–3000 |
| Mammals | 300–1200 |
| Herbaceous plants | 5,000–70,000 |
| Coniferous trees | 800–3000 |
| Deciduous trees | 4,000–10,000 |

Adopted from Pfafflin JR, Ziegler EN (2006) Environ Sci Eng. Springer

The vertical primary population intensities of *galactic cosmic rays* (GCR) and *solar particles* affect the electronics of the spacecraft. As a result, military space electronics require special technology to study the primary population intensities of GCR and solar particles that cause variations in secondary neutron and proton levels [2]. During irradiation, the incident particles collide with the nucleus and transfer part of the energy. Experimental observations show that there is an exact linear relationship between ionizing energy deposited by the incident particle in lattices of the semiconductor materials and the measured degradation [39]. However, long-term exposure to ionizing energy results in nonlinearities, and these complicate the determination of displacement damage factors.

Radiation-induced defects in silicon (Si)- and germanium (Ge)-based junction devices (e.g., CMOS, bipolar transistors, photodiodes) are well established [40, 41]. Irradiated energetic particles incident on the device lose their energy to ionizing process and result in the production of electron–hole pairs and displaced atoms. The effectiveness of the radiation-induced displacement damage depends upon the bombardment conditions and on the time after irradiation [42]. A key theme that emerged from all these studies by various workers suggests that the effects on electrical properties of the semiconductor materials depend on the irradiated rays (fission or neutrons or gamma rays) [43, 44].

As a matter of fact, the *metal oxide semiconductor* (MOS) technology, particularly *complimentary* MOS (CMOS) technology, which is the dominant commercial technology, has been severely affected by ionizing radiation [27, 28]. On the other hand, *silicon-on-insulator* (SOI) technology has been developed for radiation-hardened applications for many years. However, bipolar amplification caused by floating body effects can significantly reduce the single-event upset (SEU) hardness of SOI devices [29].

In *bipolar junction transistors* (BJTs, npn, and pnp), current gain is noticed when base emitter junction is irradiated [30, 31]. Earlier study by Goben demonstrated the damage in bipolar transistor by neutrons [32].

Many medical modalities like *computer tomography* (CT), *positron emission tomography* (PET), *single-photon emission tomography* (SPECT), *magnetic resonance imaging* (MRI), and *ultrasound* are inherently digital. These digital technologies offer significant improvements in image quality and dose utilization

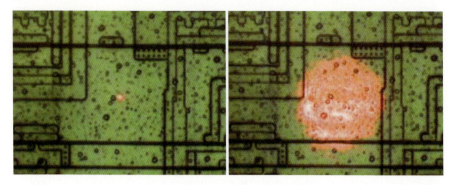

**Fig. 2.10** A large-diameter laser beam that allows for rapid identification of regions in complex integrated circuits (ICs) that are susceptible to single-event effects (Photo courtesy of Crosslink, The Aerospace Corporation, CA)

[33]. However, most of these digital circuitry (*flat-panel detectors* (FPD), *charge-coupled devices* (CCD), and *thin-film transistor* (TFT) array) used in medical imaging system are based on solid-state *integrated circuit* (IC) technology, similar in many ways to the imaging chips used in visible wavelength digital photography and video [45]. These photonic imagers are also being used in space systems, where they are exposed to space radiation environment. Ionizing radiation during X-ray and γ-ray exposure affects the normal behavior of these devices [46].

Aerospace is involved in collaborative research efforts to study the novel approaches for hardening commercially available integrated circuits (ICs) against single-event latch-up. The picosecond laser facility is also being used to study the effectiveness of various design strategies for mitigating the effects of single-event transients in digital electronic circuits. Figure 2.10 shows the use of a large-diameter laser beam allowing for rapid identification of regions of complex integrated circuits (ICs) that are susceptible to single-event effect.

It has been observed that when a cosmic ray passes through the drain region of an NMOS transistor, a short circuit is momentarily created between the substrate (normally grounded) and the drain terminal (normally connected to the positive power of the supply voltage) resulting a spike of current. The amount of charge that is collected from the ion track can burn out the device or may cause malfunctioning of the device. Figure 2.11 shows the path of the cosmic ray through the drain of an NMOS transistor.

During 1990s, many groups were involved in studying the radiation effects in charge-coupled devices (CCDs, arrays of MOS capacitors) including CMOS [47] thin-film transistors (TFTs), passive pixel arrays such as charge-induced devices (CID), photodiodes (including avalanche), infrared (IR) detectors, and other semiconductor devices that are being used in space program or in medical imaging system [48–51].

High dose of irradiation causes radiation damage and general alteration of the operational and detection properties of a detector (scintillator). There are three main aspects of radiation stability in plastic scintillator such as polymer hardness (optical

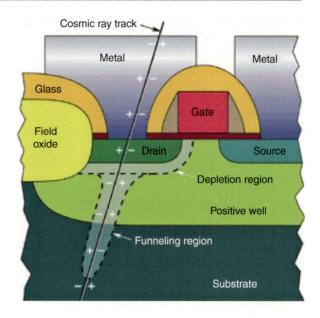

**Fig. 2.11** The path of a cosmic ray through the drain of an NMOS transistor (Courtesy of Cross-link, The Aerospace Corporation, CA)

stability), dopant stability, and stability in fiber waveguide structures. The amount of light intensity losses in the scintillator (which is a bulk effect of the polymer) depends on the dose and the building material of the scintillator. The light yield loss and the transmittance loss in plastic scintillators are about ~20 % and 2–6 %, respectively, for a dose of $10^5$Gy [52]. In scintillation fibers, radiation degrades scintillation core, polymer cladding, and core/cladding interface [53]. On the other hand, garnet-containing inorganic scintillators of terbium (Tb) and lutetium (Lu) show increased resistance to radiation damage until the dose reaches $10^5$Gy or higher [54].

## 2.6 Detection of Radiation

Ionizing radiation is ubiquitous in the environment and also comes from radioactive materials, X-ray tubes, and particle accelerators. It is invisible and not directly detectable by human senses. Therefore, instrument for the evaluation of minimum levels of delectability is of prime interest. As a matter of fact, nuclear detection is always a multistep, highly indirect process. For example, in a scintillation detector, incident radiation excites a florescence material that de-excites by emitting photons of light. The light is focused onto the photocathode of a photomultiplier (PMT) tube that triggers an electron avalanche. The most common handheld or portable instruments are ionization chamber, proportional counters, Geiger counter, optically simulated luminescence (OSL) dosimeters, and multichannel analyzer system.

Four major forms of radiation are commonly found emanating from radioactive matter: *alpha* ($\alpha$), *beta* ($\beta$), *gamma* ($\gamma$), and *X-ray* radiation. *Alpha particles*, the heaviest and the most highly charged of common nuclear radiation, are the most difficult to detect. Since these particles can easily be trapped by other materials, quantitative measurement of alpha radiation is impossible outside of a laboratory environment. The field instruments, such as the AN/PDR-56, AN/PDR-77, and ADM-300, use an extremely thin piece of aluminized Mylar® film on the face of the detector probe. It has been found that counting efficiencies of these alpha detectors are between 20 and 60 %.

## 2.6.1 Detection of Alpha ($\alpha$) and Beta ($\beta$) Particles

*Beta particles* are energetic electrons emitted from the nuclei of many natural and man-made radioactive materials. Low-energy beta particles are less penetrating than the alpha particles, requiring special techniques for detection. From the detection standpoint, unfortunately, high-energy beta and gamma radiations are not the primary decay products of the most likely radioactive materials. Rather, the major potential source of beta and/or gamma radiations is from fission products.

Alpha ($\alpha$) and beta ($\beta$) particles are detected and measured by *ionization chamber*. Absolute measurements of ($\alpha$) particle are straightforward and involve evaluation of the effective solid angle subtended by the counteractive volume (which is close to $2\pi$ for initial source flow counts). On the other hand, absolute ($\beta$) activity measurements are often carried out in a $4\pi$ flow counter. When both alpha ($\alpha$) and beta ($\beta$) particles are present in the emitting source, the spectra of $\alpha$- and $\beta$-particles will be different [55–57]. The $\beta$-spectrum is generally broader and shorter in pulse height. The corresponding count curve will show two distinct plateaus: In the first plateau, $\alpha$s are counted, and in the second, both $\alpha$s and $\beta$s are counted. It has been observed that the *beta plateau* is generally shorter and shows a greater slope than *alpha plateau*. The obvious reason is that the *beta particle* pulse height distribution is broader and less well separated from the low-amplitude noise.

Figure 2.12 shows the spectra of alpha and beta particles, when the pulse height ($H$) is plotted against differential number of pulses within the differential amplitude increment ($dN/dH$). The right-hand-side curve shows the plateaus when counting rate is plotted against the applied voltage ($V$).

Experimental detection of $\alpha$-particles from the radioactive decay of natural bismuth ($^{209}$Bi) has been reported by P. de Marcillac et al. [55]. They have detected an energy release of $3,137 \pm 1$ (statistical) $\pm 2$ (systematic) keV and a half-life of $(1.9 \pm 0.2) \sim 1,019$ years, which are in agreement with expected values. It has been observed that the nuclear structure of naturally occurring $^{209}$Bi isotope gives rise to an extremely low decay probability and it generates low-energy particles that are difficult to detect. Therefore, their detection has no quantitative use but is used as a safety precaution.

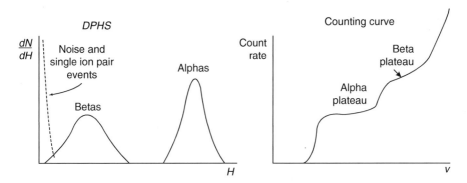

**Fig. 2.12** The spectra and plateaus of alpha and beta particles (Reprinted with permission Wiley)

## 2.6.2 Detection of Gamma Ray

The key process of detecting gamma ray is by ionization. During ionization, it gives up part or all of its energy to an electron. The energized electrons then collide with other atoms and liberate electrons. The liberated charge is collected either directly, using *gas ionization type* (ion chambers, proportional counters, or Geiger–Muller (GM) counters) counters, or indirectly (as with scintillation detector), in order to record the presence of gamma ray and measure its energy. Details of these counters and the scintillators will be discussed later.

## 2.6.3 Detection of X-Ray

*X-rays* are emitted by electrons outside the nucleus, while *gamma rays* are emitted by the nucleus. To detect X-rays, we can use either a photographic plate or photostimuable phosphor. Proportional counters are the most common X-ray detectors used, although CCD chips are rapidly gaining popularity. Besides, Geiger counter based on the ionization of gases is also being used.

Scintillators like NaI:Tl can convert X-ray photons into visible photons and can be detected by a photomultiplier tube (PMT). Other solid-state direct detectors like lithium-doped silicon (Si(Li)) and lithium-doped germanium (Ge(Li)) have also been used to detect X-rays. However, detectors like CdTe, CZT, and $HgI_2$ have also been used to detect X-rays. Recently, these materials are deposited on a TFT array or flat-panel array for medical imaging using X-rays.

## 2.7 Neutron Detection

The process of neutron detection begins when neutrons, interacting with various nuclei, initiate the release of one or more charged particles. Two basic types of neutron interactions with matter are available: recoil type and interaction type. Recoil-type counters measure only the first interaction event. Reaction type, however, takes advantage of the increased reaction probability at low neutron energies by moderating the incoming neutrons.

*Neutron* is electrically *neutral*. Its temperature is called the *neutron energy*, or a free neutron's *kinetic energy*, usually measured in electron volts (eV). *Kinetic energy (KE)*, speed, and wavelength of the neutron are related to the de Broglie relation and are classified according to the distribution of the energies. As for example, a *slow neutron* has energy less than or equal to 0.4 eV, whereas the *fast neutrons* have energy greater than 0.1 eV [58, 59]. When a neutron has energy of about 0.025 eV, it is identified as *thermal neutron*.

Advances in materials and methods have enabled the detection of neutrons with cryogenic neutron detectors to room-temperature semiconductor detectors [60–62]. A cryogenic neutron detector operates by detecting the heat pulses caused by neutron capture and scattering [37, 63, 64]. On the other hand, a semiconductor detector is able to detect neutrons from a neutron-sensitive material like uranium oxide boron, lithium, or polyethylene on aluminum layer, placed in contact with silicon diode or a GaAs detector, which finally detects the nuclear reaction products [65–67]. However, coated detectors are seen to suffer from a lack of sensitivity [68].

Gupta, Schieber, and Knittel et al. [69–71] have reported the results for a number of inorganic scintillators fabricated from various combinations of inorganic elements like Gd, B, and Li. On the other hand, Chung and Chen [72, 73] have described a method where they have used a germanium (Ge) detector where reactions of neutrons with $^{72, 74}$ Ge (n, n', $\gamma$) produce gamma ($\gamma$) rays at 596 and 691 keVs to detect gamma rays. A new-type polycrystalline CVD (chemical-vapor-deposited) diamond detector with high gain for neutron detection has been reported [74–76]. Use of a dosimeter for the detection and measurement of neutrons has also been reported by Angelone et al. [77, 78]. Detectors have also been fabricated from compounds of inorganic materials like $Li_2B_4O_7$ and $BaB_4O_7$ to study neutrons [79]. Siffert et al. on the other hand had exploited CdZnTe (CZT) and CdTe to study thermal neutrons [80].

Some organic scintillators consisting of *phosphors* are prepared by dissolving (generally) the phosphor in a hydrocarbon plastic like polystyrene or polyvinyl toluene or in liquid such as benzene, xylene, or toluene and have been used to detect neutrons. The hydrocarbon matrix with phenyl ring provided by the *phosphor* helps to hop the deposited ionization energy from one ring to another until it finally transfers to a phosphor molecule. In many cases, *silicon rubber* (Si–O–Si) [81, 82] has been used as a suitable matrix. Recently, *anthracene* and *stilbene* have achieved widespread popularity as pure organic scintillators for neutron detection [83]. At the same time, semiconducting polymers (such as poly tri-aryl amine) have offered many advantages as organic detectors over their conventional inorganic counterpart for detection and measurement of neutron [84, 85].

### 2.7.1  Thermal Neutrons

*Thermal neutrons* are used in nondestructive testing (NDT) because of their advantages over X-ray and gamma-ray testing [86, 87]. As a matter of fact, NDT

## 2.8 Boron Reaction

testing by thermal neutrons provides complementary information about the interactions of photons and neutrons with matter. Photons interact with the atomic electrons, while neutrons interact with the atomic nucleus itself. Neutrons can penetrate into the high-density materials that are opaque to X-rays thus allowing the inspection of objects obstructed by a dense material. Moreover, the attenuation of thermal neutrons (relative to X-rays) is pronounced for lighter elements such as hydrogen, carbon, and their compounds that comprise the base for many organic materials. The nuclides that are frequently used for thermal neutron detection are helium ($He^3$), lithium ($Li^6$), boron ($B^{10}$), cadmium ($Cd^{113}$), and gadolinium ($Gd^{155}$ and $Gd^{157}$) [88–90].

### 2.7.2 Slow Neutrons

The energy of a *slow neutron* is extremely low (energy below the cadmium cutoff of about 0.5 eV) with the reaction Q-value, and it loses its initial kinetic energy (KE) during conversion process. *Slow neutrons* are detected through nuclear reactions which result in energetic charged particles such as protons and alpha ($\alpha$) particles [91]. The neutron detectors are composed of two detectors—the target detector and the detector for detection and measurements. Since cross section for neutron interactions in most materials is a strong function of the energy of the incoming neutrons, different techniques have been developed to detect and measure neutrons in different energy regions. For the detection and measurements of *slow neutrons* in an efficient way, we have to consider several factors:

1. The cross section for the reaction must be very large, especially when the target material is incorporated as gas.
2. The target nuclide should either be of high isotopic abundance in natural element or artificially enriched sample.
3. The detection process should be able to discriminate gamma ($\gamma$) rays.
4. Q-value of reaction should be large.

## 2.8 Boron Reaction

Boron $^{10}B$ ($n,\alpha$) has been identified as a popular reaction material for the conversion of neutrons [92, 93]. The reactions that followed due to the reactions of the *slow neutron* and the target material are as follows:

$$Q\text{-value}$$

$$^{10}_{5}B + ^{1}_{0}n \qquad \left\{ ^{7}_{3}Li + ^{4}_{2}\alpha \quad (2.792 \text{ MeV}) \text{ for ground state} \right. \tag{2.1}$$

$$\left\{ ^{7}_{3}Li + ^{4}_{2}\alpha \quad (2.310 \text{ MeV}) \text{ for excited state} \right. \tag{2.2}$$

When the incoming neutron is a *thermal neutron* ($E_n = 0.025$ eV), 94 % of all reactions lead to the excited state [94, 95]. The Q-value of reaction becomes very large between 2.310 and 2.794 MeV compared to the *slow neutron*, and the reaction product is $^{7}_{3}$Li + $^{4}_{2}\alpha$. The energy of the individual product can be calculated from the conservation of energy and momentum as

$$E_{Li} + E_\alpha = Q = 2.31 \text{ MeV} \tag{2.3}$$

The momentum $p$ (product of mass $m$ and velocity $v$) of lithium is equal to the momentum ($p$) of alpha particle, that is, $p_{Li} = p_\alpha$. So we can write $(2m_{Li}E_{Li})^{1/2} = (2m_\alpha E_\alpha)^{1/2}$. From these equations, we can get the values of $E_{Li} = 0.84$ MeV and $E_\alpha = 1.47$ MeV. The thermal neutron cross section for $^{10}$B (n, $\alpha$) reaction is 3,840 barns. The cross-section value drops rapidly with increasing neutron energy and is proportional to $(1/v)$.

## 2.9 Lithium Reaction

The reaction between slow neutron and lithium (Li) is as follows:

Q-value

$$^{6}_{3}\text{Li} + ^{1}_{0}\text{n} \rightarrow ^{3}_{1}\text{H} + ^{4}_{2}\alpha \quad 4.78 \text{ MeV} \tag{2.4}$$

Similarly following the equations presented with boron reactions, we can arrive to the values of $E^{3}_{H} = 2.73$ MeV and $E_\alpha = 2.05$ MeV. The thermal neutron cross section for $^{6}_{3}$Li is 940 barns. Recently, commercially available pyrolytic boron nitride (BN) (5-mm$^2$ area and 1 mm thick) coated with an epitaxial layer of GaAs has been used as thermal neutron source [96].

*Gas-filled detectors* are used to detect either *thermal neutrons* via *nuclear reactions* or *fast neutrons* via *recoil interactions*. Gas-filled detecting chambers typically employ $^{3}$He, $^{4}$He, BF$_3$, or CH$_4$ as the primary gases, at a pressure varying from less than 1 atm to about 20 atm depending on the application. As neutron detectors are also sensitive to gamma ray, therefore, the gamma-ray sensitivity of a neutron detector is an important criterion in its selection.

## 2.10 Instrumentation

The instruments that are used to detect the ionizing radiation are based on ionization of the gas inside a chamber, and they are categorized as (1) *ion chamber*, (2) *proportional counter*, and (3) *Geiger tubes*.

**Fig. 2.13** The picture of an ionization chamber (Photo courtesy of Overhoff company)

## 2.10.1 Ion Chamber

The instrument, in principle, is the simplest of all the gas-filled detectors, where ions are formed by direct ionization and measured by the application of electric field. Unlike proportional counter and the Geiger tubes that are operated in pulse mode, ion chamber is operated either in *current mode* or *pulse mode* [97, 98].

Figure 2.13 shows the picture of an ionization chamber, which is an electrically closed vessel containing an internal electrode.

*Working Principle*: Ions are formed in the gas chamber either by direct interaction with the incident particle or through a secondary process in which some of the particle energy is first transferred to an energetic electron. Experimental observation shows that when a particle having energy of 1 MeV is stopped within the gas, it creates almost 30,000 ion pairs. The deposited energy is proportional to the number of ion pairs formed provided the average energy lost by the incident particle per ion pair formed ($W$) is constant [99].

The total number of energy of the incident particle, which is converted into the information carriers, is seen to fluctuate. An empirical constant known as *Fano factor* is multiplied with the predicted variance to convert it into the experimentally observed variance. The empirical constant is only significant when the detector is operated in pulse mode [100, 101].

The incident particle having sufficient energy ionizes the gas inside the chamber and creates positive ion and free electron. In course of time, the electron and the positive ion gain *thermal energy* and can diffuse away from the regions of higher density. During *thermal motion*, collision between *positive ion* and *free electron* results in *recombination*, and a state of *charge neutrality* exists. However, in presence of electric field, the free electron and positive ion gain momentum, which is known as *drift velocity* [102]. The drifting of the charges actually constitutes an *electric current*, and a *dc ion chamber* is used to measure the *ionizing current*.

At high voltage, the *ionization current* will be saturated, the electric field will be large everywhere, and the recombination will be negligible. The percent by which the measured ion current falls short of true saturation is seen to increase as a function of the measured ion current. It is not unusual to observe some imbalance in the steady-state concentrations of the two charges when electron drifts toward anode followed by the drift of positive charge toward cathode. As a result, a gradient will exist for a species that is free to migrate, and some *net diffusion* must take place in the direction of decreasing concentration. The *direction of diffusion* will be opposite to the direction of charge carrier flow by the electric field, and the effect will reduce the measured ion current. The ratio of the change in ionizing current to the total ionizing current can be modeled after Rossi and Staub as [103]

$$(\Delta I / I) = (\epsilon k T / e V) \tag{2.5}$$

where $\epsilon$ is the ratio of the average energy of charge carrier with electric field present to that without electric field, $k$ is Boltzmann constant, $T$ is temperature in absolute, $e$ is electronic charge, and $V$ is the applied voltage.

## 2.10.2 Free-Air Ionization Chamber

The *free-air* ionization chamber has become the most popular instrument to measure ionization of gamma rays [104–106]. The parallel plate electrodes inside the chamber receive the collimated gamma ($\gamma$)-ray radiation. There are three small electrodes on the top. The central electrode on the top receives the signal, and it is kept at negative potential with respect to the bottom electrode. The two other small electrodes on the top are grounded. Figure 2.14 shows the schematic of a free-air ionization chamber.

The drawback of this system is the loss of ionization due to the creation of secondary electrons at high energy ($>100$ keV) which creates some difficulties during measurements. In order to minimize the loss of ionization measurements, gamma-ray ($\gamma$) energy is kept below 100 keV. At higher energies, $\gamma$-ray exposure measurements are carried out in *cavity chambers* surrounded by a small volume of air. The secondary electrons created by collisions make the surrounding volume of air large. To improve the situation, the volume of air is compressed, and the walls are made sufficiently thick—a condition called *electronic equilibrium* is established. It has been observed that the exposure rate $R$ (C/kg-s) of an air

## 2.10 Instrumentation

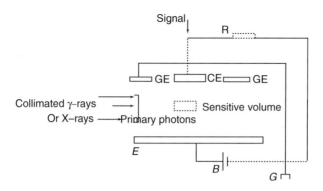

**Fig. 2.14** The schematic of a free-air ionization chamber. *B* battery, *G* ground, *GE* guard electrode, *CE* central electrode, *R* resistance

equivalent ion chamber is directly proportional to the saturated ion current ($I_s$) in amps and inversely proportional to the mass (m) in kilogram (kg), and the relation can be expressed mathematically as

$$R = (d/dt)\{(dQ/dm)\} = \{(I_s)/m\}$$
$$= \{(I_s)/(1.293 \text{ kg/m}^3)\} \cdot \{V(P/P_0)\} \cdot (T/T_0) \quad (2.6)$$

where $Q$ is the charge due to ionization, $V$ is chamber volume (m$^3$), $P$ = air pressure in the chamber, $P_0$ is the saturated pressure (760 mmHg), $T$ air temperature inside the chamber, and $T_0$ is standard temperature (273.15 K).

Again, $Q$, the ionization charge, is equal to $(eT)/(W)$, where $e$ is the electronic charge; $T$ is the kinetic energy of the charged particle; and $W$ is the mean energy required to create an ion pair in a gas. Now, we will introduce a new term called *kerma* ($\Re$), which is defined as the kinetic energy released per unit mass, and we will define it as

$$\Re_{air} = (W_{air}/e)/(1-g) = \{T\}/\{Q(1-g)\} \quad (2.7)$$

where $(1 - g)$ is the mean correction for energy given to the radiated photon. The value of $W_{air}$ = 33.97 eV with a relative standard uncertainty of 0.15 % [107].

Mostly, the ionization chambers are used in current mode, and the average rate of ion formation within the chamber is measured. However, an ionization chamber can also be operated in pulse mode in which separate radiation quantum gives rise to distinguishable single pulse. The equivalent circuit of an ion chamber operated in pulse mode is shown in Fig. 2.15.

### 2.10.3 Proportional Counter (PC)

*Proportional counter* is a gas-filled (mostly 90 % argon and 10 % methane) ionization counter that works on the same principle as the Geiger–Muller (GM) counter. One of the most important applications of PCs as detection and

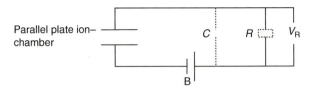

**Fig. 2.15** The schematic of the equivalent circuit of an ion chamber operated in pulse mode. $C$ capacitance, $R$ resistance, $B$ battery/external power, $V_R$ output voltage

**Fig. 2.16** An earlier version (1952) of a 4-pi gas flow proportional counter (Photo credit of Oak Ridge Associated Universities)

spectroscopic instruments is for low-energy radiation measurements ($\alpha$-counting, $\beta$-counting, mixed $\alpha$- and $\beta$-counting, and for the measurements of $\gamma$- and X-rays). They are widely used to detect neutrons. Figure 2.16 shows an earlier version (1952) of a 4-pi gas flow proportional counter. Schutmeister and Meyer are generally credited for the construction of the first true 4-pi detectors in 1947–1948.

According to the design, the PC can be *position-sensitive, parallel-plate avalanche, multi-wire proportional, or gas proportional scintillation type*. It relies on the phenomenon of *gas multiplication* due to secondary ionization, which is a consequence of increasing electric field, and acceleration of the electron liberated by the secondary ionization process. During its subsequent drift, it undergoes collisions with other neutral gas molecules and creates additional ionization.

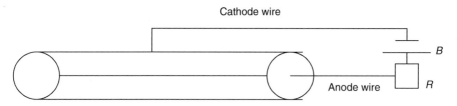

**Fig. 2.17** The schematic of a simplified proportional counter circuit. $B$ is the battery, and $R$ is the load resistance

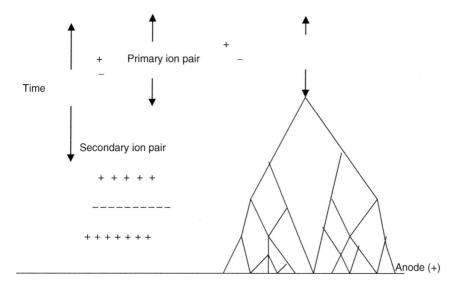

**Fig. 2.18** The schematic shows that the primary charged ion pair produces many secondary ion pairs

However, positive and negative ions achieve very little energy between collisions and do not take part in secondary ionization.

Figure 2.17 shows the schematic of a simplified proportional counter circuit, which operates in the proportional region. The circuit illustrates that the walls act as one of the electrodes (cathode) and a long fine wire at the center of the chamber acts as an another electrode (anode). The geometry of the electrodes and the voltage on them are chosen in such a way that in most of the volume of the counter, the electric field strength is not enough to produce a *Townsend avalanche* [108].

The gas multiplication process takes the form of a cascade and is known as *Townsend avalanche* (Fig. 2.18). Now, we can define the *Townsend avalanche* coefficient ($\alpha$) as the fractional increase in the number of electrons ($n$) per unit path length ($x$). Mathematically, it is expressed as

$$\alpha = (dn/n)\,(1/dx) \tag{2.8}$$

As the applied voltage is increased, a nonlinearity in the pulse form is noticed. On the other hand, when the applied voltage is low, the field is insufficient to prevent recombination of the original ion pairs. However, over some region of the voltage–pulse amplitude curve, gas multiplication is linear, which means that the collected charges are proportional to the original numbers of ion pairs created by the incident radiation. This is the true proportionality region and represents the mode of operation of conventional proportional counter. Assuming the linearity between the *Townsend coefficient* ($\alpha$) and the electric field ($E$), the multiplication factor ($M$) can be written as

$$M = [\{V/(c/a)\}] \cdot [\{\ln\ 2/(\Delta V)\}]\ [\{\ln\ (V)\}/\{(pa\ \ln\ (b/a)\} - \ln\ K\}] \quad (2.9)$$

where $K$ is constant, $b$ = cathode radius, $a$ = anode radius, $V$ = applied voltage, and $p$ = gas pressure [109].

The charge, which is developed in a pulse during measurement (assuming no nonlinearity), is proportional to the number of ion pairs created ($n_0$) by the incident radiation. The amplitude of the pulse, however, is seen to fluctuate from pulse to pulse, and the expression for variance in amplitude of the pulse is $(Qq/Q)^2$ and can be written as

$$(Qq/Q)^2 = (\sigma_{n0}/n_0)^2 + 1/n_0(\sigma_A/\bar{A})^2 \quad (2.10)$$

where $(\sigma_A/\bar{A})^2$ is the single-electron multiplication variation and $(\sigma_{n0}/n_0)^2$ is the ion pair fluctuation. Following Furry distribution and assuming that $A$ is reasonably large, the expected distribution in the number of electrons produced in a given avalanche reduces to a simple exponential form as

$$P(A) \approx \exp.(-A/\bar{A})/\bar{A} \quad (2.11)$$

when the multiplication factor $A$ (the mean value of $A$) is large due to electric field $E$, $(\sigma_A/A)^2 \approx 1/(1 + \theta) \approx c$, where $\theta$ is a parameter related to the fraction of electrons where energy exceeds a threshold energy for ionization ($0 < \theta < 1$) and $c$ is cathode radius [60].

By proper functional arrangement, modifications, and biasing, proportional counter can be used as spectroscopic instruments for soft X-rays and gamma ($\gamma$) rays. It can also be used for neutron radiation detection in mixed radiation fields. As a matter of fact, low-energy X-ray spectroscopy is one of the most important applications of proportional counter. The energy of the absorbed photoelectrons formed by photon interaction within the gas is directly related to X-ray energy. However, proportional counter is not used when the gamma rays are having higher photon energies.

### 2.10.4 Position-Sensitive Proportional Counter

*Position-sensitive proportional counter* (PSPC) is based on the principle of charge division. The anode is a high-resistive wire and is connected to two amplifiers

## 2.10 Instrumentation

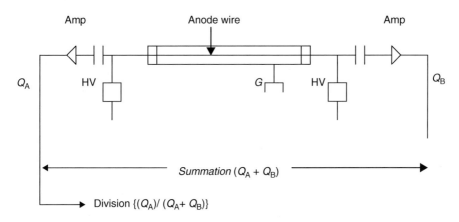

**Fig. 2.19** The diagram of position localization in proportional counters using charge division

at two ends. Electrons drift along radial field lines from their place of formation. The position of the avalanche is a good indicator of the axial position at which the original pair is formed. By summing the amplitude of the two amplifiers, a conventional output pulse is produced with amplitude proportional to the total charge ($Q_A + Q_B$). A position signal is generated by dividing the output signal of a single amplifier by summed signal $\{(Q_A)/(Q_A + Q_B)\}$ to give a pulse that will indicate relative position along length.

Figure 2.19 shows the diagram of position localization in proportional counters using charge division $\{(Q_A)/(Q_A + Q_B)\}$. It is proportional to the position of the particle track. In the Fig. 2.19, HV is the high voltage and $G$ is the ground. The input impedance $I_{imp}$ of the amplifiers is made high, compared to the resistance of the anode wire, $R$ [61], which is made sufficiently large so that the collected charge on one side divided by the total charge on the two opposite sides of the amplifiers follows a simple relation to the position of the interaction [37, 62, 63]. An alternative method is developed on the basis of the relative rise time from the preamplifiers placed at either end of the resistive wire [64].

### 2.10.5 Parallel-Plate Avalanche Counter

It consists of two parallel plate electrodes separated by a distance inside a closed chamber with a low-pressure gas inside. When a high voltage is applied, a charge particle that traverses the gap between the plates is multiplied through the usual gas amplification process and leaves a trail of ions and electrons. The numbers of ions that are formed and traveled through the gap reflect the amount

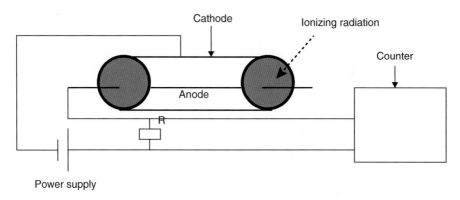

**Fig. 2.20** The experimental setup with a Geiger–Muller counter (*R* resistor)

of energy lost by the particle in transit. Experimental observation shows that for gaps of the order of 1 mm, the fast component rise time is about 2 ns under high field [65].

### 2.10.6 Geiger–Muller (GM) Counter

It is an instrument that can detect a single particle of ionizing radiation with an operating voltage in the Geiger plateau. It consists of a tube filled with low-pressure (0.1 atm) inert gas and an organic vapor or a halogen gas. Inside the chamber, there are two electrodes, between which a potential difference of several hundred volts is created without any flow of current. Figure 2.20 shows the schematic of an experimental setup with a Geiger–Muller counter.

The electrons from the gas discharge are collected as pulses within a microsecond. The pulses should have two slopes, the fast one corresponding to the collection of electrons and the slower one will correspond the drifting time of the positive ions. A single ion pair formed within the fill gas of the GM tube can trigger a full Geiger discharge, and the counting efficiency can reach 100 %.

A portable GM meter used to survey gamma ($\gamma$) radiation consists of high-voltage supply and pulse-counting meter. The higher field in *GM tubes* enhances the intensity of avalanche process. The most important achievement of a *GM counter* is that all pulses from a *GM tube* are of the same amplitude regardless of the number of original ion pairs that initiated *avalanche* process. Therefore, a *GM tube* cannot be used for direct radiation spectroscopy rather than as a simple counter of radiation-induced events [66]. Another disadvantage of *GM tube* is that besides lack of energy information, it has a large dead time that limits its application as radiation detector. Figure 2.21 shows a picture of a *Geiger–Muller (GM) counter with an attached probe head*.

Multiple pulsing is potentially much more severe in *GM tubes* than proportional counters. Therefore, special precaution must be taken in *GM counters* to prevent the

**Fig. 2.21** A picture of a Geiger–Muller (GM) counter with an attached probe head (Courtesy: Biodex Med. Syst., NY)

possibility of excessive multiple pulsing which can be prevented by internal quenching by adding a second component called the *quenching* gas. Sometimes, external quenching is performed by reducing the applied voltage with an addition of a resistor to the supply voltage [67]. Most of the *GM tubes* are operated at atmospheric pressure where windows have to support a substantial differential pressure.

For thermal neutrons, the conventional Geiger gases are not suitable because of low capture cross section that result low counting efficiency. On the other hand, for the detection of the fast neutrons, GM tubes are not suitable because they produce recoil nuclei in a Geiger gas, which will generate ion pairs and subsequent discharge.

The efficiency of counting γ-rays by GM tube depends on the probability of the interaction of the rays and the secondary electrons reaching the fill gas, which increases with atomic number of the wall material. Therefore, in order to increase the efficiency, cathode material with high $Z$ (like bismuth $Z = 83$) has been used widely in GM tubes for γ-ray detection.

## 2.11 Photomultiplier

### 2.11.1 Tube

It is a very sensitive and versatile device of radiant energy in the ultraviolet, visible, and near-infrared regions of the electromagnetic spectrum. The structure of a photomultiplier consists of a photosensitive layer called *photocathode* coupled to an electron multiplier structure. The electron multiplier serves as an ideal amplifier,

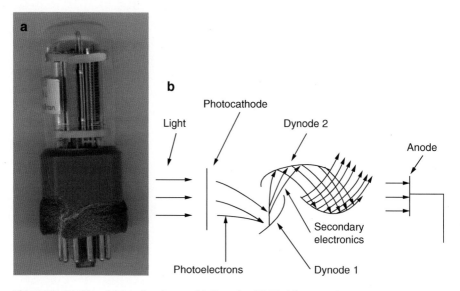

**Fig. 2.22** (a) The picture of a photomultiplier tube (PMT) (Courtesy: Post Industrial, NJ) and (b) schematic representation of a PMT and its operation

which greatly increases the number of electrons and the energy of the output signal. After multiplication through the multiplier structure, a scintillation pulse will give rise to $10^7$–$10^{10}$ electrons, sufficient to serve as charge signal. The charge is efficiently collected at the anode. Most of the PM tubes amplify the signal in a linear manner, producing an output pulse that remains proportional to the number of original photoelectrons over a wide range of amplitude [68]. Figure 2.22a shows the picture of a photomultiplier tube (PMT) and Fig. 2.22b schematic representation of a PMT and its operation.

The surface potential barrier influences another important property of photocathode known as *thermionic* noise. The photocathode is normally coated with multialkali material based on $Na_2KSb$ or $Na_2CsSb$ activated with oxygen ($O_2$). The quantum efficiency (QE) of a PM tube is defined as the number of photoelectrons emitted to the number of incident photons and can be expressed as

$$QE = (\text{Number of photoelectron emitted})/(\text{Number of incident photons}) \quad (2.12)$$

The ideal QE should be 100 %, but because of limitations, practical photocathodes show maximum QE approximately 20–30 %. The QE is a function of the frequency of wavelength or quantum energy of the incident light.

Electrons leaving the photocathode have kinetic energy (KE) approximately on the order of 1 eV or less. The dynode has a positive potential of several hundred volts. The creation of secondary electron is a sensitive function of incident electron energy. The incident electrons with higher energy will create more excited secondary electrons. The overall multiplication factor for single dynode is given by

## 2.11 Photomultiplier

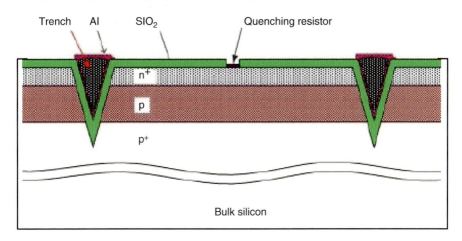

**Fig. 2.23** The cross-sectional view of a V-grooved n⁺ p p⁺ CMOS structure with a quenching resistor which represents an SSPM

$$M_f = (\text{Number of secondary electrons emitted})/(\text{primary incident electron}) \tag{2.13}$$

The secondary yield of the dynodes is seen to increase when the negative electron affinity (NEA) materials are used [69, 70]. The time required for photoemission in the photocathode or the secondary emission from dynodes is very short (0.1 ns or less). The electron transit time of a PM tube is defined as the average time difference between the arrival of a photon at the photocathode and the collection of the subsequent electron burst at the anode, and it ranges from 20 to 80 ns. The amount of transit time spread observed for a special pulse depends on the number of initial photoelectrons per pulse.

### 2.11.2 Solid-State Photomultiplier (SSPM)

*A solid-state photomultiplier* (SSPM) is an array of avalanche photodiode operating in a Geiger mode (G-APD). It was first proposed in the year 1960 as a versatile device and a low-cost solution to optimize the functionality of scintillation materials. It is generally used in medical imaging system [71, 72]. It is hoped that SSPM would achieve low noise and high quantum efficiency compared to a photomultiplier tube (PMT) at a lower cost. Sometimes, the SSPM needs special surface treatments to increase the blue light sensitivity.

Figure 2.23 shows the cross section of a V-grooved isolation-schemed design CMOS structure. The microcell is isolated from both of its common substrate and its surrounding neighbors by a dielectric layer [73, 74]. Essentially, the device is a complementary silicon metal oxide semiconductor (CMOS). The n⁺ p p⁺ doping structure results in a high field gradient in the region around n⁺ p junction enabling

90      2 Radiation Exposure: Consequences, Detection, and Measurements

avalanche multiplication of photoelectrons originating primarily in the $p$ but also in the n+ region.

The SSPM's performance is very much dependent on a quenching doped polysilicon resistor (1–100 M$\Omega$). Lower resistor values provide fast microcell recovery [75]. The device has a low *temperature coefficient* and *optical cross talk*. It has been used as a device for gamma-ray detection and for medical imaging in nuclear medicine [76]. In presence of cross talk, the energy resolution is proportional to the total number of working pixels in the SSPM that participates in the measurements. The cross talk gain factor depends on the excess bias voltage and the pixel spacing. The *detection efficiency* (DE) of the pixel depends on the *quantum efficiency* (QE) and bias-dependent *Geiger probability*. During the measurements of pixel DE, we must correct for the effects of after pulsing on the DE measurements [64].

The build matrices of SSPMs coupled with radiation detectors are capable of providing an MRI (magnetic resonance imaging) compatible readout for PET (positron emission tomography) with FWHM energy at 511 keV and timing below 15 % and 500 ps, respectively [77]. Recently, nuclear spectroscopy with LaBr$_3$:Ce scintillator coupled to a silicon photomultiplier (SSPM) has been used for positron emission tomography (PET) [78].

## 2.12 Modes of Detector Operation

### 2.12.1 Current Mode

A radiation detector can be operated in three different modes, namely, the *current mode* [79], the *mean square voltage mode* [80], and *pulse mode* [81]. In *current mode*, the output current of the detector (after being exposed to radiation) is measured. In this mode, the time-dependent current $I(t)$ can be written as (assuming that the measuring device has a fixed time $T$ (response time), which is typically a fraction of a second)

$$I(t) = (1/T) \int_{t-T}^{t} I(t')dt' \qquad (2.14)$$

The upper sketch of Fig. 2.24 shows the schematic of a *current-mode* circuitry, and the bottom is a schematic of the time-dependent output current $I(t)$ when plotted against time. Experimental observations show that at any instant of time, there was a statistical uncertainty in the signal due to the random fluctuations in the arrival time of the event. In order to minimize statistical fluctuations in signal, the response time $T$ is made large. The average current $(I_0)$ is a product of even rate and the charge produced at each event, that is,

$$I_0 = rQ = r(E/W)q \qquad (2.15)$$

where $E$ is the average energy deposited per event, $W$ is the average energy required to produce a unit charge pair, $q$ is electronic charge, and $r$ is the event rate [82].

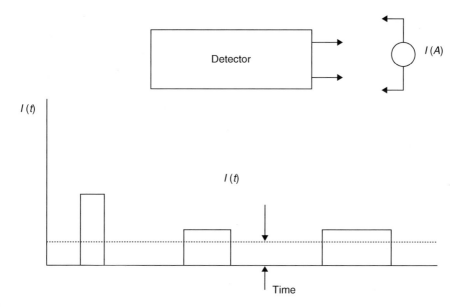

**Fig. 2.24** Schematic of current-mode circuitry and the output amplitude of the time-dependent current

## 2.12.2 Mean-Square Voltage Mode (MSV)

In MSV mode of operation, the derived signal is proportional to the square of the charge ($Q^2$) per event. MSV mode of operation is also called *Campbelling technique*. The magnitude of the signal in the MSV mode can be written as

$$\text{Mean square signal} <\sigma^2(t)> = \{(rQ^2)/T\} \quad (2.16)$$

where $r$ is the event rate, $Q$ is the charge, and $T$ is the effective time. In case of neutron-induced pulse, the mean square signal will weight the neutron component by the square of the ratio of neutron to γ-ray-induced charge because neutron-induced pulses are having much greater charge than pulses from gamma (γ) rays. It has been observed that MSV-mode operation is most useful when making measurements in mixed radiation environment.

## 2.12.3 Pulse Mode

*Pulse-mode* operation is most suitable when the event rates are high or someone has to enhance the relative response to large amplitude. However, in many applications, *MSV*-mode operations are found to be most suitable because they can preserve better information on the amplitude and timing of individual events. It is seen that the signal pulse produce from a single event depends mostly on the input

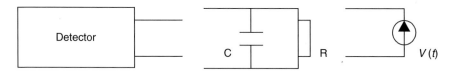

**Fig. 2.25** The schematic of the pulse-mode circuit for detector measurement. C, the capacitor, R, the resistor, V(t), time dependent voltage

characteristics of the circuit to which the detector is connected. The circuit below represents a schematic of an electronic circuit connected to a detector. The circuit has three parts: (a) the detector, (b) the electronic circuit (including preamplifier), and (c) the device to measure the signal voltage. Figure 2.25 shows the schematic of the pulse-mode circuit for detector measurement.

The preamplifier circuit has an output resistor, and the equivalent circuit capacitance is $C$. The time-dependent signal voltage $V(t)$ is developed across the resistor $R$. The time constant RC value of the circuit can be large. When RC value is small, the amount of current that flows through $R$ will be equal to the instantaneous value of the current that will flow through the detector. On the other hand, when the time constant (RC) is large, very little current will flow through the resistor $R$ during the charge collection time, and the voltage $V(t)$ developed can be traced (shown in Fig. 2.26).

Figure 2.26 (a) shows the voltage developed across the load resistor $R$ from a hypothetical detector, (b) when RC (time constant) is small and (c) when RC is large. As a matter of fact, the pulse-type operation with large RC ($\tau \gg t_C$) is by far the most common mode of operation for the detectors. For $\tau \gg t_C$, (a) the charge collection time within the detector is determined by the value of the RC, and (b) the load circuit does not influence the rise time of the pulse, but the decay time (the restoration time for the signal voltage to zero) is determined by the time constant of the load circuit. The circuit capacitance $C$ is kept constant, and the total charge $Q$ divided by the circuit capacitance gives the maximum amplitude of the pulse, or in other words, $V_{max} = (Q/C)$. The charge $Q$ is again proportional to energy of the incident quantum of radiation. Thus, when $C$ is constant, $V_{max}$ is proportional to the charge developed in the detector during radiation.

Pulse-mode operation is the most common choice for most radiation detector applications because (1) higher sensitivity is achievable than MSV- or current-mode operation and (2) it can provide more information which is lost in both MSV- and current-mode operations. In pulse-mode operation, pulse amplitude distribution is a fundamental property of the detector output. When large number of pulses is studied by a particular radiation interaction in a detector, one can find variations in the amplitudes. These variations are due to the differences in radiation energy or due to the fluctuations in the inherent response of a detector to monoenergetic radiation [83]. The most common method of displaying a pulse amplitude information is through different pulse height distribution. The abscissa represents the pulse height ($H$), and the ordinate represents the differential number d$N$ of pulses observed with the differential amplitude increment d$H$. The total number of pulses

## 2.12 Modes of Detector Operation

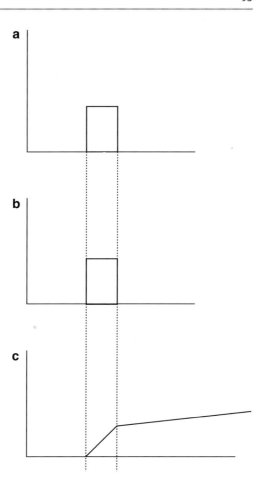

**Fig. 2.26** A sketch (a) from a hypothetical detector. (b) The signal voltage ($V(t)$) plotted against time ($t$) when RC is small and (c) is same as (b) when RC is large

$N_0$ can be obtained by integrating the area under entire spectrum and can be expressed mathematically as

$$N_0 = \int_0^\infty (dN/dH) dH \qquad (2.17)$$

The differential and the integral distributions convey exactly the same information, and one can be derived from the other. It should be noted that the plateaus in the integral spectrum correspond to the valleys in the differential distribution.

In pulse-mode operation, the output from the detector is fed to a counting device, and it is often desirable to select a stable operating point where the drift is minimum. When the slope of the integral has its minimum, one can expect small changes in the distribution level, and ultimately, there will be minimum impact on the total number of pulses recorded [84].

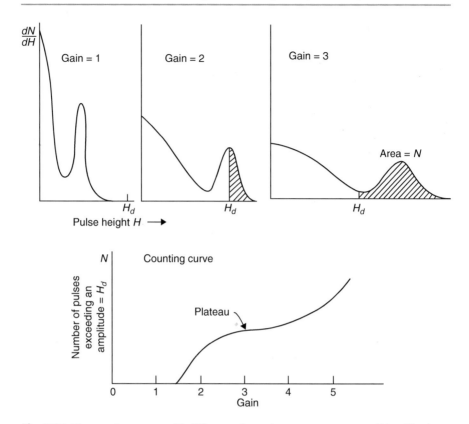

**Fig. 2.27** The counting curves with different gains under constant source condition. The *lower curve* is a plot of number of pulses exceeding amplitude $H_d$ plotted against three gains as shown on the three *top curves* (Reprinted with permission, Wiley [103])

The regions of minimum slope on the integral distribution are called *counting plateau*, and it represents area of operation in which minimum sensitivity to drift in discrimination level is achieved. When the voltage gain applied to the same source of pulses is increased, we will expect increased in pulse height amplitude, but the areas within the pulses will remain constant [85].

The top curves of Fig. 2.27 represent the differential number of pulses (d$N$) to differential amplitude height d$H$ when plotted against the pulse height ($H$) with different gains (1, 2, and 3). The pulses with gain 1 are smaller than $H_d$ so we do not find any count beyond $H_d$. However, pulses recorded in 2 show the gain exceeding $H_d$, and the curve with gain 3 has extended to a larger area beyond $H_d$.

The abscissa of the lower curve represents the gain where the ordinate represents the number of pulses exceeding amplitude $H_d$. The single curve is a mirror image of the integral distribution. Therefore, we can anticipate plateaus in the counting curve with the effective discrimination pulse height $H_d$, and it will pass through minima in the differential pulse height distribution [110].

## 2.13 Recording and Measurement Techniques

### 2.13.1 Spectroscopy and Spectrometry

The term *spectrometry* and *spectroscopy* become *synonymous* in the sense that both were used to refer to *recording* and *measurement* techniques. Spectrometric methods are subdivided, as a whole, into two main divisions: the first one is (a) radiation spectrometry, which comprises *absorption spectrometry* [111, 112], *emission spectrometry* [113, 114], *Raman scattering spectrometry* [115], *magnetic resonance spectrometry* [116, 117], and the second one is (b) *mass spectrometry* [118, 119].

*Radiation spectrometry* and *spectroscopy* cover an ensemble of analytical methods allowing the composition and structure of matter to be ascertained, based on investigation of the spectra yielded by the interaction between atoms and molecules and various types of electromagnetic radiation, emitted, absorbed, or scattered. Therefore, radiation spectroscopy has been found to be a powerful tool for rapid analysis and measurements of *energy resolution, photo-peak positions, photo-peak efficiency*, and *background* of radioactive samples.

### 2.13.2 Gamma-Ray Spectroscopy

We know that a Geiger counter can determine count rates for radiation, but the spectroscopy has the ability to determine both the count rate and the energy of the radiation. Many radioactive sources emit many gamma rays having different energies. X-ray or gamma-ray photons do not have an intrinsic charge. Therefore, they create ionization excitation due to their interactions in medium and ultimately produce photons. Thus, the measurement of these photons is dependent on the interaction with the electrons of the media.

A gamma-ray spectrum is created by taking measurements of emitted gamma rays and processing them [120]. An unknown radioisotope is identified by comparing the features on the gamma-ray spectrum with a known spectrum. However, there are several factors that complicate the spectrum, namely, *secondary electron escape, bremsstrahlung escape, characteristic X-ray escape, secondary radiations created near the source, the effect of the surrounding materials*, and *coincidence*.

For high-energy gamma rays, the *secondary electrons* will have high energy with a high probability of penetration inside the detector and leakage of electrons out of the system. The leakage of electrons will distort the response function. Further, Compton continuum along with other low-amplitude energies will be shifted to favor lower amplitudes. As a result of the escaping events, the photopeak and the photofraction will be reduced. These phenomena are observed due to *secondary electron escape*.

The second effect of the list given above is the *bremsstrahlung escape* effect, which will alter the $Z^2$ factor of the detector. It will ultimately alter the response function in the same manner that the secondary electrons alter the response function. Detector material with low atomic number may improve the effect.

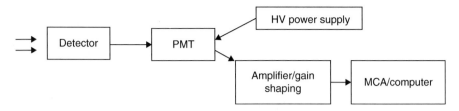

**Fig. 2.28** The schematic of the accessories and the instruments used in gamma-ray spectroscopy

The third effect mentioned above is the *X-ray escape*. It can create a new peak at a distance of the main photo-peak. The phenomenon is prevalent in low-energy incident photons and when the detector has large surface to volume ratio.

The next item in the above series which complicates the spectrum is the secondary radiation created near the source material, which emits $\beta^+$ and $\beta^-$ particles during decay and are received by the detector. Both of these decay products increase overall radiation energy that is detected in the system. As a result, we can expect a shift of the energy spectrum to favor higher amplitudes.

The emitted gamma ray from the radiation source will interact with the detector as well as with the materials that house the detector or any other surrounding material/s. It is not unlikely that the creation of radiation from the surrounding material might have a characteristic of X-rays, and as a result, we can notice an additional peak. In addition to this, we might be able to observe a wider peak of the spectrum due to the backscattering of the photons.

In the measurement of gamma-ray energies with several hundred keV, several inorganic and semiconductor scintillators have been used [121]. Among these, NaI (Tl) is by far the most popular and the germanium semiconductor predominantly of the Ge(Li) type.

Figure 2.28 shows the schematic of the accessories and the instruments used in gamma-ray spectroscopy.

When gamma ray passes through the detector (scintillator), one of the three processes, namely, *photoelectric absorption, Compton scattering,* and *pair production,* can occur. These three processes produce excited electrons that transfer their energy to the detector crystal and excite the atoms in the lattice. These atoms decay to the ground state in course of time and emit photons whose energy is proportional to the deposited energy of the gamma ray. The photomultiplier tube (PMT) then uses these photons to produce photoelectrons, and these charges are converted to voltage pulses and amplified without losing any information. The signal is then fed to a multichannel analyzer (MCA) and then to a computer that registers a count for the channel corresponding to the specific voltage fed through. Figure 2.29 shows a typical MCA spectrum for $^{137}$Cs source with a NaI (Tl) detector. The photoelectric effect results in a peak, called photo-peak, in the photomultiplier spectrum. Besides photo-peak, there are two broad peaks with smaller intensities that appear before the photo-peak. The channel numbers of these two peaks are also smaller than the photo-peak.

## 2.13 Recording and Measurement Techniques

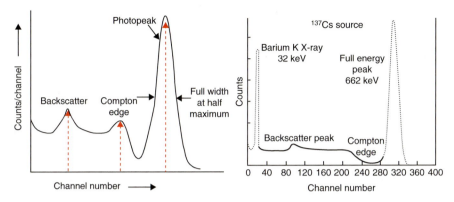

**Fig. 2.29** $^{137}$Cs spectrum with NaI (Tl) detector (Courtesy: Rutgers University, NY)

The photo-peak in $^{137}$Cs is around 315, and that of the Compton edge and backscatter peak are at 210 and 90, respectively. The theoretical energy of a Compton edge $E'$ is given by [122]

$$E' = \{(E)\}/\{(1 + mc^2)/2E\} \qquad (2.18)$$

And the theoretical formula for the backscattered peak is

$$E' = \{(E)\}/\{(2 + mc^2)/2E\} \qquad (2.19)$$

When the energies for the Compton edge and backscattered peak are calculated from the theoretical equations (Eqs. 2.18 and 2.19), we get their respective values as 410 and 180 keV. The energy related to the photo-peak is found to be 662 keV.

The photo-peaks of $^{60}$Co are observed at channel 860 and 961 and that correspond to the energies of 1.17 and 1.33 MeV, respectively (Fig. 2.30). Derived from the point slope formula, the equation below is used to calculate the energy corresponding to each channel:

$$\text{Energy} = (1.33 - 1.17 \text{ MeV})/(961 - 860) \times \text{channel} - 192.24 \qquad (2.20)$$

The energy spectrum for Co-60 using a NaI(Tl) detector is shown in Fig. 2.30. Availability of thallium-doped sodium iodide (NaI:Tl) scintillator in large size together with its high density results in very high interaction probabilities of gamma rays. As a result, the energy resolution of germanium (Ge) doped with lithium (Li) has a typical value less than 1 % compared to 5–10 % of NaI:Tl. Not only that. The combined smaller size availability and lower atomic number of germanium give photo-peak efficiencies of at least an order of magnitude lower than a NaI(Tl). The lower atomic number results also in smaller photoelectric cross section by a factor of 10–20 compared with NaI:Tl. Therefore, photoelectric absorption in a single

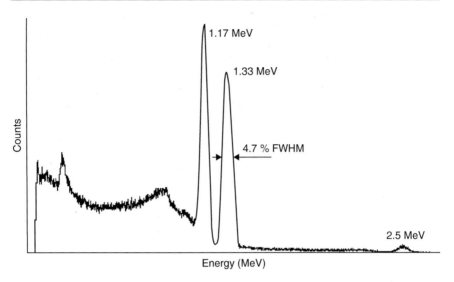

**Fig. 2.30** The energy spectrum for Co-60 with NaI(Tl) as detector (Courtesy: ORTEC)

interaction is much less probable in Ge(Li). The other disadvantage of germanium (Ge:Li) detector is that it requires very low temperature to operate.

Another important criterion of Ge(Li) detector is Compton continuum. Because the ratio of the Compton to photoelectric cross section is much larger in Ge than NaI, a much greater fraction of all detected events lie within the continuum rather than under the photo-peak. Besides, Ge(Li) detector has greater transparency to secondary gamma rays compared to NaI, which plays an important role in spectroscopy The escape of the characteristic X-rays from germanium following photoelectric absorption can be significant especially with the detector having small size with a large surface to volume ratio. When the gamma rays have high energies, the escape of annihilation radiation following pair production within the detector is very significant.

Figure 2.31 shows the relative intensity of the full-energy peak, the single-escape, and double-escape peaks. The single-escape peak occurs at 0.511 MeV or at 2.00 MeV, and the double-escape peak occurs at 1.02 MeV or 1.49 MeV. The full-energy peak, on the other hand, occurs at 2.511 MeV, and it represents those events where there is a combination of pair production and photoelectric effect in which all the energy absorbed in the detector.

### 2.13.3 Neutron Spectroscopy

Measurements of the high-energy neutron component are needed to assess the increased probability for carcinogenesis and DNA and central nervous damage—some of which may be caused by single hit of energetic ion or neutron. In addition,

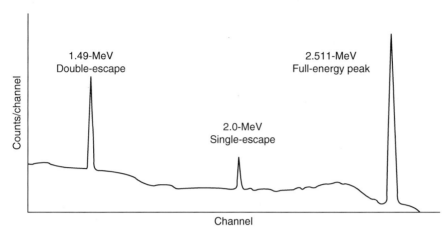

**Fig. 2.31** The relative intensity of full-energy peak and single-escape and double-escape peaks are shown (Courtesy: ORTEC)

neutron spectrometry can provide a wealth of information about the structure of the nuclei and the characteristics of nuclear reaction. Detectors employing organic scintillators, both liquids and solids with the ability to use pulse shape discrimination to separate neutron events from photon events, have found some use in the laboratory [123, 124]. The techniques to measure neutron spectrum includes proton-recoil spectrometers, neutron time of flight measurements, and $^3$He spectrometers.

The $^3$He spectrometer is a gas-filled *proportional counter* (discussed earlier about the construction and working principle) containing $^3$He, argon, and some methane. Detector based on $^3$He was first developed by Cuttler and Shalev and is known as C-S spectrometer. The energy spectrum of a $^3$He spectrometer includes a full-energy peak at the neutron energy $E_n$ + 764 keV, a thermal neutron capture peak at 764 keV, and a $^3$He (n, n') elastic recoil spectrum with a minimum at 0.754 $E_n$. (Obtained from the equation $E_{max}$ (maximum energy transferred) = {(4AE)/(A + 1)2}, where $A$ is the atomic weight and $E$ is the maximum energy). Under ideal conditions, neutron energy measurements have been made with a resolution of <15 keV. However, the energy resolution of a C-S spectrometer degrades at higher energies. Continued development in electronics and careful analysis have resulted an energy resolution of approximately 0.040 MeV for thermal peak and approximately 0.13 MeV at the deuterium–deuterium neutron energy [125, 126].

Figure 2.32 shows the differential energy spectrum and the peaks from fast neutrons incident on a $^3$He spectrometer. A compact high-energy neutron spectrometer (COLONS) has been used with a single NE-213 liquid scintillator in neutron spectrometry. The approach is based on the use of both pulse height (*L*) and pulse shape (*S*) measurements. The spectra obtained by unfolding the *L-S* signature are comparable to the time of flight spectra at low energies ($E$ < 60 MeV) [127, 128]. During magnetic fusion experiments with NE-213 scintillator, neutron flux monitors

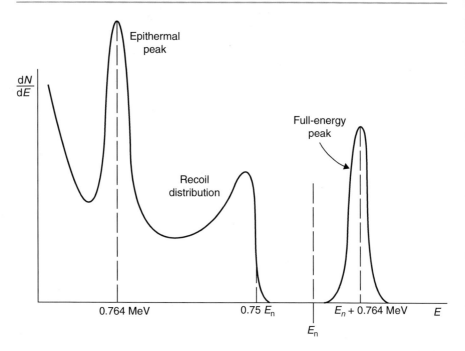

**Fig. 2.32** The differential energy spectrum and the peaks from fast neutrons incident on a $^3$He spectrometer (Reproduced with permission: John Wiley)

are calibrated by means of a standard neutron source (e.g., $^{252}$Cf) of known intensity. Another very important calibration component is the energy scale determination. Generally, for neutron spectrometer calibration and light output testing in energy range up to 8 MeV, the reaction of Be (d,n) might be useful. An NE-213 detector has also been used to measure neutron spectra for deuterium plasma during ohmic discharges of Frascati Tokamak Upgrade (FTU) [128].

Another single-crystal semiconductor detector that has been used in high-energy neutron spectroscopy is silicon (Si). Neutrons do not directly ionize target atoms of Si; therefore, elastic and nonelastic reactions that produce secondary charged particles have been used to measure neutron energy. When the source (radiation) spectrum is monoenergetic, the detector response function can be evaluated at the incident neutron energy and can be presented mathematically as [129]

$$C(E_D) = A(E_D, E_N)\Phi(E_V)\delta E_N \tag{2.21}$$

where $C$ is the number of counts detected at a given deposited energy $E_D$, $\Phi(E_V)\delta E_N$ is the integral neutron fluence centered at $E_N$, and $A$ is the response function of the detector, which depends upon both the incident and the deposited energies. When the source spectrum is not monoenergetic, the above equation can be generalized as

## 2.13 Recording and Measurement Techniques

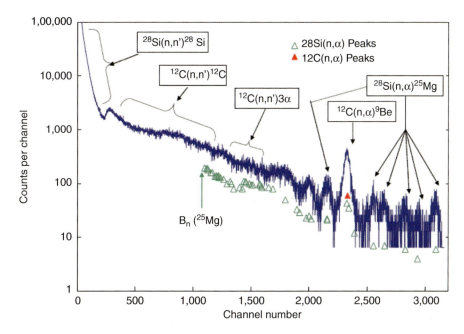

**Fig. 2.33** A 14-MeV neutron response data of SiC detector (Reprinted with permission from World Publishing)

$$C(E_D) = \int A(E_D, E_N)\Phi(E_V)dE_N \qquad (2.22)$$

or in discrete form as

$$C_i = \Sigma_j C_{ij} = \Sigma_j A_{ij}\Phi\Delta En, \qquad (2.23)$$

where $C_{ij}$ is the portion of the total counts in the $i$th energy deposition bin generated by the $j$th neutron energy bin and $A_{ij}$ is the response matrix element for the $i$th energy deposition bin and $j$th neutron energy bin. In matrix notation, this can be written as $C = A\Phi$.

If the response matrix, $A$, is known, Eq. 2.23 becomes a system of linear equations, and one can estimate the fluence in each neutron energy bin. The model that we have discussed here is useful for neutron spectroscopy when silicon will be used as a detector.

Silicon diode sandwiched with a thin layer of lithium fluoride (LiF) has also been used as a neutron spectrometer. Neutrons incident on the assembly interact with $^6$LiF layer, giving rise to tritons and alpha particles, and are detected in silicon semiconductor counters. The silicon counter pulses are ultimately recorded by a pulse height analyzer. Besides silicon diode, p-i-n silicon device coupled to a polythene radiator has also been used in neutron spectrometry.

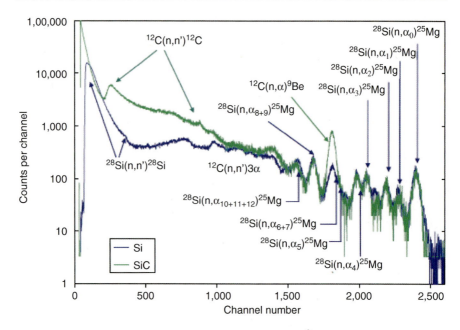

**Fig. 2.34** Comparison of the neutron response of a 28.3 mm$^2$ × 100 μm SiC detector and 450 mm$^2$ × 100 μm Si detector (Reproduced with permission, World Science Publishing, Netherland)

The neutron spectra are derived from the measurement of ionization induced by recoil protons in silicon. The silicon detector provides a pulse whose amplitude is proportional to the collected charge. The energy $E_d$ deposited by the recoil protons in the spectrometer irradiated with a monoenergetic energy source (between 0.5 and 5 MeV) has been calculated both by analytical method and by Monte Carlo simulations. The calculated results have been verified experimentally [130].

Recently, silicon carbide (SiC) Schottky diode has been used to measure the response peaks of 14-MeV fast neutron without a proton-recoil converter layer. The resulting 14-MeV pulse height response data are shown in Fig. 2.33.

Figure 2.33 shows a 14-MeV neutron response data of SiC detector. At low-energy portion of the spectrum, the continua for $^{28}$Si and $^{12}$C elastic and inelastic scattering dominate the detector response. At higher energies, specific reaction peaks dominate. The most prominent peak of $^{12}$C(n, α)$^9$Be is a product of 8.3 MeV in recoil-ion energy. Five peaks of $^{28}$Si(n, α)$^{25}$Mg are observed in the spectrum which have been plotted from the collected data.

The photo-response of SiC at 14MeV energy is compared with a passivated ion implant Si detector with the same active volume and is shown in Fig. 2.34. From the figure, we can see that the peak heights, peak widths, and the energy positions of SiC and Si closely match each other [131, 132].

## 2.13 Recording and Measurement Techniques

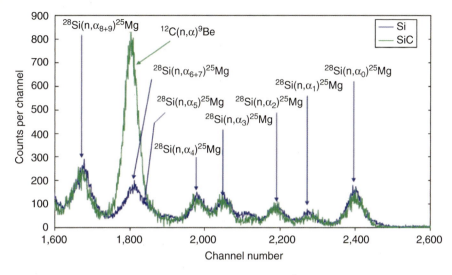

**Fig. 2.35** Comparison of the high-energy peaks of a 28.3 mm$^2$ × 100 μm SiC detector and 450 mm$^2$ × 100 μm Si detector (Reprinted with permission from World Publishing)

Figure 2.35 shows the comparison between the high-energy peaks of a 28.3 mm$^2$ × 100 μm SiC detector with that of a 450 mm$^2$ × 100 μm Si detector.

### 2.13.4 X-Ray Spectroscopy

X-ray spectrum analysis provides a qualitative result of the elemental composition of a material. In energy X-ray spectrometer (XDS), semiconductor detector measures energy of the incoming photons, while in wavelength dispersive X-ray spectrometer (WDS), the single crystal diffracts the photons which are collected by the detector and recorded for analysis. Synchrotron-generated X-ray typically has high intensity, and one can tune the wavelength of the incoming ray. The powerful X-ray plays a dominant role in nuclear medicine because of its shorter pulse duration, synchronizability of the ultrashort pulses, and availability of useful fluxes [133, 134].

X-ray emission spectroscopy has been used in nuclear medicine for the quantitative analysis of calcium, sulfur, and phosphorus in urinary calculi. The other areas of the use of X-ray spectroscopy is XRF (X-ray fluorescence spectroscopy). It is used to determine the primary distribution of different organic and inorganic elements in human bone joints, presence of biologically important metals in soft and classified tissues, the iodine content in thyroid, and the existing amount of strontium and lead in bone [135].

The daily partial workload in a nuclear medicine department consists of functional imaging of organs—for example, thyroid, brain, cardiac, liver, kidney—using X-ray and a detector coupled to electronic circuitry [136]. From the soft

X-ray to the hard X-ray regions, gallium arsenide (GaAs) detector has been very useful because of its *Fano* energy resolution [137]. However, in the hard X-ray band, cadmium telluride (CdTe) and cadmium zinc telluride CdZnTe (CZT) are routinely used. For nuclear medicine application, radiation dose received by a patient during imaging or therapy is an important issue. From our practical experience, we know that for a safe dose, the upper energy range of the radiation energy is kept between 200 and 250 keV, where in general, the detector has efficiency between 30 and 35 %. There are several inorganic scintillators (*detectors*) that are being used in nuclear medicine for imaging spectroscopy and will be elaborately discussed in a separate chapter of the book.

One important property of a *detector* in *radiation spectroscopy* can be examined by noting its response to a monoenergetic source of radiation. The response to the radiation is in the form of a curve obtained through an electronic circuitry, and the response curve is identified from its different parameters, for example, the *energy resolution, photo-peak positions*, and *photo-peak efficiency*. Depending on the energy of the incident radiation, photons interact selectively with various electron shells or levels, making up the electronic structure of the atoms or molecule. In order to get some knowledge about the interactions of the incident radiation with the various tissues and cells during the irradiation and with the detector, the analysis of the above parameters is important.

## 2.13.5 Characteristic Parameters of the Spectrum

### 2.13.5.1 Energy Resolution

The most important parameter that is measured in radiation detectors is the *energy distribution* of the incident radiation. The distribution is called the response function of the detector for the energy used during the experiment. This is actually done by noting its response to a monoenergetic source of radiation [138–140].

The energy resolution curve is said to have *good resolution* when the curve is *steep* and does not have much *broadening* [141, 142]. The full width at half maximum (FWHM) of the resolution curve is a measure of the *quality* of a detector and is defined as the width distribution at a level that is just half the maximum ordinate of the peak (Fig. 2.36) [143, 144]. The *standard deviation* $\sigma$ of the peak in the pulse height spectrum is proportional to the square root of the total number of charge carriers $N$ ($\sigma = K\sqrt{N}$ and FWHM $= 2.35/\sqrt{N}$, where $K$ is a proportionality constant). The *resolution R* due statistical fluctuation can be expressed mathematically as

$$R = (\text{FWHM}/H_0) = \{(2.35K\sqrt{N})/(KN)\} \qquad (2.24)$$

The energy resolution is a dimensionless fraction conveniently expressed as a percentage. It can be defined as the FWHM divided by the location of the peak centroid $H_0$. From the above equation, it is clear that $R$ is dependent on the number

## 2.13 Recording and Measurement Techniques

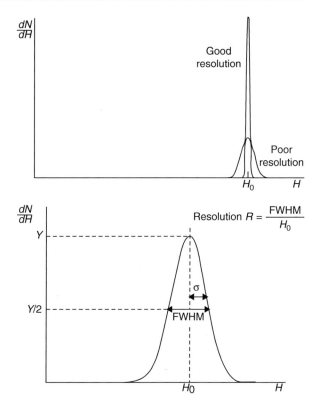

**Fig. 2.36** (a) The peak of two spectra with good and poor resolution (R) and (b) a Gaussian-shaped curve with standard deviation $\sigma$ and FWHM is given by 2.35b

of charge carriers $N$, and it is obvious that to improve $R$, one has to increase the number of charge carriers.

### 2.13.5.2 Efficiency

The other important parameters that one should be concerned with are (1) the efficiency of the detector and (2) the correct measurement of the output pulse that is received for a particular incident quantum radiation, which interacts with the active volume of the detector. Thus, it is the primary goal of an experiment to see that the radiation particles like alpha ($\alpha$) and beta ($\beta$) are properly received by the active volume of the detector. When the conditions are fulfilled, the detector is said to have a *counting efficiency* of 100 %. On the other hand, uncharged radiation like gamma ($\gamma$) rays or neutrons must first undergo a significant interaction in the detector before detection is possible. These radiations travel large distances between interactions, and the efficiency of detector is always less than 100 % [145, 146].

*Counting efficiency* can be subdivided into two categories, namely, the *absolute* and *intrinsic*. The *absolute efficiency* is defined as the ratio of the number of pulses recorded to the number of radiation quanta emitted by the source. The *intrinsic*

**Fig. 2.37** Solid angle ($\Omega$) subtends by the point source to the detector plane

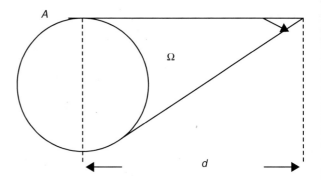

*efficiency*, on the other hand, is defined as the ratio of the number of pulses recorded to the number of quanta incident on the detector. For isotopic sources, the two efficiencies are related to ($4\pi/\Omega$), where $\Omega$ is the solid angle of the detector seen from actual source position, and can be expressed mathematically as

$$\Omega = \int_A (\cos \alpha)/r^2 \, dA \qquad (2.27)$$

Again, a point source when located along the axis of a right circular cylindrical detector and $\Omega$ can be written as

$$\Omega = 2\pi[1 - \{d/(d^2 + a^2)\}] \qquad (2.28)$$

When $d \gg a$, we can replace Eq. 2.20 as

$$\Omega = \pi\{a^2/d^2\} \qquad (2.29)$$

Figure 2.37 Solid angle ($\Omega$) subtends by the point source to the detector plane [147]

### 2.13.5.3 Dead Time

When two separate pulses are recorded as two events, there should be a minimum amount of time lag between the two events. The time lag or the time interval between the two events is known as *dead time* [148]. The minimum time of separation can depend on the detector itself or on the associated electronics.

We can set up two models, namely, the *paralyzable* and *non-paralyzable* to represent *dead time* [149–151]. In the former case, true interaction rate ($n$) is proportional to $m$ times the system dead time ($\tau$) where $m$ is the recorded time rate. In the later case, $n$ is proportional to recorded count rate ($m$) and inversely proportional to $(1-m\tau)$, where $\tau$ is system *dead time* [152].

The most common method of measuring dead time is the two-source method based on observing the counting rate from two sources individually and in combination. It is to be noted that the method is based on observing the differences between two nearly equal large numbers; therefore, it is very hard to get a reliable

## 2.13 Recording and Measurement Techniques

value of dead time unless one is very careful enough during the measurements. The second method that is applied to measure the dead time is the measurement of the departure of the counting rate from the known exponential decay of the source. The technique is known as *decaying source* method [152].

### 2.13.5.4 Dead-Time (Coincidence) Correction

Let us take the total insensitive time per unit time as $\Re\tau$, when the counter has a recovery time $\tau$ (the dead time or resolving time), where $\Re$ is the observed counting rate. On the other hand, $\Re^*$ is the rate that would be recorded if there are no dead-time losses, and the number of lost counts per unit time is taken to be as $(\Re^*-\Re)$. Mathematically, the value of $(\Re^*-\Re)$ is given by the product of the rate $\Re^*$ and the fraction of the insensitive time $\Re\tau$, or in other words, we can write

$$(\Re * - \Re) = \Re * \Re\tau \tag{2.30}$$

$$\text{Or, } \Re* = (\Re/1 - \Re\tau) \tag{2.31}$$

Now, we can set up the *Schief formula* from the above Eq. 2.31, which can be expressed mathematically as $\Re^*=\Re \exp(\Re^*\tau)$. As a matter of fact, it is more closely to the conditions of dead-time loss in which a new pulse within a dead time can initiate a new dead-time period although it is not being recorded. But the fact of the matter is that whether an event is recorded or not, it still prevents the recording of the second event occurring within the time $\tau$. When binomial expansion is applied to $\{1/(1-\Re\tau)\}$, the Eq. 2.31 transform to

$$\Re* = \Re (1 + \Re\tau) = \Re + \Re^2\tau \tag{2.32}$$

The Eq. 2.32 is more convenient for the interpretation of an event designed to measure $\tau$. The procedural method is to measure the rates of $\Re_1$ and $\Re_2$ produced by two separate sources and the rate $\Re_t$ produced by two sources together with rates of the back ground effect $\Re_b$. Obviously,

$$\Re_1^* \text{ and } \Re_2^* = \Re_t^* \text{ and } \Re_b^* \tag{2.33}$$

In the above equation, we have neglected the dead-time loss in the measurement of the low background rate. Replacing $\Re_1^*$ by $(\Re_1^* + \Re_1^2 \tau)$ and rearranging, we can write the value of $\tau$ as

$$\tau = \{(\Re_1 + \Re_2) - (\Re_1 - \Re_b)\}/\{(\Re_t^2 - \Re_1^2 - \Re_2^2)\} \tag{2.34}$$

**Table 2.5** Estimated dose of 201-Th ($^{201}$Th) with an energy of 74 MBq ($2 \times 10^{-3}$ Ci)

| Tissue | MGy/74 MBq | Rad/2 mCi |
|---|---|---|
| Heart wall | 10.0 | 1.0 |
| Kidneys | 24.0 | 2.4 |
| Liver | 11.0 | 1.1 |
| Thyroid | 13.0 | 1.3 |
| Whole body | 4.2 | 0.42 |

Reprinted with permission: DuPont de Numours 1988

## 2.13.6 Dose Received During Medical Activities

Living creatures like human beings and animals are exposed to man-made radiation (from radionuclides, X-rays, and gamma rays), either during radiation therapy or during medical imaging. The medical personal involved in the operation also receives a partial amount of the dose [153]. Thus, it is essential to monitor or measure the amount of radiation dose a particular living being is receiving during the course of medical treatment or during handling of the radioactive sources. Direct measurement of the radiation received by a living creature can be described as a process where the measurement generally consists of *whole bodycounting*, *organ counting*, or *in vivo* measurement. In principle, direct measurements of radionuclides include detection of photons (X-ray, gamma ($\gamma$) ray, or bremsstrahlung radiation), beta ($\beta$) particles, or alpha ($\alpha$) particles.

In X-ray tube, when fast electrons are bombarded with matter, a part of the energy is converted into electromagnetic (EM) radiation in the form of *bremsstrahlung*. The technique for direct measurement of X-ray is in principle the same as that determining the content of $\gamma$-ray emitting radionuclides. The exposure rate can be expressed mathematically as

$$\text{Exposure Dose } (E_{\text{D}}) = K \ (A_{\text{s}}/d^2) \tag{2.35}$$

where $K$ is exposure rate constant, $A_S$ the activity of the source, and $d$ the distance of the source to the target. Research efforts are underway to use Monte Carlo techniques to calculate patient dose. Using computer models, researchers are generating millions of virtual data for different radiation exposures. It has been estimated that miscalculations result in almost 20–30 % excess radiation dose than it is necessary for diagnosis of the affected cells.

In the field of medicine, radiation is used for diagnostic studies, to destroy some abnormal tissues and cells, and for medical imaging. As for example, thorium-201 is generally used in diagnostic studies of coronary artery diseases and location of lymphomas. A typical estimated dose of thorium-201 ($^{201}$Th) is shown in Table 2.5.

In the field of imaging, visualization of tissue perfusion with radionuclides such as $^{201}$Th has been successful for coronary artery disease [154]. $^{201}$Th is the form of thallous chloride and has been the principle tracer used in myocardial perfusion imaging.

Another isotope of thorium is $^{232}$Th which has been used for specific imaging of liver and spleen, but it has been mostly used in cerebral angiographies. Approximately, 10 years after its introduction, there are some reports of possible carcinogenic effects of the isotope. Despite of these publications, the use of $^{232}$Th has been increased because of the lack of acute toxicity and excellent radiological results compared with other contrast media [155].

The other isotope that has been frequently used is $^{99m}$Tc. It has been found that with higher dose, $^{99m}$Tc yields better images with higher count density. Although the isotope chemistry of iodine ($^{131}$I) is much more simple than isotope chemistry of $^{99m}$Tc, the long physical half-life, emission energy, biological activity, and cost have made it less useful as *single-photon emission agent*.

*Single-photon emission computed tomography* (SPECT) is the mainstream perfusion imaging, while *positron emission tomography* (PET) continues to be the modality of choice for agents that track metabolism. Limitations inherent to the use of nonphysiologic single-photon emitting radiotracers have limited the quantitative power of SPECT. On the other hand, the use of positron emission radioactive tracers offers the potential for in vivo quantification of specific biological processes. In addition to this, the *positron emission tomography* (PET) is uniquely quantitative and allows for accurate attenuation, correction of emission data, higher efficiency, and better contrast resolution than SPECT. We will discuss all of these in a separate chapter.

### 2.13.7 Measurement of Dose Received

*Dosimeter* measures the amount of *radiation* energy absorbed over a given period of time by an object (e.g., living animals) or part of the body object (e.g., an internal organ or tumor). It also can measure the amount of nuclear medicine or the radiation dose administered into the body. Here, *radiation* does not refer only to ionization energy but to light, radio waves, or ultrasound. In other words, we can say that it is a device that measures cumulative radiation exposure. Therefore, a *Geiger counter*, which is a *radiation detector*, cannot be identified as a *dosimeter*, because it gives a moment to moment reading of radiation intensity.

The calculation in *dosimeter* is done on the basis of the *half-life* ($T_p$ or $T_{1/2}$) which is defined as the time after which the radioactive nucleus and the decay time are reduced to half. The *effective lifetime* ($T_e$), on the other hand, is associated with the *biological half-time* ($T_b$) and *radioactive lifetime* as ($T_p$)

$$1/T_e = 1/T_b + 1/T_p \qquad (2.36)$$

The *biological half-time* ($T_b$) is defined as the total time taken to excrete half of the radioactive material (the nuclear medicinal nuclide administered) from the body or organ of the body. On the basis of the above definition, $T_b$ for $^{99m}$Tc-MDP is

**Table 2.6** The half-life of the radiopharmaceuticals

| Radiopharmaceuticals | $T_p$ | $T_b$ | $T_e$ |
| --- | --- | --- | --- |
| $^{99m}$Tc-sulfur colloid | 6 h | $\propto$ | 6 h |
| $^{99m}$Tc-MDP | 6 h | 4 h | 2 h |
| $^{67}$Ga-citrate | 78 h | 530 h | 68 h |
| $^{123}$I | 13 h | 26 h | 8.7 h |
| $^{131}$I (30 % uptake) | 8 d | 70 d | 7 d |

Reprinted with permission: Powsner and Powsner [136]

defined as the time it takes for one-half of the radiopharmaceutical to be filtered and excreted by the kidneys and bladder. The dose ($D$) delivered to the whole body or to a particular organ can be written as

$$D = \tilde{A} \times C \tag{2.37}$$

where dose ($D$) = total activity ($\tilde{A}$) of the nuclear medicine within the time of stay inside the body multiplied by the correction factor ($C$) and $\tilde{A}$ is defined as the residence time ($\tau$) over which the organ will receive the dose from the initial activity $A_0$. Thus, mathematically, $D$ can be written as

$$D = \tau \times A_0 \times C. \tag{2.38}$$

where $\tau = 1.44$ Te and $\tilde{A} = 1.44 \times T_e \times A_0$. The value of $C$ can be obtained from a standard table (Table 2.6).

### 2.13.8 Dosimeters

*Dosimeters* measure the amount of radiation energy absorbed over a given period by an object that can be a living being or any other object on Earth. An adaptive response may be harmful even if it is a small dose and does not have the ionizing effect. Adaptation is not observed in every case of exposure, but when it does occur, the reaction may be harmful. In living creatures, 70 % of the damage to DNA can be caused by water radiolysis. The interest for low radiation dose effects in the last decades has been increased dramatically because of the increased number of artificial radiation sources that are generating exposure of the Earth's biosphere [156, 157]. Thus, there is a growing need to detect the amount of radiation received by a living body or the part of the living body when it is exposed to any amount of radiation.

**Fig. 2.38** (**a**) The picture of a commercial dosimeter with (Courtesy of DCA Dosimeter, NJ) (**b**) a probe for commercial use of monitoring radiation (Courtesy: Arrow Tech. ND) (Courtesy of DCA Dosimeter, NJ)

370 End Window Probe

## 2.13.9 Commercial Dosimeters

These *dosimeters* are available in different sizes, and the accuracy of these instruments can be energy dependent or rate dependent. As the dosimeter is exposed to radiation, ionization occurs in the surrounding chamber decreasing the charge on the electrodes in proportion to the exposure. Figure 2.38 shows the picture of a dosimeter with probe.

## 2.13.10 MOSFET Dosimeter

*MOSFET* (metal oxide silicon field effect transistor) dosimeters are used to record the amount of dose received by a body, where anthropomorphic phantoms with 20 high-sensitivity diagnostic detectors record the data and transfer all the data to a laptop computer to analyze the dose. It is to be noted that radiation doses received by different parts of the body are different. Thus, the dosimeter has to be set up according to the process technology of the particular scanner. However, newer-generation scanners include *automated dose modulation system* (Fig. 2.39) [158, 159].

Recently introduced combined modulation (CARE Dose 4D: Siemens Medical Solutions) automatically determines whether the patient is slim or obese from a

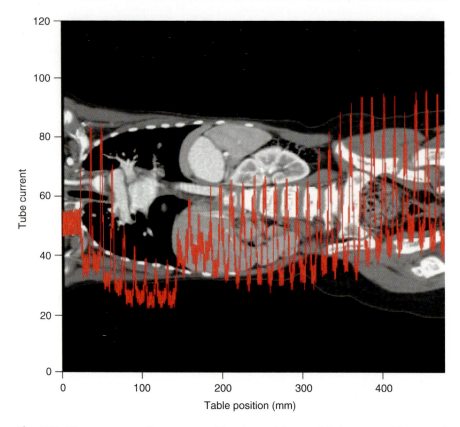

**Fig. 2.39** The newer-generation scanners with automated dose modulation system (Courtesy of Toshiba, Diagnostic imaging, Aug 2005, p 64)

single localizer radiograph and adopts the dose to the user-specified image-quality reference tube current-time product and strength of modulation [160].

The complexity of the measurement task therefore primarily depends upon the photon energies to be measured, anticipated distribution in the cells, sensitivity and accuracy required, and the intended application of the information obtained from the measurement. Blumart and Weiss made the first measurements of radionuclides in the human body with ionization chambers in 1927 and later by Schlundt in 1929. In 1937, Evans made a significant improvement in sensitivity using a Geiger–Muller (GM) tube.

### 2.13.11 Miniature-Type Dosimeters

Recently, several miniature-type dosimeters are available in the market. They are small and convenient to carry in the pocket. One of the most frequently used

## 2.13 Recording and Measurement Techniques

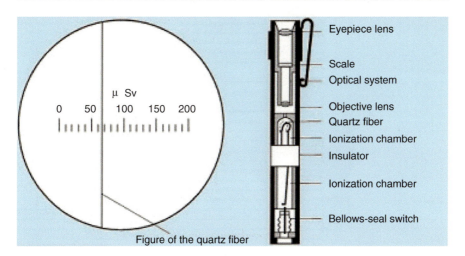

**Fig. 2.40** The picture of a pen ionization meter with an illustrative sketch of the quartz fiber (Courtesy of Informationskreis KernEnergie, Berlin)

dosimeter is the *pen ionization dosimeter*. It carries a long narrow chamber filled with nonconducting gas. Two metal electrodes are placed on two sides that form the anode and the cathode. An initial charge is established between the anode and the cathode. Figure 2.40 shows the picture of a *pen ionization dosimeter* with an illustrative sketch of the quartz fiber.

Radiation passing through the gas (normally nonconducting) chamber partly ionizes the gas and produces free electrons. These negatively charged electrons are free to move at the end of the tube (anode) because of an existing positive charge. As a result, the initial neutral gas inside the chamber loses electrons. The amount of charge lost from the gas is a measure of the amount of radiation that has passed through the gas chamber. The amount of charge that has been passed through the chamber is again proportional to the current that passed through the circuit. The current is read by the pen dosimeter, which ultimately measures of the charge that has been passed through the chamber. However, the readings with the pen dosimeters can be severely affected by the vibration of the dosimeters during measurements.

Similar to the *pen ionization dosimeter*, there is an another direct-reading pocket dosimeter, which is called *pocket dosimeter*. The dosimeter contains a small ionization tube filled with nonconducting gas approximately 2 mm in volume. Inside the ionization chamber, a central metal wire forms the anode. The anode wire is again attached to a metal-coated quartz fiber. When the anode is charged to a positive potential, the charge is distributed between the anode wire and the quartz fiber. Electrostatic repulsion deflects the quartz fiber, which is proportional to the amount of charge existing between the electrodes.

When the chamber is exposed to radiation, electrons are produced and attracted toward anode, which is positively charged. As a result, the deflected quartz fiber moves toward the original position. The shifting of the fiber from its deflected position is a direct measure of the ionization of the gas.

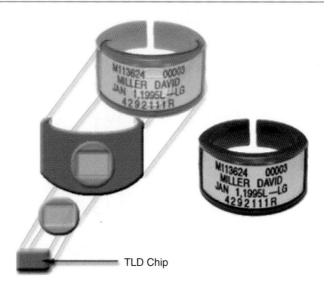

**Fig. 2.41** The schematic of a TLD ring (Courtesy: Harvard University, MA)

A more modern dosimeter is developed in the market, which is based on the luminescence properties of a crystal-like lithium fluoride (LiF) or calcium fluoride (CaF$_2$), and the instrument is known as *thermoluminescent dosimeter*. When the crystal is exposed to radiation, it is heated giving off an amount of energy that is proportional to the amount of radiation energy it is exposed. This light is detected by a detector and recorded digitally. *Thermoluminescent dosimeter* (TLD) can detect doses as low as 1 mrem. The advantage of the device is its linearity of response to a dose, which is relatively energy independent. It is also sensitive to low doses. Figure 2.41 shows the schematic of a TLD ring. *TLD rings* are used as dosimeter to measure whole-body exposure to radiation. The lithium fluoride (LiF) chip is encapsulated inside the ring. These rings are small and very convenient to use.

Recently, *optically simulated luminescence dosimeters* (OSLD) is finding a widespread application in a variety of radiation dosimetry fields. The meter is used in personal monitoring, environmental monitoring, retrospective dosimetry, (including geological dating and accident dosimetry), and space dosimetry [161]. The device consists of a crystalline thin film of aluminum oxide (Al$_2$O$_3$) which undergoes cumulative structural changes as it receives the external radiation. In order to measure the amount of radiation, the film is exposed to a green laser light. The amount of blue light emitted by the film in response is proportional to its radiation exposure. Unlike a TLD, an OSLD can supply an instant readout that can be repeated if necessary.

Figure 2.42 shows the schematic a commercial OSL whole-body dosimeter.

There are some solid-state devices (transistors) which detect the ionization current when exposed to an external radiation. The measured ionization current is proportional to the radiation the device is exposed. These devices are available in

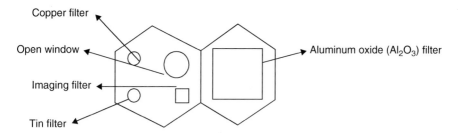

**Fig. 2.42** The schematic of a commercial OSL, a whole body dosimeter

the market since 1980. But compared to the other dosimeters, these solid-state device backed dosimeters are more expensive, and sensitivity is also not very attractive.

In spacecraft and in accelerator particle physics experiments, organic scintillating fibers are capable to cover surfaces of relatively large volumes. The benefit of these detectors is that they do not need regular crystalline lattices to form electron–hole (e–h) pairs. The base component of the scintillators is typically polystyrene or polyvinyltoluene, which is the primary absorber of radiation. The fibers are typically between 0.5 and 1 mm in diameter with 25-μm-thick cladding of acrylic around the scintillating core [162]. The fibers are optically coupled to a waveguide block or fiber ribbon (with or without a wave shifter) for readout by a photomultiplier tube (PMT) or a visible light photon counter (VLPC). These dosimeters are able to monitor the impact of radiation conditions on astronauts and support equipment during extravehicular activity (EVA).

The ionizing radiation is measured by dosimeters. But the radiation received from laser light, radio waves, and ultrasound that are often required in medical context are hard to measure by the regular commercial dosimeters. One method of detecting these doses delivered to a volume of tissue is to measure the temperature increase of the tissue. However, this technique does not work for tissues embedded in living organisms or for whole-body exposure. Dosimetry, therefore, for radio and ultrasound relies heavily on computational models rather than direct measurements. Recently, the use of metabolomics for the discovery of biomarkers for radiation exposure is one of the promising new approaches to radiation biodosimetry. To reduce the time necessary to detect and measure these biomarkers, differential mobility spectrometry–mass spectrometry (DMS–MS) systems have been developed and tested [119].

## 2.14 Statistical Fluctuations in Nuclear Process

It is to be admitted that it is not possible to measure any physical magnitude of a scientific process without any error. This is especially valid for nuclear processes, where the *radioactive decay* is a *random process* due to *statistical fluctuation* of the

**Table 2.7** Data distribution function

| Data | | Frequency distribution function |
|---|---|---|
| 5 | 8 | $F(5) = (2/20) = 0.1$ |
| 6 | 7 | $F(6) = (2/20) = 0.1$ |
| 8 | 6 | $F(7) = 0.1$ |
| 9 | 5 | $F(8) = 0.1$ |
| 7 | 9 | $F(9) = 0.1$ |
| 10 | 11 | $F(10) = 0.1$ |
| 11 | 14 | $F(11) = (3/20) = 0.15$ |
| 12 | 12 | $F(12) = 0.1$ |
| 13 | 13 | $F(13) = 0.1$ |
| 14 | 11 | $F(14) = (1/20) = 0.5$ |
| | | $\sum_0^\infty F(x) = 1$ |

*physical process.* As a result, there are random errors involved in the experimental data because they are different from the true value of the data [163, 164].

In nuclear physics experiment, random sample of events is distributed randomly in time, and it contains a number of events that fluctuate from sample to sample. These fluctuations in the observations result from the statistical nature of the physical process itself (such as small radioactive decay of a radioactive material, $dN/dt = -\lambda N$, $N$ is the number of nuclei present, and $\lambda$ is decay constant).

The application of statistical theory to such measurements is therefore doubly important because it contributes to our understanding of nuclear processes and it gives insight into statistical distributions, which describes other random processes whose individual events are not observable [165].

The value of counting statistics can be grouped into two main categories. The first one is to serve as a check on normal functioning of the equipment, and the second one deals with the situation in which a series of measurements are taken. In any series of such measurements, the frequency of occurrence of a particular value is expected to follow some *probability distribution laws* or *frequency distributions.*

Let us consider the frequency distribution function of an experimental procedure is $F(x)$, which is defined as the ratio of the number of occurrences of the value $x$ to the number ($N$) of measurements taken . Thus, the distribution, when normalized, gives rise to

$$\sum_0^\infty F(x) = 1 \tag{2.39}$$

For the purpose of illustration, we have presented our hypothetical set of data consisting of 20 entries in Table 2.7.

The data distribution function that has been displayed in Table 2.7 is shown as bar graph in Fig. 2.43.

Figure 2.43 shows the distribution function for the data shown in Table 2.7. Along x-axis, the function $F(x)$ is plotted, and along y-axis, $\bar{x}_e$ is plotted where

## 2.14 Statistical Fluctuations in Nuclear Process

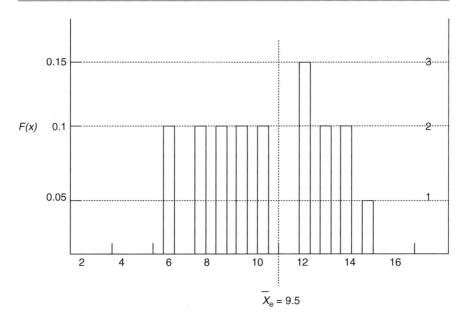

**Fig. 2.43** The distribution function for the data shown in Table 2.7. Numbers 1, 2, and 3 denote the number of occurrences

$\bar{x}_e = \sum_0^\infty F(x)$. Now, for a series of $N$ measurements $x_i$ (a single typical value from a set), the most probable estimates of the true mean $\bar{x}_e$ and of the sample variance $s^2$ are

$$\bar{x}_e = \sum_{i=1}^{N} \{(x_i)/N\} \tag{2.40}$$

$$\text{and} \quad s^2 = \sum_{i=1}^{N} \{(x_i - \bar{x}_e)^2\}/\{(N-1)\} \tag{2.41}$$

Indeed, the value $\bar{x}_e$ is the most probable estimate of the mean, that is, one of which minimizes the sum of the squares of the derivations $d_i = (x_i - x_e)^2$ (the least square criterion). When we expand $\sum_{i=1}^{N} \{(x_i - \bar{x}_e)^2\}$, we will get [166]:

$$\sum_{i=1}^{N} \{(x_i - \bar{x}_e)^2\} = \sum_{i=1}^{N} x_i^2 - 2\bar{x}_e \sum_{i=1}^{N} x_i + N\bar{x}_e^2 \tag{2.42}$$

The derivative of $\sum_{i=1}^{N} \{(x_i - \bar{x}_e)^2\} = -2\bar{x}_e \sum_{i=1}^{N} x_i + 2N\bar{x}_e = 0 \tag{2.43}$

By definition, $\bar{x}_e = x_i$. As number of observation ($N$) increases, the estimator $\bar{x}_e$ becomes more accurate: $x_e = \bar{x}_e$.

**Table 2.8** Count rate data from a radioactive source

| Minute | Counts ($x_i$) | $\Delta_i$ | $\Delta_i^2$ |
|---|---|---|---|
| 1 | 99 | −5 | 25 |
| 2 | 115 | +11 | 121 |
| 3 | 110 | +6 | 36 |
| 4 | 95 | −9 | 81 |
| 5 | 107 | +3 | 9 |
| 6 | 109 | +5 | 25 |
| 7 | 98 | −6 | 36 |
| 8 | 105 | +1 | 1 |
| 9 | 90 | −14 | 196 |
| 10 | 112 | +8 | 64 |

The average count $\bar{x}_e = 104$, and $\Delta_i = x_i - \bar{x}_e$

The Eq. 2.33 can be written in computational version expanding Eqs. 2.42 and 2.43, and we can write

$$
\begin{aligned}
s^2 &= \sum_{i=1}^{N} [\{x_i - \bar{x}_e)^2\}/\{N - 1\}] \\
&= \Sigma_i x_i^2/(N-1) - 2x_e \Sigma x_i/(N-1) + \{N/(N-1)\}\bar{x}_e^2 \qquad (2.44) \\
&= N(\bar{x}_e^2)/(N-1) - 2\bar{x}_e N(\bar{x}_e)/(N-1) + N/(N-1)x_e^2 \\
&= N/(N-1)(x_e - \bar{x}_e^2) \qquad (2.45)
\end{aligned}
$$

where $x_e^2$ and $x_e$ are the mean estimator for $x$ and $x^2$, respectively, in analogy with Eq. 2.43.

Let us now consider a set of data that are collected from a Geiger counter measuring a long-lived (steady) radioactive source and are shown in tabulated form in Table 2.8. It has been observed that the number of counts recorded per minute fluctuates almost all the time during the measurements.

Now, we are going to introduce three specific statistical models to identify the probability success of any one trial as $p$. In case of radioactive decay, the probability is equal to $(1 - e^{-\lambda t})$ where $\lambda$ is called the decay constant of the radioactive sample.

### 2.14.1 The Binomial Distribution

The binomial distribution is one of the most frequently used statistical models used in probability success of any trial. In a random group of $z$ independent trials, the probability that the event will occur $x$ times is represented by $P_x$ as [167–179]

$$
P_x = \{(z!)/(x!(z-x)!\}\{p_x (1-p)^{z-x}\} \qquad (2.46)
$$

## 2.14 Statistical Fluctuations in Nuclear Process

Now, if we carry out the summation of the above equation, we will reach to a simple result as

$$\bar{x} = zp \tag{2.47}$$

The example of binomial process is a trial and error process either rolling a die or tossing a coin. Rolling a die can have six numbers: 1–6. Let us define a successful roll as one which has any of the numbers 4, 5, and 6. Then the individual probability of success will be (3/6) or 50 %. Therefore, we can calculate the expected average number of success by multiplying the number of trial $z$ (10) by the probability $p$ that any one trial will result in a success. By substituting the value of $z$ and $p$ in Eq. 2.47, we can get

$$\bar{x} = 5.0 \tag{2.48}$$

Now, we can define the predicted variance $\sigma^2$ as the measure of the scatter about the mean predicted by a specific statistical model $P(x)$ as

$$\sigma^2 = \sum_{x=0}^{z} (x - \bar{x})^2 P(x) \tag{2.49}$$

If we carry out the summation of the above equation for specific case of $P(x)$ given by the binomial distribution, we will get the following result:

$$\sigma^2 = zp(1 - p) \tag{2.50}$$

By substituting the values of $z$ and $p$, $\sigma^2$ value becomes

$$\sigma^2 = (10) \times (0.5)\,(1 - 0.5) = 5 \times 0.5 = 2.5 \tag{2.51}$$

because $\bar{x} = zp$, and we can also write

$$\sigma^2 = \bar{x}(1 - p) \tag{2.52}$$

$$\text{and} \quad \sigma - \{\bar{x}(1 - p)\}^{1/2} \tag{2.53}$$

where $\sigma$ is called the *standard deviation* and its value is equal to $(2.5)^{1/2} = 1.58$. Indeed, *standard deviation* ($\sigma$) can be defined as the breadth of the statistical fluctuations about the true mean value.

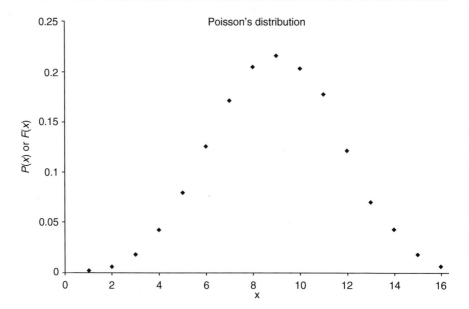

**Fig. 2.44** The Poisson distribution for a mean value $\bar{x} = 9.5$

### 2.14.2 The Poisson Distribution (PD)

PD is a discrete distribution of the binomial distribution (BD) for the case when average number of success is very much smaller than the possible number, that is, $<< z$ because $p << 1$ [169]. As a matter of fact, PD describes all random processes like disintegration of atomic nuclei or the appearance of cosmic ray bursts, whose probability of occurrence is small and constant. Figure 2.44 represents graphical presentation of *Poisson distribution*.

The *Poisson* distribution (*PD*) applies to substantially all observations made in experimental nuclear physics. The *PD* can be deduced as the limiting case of the binomial distribution for those random processes in which the probability of occurrence is very small, $p << 1$, while the number of trials $z$ becomes very large, and the mean value $m = pz$ remains fixed. It can be shown that for a small and constant probability of success, the binomial distribution reduces to

$$P(x) = \{(z^x p^x)/(x!)\}\{(e^{-Pz})\} = \{(m^x)/(x!)e^{-m}\} \quad (254)$$

where $m = \bar{x}$. The normalization of the Poisson distribution can be written as

$$\sum_{x=0}^{n} P(x) = 1 \quad (2.55)$$

## 2.14 Statistical Fluctuations in Nuclear Process

**Table 2.9** Binomial distribution and its Poisson approximation

| Binomial distribution | Poisson distribution |
|---|---|
| $z = 20$ and $p = 0.08$ | |
| $m = zp = 1.6$ | |
| $P(x = 0) = 0.1887$ | $P(x = 0) = 0.2019$ |
| $P(x = 1) = 0.3282$ | $P(x = 1) = 0.3230$ |
| $P(x = 2) = 0.2711$ | $P(x = 2) = 0.2584$ |
| $P(x = 3) = 0.1414$ | $P(x = 3) = 0.1378$ |
| $P(x = 4) = 0.0523$ | $P(x = 4) = 0.0551$ |
| $P(x = 5) = 0.0145$ | $P(x = 5) = 0.0176$ |
| $P(x = 6) = 0.0032$ | $P(x = 6) = 0.0047$ |
| $P(x = 7) = 0.0005$ | $P(x = 7) = 0.0011$ |
| $P(x = 8) = 0.001$ | $P(x = 8) = 0.0002$ |
| $P(x = 9) = 0.000$ | $P(x = 9) = 0.0000$ |
| $P(x = 10) = 0.000$ | $P(x = 10) = 0.0000$ |
| $P(x = 20) = 0.0000$ | $P(x = 20) = 0.0000$ |

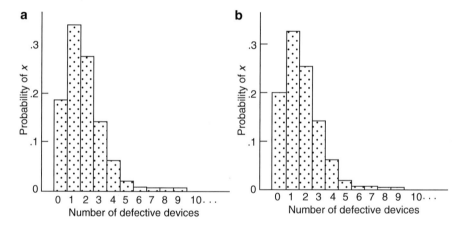

**Fig. 2.45** The graphical representation of the (**a**) binomial and (**b**) the Poisson distributions

And the mean value of the distribution $m = \bar{x}$ can be expressed mathematically as

$$m = \bar{x} = \sum_{x=0}^{n} xP(x) = pz \quad (2.56)$$

and the predicted variance $\sigma^2 = \bar{x}$ and the standard deviation $\sigma = (\bar{x})^{1/2}$. Thus, we can see that the predicted standard deviation of any *Poisson* distribution is the square root of the mean value that characterizes the same distribution. We will also introduce the general conditions (which are listed below) under which *Poisson* distribution will work for radioactive decay or radioactive disintegration:

(a) The chance for an atom to disintegrate in any particular time interval is the same for all identical atoms in the group.

122      2 Radiation Exposure: Consequences, Detection, and Measurements

(b) The chance for an atom to disintegrate during a certain time interval is the same for all time intervals of equal size.
(c) The total number of atoms and the total number equal time interval are large.

Now, if we compute the equation $P(x) = \{(z^x p^x)/(x!)\}\{(e^{-pz})\} = \{(m^x)/(x!)E - m\}$ using Sterling's approximation to the factorial $x!$, we can write

$$P(x) \cong \{(1/2\pi x)^{1/2}(m/x)^x(1/e^{(m-x)}\}. \tag{2.57}$$

Then for $m = 10$, 100, and 1,000, the value of $P(x)$ becomes 0.127, 0.040, and 0.013. Indeed, for small values of $m$, the Poisson distribution is very asymmetric and is not well approximated by the normal distribution.

Suppose in a *light-emitting diode* (LED) manufacturing plant, 8 % of the manufactured devices are defective. To illustrate the use of the Poisson approximation for the binomial, the probability of obtaining exactly one defective *LED* from the sample of 20 using Eq. 2.57, we will get the value of $P(x) \approx 0.3230$. In order to compare the values of $P(x)$ for different values of $x$ when $z = 20$ and $p = (8/100) = 0.08$, we have presented the values in the following table (Table 2.9). Figure 2.45 shows the graphical representation of the binomial and the Poisson distributions.

### 2.14.3 Gaussian Distribution (GD)

The *Gaussian distribution* (GD) function is a continuous distribution, and mathematically, it is an approximation to the *binomial* (BD) for the limiting case, where the number of observations z becomes infinitely large and the probability $p$ of success for each is finite so that $zp >> 1$. The *Gaussian* probability distribution is sometimes called the *bell-shaped curve* or *normal distribution*. It is perhaps the most used distribution in all sciences. Unlike the *binomial* and *Poisson* distributions, the *Gaussian* is a continuous distribution.

The *GD* is an analytical approximation of binomial (or Poisson) distribution when the value of z (number of trial) is very large. It is applicable to distributions in which the observed variable is not confined to integer values but can take on any value from $-\infty$ to $+\infty$. The *GD* thus applies to continuously variable observed magnitude whereas the binomial and Poisson distributions are applied to discontinuous variables. Thus, when the mean value of the distribution is large, the binomial theorem can lead to Gaussian distribution:

$$P(x) = \{1/(2\pi x)^{1/2}\}\{\exp -[(x - m)^2/2m)]\} \tag{2.58}$$

The normalized form is $\sum_{x=0}^{\infty} P(x) = 1$, and the distribution is characterized by a single parameter $m$, which is given by the product $zp$, and the predicted variance $\sigma^2$ is equal to $m$.

Consider tossing a coin for 10,000 times. $P(\text{head}) = 0.5$ and $z = 10,000$. For binomial distribution, mean number of heads $= zp = 5,000$. Standard deviation $= 50$. The probability to be within $z \pm 1\sigma$ for binomial distribution will be 0.69, and for Gaussian distribution, it will be 0.68. Table 2.10 shows comparison of an area

## 2.15 Chi-Square Distribution

**Table 2.10** Comparison of an area of ±1σ between Gaussian and Poisson

| Mean | Gaussian | Poisson | % Difference |
|---|---|---|---|
| 10 | 0.6827 | 0.74 | 7.8 |
| 25 | 0.6827 | 0.73 | 6.9 |
| 100 | 0.6827 | 0.707 | 3.5 |
| 250 | 0.6827 | 0.689 | 0.87 |
| 5,000 | 0.6827 | 0.6847 | 0.29 |

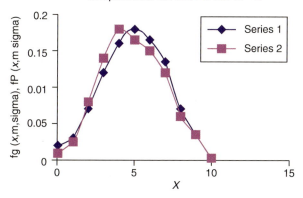

**Fig. 2.46** The comparison of Gaussian (series 1) and Poisson distributions (series 2), and $m$ is mean value of distribution

of ±1σ between Gaussian and Poisson, and Fig. 2.46 shows the Gaussian and Poisson distributions (GD and PD) with $m = 2$.

It has been observed that the counting particles in atomic and nuclear physics follow *Poisson distribution*. *Gaussian distribution* is easy to handle algebraically and reasonable. Unless the measurements provide clear evidence to the contrary, we assume the distribution Gaussian. But when the measurements are discrete, we have to apply non-Gaussian distribution [170].

The adaptation of the theoretical Poison and Gaussian distributions is carried out by soft wire and can be easily verified with the aid of a programmable calculator. An even more comfortable way is to use a spreadsheet analysis program which in general can provide the graphical representation, too. The comparison of the GD and PD shown in Fig. 2.32 is produced by means of the program Excel® by Microsoft®. Furthermore, it is possible to introduce the real measured values into the theoretical curves as well.

## 2.15 Chi-Square Distribution

Chi-square distribution (CSD) ($\chi^2$) is another best-known goodness of fit statistics that numerically compares two sampled distribution. It can be defined as

**Fig. 2.47** Chi-square probability distribution function (pdf) graph with 3-D freedom (df) for a two-sided test at significance level $\alpha = 0.5$

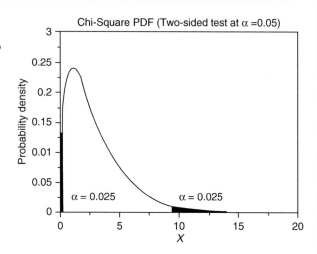

**Table 2.11** Portion of a chi-square distribution table

| Statistical degrees of freedom | Number of measurements | $z = 0.8$ | $z = 0.7$ | $z = 0.6$ | $z = 0.5$ |
|---|---|---|---|---|---|
| 18 | 19 | 12.85 | 14.44 | 15.89 | 17.33 |
| 19 | 20 | 13.72 | 15.35 | 16.85 | 18.33 |
| 20 | 21 | 14.58 | 16.26 | 17.80 | 19.34 |

$$\chi^2 = \{(N-1)s^2\}/(\bar{x}_e) \tag{2.59}$$

For Poisson distribution, we can write $s^2 \equiv \sigma^2$ and $\sigma^2 = \bar{x} = \bar{x}_e$. When we have $N = 20$, $s^2 = 7.94$ and $x_e = 9.5$. These values will give rise to $\chi^2 = 15.88$.

The sum of all the square normalized differences is the chi-square statistic, and the distribution depends upon the number of bins through the degree of freedom (df). The value of df is normally one less than the number of bins.

The probability density function (PDF) of chi-square ($\chi^2$) distribution with $r$ degrees of freedom (df) is given by [171]

$$p_r(X) = \{(X^{r/2-1}e^{-X/2})\}/\{\Gamma(1/2r)_2^{r/2}\} \tag{2.60}$$

where $r$ is the degree of freedom and $\Gamma(x)$ is a gamma function.

The Fig. 2.47 shows chi-square probability distribution function (pdf) graph with 3 degrees of freedom (df) for a two-sided test at a significance level $\alpha = 0.5$.

Table 2.11 shows the portion of a chi-square distribution table.

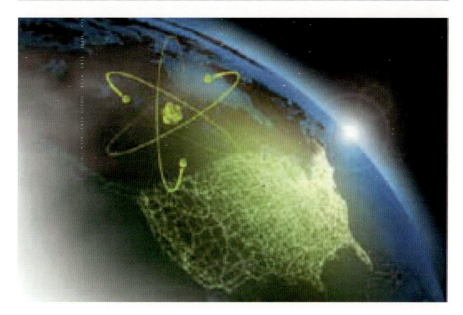

**Fig. 2.48** An expanded use of nuclear energy (Photo courtesy of Idaho National Laboratory)

## 2.16 Pros and Cons of Radiation Energy

*Nuclear energy* is a promising energy source for mankind, and it is argued that it is the safest, economical, and clean energy source [172]. It gives certainty that the *nuclear energy* will never be halted in the industrial developments due to lack of sufficient power and it guarantees the continuance of expansion.

Figure 2.48 shows the expanded use of nuclear energy. It is expected that the clean and efficient nuclear-assisted hybrid energy will be used in every sphere of our daily life and will reduce greenhouse gas emission considerably. However, there have been several analyses for the severe discordance in perceptions regarding nuclear energy [173, 174]. The studies reveal that for next decades, there are only a few realistic options for reducing carbon dioxide emission from electricity generation.

On the field of nuclear medicine, an X-ray is used to *kill cancerous cells*, and radiography is performed by means of gamma ($\gamma$) rays [173, 174]. In bioscience, radiation is mainly used for sterilization and enhancing mutations. It is also used in sterilizing medical hardware or food [5]. However, over the decades, its growth has been severely retarded due to some of the recent nuclear accidents [175–178] and experimental findings about the cause of several long-lasting diseases. Thus there has also been a view, rather vocal at that, that it is environmentally disastrous and endangers the health of the operating personnel and the public of the present generation.

## 2.17 Summary

Detecting and quantitating nuclear radiation either from man-made exposure or environmental exposure is the most important field of study for the nuclear scientists. Following the September 11 attacks, the threat of a potential for a radiological attack has made the detection of radiation sources the top priority. The radiation received by the plants and animals and the consumer products can be ionization radiation or nonionizing radiation. The first one is more severe than the later. Because *ionizing radiation* (shorter wavelengths) is capable of disrupting and separating electrons from atoms. Therefore, detection of the radiation sources is performed in the laboratories, with the help of different ion-detecting instruments, like *ion chambers, proportional counters, Geiger–Muller counter, and photomultiplier tube*.

The nuclear radiation from man-made source or from environment causes nuclear reactions with the living tissue in many ways depending on the type of radiation. This radiation includes neutrons of various energies and photons (gamma and X-rays). In addition to the primary radiation, fission also produces radioactive isotopes of many elements, which in turn can emit particles and photons, known as secondary radiation. Some of the radiation sources that the living creatures and plants received are difficult to avoid. It is also true that effects of radiation exposures sometimes are not harmful because the radiation dose is too small. Thus, the primary issue became the measurement of the amount of dose a particular body or a plant is receiving. Therefore, the need to measure the amount of dose during nuclear fuel fabrication and processing, nuclear reactor operation, radioactive ore mining operations, waste storage facilities, and environmental monitoring laboratories became a priority issue.

As a result, several types of dosimeters are invented and ultimately came into the commercial arena. These radiation detectors can be operated in three different modes, namely, the *current mode*, the *mean square voltage mode*, and *pulse mode*. Another important parameter that is measured in radiation detectors is the *energy distribution* of the *incident radiation*. The distribution is called the *response function* of the detector for the energy used in determination. This is actually done by noting its response to a monoenergetic source of radiation.

## References

1. Barth JL, Dyer CS, Stassinopoulos EG (2003) Space, atmospheric and terrestrial radiation events. IEEE Trans Nucl Sci 50(3):466 and also Streffer C (1997) Health impact of large releases of radionuclides, biological effects of parental irradiation. Ciba Foundation Symp 203:155 and Biological effects of inhaled radionuclides ICRP Pu, 31 July 2007 and Carlson TM et al (2001) Radionuclide contamination at Kazakhstan's Semipalatinsk Test Site: implications on human and ecological health, 2001 (US Dept. of Energy) and Larsson E, Meerkhan SA, Strand SE, Jonsson BA (2021) A small scale anatomic model for testicular radiation dosimetry for radionuclides localized in human testes. J Nucl Med 53(10):72

# References

2. Velinov PIY, Mateev LN (2008) Improved cosmic ray ionization model for the system ionosphere—atmosphere—calculation of electron production rate profiles. J Atmos Solar-Terrest Phys 70(2–4):574 and also ONeill PM (2010) Badhwar-ONeill 2010 galactic cosmic ray flux model—revised. IEEE Trans Nucl Sci 57(6):3148
3. Rando R et al (2004) Radiation testing of GLAST LAT tracker ASICs. IEEE Trans Nucl Sci 51(3):1067
4. Health Risks from Exposure to low level Ionizing Radiation (2006) BEIR VII, Phase 2, The National Academy Press, Washington, D.C. and also U.S. Nuclear Regulatory Commission (2011) Principle people and the environment. 17 Oct 2011
5. Eisberg RM (1961) Excited states of atom, Chapter 13. In: Fundamentals of modern physics. Wiley, New York and also Loaharanu P, Thomas P (2001) Irradiation for food safety and quality. Technomic Pub. Co., Lancaster
6. Knoll GF (2000) Radiation sources, Chapter 1. In: Radiation detection and measurements, 3rd ed. Wiley, New York and also Bertell R (1985) Nuclear radiation and its biological effects. The Book Pub. Co., Tennessee, p 15
7. Halliday D, Resnick R (1981) Electromagnetic waves, Chapter 38. In: Fundamentals of physics, 2nd edn. Wiley, New York and also Bethe H, Ashkin J (1953) Passage of radiation through matter. In: Segre E (ed) Experimental nuclear physics, vol I. Wiley, New York, pp 167–357
8. Crosbie WA, Gittus JH (1989) Medical response to effects of Ionizing radiation. Elsevier Appld. Sci. and also Biological effects of radiation and units of dose. In: Radiation safety manual. Stanford Univ. Palo Alto Health Care Syst. 20910, CA, p 19
9. Mettler FA, Upton AC (2008) Medical effects of ionizing radiation. Elsevier Sci. Pub., New York and also Boice Jr JD, Fraumeni Jr JF (1984) Radiation carcinogenesis. Raven Press, New York
10. Wolbarst A (2000) Physics of radiology. Medical Physics Pub., Madison and also Wilson R (1995) Effects of ionizing radiation at low doses. In: Metter FA, Upton AC. 2nd edn. W.B. Saunders, Philadelphia
11. Salvato JA, Leonard N (2003) Environmental engineering. Wiley, Hoboken and also Advanced nuclear reactor safety issues and research needs. In: Workshop Proc., Paris, 18–20 Feb 2002. Int. Atomic Energy Agency (AIEA and IAEA Pub.)
12. Lloyd DC, Dolphin GW (1977) Radiation induced chromosome damage in human lymphocytes. Br J Ind Med 34(4):261
13. Jagetia GC, Venkatesha VA, Reddy TK (2003) Naringin a citrus flavonone protects against radiation induced chromosome damage in mouse bone marrow. Mutagenesis 18(4):337
14. Pshenichnov I, Mishustin I, Greiner W (2005) Neutron from fragmentation of light nuclei in tissue like media. Phys Med Biol 50:5493
15. Nagamine K, Torakai E, Shirnomura K, Ikedo Y, Schultz JS (2009) Molecular radiation biological effect in wet protein and DNA observed in the measurements of labeled electron with muons. Phys B Cond Matt 494(5–7):553
16. Stisova V, Goffmont S, Maurizot MS, Davidkova M (2006) Radiation damage to DNA protein specific complexes. Radiat Prot Dosimetry 122(1–4):106
17. Ben-Ishay Z, Prindull G, Yakelev S, Sharon S (2008) Cumulative bone marrow stromal damage caused by X-ray irradiation. Med Oncol 7(1):55
18. Valentin J (2003) RBE, quality factor, radiation weighting factor. Ann ICRP 33(4):1
19. Leo WR (2005) Techniques for nuclear and particle physics experiments. Springer, Germany, p 71
20. Ignarro LJ (ed) (2000) Nitric oxide biology and pathobiology. Academic, San Diego
21. Gordon MY, Amos TAS (1996) Stochastic effects in hemopoiesis. Stem Cells 12(2):175
22. Nowakowski B, Lemarchand A (2001) Stochastic effects in a thermochemical system with Newtonian heat exchange. Phys Rev 64(6):061108
23. Darnell J, Lodsh H, Baltimore D (1986) Molecular cell biology. Scientific American Books, New York, p 80

24. Arlett CF et al (2006) Chemical and cellular ionizing radiation sensitivity in patient with XP. Br J Radiol 79(942):510
25. Kim GJ, Chandrasekaran K, Morgan WF (2006) Mitochondrial dysfunction persistently elevated levels of reactive oxygen species and radiation induced genomic instability: a review. Mutagenesis 21(6):361
26. Limoli CL et al (2003) Persistent oxidative stress in chromosomally unstable cells. Cancer Res 63:3107
27. Bennard M et al (2008) Review analysis of radiation induced degradation observed for input bias current of linear integrated circuit. IEEE Trans Nucl Sci 55(6):3174
28. Schwank JR, Cavrois VF, Shaneyfelt MR, Piallet P, Dodd PE (2003) Radiation effects in SOI technologies. IEEE Trans Nucl Sci 50(3):522
29. Shapiro J (2002) Radiation protection, 4th edn. Harvard Press, Cambridge
30. Re V et al (2010) Mechanisms of noise degradation in low power 65 nm CMOS transistors exposed to ionizing radiation. IEEE Trans Nucl Sci 57(6):3071 and also Cooper WJ, Curry RD, O'Shea KE (1998) Environmental applications of ionizing radiation. Wiley, New York
31. Cress CD et al (2010) Radiation effects in single walled carbon nanotube thin-film transistors. IEEE Trans Nucl Sci 57(6):3040 and also Real A et al (2004) Effects of ionizing radiation exposure to plants. J Radiol Phys Prot 24:A123 and Dengel S, Aeby D, Grace J (2009) A relationship between galactic cosmic radiation and tree rings. New Phytologist 184(3):545
32. Mulvey S. Wild life defects Chernobyl radiation. BBC News
33. Kryshev II (1995) Radioactive contamination of aquatic tic eco systems following Chernobyl accident. J Environ Radiol 27:207–219
34. Mould RF (2000) Chernobyl record: the definitive history of the Chernobyl catastrophe. CRC Press, Boca Raton
35. Seifriz W (2005) Reaction of protoplasm radiation. 25(1):196. Springer-Wien, Pub., Germany
36. Gopinath DV (2007) Radiation effects, nuclear energy, and comparative risks. Curr Sci 93(9): 1230
37. Goans RE et al (1996) Dose estimation using Lymphocyte depletion kinetics, Armed Forces Radiobiology Research Institute Workshop, Washington, 25–27 Sept 1996
38. Coans CE, Holloway CE, Bayer EM, Ricks RC (1997) Early dose assessment following severe radiation accidents. Health Phys 72(4):513–518
39. Oldham TR, Mclean FB (2003) Total ionizing dose effects in MOS oxides and devices. IEEE Trans Nucl Sci 50(3):483
40. Li Z (1995) Experimental comparisons among various models for reverse annealing of the effective concentration of ionized space charges of neutron irradiated silicon detectors. IEEE Trans Nucl Sci 36:1825 and also Gerardin S, Paccagnella A (2010) Present and future non-volatile memories for space. IEEE Trans Nucl Sci 57(6):3016
41. Srour JR, Marshall CJ, Marshall PW (2003) Review of displacement damage effects in silicon devices. IEEE Trans Nucl Sci 50(3):653
42. Chaudhari P, Bhoraskar SV, Padgavkar S, Bhoraskar VN (1991) Comparison of defects produced by 14-MeV neutrons and 1-MeV electrons in n-type silicon. J Appl Phys 70(3):1261
43. Jones R, Carvalho A, Goss JP, Bidden PR (2009) The self interstitial in silicon germanium. Mater Sci Eng 159–160:112
44. Summers GP, Burke EA, Dale CJ, Wolicki FA, Marshall PW, Gehlhausen MA (1987) Correlation of particle induced displacement in silicon. IEEE Trans Nucl Sci NS-34 (6):1134 and also Keiter ER, Russo TV, Hembree CE, and Kambour KE (2010) A Physics based device model of transit neutron damage in bipolar transistor. IEEE Trans Nucl Sci 57(6):3305
45. Myjack MJ, Seifert CE (2008) Real time Compton imaging for gamma ray tracker handheld CZT detector. IEEE Trans Nucl Sci 55(2):769
46. Pickel JC, Kalma AH, Hopkinson GR, Marshall CJ (2003) Radiation effects on photonic images. IEEE Trans Nucl Sci 50(3):671

# References

47. Hopkinson GR, Dale CJ, Marshall P (1996) Proton effects in CCDs. IEEE Trans Nucl Sci 43:614 and also Philbrick RH (2002) Modeling the impact of preflushing on CTE in proton irradiated CCD-based detector. IEEE Trans Nucl Sci 49(2):559
48. Prigozhin G et al (2000) Characterization of radiation damage in Chandra X-ray CCDs. Proc SPIE 4140:123
49. Akkerman JB, Chadwick MB, Levinson J, Murat M, Lifshitz Y (2001) Updated NIEL calculations for estimating the damage induced by particles and $\gamma$-rays in Si and GaAS. Radiat Phys Chem 62:301
50. Dowsett DJ, Kenny PA, Eugene R (1998) The physics of diagnostic imaging. Chapman & Hall Medical, Gt. Britain
51. Theuwissen JP (1995) Solid state imaging with charge coupled devices. Kluwer, Dordrecht/London
52. Bross AD, Pla-Dalmau A (1994) Radiation effects in plastic scintillators and fibers. Int Conf Cal. In HEP, Fermi Nat. Lab., Oct 1994, Final Rept. 91/74
53. Marini G et al (1985) Radiation damage to organic scintillation materials. CERN Int. Rept. 85–08
54. Kozma P, Kozma P Jr (2004) Radiation resistance of heavy scintillators to low energy. Radiat Phys Chem 71(3–4):705–707
55. Marcillac PD, Corn N, Dambier G, Leblanc J, Moalic J-P (2003) Experimental detection of alpha particles from radioactive decay of natural bismuth, Letts. To Nature 422:876 and also Mann WB (1978) NCRP Report 58, A handbook of radioactivity measurements procedures, section 3.7, National Council on Radiation Protection and Measurements, Washington, D.C.
56. Baschenko SM (2004) Remote optical detection of alpha particle source. J Radiol Prot 24:75 and also Semkow TM, Parekh PP (2001) Principle of gross alpha and beta radioactivity detection in water. Health Phys 81(5):587
57. Hobzova L, Patel M, Spuray Z (1983) TSEE detection of alpha and beta radiation. Radiol Prot Dosimetry 4(3–4):137 and also Jesse WP, Sadauskis J (1995) Ionization impure gases and the average energy to make pair for alpha and beta particles. Phys Rev 97(4):1668
58. L' Annunziata MF, ElBaradei HM, Burkart W (2003) Hand book of radioactivity analysis. Elsevier Sci. & Technol., New York and also Toh K, Yamagishi H, Sakasai N, Nakamura TN, Soyama K (2009) Development of two dimensional micropixel gas chamber capable of individual line readout for neutron measurements. In: IEEE nuclear science symposium and medical imaging, Orlando, 25–31 Oct 2009
59. Cullum BM, Mobley J, Bogard JS, Moscovitch M, Phlys GW, vo-Dinh T (2001) Detection of neutrons using novel three-dimensional optical random access memory technology. SPIE Proc 4199:165–172 and also Ebisu T, Watanabe T (2003) Cryogenic detection of neutron using superheated superconducting tin granules. Nucl Instrum Methods A503(3):589
60. Campion PJ (1971) The operation of proportional counters at low pressures for micro density. Phys Med Biol 16:611
61. Fischer BE (1977) Nucl Instrum Methods 141:173
62. Westphal GP (1976) Nucl Instrum Methods 134:387
63. Knoll GF (2000) Radiation detection and measurement, 3rd edn. Wiley, Hoboken, p 189
64. Borkowski CJ, Kopp MK (1970) Some applications and properties of one and two dimensional position sensitive proportional counters. IEEE Trans Nucl Sci NS-17(3):340
65. Stelzer H (1976) Nucl Instrum Methods 133:409
66. Emery EW (1966) Geiger-Muller and proportional counters. In: Attix FH, Rosech WC (eds) Radiation dosimetry, vol II. Academic, New York
67. Wilkinson DH (1950) Ionization chambers and counters. Cambridge University Press, London
68. RCA Photomultiplier Manual Technical series (1970) PT-61, RCA Solid State Division, Electro optics and Devices, Lancaster, 1970 and also Photomultiplier tube, principle to application (1994) Hamamatsu Photonics, K.K.
69. Knall HR, Persy DE (1972) Recent work on fast photomultipliers utilising GaP(Cs) dynodes. IEEE Trans Nucl Sci NS-19(3):45

70. Simon RE, Williams BF (1968) Secondary electron emission. IEEE Trans Nucl Sci NS-15:167
71. Melntyre RJ (1961) Theory of micro-plasma instability in silicon. J Appl Phys 32(6):983 and also Goetberger A et al (1963) Avalanche effects in silicon $p$-$n$ junctions II. Structurally perfect junctions. J Appl Phys 34(6):1591
72. Gulinatti A et al (2005) Time resolution at room temperature with large area single photon avalanche diode. Electr Lett 41(5):272 and also Dautet H et al (1993) Avalanche effects in silicon p-n junctions II. Structurally perfect junctions. J Appl Opt 32(21):3894
73. McNally D, Golovin V (2008) Review of solid state photomultiplier developments by CPTA and photonique SA. Nucl Instrum Methods A-140:5
74. Buzhan P et al (2003) Silicon photomultiplier and its possible applications. Nuclr Instrum Method 504:48
75. Sciacca E et al (2003) Silicon planar technology for single photon detectors. IEEE Trans Electron Dev 50:918
76. Ignatov SM, Maneuski DA, Potapov VN, Chirkin VM (2007) A scintillation γ-ray detector based on a solid state photomultiplier. Instrum Exp Technol 50(4):474
77. Seiferr S et al (2009) Ultra precise timing with SiPM-based TOF PET scintillator detector. IEEE Nuclear Science symposium medical imaging, Orlando, 25–31 Oct 2009
78. Roberts OJ, Jenkins DG, Joshi P (2009) Nuclear Spectroscopy with novel LaBr3:Ce scintillator and Si-PM detector. In: IEEE transactions on nuclear science symposium medical imaging, Orlando, 25–31 Oct 2009 and also Degenhardt C et al (2009)The digital silicon photomultiplier—a novel sensor for the detection of scintillation light. In: IEEE transactions on nuclear science symposium medical imaging, Orlando, 25–31 Oct 2009
79. Purghel N, Valcov N (1995) Nucl Instrum Methods B95:7 and also Knoll GF (2000) Radiation detection and measurement, 3rd edn. Chapter 4. Wiley, Hoboken, p 105
80. Harrer JM, Beckerley JG (1973) Nuclear power reactor instrumentation systems handbook, vol 1, Chapt 5, TID-25952-PI and also Campbell NR, Francis VJ (1946) IEEE 93, Part III
81. Frame PW (2005) A history of radiation detection instrumentation. Health Phys 88(6):613 and also Knoll GF (2000) Radiation detection and measurement, 3rd edn. Chapter 4. Wiley, Hoboken, NJ, p 109
82. Kanno I et al (2008) A current mode detector for unfolding X-ray energy distribution. J Nucl Sci Technol 45(11):1165
83. Ranger NT (1999) Radiation detector in nuclear medicine. Radiographics 19:48
84. Ruby L (1994) Further comments on the geometrical efficiency of a parallel disk source and detector system. Nucl Instrum Methods A337:531 and also Purghel L, Valcov N (1995) Particle and energy dependence of the statistical fluctuations of an ionized chamber current. Nucl Instrum Methods B95:7
85. Knoll GF (2000) Radiation detection and measurement, Chapter 4, 3rd edn. Wiley, New York
86. Conway AM et al (2009) Si-based pillar structural thermal neutron detectors. In: IEEE nuclear science symposium medical imaging, Orlando, 25–31 Oct 2009
87. Tremsin AS, Feller WB, Downing RG, Mildner DFR (2004) The efficiency of thermal neutron detection and collimation with microchannel plates of square and circular geometry. IEEE Trans Nucl Sci 512:1020
88. McGregor DS et al (2004) Design considerations for thin film coated semiconductor thermal neutron detectors. Nucl Instrum Methods A500:272
89. Friedrich H, Dangendrof V, Demian AB (2002) Position sensitive thermal neutron detector with Li-6-foil converter coupled to wire chambers. Appl Phys A Mater Sci Proc 74:S124
90. Izumi N et al (2003) Development of a gated scintillation fiber neutron detector for areal density measurements of inertial confinement fusion capsules. Rev Sci Instrum 74:1722
91. Price WJ (1964) Nuclear radiation detection, 2nd edn. Chapt-10, McGraw Hill, New York, and Allen WD (1960) Neutron detection. George Newnes, Ltd., London
92. Tremsin AS, Feller WB, Downing RG (2004) Efficiency optimization of neutron imaging detectors with $^{10}$B doped MCPs. Nucl Instrum Methods A500:269

# References

93. Boyace NO, Kowash BR, Wehe D (2009) Thermal neutron imaging with a rotationally modulated collimator. In: IEEE nuclear science symposium, Orlando, 25–31 Oct 2009
94. Gersch HK, McGregor DS, Simpson PA (2002) A study of the effect of incremental gamma ray doses and incremental neutron fluences upon the performance of self-biased 10B coated high purity epitaxial GaAs thermal neutron detector. Nucl Instrum Methods A489:85
95. Fuller WB, Downing RG, White PL (2000) Neutron field imaging with microchannel plates. Proc SPIE 4141:291
96. Abdushukurov DA et al (1994) Model calculation of efficiency of gadolinium based converters of thermal neutrons. Nucl Instrum Methods B84:400
97. Knoll GF (2000) Radiation detection and measurements, 3rd edn. Chapter 5, Wiley, New York, p 148
98. Boag JW (1966) Ionization chambers. In: Attix FH, Roesch WC (eds) Radiation dosimetry, vol II. Academic, New York
99. Torii T (1995) Ionization efficiency of a gas flow ion chamber used for measuring radioactive gases by Monte Carlo simulation. Nucl Instrum Methods A-356:255 and Knoll GF (2000) Radiation detection and measurements, 3rd edn. Chapter 5, Wiley, New York, p 152
100. Friedlander G, Kennedy JW, Macias ES, Miller JM (1981) Nuclear and radiochemistry, 3rd edn. Wiley, New York, p 363
101. Goulding FS, Landis DA (1974) Semiconductor detector spectrometer electronics. In: Cerny J (ed) Nuclear spectroscopy and reactions part a. Academic, New York
102. Sze SM (ed) (1998) Modern semiconductor device physics. Wiley, New York, pp 353–382
103. Rossi BB, Staub HH (1949) Ionization chambers and counters. McGraw-Hill, New York
104. Burns DT, Buermann L (2009) Free air ionization chamber. Metrologia 46:S-9
105. Boag JW (1966) Introduction to radiological physics and radiation domsimetry. Ionization chambers. Rad. Dosi. vol II. In: Ahix FH, Roech WC (eds) Academic Press, New York, pp 1–72
106. Greening JR (1960) A compact free air chamber for use in the range 10–50 kV. Br J Radiol 33:178
107. Buermann L, Grosswent B, Kramer H-M, Selbach H-J, Gerlach M, Hoffmann M, Krumrey M (2006) Measurement of the X-ray mass energy absorption coefficient of air using 3 keV to 10 keV synchrotron radiation. Phys Med Biol 51(5125)
108. Hasegawa M et al (2008) Initial plasma produced by Townsend avalanche breakdown on QUEST tokamak. Jpn J Appl Phys 47:287
109. Knoll GF (2000) Radiation detection and measurement, 3rd edn. Wiley, New York, p 173, Chapter-6
110. Knoll GF (2000) Radiation detection and measurement, 3rd edn, Chapter 4. Wiley, New York, p 112
111. Platt U, Stutz J (2010) Differential optical absorption spectroscopy. Springer, New York
112. Welz B, Sperling M (1999) Atomic absorption spectroscopy. Wiley VCH, Weinheim
113. Twyman RM (2000) Atomic emission spectroscopy. Elsevier, New York
114. Ross CB, Frdeen KJ (1997) Concepts, instrumentation and techniques in inductively coupled plasma optical spectroscopy, 2nd ed. Perkin Elmer, Cambridge
115. Smith ZJ, Berger AJ (2008) Integrated Raman angular scattering magnetic angular microscopy. Opt Lett 3(7):714 and also Gardiner DJ (1989) Practical Raman spectroscopy. Springer, New York
116. Vink R (1997) Magnetic resonance spectroscopy. In: Reilly P, Bullock R (eds) Head injury. Chapman & Hall, London
117. Emsley JW, Feeney J (2012) Progress in nuclear magnetic resonance spectroscopy. Elsevier, New York
118. Downard KM (2007) Historical account: Francis William Aston, the man behind the mass spectroscopy. Eur J Mass Spectrom 13(3):177
119. Coy SL et al (2010) Detection of radiation – exposure biomarkers by differential mobility prefiltered mass spectrometry. Int J Mass Spectrom 291:108

120. Gupta TK et al (2004) LuI$_3$:Ce—a new scintillator for gamma ray spectroscopy. IEEE Trans Nucl Sci 51(5):2302
121. Gupta TK et al (2002) RbGd$_2$Br$_7$:Ce scintillators for gamma ray and thermal neutron. IEEE Trans Nucl Sci 49(4):1655
122. Rapach TA, Pelcher MA. Gamma ray spectroscopy. Department of Physics and Astronomy, University of Rochester, Rochester, NY 14627
123. Brooks FD, Klein H (2002) Neutron spectrometry–historical review and present status. Nucl Instrum Methods Phys Res A 476:1
124. Kaschuck YA, Esposito B, Trykov LA, Semenov VP (2002) Fast neutron spectrometry with organic scintillators applied to magnetic fusion experiments. Nucl Instrum Methods Phys Res A 476:511
125. Chichester DL, Johnson JT, Seabury EH (2010) Measurement of the neutron spectrum of a DD electronic neutron generator'. In: 21st international conference on application of accelerators in research and industry, 2010, INL/CON-10-18572
126. Chichester DL et al (2009) Active neutron interrogation to detect shielded fissionable material. Appl Radiat Isot 67:1013
127. Brooks FD et al (2007) A compact high energy neutron spectrometer. Radiat Prot Dosimetry 126(1–4):218
128. Kaschuck YA, Espositi B, Trykov LA, Semenov VP (2002) Fast neutron spectroscopy with organic scintillators applied to magnetic fusion experiments. Nucl Instrum Methods Phys Res A 476:511
129. Kinison JD, Maurer RH, Roth DR, Haight RC (2003) High energy neutron spectroscopy with thick silicon detectors. Radiat Res 159:154
130. Agosteo S et al (2003) Neutron spectrometry with a recoil radiator – silicon detector devices. Nucl Instrum Methods Res A 515:589
131. Franceschini F, Ruddy FH, Petrovic B (2008) Simulation of the response of Silicon Carbide (SiC) fast neutron detectors. Reactor Dosimetry State of the Art 2008, p 128
132. Franceschini F, Ruddy FH. Silicon carbide neutron detectors. and Strokan N, Ivanov A, Levedev A (2009) Silicon carbide nuclear radiation detectors, SiC power materials, devices and applications. In: Feng Z (ed) Springer, New York, Chapt 11, p 441
133. Tsuji K, Injuk J, Grieken RV (eds) (2004) X-ray spectrometry: recent technological advances. Wiley, Chichester
134. McDonald CA, Gibson WM, Peppler WW (2002) X-ray optics for better diagnostic imaging. Technol Cancer Res Treat 1:111
135. Chettle D (2008) X-ray spectroscopy in medicine. X-Ray Spectrom 37:1–2
136. Poesner RA, Powsner ER (2006) Essential nuclear medicine physics, 2nd edn. Blackwell Pub, Malden
137. Owens A et al (2003) Hard X-ray spectroscopy using small-format TlBr array. Nucl Instrum Methods Phy. Res A 497A:359–369 and also Owens et al (2001) Hard X-ray spectroscopy using small format GaAs arrays. Instrum Method Res A 466:168–173
138. Nirula M et al (2005) Development of nuclear radiation detectors with energy resolution capability on CdTe n+−GaAs heterojunction diodes. IEEE Electron Dev 26(1):8
139. Tavora LMN et al (2004) Intrinsic limitations in the energy resolution of drift field based radiation detectors. Radiat Phys Chem 71(3–4):723
140. Saha GB (2003) Physics and radiology of nuclear medicine. Springer, New York, p 88
141. Menon VM, Tong W, Forest SR (2004) Control quality factor and critical coupling in microring resonators through interaction of a semiconductor optical amplifier. IEEE Photonic Tech Cells 16(5):1343
142. Wheeler JL, Wang W, Tang M (2002) A comparison methods for the measurements of spatial resolution in two-dimensional circular EIT images. Physiol Meas 23:169
143. Steiner G, Watzenin D, Zangl H, Wegleiter H, Fuchs A (2007) In: 13th International conference on electrical bioimpedance and 8th conference on electrical impedance tomography, Graz, Aug 29th–Sept 2nd 2007

# References

144. Clement GT (2005) Spectral image reconstruction for transcranial ultrasound measurement. Phy Radiat Biol 50(23):5557
145. Wielopolski L (1994) Monte Carlo calculation of the average solid angle subtended by a parallelepiped. Nuclr Instrum Sci 226:436
146. Rizk RA, Hathout AM, Hussein ARZ (1986) Solid angle calculation. Nucl Instrum Methods A245:162
147. Cook J (1980) Solid angle subtended by two rectangle. Nucl Instrum Methods 178:561
148. Particle counting in radioactivity measurements (1994) ICRU Report 52, ICRU Bethesda
149. Hasegawa T et al (2004) On clock non-paralyzable count-loss model. Phys Med Biol 49:547 and also Diethorn WS (1974) Int J Appl Radiat Isot 25:55 and also Rogers DJ, Bienfang JC, Nakassis A, Xu H, Clark CW (2007) Detector dead-time effects and paralyzability in high speed quantum key distribution. New J Phys 9:319
150. Apanasovich VV, Paltsev SV. Distortion of photon correlation functions in detection systems with paralyzable dead-time effects and also Muller JW (1974) Nucl Instrum Methods 117:401 and also Ibid 117:401 (1974)
151. Faraci G, Pennisi AR (1983) Nucl Instrum Methods 212:307 and also Marcikic I, Linares AL, Kurtsiefer C (2006) Free space quantum key distribution with entangled photons. Appl Phys Lett 89:101122
152. McCormac AM (1962) Nucl Instrum Meth 15:268 and also Currie LA (2004) Detection and quantification limits: basic concepts, international harmonization, and outstanding (low level) issues. Appl Radiat Isot 61:145
153. Kortov VS, Milman II, Nikiforov SV, Gorelova EA (2000) The use of thermoluminescent detectors for radiation monitoring on territories of atomic power plants. J Int Res Pub ISSN 1311–8978 (1) and also McKeever SWS (1985) Thermoluminescence of solids. Cambridge University Press, Cambridge/London, p 376
154. Gerson M (1997) Cardiac nuclear medicine, 3rd ed. McGraw Hill, New York
155. van Kampen RJW, Erdkamp FLG, Peters FPJ (2007) Thorium dioxide related haemangiosarcoma of the liver. J Med 65(8)
156. Kuzin AM (1963) On the role of DNA in the radiation damage of the cell. Int J Radiat Biol 6 (3):201
157. Tubiana M, Dutreix A (1990) Introduction to radiobiology. Taylor & Francis Ltd., London
158. Mulkens TH et al (2005) Use of an automatic exposure control mechanism for dose optimization in multidetector row CT examinations. Radiology 237:213
159. Rubin GD, Rofsky NM (2008) Computer tomography and magnetic resonance angiography. Lippincott Williams and Wilkins, New York, 32
160. Rizzo SMR, Karla MK, Schmidt B (2005) Automatic exposure control techniques for individual dose adaptation. Radiology 235:335
161. Boetter-Jensen L, McKeever SWS, Wintle AG (2003) Optically simulated luminescence dosimetry. Elsevier Pub, Amsterdam
162. Wrbanek JD, Wrbanek SY, Fralick GC, Chen LY (2007) Microfabricated solid state radiation detectors for active personal dosimetry. NASA/TM 2007–214674
163. Heyde KL, Heyde KL, Heyde H (2008) Basic ideas and concept in nuclear physics. Taylor & Francis, New York
164. Evans DR (1963) Statistical fluctuations in nuclear process. Academic, New York, p 761
165. Hendee VR, Ritenour ER (2002) Probability statistics, medical imaging. Wiley, New York, p 180
166. Poenaru DN, Greiner W (1977) Experimental techniques in nuclear physics. Walter de Gruyter Inc., Hawthorne, New York
167. Hoel PG (1954) Introduction to mathematical statistics, 2nd edn. Wiley, New York
168. Evans RD (1955) The atomic nucleus. Robert Krieger Pub, Malabar
169. Freedman D, Pisani R, Purves R (1978) Statistics. W.W. Norton Co. Pub, New York
170. Squires GL (2001) Practical physics. Cambridge University Press, London

171. Lomax RG (2007) An introduction to statistical concepts, 2nd edn. Routledge Publication, New York
172. Ruby L et al (1994) Further comments on the geometrical efficiency of a parallel-disk source and detector system. Nucl Instrum Methods A337:531
173. Mayles P, Rosewald JC, Nahum A (2007) Hand book of radiation therapy physics, theory and practice. Taylor and Francis, Boca Raton
174. Williams R, Thwaites DI Radiation therapy. Oxford University Press, Oxford, London
175. Matson J (2011) Fast facts about radiation from the Fukushima Daiichi Nuclear Reactors, Scientific American, 16 Mar 2011
176. A rare look inside Fukushima Daiichi Nuclear Power Plant, National Geographic, 23 May 2011
177. Lost City of Chernobyl, English Russia, 13 Sept 2006
178. Chernobyl: 25 years after nuclear disaster, Huff Post Green, 2 Feb 2012

# Mathematical Modeling of Radiation

**3**

## Contents

3.1    Introduction .................................................................. 135
3.2    Trapping and De-trapping ...................................................... 137
3.3    Polarization .................................................................. 141
3.4    Electrode Design ............................................................. 143
3.5    Frisch Grid Design ........................................................... 146
3.6    Coplanar Design ............................................................. 150
3.7    Pixelated Design ............................................................. 153
3.8    Digital Radiation Detector ..................................................... 157
       3.8.1    The Detective Quantum Efficiency (DQE) ................................. 158
3.9    Direct Conversion Efficiency ................................................... 162
3.10   Measurement of Alpha, Beta, and Gamma Radiation ............................. 166
3.11   Noise in a Radiation Detector .................................................. 168
       3.11.1    Shot Noise ............................................................. 169
       3.11.2    Flicker Noise .......................................................... 173
       3.11.3    Burst Noise ............................................................ 173
3.12   Noise and Its Effect on Medical Imaging ......................................... 175
3.13   Dead Time ................................................................... 177
       3.13.1    Paralyzable and Non-paralyzable Models ................................. 177
References ......................................................................... 180

## 3.1   Introduction

Mathematical modeling has been used routinely in the design and analysis of semiconductor radiation detectors because it saves development time in the initial stages and saves the manufacturing cost as a whole. Moreover, one can easily change parameters, such as *trapping and de-trapping times* ($\tau_{t}$, $\tau_{D}$), *electric field strengths* ($\varepsilon$), *electron–hole mobility* ($\mu_{n}$, $\mu_{h}$), and *electrode designs*, during computer simulation and can minimize the *polarization* and maximize overall detection efficiency ($\eta$) of the detector without spending much time in the laboratory. Thus,

T.K. Gupta, *Radiation, Ionization, and Detection in Nuclear Medicine,*      135
DOI 10.1007/978-3-642-34076-5_3, © Springer-Verlag Berlin Heidelberg 2013

during computer simulation, these parameters are changed partially or fully until the model spectrum matches closely to the real spectrum [1, 2].

The numerical simulation as described above is of great interest during the analysis, design, and modeling of a radiation detector. Recently an accurate model for determining the three-dimensional (3-D) distribution of charge pulses produced in a semiconductor detector has been developed [3, 4]. The simulation presents a significant reduction in computational timing and has become a fruitful tool for 3-D configuration. The model also takes account of different parameters of the detector, for example, *the pulse processing, the charge collection efficiency*, and *the associated electronic noise* [5]. However, choosing parameters in a system is a challenge, especially in the network design. As for example, to reduce the statistical noise to an acceptable level and to accurately correlate with experimental work require heavy computational maneuvering due to the large number of photons and their history [6].

The next key question for mathematical modeling is the choice of *algorithm*, which directly impacts key parameters in a system. For example, a simple threshold-based detection algorithm may consider data from each sensor and therefore requires no communication. However, simulations are necessary to understand the behavior of a system, the characteristics and limitations of the algorithms, and the impacts on the system from changes in parameters. It requires thorough knowledge of the radiation detector and the related physics involving the source of generation of X-ray, the interaction of the X-ray beam with the detector and finally the imaging process. The Monte Carlo method is the most regularly used numerical simulation adopted to address the interaction problem of X-rays with the scintillators.

Experimental observations show that the efficiency of charge collection in the detector was influenced by the ratio of charge collection and *trapping* and *de-trapping* times of the charge carriers, and one should make sure that simulations should have taken into account these factors. In principle, the findings from the simulations depend on the parameters of the deep levels; hence, it is important to work with parametric quantities representing the conditions present in state of the art of the detectors. However, using a model with one mid-gap level, one can simulate both acceptors and donors in a simpler way than the deep-level model [6]. However, Petasecca et al. have adopted a model, which includes radiation-induced deep level recombination centers, and the study of Shockley–Read–Hall (SRH) statistics has been successful to evaluate the radiation damage effects in p-type silicon [7].

Simulation has been used for a long time for modeling the medical imaging on a domain composed of an electrode surrounded by its eight neighbors in a 3-D geometry. During simulation, the charge induction efficiency is computed assuming standard *electron lifetime* ($\tau$) and *mobility* ($\mu$). The bi-parametric spectrum and pulse height spectra are then simulated, taking into account a realistic noise model. The pulse height and rise time standard deviations are then computed using spectral density analysis of the detector. In case of pixelation geometry, numerical simulation has been able to optimize energy resolution of a medical imager with wide pixel and small interpixel spacing [8].

## 3.2 Trapping and De-trapping

In an imaging detector, *charge trapping* and *de-trapping* effects appear in many compounds (CdTe, CdZnTe, $HgI_2$, $PbI_2$, $PbF_2$, CsI, etc.) and elemental semiconductors (Si and Ge) used in radiation (imaging) detector [9]. As a matter of fact, *polarization effect* [10] is believed to be the possible cause of *trapping* and *de-trapping* of carriers including change of *defect structure* of the detector. The *trapping* and *de-trapping* of carriers will change the *space charge* distribution in the radiation detector and will modify the *induced charge* due to radiation.

In presence of traps, the electric field within the detector becomes weaker with time and shows *polarization*, which ultimately decreases the counting rate with time. But in absence of traps, the current density ($J$) of an insulator (semiconductor with high resistivity at room temperature) will follow the relation of Mott and Gurney [11] and can be presented mathematically as

$$J = \{(9K\mu V^2)/(8d^3)\} \tag{3.1}$$

In the presence of traps, the above equation will transform into an exponential equation as follows:

$$J = (N/N_t) \exp(E_t/kT) \tag{3.2}$$

where $N$ is the number of total charges produced due to external radiation, $N_t$ is the number of trapping centers at energy level $E_t$, $k$ is Boltzmann constant, and $T$ is the absolute temperature. When $E_t$ is very large (several eVs), the space charge limited current does not induce significant conductivity and consequently does not increase the noise level.

The charge carriers ($N$) under the influence of the bias voltage will decrease in the presence of traps, and the number of the charge carriers that will be trapped will depend upon the number of traps and the energy level of the trapping centers. As a result, the number of free charge carriers will decrease and can be expressed mathematically as

$$(N - N_t) = N \exp(t/\tau_d) = N \exp(-x/\lambda) \tag{3.3}$$

where $\tau_d$ is the mean drift time, $x$ is the distance traveled by the carriers ($x = v_d t$), $\lambda$ is the mean free path, and $v_d$ is the drift velocity. Now if the carriers are reemitted due to thermal process, the probability of such process will give rise to a value reciprocal to the de-trapping time as

$$(1/\tau_D) = \Im \exp(-E_t/kT) \tag{3.4}$$

where $\tau_D$ is the de-trapping time, $E_t$ is the energy level of the *trapping center*, and $\Im$ is the frequency factor.

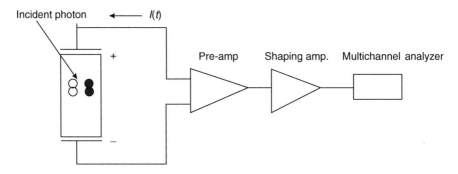

**Fig. 3.1** Schematic diagram of detector and the associated electronics

When periodicity of a perfect single crystal is perturbed due to the presence of foreign materials or crystal defects, discrete energy levels ($E_t$) are introduced into the band gap that represent some *defect centers* or *trapping* centers. These defect centers are also called *generation–recombination* (G-R) centers. These recombination centers can act as traps, where a carrier can be captured and subsequently be injected back into the band. Although not contributing directly to *recombination*, *trapping* and *de-trapping* can impact heavily on the overall carrier dynamics and can affect the detector performance. In absence of traps, the induced charge due to the external radiation is a function of the interaction location ($x$) and can be defined as

$$= qN_0[(v_h\tau_h)/(d)\{(1-\exp[(-x/v_h\tau_h)]\} + [(v_e\tau_e)/d\{(1-\exp[(-x/v_e\tau_e)]\} \quad (3.5)$$

However, in the presence of traps, the lifetime ($\tau$) of the carriers will be modified to $\tau_h = (1/N_t v_{th}\sigma h)$ and $\tau_e = (1/N_t v_{th}\sigma_e)$, where $\tau_h$, and $\tau_e$ are the hole and electron lifetimes, $N_t$ is the number of trapping centers at an energy level $E_T$, $v_{th}$ is the thermal velocity of the carriers, and $\sigma_h$ and $\sigma_e$ are capture cross sections of the holes and electrons. Now the time-dependent current $I(t)$ due to trapping and de-trapping can be written as

$$I(t) = \{(N_0 q)/(\tau_r)\}\{1 - \exp(t/\tau_D)\} \quad \text{for } t \leqslant \tau_r \quad (3.6)$$

and $I(t) = 0$ when $t \geq \tau_r$, where $\tau_r$ is the transit time of the carrier.

Following Hecht formula the above Eq. 3.6 can be written as

$$V(t) = \{(N_0 q)/(C)(\tau_D/\tau_r)\}\{1 - \exp(t/\tau_D)\} \quad (3.7)$$

where $C$ is the preamplifier (Fig. 3.1) capacitance. If the de-trapping time is not negligible, the voltage pulse will show fast and slow components with time and the above Eq. 3.7 will be modified to:

## 3.2 Trapping and De-trapping

For fast component : $V(t)$
$$= [\{(N_0q)/(C)\}\{\tau_e/\tau_r\}][\{(1/\tau_D + \tau_e/\tau_d)\}\{1 - \exp(t/\tau_e)\}]$$
(3.8)

where $\tau_e = \{(\tau_d\tau_D)/(\tau_d + \tau_D)\}$. For slow component, however, it is difficult to reach a satisfactory analytical solution. If the charges are generated near to one of the electrodes and *surface recombination is neglected*, the efficiency ($\eta$) of the detector will be given by

$$\eta = \{(\mu\tau\varepsilon/d)\}\{1 - \exp(-d/\mu\tau\varepsilon)\}$$
(3.9)

When *surface recombination* is considered to be present, the above equation will transform into

$$\eta = \{(\mu\varepsilon)/(\mu\varepsilon + s_v)\}\{(\mu\tau\varepsilon/d)\}\{1 - \exp(-d/\mu\tau\varepsilon)\}$$
(3.10)

where $s_v$ is the *surface recombination velocity*, $\mu$ is the mobility, $\tau$ is the lifetime of the carriers, and $\varepsilon$ is the electric field.

The *surface recombination velocity* ($s_v$) is defined as the number of carriers recombining per second per unit area divided by the excess concentration over the equilibrium value at the surface. It is a function of density and energy of interface states, capture probabilities, carrier concentrations, and degree of occupancy of the interface states and can be defined mathematically as

$$s_v = N_{ts}v_{th}\sigma$$
(3.11)

where $N_{ts}$ is the density of surface states per unit area, $\sigma$ is the capture cross section, and $v_{th}$ is the thermal velocity of the carriers. *Surface recombination velocity* ($s_v$) plays an important role in determining the *mobility-lifetime* ($\mu\tau$) product—the *quality factor* of the radiation detector. Surface recombination is a special case of recombination, which is analyzed in terms of surface recombination velocity (SRV) instead of lifetimes ($\tau_h$ and $\tau_e$). The X-ray spectra and photocurrent methods are able to give a rough estimate of mobility-lifetime product ($\mu\tau$), by analyzing either the charge collection efficiency as a function of the applied electric field or the shape of the X-ray photo peak [12].

Frequently mobility-lifetime products for electrons and holes are measured rather than separately determined. For a given detector geometry and field strength, the mobility-lifetime products determine the charge collection efficiency and are often quoted when evaluating nuclear detector materials [13].

At a semiconductor surface there are *surface energy states* with energies within the energy gap of semiconductor. These energy states can be divided into layer states (owing to the characteristics of the oxide layer; they arise from the absorbed ions, molecules, or imperfections). The interface states owing to the characteristics of the semiconductor-oxide interface arise from initial semiconductor surface

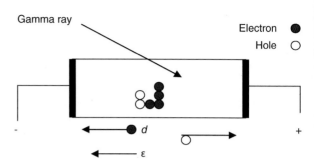

**Fig. 3.2** Schematic of a basic planar detector where charge carriers are created by external radiation and separated by an external bias. $d$ is the distance between electrodes, and $\varepsilon$ is the electric field

treatment before oxide formation, and their behavior is similar to the bulk recombination centers.

It has been reported that the self-trapped hole in cesium iodide (CsI) crystal consists of two bound iodine (I) atoms (the $V_k$ center, $I_2^-$) aligned along <100> axis of the body-centered cubic (bcc) lattice. However, in alkali halide crystals the situation is quite different. Here one can find two-atom $V_k$ centers. On the other hand, in lead fluoride crystal ($PbF_2$), the hole can be shifted to one of the 12 nearest neighbor lead (Pb) atoms by displacement of two fluorine (F) atoms between them [14].

The band gaps of certain materials like diamond and mercuric iodide ($HgI_2$) are large, and these materials show high resistivity. As a result, the neutrality of the crystal after the passage of ionizing radiation is not restored in the time interval between two events. Therefore, one can expect space charge accumulation within the crystal. These charges might be trapped partially for a period of time depending on the position of the trapping levels introduced in the forbidden gap. These immobile trapped charges will constitute an electric field opposite to the direction of the applied field. Thus, the field distribution due to the external field will no longer remain constant. The modified field in presence of the traps can be expressed mathematically as

$$\varepsilon(x) = (V/d) - (2\pi\rho/K)(d - 2x) \tag{3.12}$$

where $\rho$ is the space charge per unit volume and is equal to $(N_0 E_0 q/\epsilon v)$, $N_0$ is the number of radiations counted by the crystal, $q$ is the electronic charge, $E_0$ is the energy of the incoming radiation of volume $v$, $V$ is the applied bias voltage, $K$ ($K = \epsilon/\epsilon_0$, where $\epsilon$ is the permittivity of the material and $\epsilon_0$ is the permittivity in vacuum and is equal to $8.85 \times 10^{-12}$ F/m) is dielectric constant of the material, and $d$ is the distance between the electrodes. Figure 3.1 represents a schematic of a radiation detector with associated electronics, and Fig. 3.2 shows the physical activities inside the detector once it is irradiated by the incoming radiation.

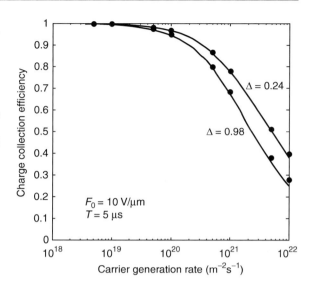

**Fig. 3.3** Monte Carlo simulation of charge collection efficiency vs. carrier generation rate when trapping and de-trapping are taken into considerations. $F_0$ is the electric field, $\Delta$ is the absorption depth, and $T$ is the exposure time (Reproduced with permission: IEEE, [115] and Elsevier Pub.)

## 3.3 Polarization

*Polarization* is defined as the time-dependent performance of the detector. Possible causes of polarization include trapping, de-trapping, and change of defect structure in the detector. The trapping and de-trapping of charge carrier change the space charge distribution in the detector and thus modify the electric field profile. The charge collection efficiency will be altered correspondingly through the change in the average drift length of the carrier, as we have already noticed from the relation established in Eq. 3.10.

Figure 3.3 shows the photo-generated charge rate vs. total charges collected when recombination due to trapping and de-trapping is taken into consideration. The upper curve represents the total charges for mammographic application, and the lower one represents for radiographic applications [15]. Apart from the trapping and de-trapping, the effectiveness of moving charge carriers away from surface recombination centers can also affect the image resolution of the detector as we can see from the relation

$$\eta = [\{1/(1 + s/\mu\varepsilon) \cdot (\tau\varepsilon\mu/d)\}\{1 - \exp(-d/\mu\tau\varepsilon)\}] \quad (3.13)$$

where $s$ is the surface recombination velocity. It has been pointed out by Levi et al. [16] that whenever charge carriers are generated near the surface (electrode), surface recombination in addition to the bulk trapping will change the *Hecht equation*.

Semiconductor like cadmium zinc telluride CdZnTe (CZT) has poor charge collection efficiency because of their large amount of trapping centers. As a result,

**Fig. 3.4** The $^{57}$Co spectrum for a CZT detector [11] (Courtesy: Fermiionics and BNL, DTRA prog)

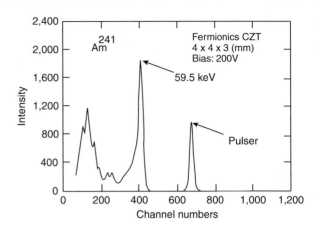

peaks in a spectrum have broad shoulder to the lower energy side, a typical sign of high hole trapping [17, 18]. Figure 3.4 shows the $^{57}$Co spectrum for a CZT detector.

As a result, FWHM value for the particular peak is difficult to measure, and we can use *Hecht equation* to take into account all these trapping centers. Accordingly, Eq. 3.13 will be modified to

$$\eta = (Q/Q_0) = [\{(\lambda_e/d)\}\{1 - \exp(-d-x)/\lambda_e)\}][\{(\lambda_h/d)\}\{1 - \exp(-d-x)/\lambda_h)\}] \quad (3.14)$$

where $x$ is the distance from interaction site to cathode, $d$ is the thickness of the detector, $\lambda$ is the mean free path, and e and h in the subscripts denote electron and hole. The value of $\lambda_e$ is almost 200 times larger than $\lambda_h$, so the trapping incidents have to account for $\lambda_h$ only. In this case, the induced charge due to energy deposition of X-ray (or γ-ray) is equal to the initial charge multiplied by $\eta$. The parameters $\lambda_e$ and $\lambda_h$ are determined by comparing the shape of the peaks (tails) in response functions with the ones in measured X-ray or gamma-ray spectra.

The *trapping* and *de-trapping times* may vary considerably among different traps, so we can expect the time-dependent detector performance with *different time scale*. The polarization may have drastically different effects on the X-ray spectra. The difference in time scale may be related to the fabrication procedures of the detector, the impurities and defects that are introduced during single crystal growth, and the behavior of the contacts due to different deposition techniques. Other factors that contribute to the polarization are the incident radiation energy, flux, penetration depth, and detector impurities.

Gradual change in energy resolution, count rate, or peak position over time is addressed as the consequence of *polarization*. In mercuric iodide (HgI$_2$), polarization is generally thought to be the effect of the degradation of the exposed material in the air, and it is suggested by many that immediate encapsulation of the material after processing can prevent the degradation. Another type of polarization is observed in p-n junction diodes and also in $HgI_2$, which is due to the formation of *dead layer*. It causes a gradual reduction in X-ray peak counts or peak position over a period of several hours after application of bias voltage. The *dead layer thickness* measured in $HgI_2$ detectors varies from 0.91 μm at 5.89 eV to 1.6 μm at 22.16 keV [19]. According to some experts, the cause of these phenomena is due to the *electrodrift* and *diffusion* of impurities [20] or space charge formation [21].

*Dead layer* is formed due to the formation of surface channels. Surface channels on the other hand actually appeared due to the formation of surface states during the surface preparation of the material (device). The surface channels actually control the *I–V* characteristics of the diode and become a source of noise (leakage current) in the detector. It has been suggested that purification of the material, improvements of the growth process, increasing bias voltage, lowering flux, and/or increasing operating temperature and use of thin device have greatly reduced the *polarization effects* [22, 23]. Most of the time, the thicker device shows polarization effect the most due to the inherent properties of the bulk material.

Sometimes, a pronounced tail is observed in the high-energy side of the spectrum when $HgI_2$ is used in gamma-ray spectrometer. However, continuous bias application reduced the effect very much. The possible explanation of this is the formation of charge multiplication process associated with Auger recombination. The process involves charge trapping of electron and recombination of holes. As a result, energy released by the recombination is transferred to a remaining trapped electron, which is ultimately ejected to the conduction band [24].

In cadmium telluride (CdTe) and cadmium zinc telluride (CdZnTe), polarization effect is believed to be due to the trapping of electrons that ultimately causes an accumulation of space charge and a reduction of the depletion layer thickness. According to some experts, the chlorine doping to increase the bulk resistivity may be one of the causes of polarization in these materials. However, *careful processing*, *proper purification*, and use of proper *contact material* and *electrode design* can minimize the polarization effect considerably [25].

## 3.4 Electrode Design

Let us consider the metal contact over a semiconductor material. When a metal is brought into intimate contact with a semiconductor, the Fermi level of the metal is fixed in relation to the bands of the semiconductor. It has been observed that in most of the semiconductors, the Fermi level at the interface comes about one-third the band gap ($E_g$) above the valence band [26, 27]. When there is no metal in contact with the semiconductor, the Fermi energy level at the surface is $q\varphi_0$. Assuming an acceptor surface state density $D_s$ and equating the charge induced on the metal surface to the surface state charge and the space charge in the semiconductor, we can write the barrier height $\varphi_B$ as

$$\varphi_B = a(\varphi_m - \chi) + (1 - a)(E_g/q - \varphi_0) = (a\varphi_m + b) \qquad (3.15)$$

where $a = \{\varepsilon_i/(\varepsilon_i + \delta q^2 D_s)\}, b = [-a\chi + \{(1-a)(E_g/q - \varphi_0)\}]$, $\delta$ is the interfacial layer, $\chi$ is the electron affinity of the semiconductor, $\varphi_m$ is the metal work function, and $\varepsilon_i$ is the dielectric constant of the interfacial layer. The values of $\varphi_0$ and $D_s$ can be expressed as

$$D_s = \{(1-a)\varepsilon_i\}/\{(a\delta q^2)\} \quad \text{and} \quad \varphi_0 = \{(-Eg/q)\} - \{(a\chi + b)/(1-a)\} \tag{3.16}$$

When the number of the surface states is high, $\varphi_B = (E_g/q - \varphi_0)$, and the barrier height is independent of the metal work function $\phi_m$. On the other hand when the surface states is small and $\varphi_B = (\varphi_m - \chi)$, the metal work function plays an important role.

The estimation of barrier height is important in the sense that it determines the current flow. The current density $(J)$ can be expressed mathematically following Richardson thermionic emission as

$$J = AT^2 \exp(-q\phi_B/kT)[\exp(qV/kT) - 1] \tag{3.17}$$

where $A$ is the Richardson constant for thermoionic emission, $k$ is Boltzmann constant, $T$ is temperature in absolute, $q$ is the electronic charge, and $V$ is the applied bias voltage. The barrier height can also be determined from the width of the space charge region. For abrupt p-n junction, the depleted width is

$$W = [2\varepsilon(V_{bi} - V - kT/q)/(qN_D)]^{1/2} \tag{3.18}$$

where $V_{bi}$ is the built-in voltage, $N_D$ is the donor concentration density, and $\varepsilon$ is the dielectric constant of the material.

Electronic states at the surface of semiconductor detectors that are produced during the final surface treatment and by adsorbed vapors during the operation of the device lead to the formation of surface channels. The thickness of the surface channels is of the order of Debye length or bulk screening length ($L_D$), which is several tenths of a millimeter in good detectors. The $L_D$ can be expressed mathematically as

$$L_D = \{(\varepsilon_0\varepsilon_r kT)/(q^2 N)\}^{1/2} \tag{3.19}$$

The surface channel changes the electric field inside the detector and its surface in such a way that there are some regions where the electric field is much higher than an average electric field. It has been shown experimentally that the surface channels are the cause of *dead layer* [28, 29]. The better side of the detector is therefore the n-side for n-type surface channel and p-side for the p-type surface channel. An n-type surface channel leads to *a dead layer* beginning at the p-contact.

In coaxial Ge (Li) detector, the $(1/r)$ dependence of the electric field may partly compensate the effect of the channel on the surface of the electric field. With usual contact arrangement (n-contact at the outside and p-contact in the core), a p-channel

## 3.4 Electrode Design

in a conventional coaxial detector can lead to a lower surface electric field than a planar one. This leads to a near perfect I–V characteristic with coaxial detectors compared to a planar one. An n-type channel in a coaxial detector makes the surface electric field near the core even larger, thus leading to a poor reverse characteristic.

The metal electrodes are used to impose an electric field across the semiconductor and collect the charge created by the incident radiation. In principle these electrodes should be ohmic, meaning that the current–voltage relation through these contacts should be linear. As the current produced by the applied bias is always larger than it is created by the incident radiation, blocking contacts are generally applied in Si and Ge devices. Whereas in mercuric iodide ($HgI_2$) and cadmium telluride (CdTe) where the band gap of the materials is high, the ohmic contacts are desirable to get a lower leakage current. However, deposited metal electrodes appear to produce Schottky barrier-like contacts that can limit the charge collection efficiency for at least one type of carrier. As a result, contact metal became a subject of research in the fabrication of room temperature nuclear detectors.

The choice of a metal surface barrier for semiconductor radiation detector is in practice made on the basis of a reverse current characteristic. Metals with work functions greater than 4.4 eV are all acceptable. Evaporated gold (Au) adheres poorly to Ge surface. On the other hand, chromium (Cr) adheres well to Ge surface but brittle and easily cracked by external pressure. Platinum (Pt) adheres well but requires high temperature to melt the metal and needs electron beam evaporation. Of all the metals tried so far for ohmic contacts, nickel (Ni) and palladium (Pd) both give acceptable small reverse currents, and the metals adhere well to the surface. In addition to that, these metals can be evaporated from a tungsten filament.

In high-purity germanium (HPGe), two types of contacts are formed. For n+ type, lithium (Li) of several millimeter thick is diffused at 270–300 °C, and for p-type contact, boron (B) is implanted inside HPGe. When HPGe is p type, lithium (Li) diffusion forms a p-n junction, leaving a 1-mm undepleted Ge region that reduces the low-energy detection sensitivity. However, in coaxial HPGe detector, Li will not diffuse significantly into the crystal at low temperature. A phosphorus implantation, however, is an alternative procedure, which requires a careful post-implant annealing.

For mercuric iodide ($HgI_2$), palladium (Pd) and carbon (C) are usually used as contact materials for X-ray and gamma-ray detectors. Sputtered Pd, however, introduces undesirable surface states. The dark current technique has been used to measure the barrier height, which gives rise to a value of about 1.2 eV for both Pd and C contacts. In case of cadmium telluride (CdTe), electroless deposition of gold (Au) and platinum (Pt) has been found to be very successful [30]. During chlorine reaction (to increase the bulk resistivity), cadmium (Cd) leaves the material to the solution as $Cd^+$ and tellurium ($Te^-$) precipitate on CdTe surface as p-type film mainly at the interface. Part of the Te is found to exist in the deposited layer of Au or Pt, allowing the contaminated contact metal to react with Te. This does not happen with Cd. However, the reaction kinetics of Au and Pt with Te is different. In the presence of oxygen, the Te layer can form $TeO_x$ layer, which is a semiconductor material, and the metal complex forms a hetero-junction between $TeO_x$ and CdTe

146                                                              3 Mathematical Modeling of Radiation

with highly distributed interfacial regions [31]. In case of tunneling, it can be distributed to Au or Pt or Cl$^-$ (doped).

Diffused indium (In) forms a low resistive good contact to n-type CdTe crystal at 300 °C and for n-type CdTe material. In a neutral argon (Ar) atmosphere, it forms a good has been successfully used in good X-ray and alpha ($\alpha$)-particle detectors. The surface of CdTe is etched or polished, and a layer of indium (In) or the oxide of indium (InO$_3$) is deposited. It is then thermally diffused within the temperature limits, and a p-type contact is made with gold (Au). However, for gamma-ray detector, metal–semiconductor–metal (M-S-M) structure with Au or Pt contacts is very popular.

In lithium-drifted (Si–Li) silicon, sputtered-coated amorphous silicon ($\alpha$-Si) has been used as n-type contact. For p-type contact, boron is being implanted on the silicon surface, and both of these contacts have been successfully used in silicon radiation detector [32].

p- and n-type contacts in cadmium zinc telluride (CZT, Cd0.96Zn0.04 Te) are formed by diffusing gold (Au) and indium (In). However, chlorine is doped in the crystal to (i) increase the bulk resistivity and (ii) to compensate background impurities and defects. Due to the low leakage currents (<10 nA at room temperature), CZT detectors are usually fabricated with platinum (Pt) and gold (Au) ohmic contacts by using metal–semiconductor–metal (MSM) structures. CZT detectors with Pt contact have shown good energy resolution of 1.4 % (FWHM) at 59.5 keV (at −37 °C) [33]. However, poor hole charge transport has resulted in long tail spectra. On the other hand, p-i-n structures in CZT detectors that are used in gamma-ray detectors are formed by p- and n-type contacts. These contacts are formed by liquid phase epitaxial deposition (LPD) of p- and n-type crystals of HgCdTe.

## 3.5 Frisch Grid Design

It has been reported that the *Frisch grid* design can deliver better sensitivity to a radiation detector over planar design [34, 35]. The prototype parallelepiped design with side grids is limited to thin dimensions due to *weighting potential* restrictions. The design allows one to fabricate relatively large volume detectors with reduced contamination in the measurement region. Moreover, the device utilizes several physical effects to increase gamma-ray interaction efficiency and energy resolution including geometrical weighting, small pixel, and Frisch grid effects [36].

Figure 3.5 shows the schematics of a planar device and a *Frisch grid* device. The *Frisch grid* design and its performance indicate that trapezoid prism (TP) design is a better choice over parallelepiped design because the $\gamma$-ray interaction ($F_i$) is better in the interaction region. As a result, TP design gives rise to higher sensitivity and higher rejection ratio [37], which can be understood from the following relation:

$$F_i = \{(1/2(W_c + W_p))(H_i + 1/2H_p)\}/\{(1/2(W_a + W_c)(H_i + H_p + H_m)\}$$
$$\approx \{((W_c + W_p))(2H_i + H_p)\}/\{(2(W_a + W_c)(H_i + H_p + H_m)\} \qquad (3.20)$$

## 3.5 Frisch Grid Design

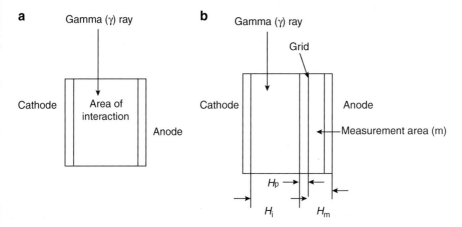

**Fig. 3.5** Schematic of (a) planar device and (b) Frisch grid device

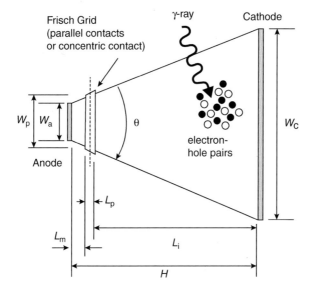

**Fig. 3.6** Schematic of the general features for a geometrically weighted semiconductor Frisch grid radiation spectrometer (Reproduced with permission, Elsevier Pub.)

where $W_c$ and $W_a$ are the width of the cathode and anode, respectively; $W_p$ is width at the previous region center; $H_i$ is the height of the interaction region; $H_p$ is the height of the previous region; and $H_m$ is the measurement region height. Figure 3.6 below shows the schematic of the general features for a geometrically weighted semiconductor Frisch grid radiation spectrometer.

The probability of distribution of the normalized $\gamma$-ray radiation under uniform irradiation can be expressed mathematically as

$$P_N(x)dx = \{2x\tan(\theta/2) + W_a\}/\{H_d^2 \tan(\theta/2) + H_d W_a\} \qquad (3.21)$$

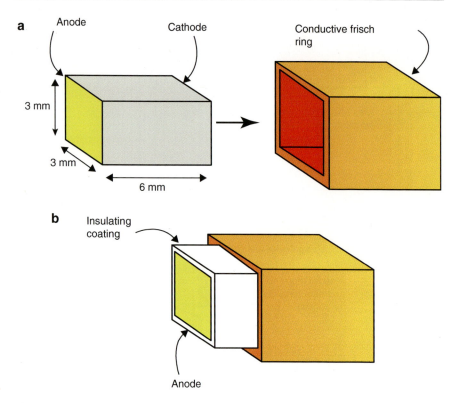

**Fig. 3.7** The different parts of a bar-shaped Frisch-grid detector. (**a**) The inner bar has been detached from the conductive ring. (**b**) Partial insetion of the bar inside the conducting ring (Reproduced with permission from Elsevier Pub.)

where $H_d$ is the total detector height, $x$ refers to the distance from the anode toward cathode, and $\theta$ is the acute angle at the anode (Fig. 3.6). The combined effects of the *geometrical weighting* and pixilated detector cause the formation of pseudo-peak which enhances the overall resolution of a geometrically weighted semiconductor device. The Frisch grid effect, on the other hand, increases the overall device performance due to the screening of the hole charge carriers' motion [38].

In Frisch grid design, a potential is applied to the device such that electrons are drifted toward the cathode. Placing a grid near the anode ensures that the induced signal forms only from the drifted electrons. Several methods of creating a Frisch grid effect without an embedded grid have been studied [39, 40]. The most effective design so far is the noncontacting Frisch grid detector, where a bar-shaped detector is inserted into a conductive ring, which also allows for an insulator filling between the Frisch ring and the detector body [41]. Figure 3.7 shows the schematic of a bar-shaped Frisch grid detector.

## 3.5 Frisch Grid Design

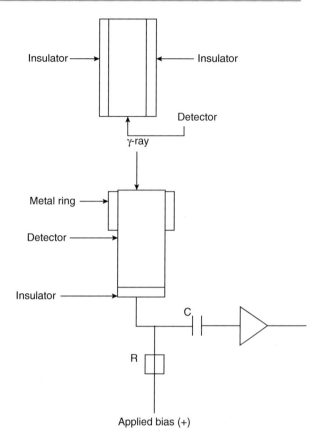

**Fig. 3.8** Schematic of a capacitive Frisch grid design with operational connections. *R* resistor, *C* capacitor

Charge induction on the anode occurs mainly in the gap region (measurement region) between the anode and Frisch ring. The highest energy resolution at 662 keV (1.7 % FWHM) has been reported with Frisch ring edge placed at 2 mm back from the anode [42].

To compensate electron trapping, Luke [43] employed a linear compensation technique using subtraction circuit with relative gain applied between channels, assuming good material uniformity. Recently, He et al. proposed a method [44] for depth correction of electron trapping. In coplanar geometry, fine tuned outer most grids and outer most gaps minimize the difference in weighting potential [45, 46]. Using radial sensing method, energy resolution of 2 and 2.4 % FWHM at 662 keV γ-ray energies has been reported.

It has been observed that coplanar design is very effective and promising to achieve energy resolution, and has provided an important breakthrough to overcome the hole-trapping problem in compound semiconductors. Unfortunately, the design is expensive, and the process is complicated. Furthermore, interstrip leakage current becomes an additional source of noise in the circuit. In order to minimize

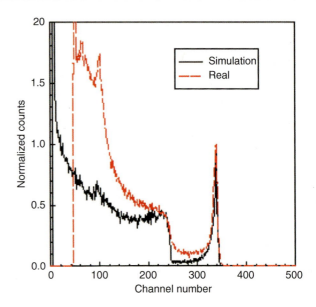

Fig. 3.9 Pulse height spectrum from a capacitive Frisch grid detector using $^{137}$Cs source (Courtesy: RMD)

these effects, a newly designed detector named *capacitive grid detector* has been proposed by Montemont et al. [47].

In the *capacitive Frisch grid design*, the bulk detector is insulated with a dielectric layer which enables one to modify the *weighting field* ($E_w$) without affecting the applied field ($E$). The *weighting field* corresponds to an instantaneous variation in charge separation and represents the transit field that follows Gauss equation:

$$\text{Div } \varepsilon E_w = 0 \quad (3.22)$$

And the dielectric relaxation time constant $\tau = \varepsilon/\sigma$, where $\sigma$ is the conductivity of the bulk detector material, $\varepsilon$ is the dielectric permittivity, and $E_w$ is the weighting field. Since the instantaneous current produced by the charge $q$ is conserved, we have div $\sigma E = 0$.

Figure 3.8 shows the schematic of a capacitively coupled Frisch grid device with operational circuit. The thin insulating layer shown in the above figure is fabricated with a high insulating material like Kapton® or Mylar®, and a metal ring is formed over the insulator.

Figure 3.9 shows pulse height spectrum from a Capacitive Frisch Grid detector using $^{137}$Cs source. For comparison a simulated spectrum is shown in the figure.

## 3.6 Coplanar Design

The recent introduction of coplanar grid techniques has led to resurgence in interest in developing large volume compound semiconductor detectors that have a γ-ray response and good spectrographic resolution [48]. As a matter of fact, the coplanar

## 3.6 Coplanar Design

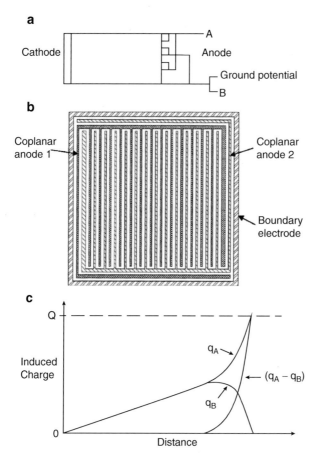

**Fig. 3.10** (**a**) Coplanar grid structure with coplanar electrodes, (**b**) coplanar grid design with two coplanar anodes and a boundary electrode, and (**c**) calculated induced charge signal on the anode A and B as a function of the position of the drifting charge –Q originating near the full area cathode. ($q_A - q_B$) denotes the external subtraction of the two signals on A and on B electrodes (Courtesy: P.N. Luke)

grid technique is essentially an evolution of the classical Frisch grid technique to reduce position-dependent charge collection effects in gas detectors [49]. Compared to simple planar contacted device, a coplanar grid detector has two anode electrodes in the form of interdigital grids. One is called the collecting electrodes, and the other is a non-collecting electrode. The collecting electrode is provided with a bias voltage ($V_g$), whereas the second one is kept at ground potential. Signals are collected from both of the electrodes and subtracted to give a net output signal. In the single-electrode configuration, signal is collected only from one grid, and optimization of detector response is accomplished by adjusting the relative areas of the two grid electrodes (Fig. 3.10).

Both the designs show no dependencies of the output signals on the hole collection. Thus, both the designs effectively eliminate the problem of poor hole transport. However, semiconductor materials also exhibit significant levels of hole trapping. This introduces a depth dependence on the net induced signal since electron trapping is dependent on the depth of the semiconductor surface (here from the grid to cathode region). With single-electrode readout method, one has to design the electrode to match the electron transport characteristics and not the gain adjustments like the conventional coplanar-grid technique.

Hole trapping in semiconductor planar detector inhibits its ability to efficiently collect total charge, because in planar geometry, good energy resolution depends upon efficient charge collection of both polarities. The charge collection in semiconductor radiation devices will follow the same principle as semiconductor devices, and the continuity equations for excess charge carriers (neglecting trapping de-trapping phenomena) can be presented mathematically as

$$\partial n / \partial t + \nabla \cdot (n \mu_n \nabla \varphi) - \nabla \cdot (D_n \nabla n) + (n / \tau_n) = G_n \qquad (3.23)$$

where $\mu_n$, $D_n$, and $\tau_n$ are mobility, diffusion coefficient, and lifetime of the electrons, respectively, and $G_n$ is set to an impulse of unit charge. As a result, the charge induced at a selected electrode of the radiation detector can be presented following Ramo's theory [50] as

$$Q_n = \int_0^t \mathrm{d}t \int_\Omega \mu_n \nabla \varphi \nabla \psi \mathrm{d}\Omega \qquad (3.24)$$

where $\nabla \psi$ is the *weighting potential*, which we have discussed earlier. For a planar device, we have to consider the contribution due to hole in addition to the contribution due to electrons, and similar model can be set forward. When we solve Eq. 3.24, we can set the photocurrent $J_{\mathrm{ph}}$ as

$$J_{\mathrm{ph}} = qd[\mu_n n(x) + \mu_p p(x)]\varepsilon(x) - q\mu_n n_0 + \mu_p p_0)V/d \qquad (3.25)$$

where $q$ is the electronic charge, $d$ is the thickness of the detector, $\mu_n$ is the mobility of electrons $n(x)$, $\mu_p$ is the mobility of hole, $p(x)\varepsilon(x)$ is the electric field, and $V$ is the bias voltage. Therefore, the charge collection efficiency (CCE) can be derived from Eq. 3.25 as

$$\mathrm{CCE} = \left( J_{\mathrm{ph}} / J_{\mathrm{ph\,max}} \right) \qquad (3.26)$$

where $J_{\mathrm{ph\,max}}$ is the maximum photocurrent which can be measured assuming no trapping and de-trapping of the photogenerated carriers. However, when the electrode structure is complex, induction efficiencies that have high-frequency spatial components and have large numbers of interaction positions can make the solution complex, leading to prohibitive computing time [51]. In large CZT crystal the nonuniformity around tellurium (Te) is responsible for nonuniformity of electron transport and deterioration of the energy resolution along coplanar detectors [52, 53].

## 3.7 Pixelated Design

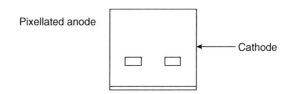

**Fig. 3.11** Schematic of a pixelated detector (*PD*)

In our previous discussion, we have mentioned that in semiconductor detector, holes are collected in a less efficient way compared to their counter charge electrons. As a result, the planar geometry has been modified in several ways where the detector signal is dependent on the efficient collection of electrons.

The two most effective designs are *coplanar* and *Frisch grid* geometry. Both *Frisch grid-* and *coplanar*-based designs have proven to be better choice over planar design because of its ability to calculate the energy resolution from the detection of electrons [54, 55]. However, all of the methods suffer from problems, which include leakage current between the anode and grid, processing difficulties, or electric field distortions.

### 3.7 Pixelated Design

In some semiconductors, single-polarity charge sensing through a coplanar grid electrode is applied to overcome the difficulties encountered due to poor hole transport in the semiconductor detector. Unfortunately, even with single-polarity charge sensing technique and methods to compensate for electron trapping, such as relative gain and depth sensing, the variations in electron trapping and material nonuniformity can still degrade the energy resolution. In *pixelated detector* (*PD*), however, the *small pixel effects* can reduce the effects of hole trapping [56].

*Pixelated detectors* are operated generally in the charge integrating mode and are dependent on the diffusion of the charge carriers from the point of its origin. These detectors can be approximated by Gaussian distribution function with a standard deviation $\sigma$ and can be defined as

$$\sigma = (2Dt)^{1/2} = \{(2kTx)/(q\varepsilon)\}^{1/2} \qquad (3.27)$$

where $D$ is the diffusion coefficient, $t$ is the time, $k$ is Boltzmann constant, $T$ is the temperature, $q$ is the electronic charge, $x$ is the drift distance, and $\varepsilon$ is the electric field.

Figure 3.11 shows the schematic of a pixelated detector (*PD*). In *PD*, the anode is generally pixelated, and the cathode remains as planar. Using Hecht equation we can write the charge induction efficiency in a pixelated device as

$$(Q/Q_0) = \eta = [\{(\mu\tau)_h \varepsilon\}/\{(d)\}][\{1 - \exp(-x/(\mu\tau)_h \varepsilon\}] + [\{(\mu\tau)_e \varepsilon\}/\{(d)\}] \\ \times [\{1 - \exp(x-d)/(\mu\tau)_e \varepsilon\}]. \qquad (3.28)$$

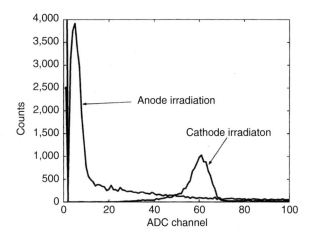

**Fig. 3.12** $^{141}$Am spectra from the cathode signal (Reproduced with permission, IEEE, Baciak and He)

where $\mu$ is the mobility, $\tau$ is lifetime of the carriers, $\varepsilon$ is the electric field, $d$ is the thickness of the detector, and $x$ is the interaction distance from the cathode.

For an event occurring near the cathode ($x \approx 0$), the above equation will transform to

$$(Q_c/Q_0) = [\{(\mu\tau)_e \varepsilon\}/\{(d^2)\}][\{1 - \exp(-d/(\mu\tau)_e\varepsilon\}] \quad (3.29)$$

For an event near the anode ($x \approx d$), it will convert the Hecht equation as

$$(Q_A/Q_0) = [\{(\mu\tau)_h \varepsilon\}/\{(d)\}][\{1 - \exp(-d/(\mu\tau)_h\varepsilon\}] \quad (3.30)$$

$Q_c$ represents the charge (channel number on the ADC) induced on the planar cathode, and $Q_A$ is the charge induced on one of the pixelated anode, and $Q_0$ is the total charge induced by the interaction of the radiated source (without any de-trapping or recombination). The value ($Q_c/Q_A$) represents the ratio between the peaks (taking into account one pixel at a time), and one can estimate the values $(\mu\tau)_h$, $(\mu\tau)_e$ from the equation ($Q_c/Q_A$). Recently, the current-mode ADC chip prototype for the readout of depleted field-effect transistor (DEPFET) particle pixel detectors with high spatial resolution has been reported. The measured signal-to-noise ratio (SNR) of the ADC, as reported by Perie et al., is about 660, where the cyclic ADC is based on current-mode memory cell [57]

Figure 3.12 shows $^{141}$Am spectra from the cathode signal. As the bias is increased, the planar spectra are more prominent because holes can travel further before they are being trapped. But the pixelated anode spectrum is more prominent than the planar spectrum due to the electron movement, which generates all of the *induced charge*. The *induced charge* produced near the pixel is dependent on the depth of interaction and the *weighting potential*. However, a thicker detector should have a weaker dependence on depth compared to a thinner detector, assuming the

## 3.7 Pixelated Design

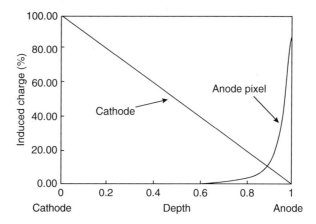

**Fig. 3.13** The weighting potential for a planar cathode and a pixelated anode

pitch depth (pixel size) is same in both cases. As a matter of fact, depth-sensing cathode to anode signal (C/A) can be expressed mathematically as

$$(C/A) = (V_C/V_A) = d_{in} \qquad (3.31)$$

where $V_C$ and $V_A$ are the voltage signal on the cathode and voltage signal to the anode and $d_{in}$ is the interaction depth. In order to correct for depth interaction, a gain adjustment must be performed. However, the measured signal is seen to vary from voxel to voxel due to *weighting potential* of the pixelated anode, electron trapping, and nonuniformity of the crystal along the length. However, at higher biases, electron trapping is lowered, and the *weighting potential* becomes a larger factor in determining the resolution [58].

Figure 3.13 shows the *weighting potential* of a planar cathode and a pixelated anode. The induced signal on the planar cathode is strongly dependent on the movement of both the charge carriers, electrons, and holes. The Shockley–Ramo theorem [59] can be applied to calculate the induced charges at different pixels due to the motion of charge carriers in the detector based on weighting potential of individual pixels. On the other hand, the induced signal on the anode is not significantly affected by the movement of the holes and the depth of interaction. Using Green's reciprocation theorem, the induced charge on a pixel $p$ by a positive unit charge at $x$ is equal to the weighting potential, $V_{wp}(x)$ of that pixel. Considering the pixel to be a square one and after interaction with the X-ray, the weighting potential can be written mathematically as [60, 61]

$$V_{wp}(x,y,z)$$
$$= (1/2\pi) \int_{y1p}^{z2p} \int_{y1p}^{y2p} \sum_{k=\infty}^{\infty} \{(1-x+2k)\} / \left\{ (1-x+2k)^2 + (y'-y)^2 (z'-z)^2 \right\}^{3/2} dy' dz'$$

$$(3.32)$$

In case of a pixelated device, the *line-spread function* (LSF) along the $x$-axis at the pixel plane is calculated by considering *induced charges* along the $y$-axis due to distributed bulk trapped charges in $x$–$z$ plane at $y = 0$. The LSF can be expressed mathematically as [62]

$$\mathrm{LSF} = (1/2\pi)$$
$$\times \int_1^0 \int_{-\infty}^{\infty} \sum_{k=\infty}^{\infty} \{(q_t(x)(1-x+2k)\}/\{(1-x+2k)^2+y+z^2)^{3/2}\}\mathrm{d}z\,\mathrm{d}x + Q_b\delta(y)$$

$$(3.33)$$

where $Q_b$ is the normalized mean number of carriers that reach the pixel electrode, $q_t$ is the normalized net space charge concentration across the photoconductor, and $k$ is an integer. On the other hand, the study of spatial resolution characteristics which is described as the *modulation transfer function* (MTF) of a medical imaging detector is an important metric of examining the quality of an image that it produces. The MTF of a detector is the relative response of the imaging system as a function of spatial frequency ($f$). The MTF is unity at zero spatial frequency. In the presence of bulk trapping, the MTF can be expressed as

$$\mathrm{MTF} = \{G(f)/G(0)\} \qquad (3.34)$$

where $\quad G(f) = (1/2\pi) \int_1^0 \int_{-\infty}^{\infty} \quad \sum_{k=\infty}^{\infty} \{(q_t(x)(1 - x + 2k)(1 - x + 2k)\}/$
$\{(1 - x + 2k)^2 + y + z^2)^{3/2}\}e^{-j2\pi fy}\mathrm{d}z \ \mathrm{d}y\,\mathrm{d}x + Q_b$

In the *pixelated device* the signal formation does not hold a linear relationship between the carrier travel distance and the induced charge [63, 64].

Edge effects in pixelated detectors having high dielectric constant show the tendency of the field lines emanating from the charge carriers (holes) to remain within the detector volume. As a result, the spectra near the edge and the corners of the pixel exhibit a longer low-energy tail [65]. In order to investigate the edge problem, we need to satisfy the boundary condition as

$$\epsilon_1 \sin \theta_1 = \epsilon_2 \sin \theta_2 \qquad (3.35)$$

where $\epsilon_1$ and $\epsilon_2$ are the dielectric constants of the medium and the detector, and $\theta_1$ and $\theta_2$ are the tangential components of the electric field lines along the detector boundary, respectively. In order to observe a uniform conduction over the both ends of the detector, an infinite number of image charges are required to extract a solution. The value of the image charges that satisfies the condition mentioned above is

$$q = h\{(\epsilon_1 - \epsilon_2)/(\epsilon_1 + \epsilon_2)\} \qquad (3.36)$$

where $q$ and $h$ are the image charges situated on the reverse side of the edge and charge of the hole carriers, respectively. Thus, the value of $q$ can be determined by knowing the values of $\epsilon_1$ and $\epsilon_2$ and the value of $h$ inside the detector [66].

From Eq. 3.24, we can see that the induced charge is dependent on the weighting potential. In *pixelated* detector the anode has a small electrode compared to the cathode. As a result, the *weighting potential* changes more abruptly near the anode than in the region near the cathode, meaning that more electrons will be drifted to the anode region than the number of holes produced at the small anode [67]. This will be further alleviated when small pixel effect will be coupled to the geometrically weighting effect.

If we consider trapping and recombination, we can write the charge induced ($Q_0$) equation from a single $\gamma$-ray event in a planar device, (following Shockley–Ramo theory) [68] as

$$Q_0 = q\{(v_h\tau_h)/d[1 - \exp(-x/v_h\tau_h) + (v_e\tau_e)/d[1 - \exp(x - d)/(v_e\tau_e)]\} \quad (3.37)$$

where $x$ is the distance between the anode and cathode, $v$ and $\tau$ are the charge carrier velocity and lifetime, sub h and e denote hole and electron, $d$ is the thickness of the detector, and $(v\tau/d)$ is defined as the carrier extraction factor (CEF) [69]. For good resolution, a detector should have CEF $> 50$.

For *Frisch grid structure*, Eq. 3.37 will be modified as

$$Q_0 = q\{(v_{hm}\tau_{hm})/d[1 - \exp((d - x - H_m - H_p/2)/(v_{hm}\tau_{hm}/d(H_p/2 + H_m)))]$$
$$+ (v_{em}\tau_{em})/d[1 - \exp((x - d)/((v_e\tau_e)/d)(H_p + H_m))]\} \quad (3.38)$$

## 3.8 Digital Radiation Detector

The development of digital radiation detectors for X-ray and gamma ($\gamma$)-ray imaging in general is motivated by many advantages of the digital imaging system over the screen–film imaging system. Digital imaging can be stored and transferred and can be used for enhancement of the picture. The physicians and the technical personnel involved can benefit from computer-assisted diagnosis, and in industry, automatic pattern recognition systems may be utilized to spot faults in devices under inspection. Moreover, transfer of images from one place to the other can be done electronically through the computer, and more contrast information can be collected due to wider dynamic range of the digital detectors including real-time imaging [70, 71].

The digital imaging radiation detectors can be of two types, direct and indirect conversion of radiation energy into electrical energy. *Direct detectors* convert the radiation energy directly to electrical charge, while the *indirect detectors* convert the radiation energy first to visible photons, which then converted to electrical charge.

**Table 3.1** Comparison between the working principles of indirect and direct radiation detectors

| Direct | Indirect |
|---|---|
| X-ray | X-ray |
| Phosphor screen | Photoconductor |
| Photons | Electron–hole pair |
| Coupled to light sensitive | Coupled to readout device |
| Readout, CCD a-Si | a-SiH active array CMOS |
| Spreading, loss of light | Collection efficiency $\propto (\mu\tau)$ |
| Poor conversion efficiency | Dark current $\propto (1/\text{resistivity})$ |
| 20–30 eV/optical photon | 4.8 eV/optical photon |
|  | Better SNR |

Let us consider that the charge collection efficiency ($\eta$) and the conversion gain ($g_1$) of both the direct and the indirect radiation detectors are the same. Then the signal received after the radiation energy is converted to either photon or electrical charge will be

$$S_i = \eta g_1 S_f \tag{3.39}$$

where $S_i$ is the initial signal received by the detector, $S_f$ is the final signal (image signal), $\eta\,(= Q/Q_0)$ is equal to charge produced by the incoming radiation ($Q$) to the actual charge measured on the electrode ($Q_0$), and $g_1$ is the conversion gain of the signal. So the zero-frequency quantum efficiency (QE) will be

$$QE = ((SNR_{out})^2\}/\{(SNR)_{in}\}^2 = (\eta)/(1 + F_1/g_1) \tag{3.40}$$

where SNR is the *signal-to-noise ratio*, $F$ is the *Fano factor* [72, 73], and $g_1$ and $g_2$ are the conversion gains after first and second conversion of the signals (first to photons and then to electrical charge).

For the indirect detector this signal is again transferred from the visible photon to electrical charge, so the above equation will be

$$QE = \{(\eta)\}/\{(1 + (F_1/g_1 + F_2/g_1g_2)\} \tag{3.41}$$

Therefore, the quantum efficiency for both the cases may not be the same. Experimental observations show that the *detective quantum efficiency* (DQE) of the indirect detectors is significantly affected by the intermediate stage, that is, during the conversion of photons to electrical charge.

## 3.8.1 The Detective Quantum Efficiency (DQE)

Medical imaging system must be designed to ensure that maximum image quality is obtained without image noise which is described by the Wiener spectrum or noise

## 3.8 Digital Radiation Detector

power spectrum (NPS). Fourier-based metrics such as modulation transfer function (MTF), NPS, noise equivalent number of spectra, and DQE are generally accepted as primary measures of image noise and detector performance. Fluorescent X-rays are absorbed at random locations in a detector, introducing spatial correlation into an image and an additional frequency-dependent effect on DQE not accounted for by *Swank factor* (DQE $= \eta S$, where $\eta$ is the quantum efficiency and $S$ is the Swank statistical factor, which affects the large area (i.e., zero frequency) DQE). As a result, it cannot justify the decrease of DQE with increase of spatial frequency. It is shown that when Swank noise is identified in a Fourier model, the Swank factor must be frequency dependent [74, 75].

Experimental observations show that fluorescence reabsorption affects both MTF and NPS in complex ways and, at low spatial frequencies, MTF can decrease up to 50 % [76, 77]. Analytical models are based on cascaded system where transfer of signal and noise is described by cascading input–output relationship for each other [78] and parallel cascaded system takes account of the reabsorption of fluorescent X-rays, for more sophisticated and complex image-process calculations [75].

Considering DQE as a function of spatial frequency ($u$) is an accurate way to look into the sensitivity of an imaging system. Particularly for a digital X-ray technology, this simple measure is a common reference point that accounts for *noise, signal loss mechanisms, sensitivity*, and *resolution*. Essentially, the DQE is equal to the square of the ratio of the imager's output *signal-to-noise ratio* (SNR) to its input SNR and can be expressed mathematically as

$$DQE(\bar{q}, u) = \{(SNR)^2_{out}\}/\{(SNR)^2_{in}\} = \{\bar{D}^2(MTF)^2(f_x)\}/\{\bar{q} \, NPS(\bar{q}, f_x)\} \quad (3.42)$$

where $D$ is the dose, $f_x$ is spatial frequency along $x$ direction, MTF is *modulation transfer function* (a measure of resolution), $q$ is the number of incident X-ray quanta, and NPS is the *noise power spectrum* produced by the imager. Higher DQE translates directly into better image quality for a given dose.

The NPS or the *Weiner spectrum* $NPS_d(f_x, f_y)$ is defined as the Fourier transform of the autocovariance $d(x, y)$, and $f_x$ and $f_y$ are the spatial frequency along $x$ and $y$ directions [79]. The $NPS_d(f_x, f_y)$ is expressed mathematically as

$$NPS_d(f_x, f_y) = {}^{Lt}X, Y, \to \propto E\{(1/2X)(1/2Y)| \int_{-x}^{x} \int_{-y}^{y} \Delta d(x, y) \exp(-2\pi(ux + vy) dx dy|^2\}$$

$$(3.43)$$

where $\{E\}$ is the expectation operator, $\Delta d(x, y) = d(x, y) - E\{d(x, y)\}$, and $f_x$ and $f_y$ are spatial frequencies in the $x$ and $y$ directions, respectively. NPS can be said to be a dimensionless signal such as a pixel value in a digital image and has the unit $mm^2$, while the NPS of a distribution of quanta has the unit $mm^{-2}$.

The light response in a scintillator (indirect conversion of energy) due to an incident ionizing radiation having energy $E_0$ is nonlinear and dependent on the atomic number $Z$ and mass number $A$ of the scintillating material [80]. The general equation for the differential light output per unit energy for a scintillator can be written as

$$(dL/dE) = \eta_s\{(1)/(1 + \P B|dE/dx|_0)\} \tag{3.44}$$

where $\eta_s$ and $\P B$ are the scintillation efficiency and quenching factors, respectively [81]. In most detectors, the response to an incident radiation shows that electrons always give the most light per deposited energy for scintillators, whereas heavier particles give less light per deposited energy. For scintillators this phenomenon is referred to as quenching [82]. As the incident ray entered into the detector, the fractional energy loss of the incident particle ($\Im$) can be expressed mathematically as

$$(dL/dE) = (1 - \Im)(dL/dE)_P + \Im(dL/dE)_\delta \tag{3.45}$$

where $(dL/dE)_P$ is the differential scintillation efficiency referring to the primary column and $(dL/dE)_\delta$ corresponds to the differential scintillation efficiency due to $\delta$-ray. The value of $\Im$, the fractional energy loss of the incident particle, again can be expressed as

$$\Im = [\{(-dE/dx)_\delta\}/\{(-dE/dx)_e\}] \tag{3.46}$$

where $\{(-dE/dx)_\delta\}$ is the energy deposited by the $\delta$-ray outside the primary column (P) and $\{(-dE/dx)_e\}$ is the electronic stopping power.

The total differential light output $\{dL(x)\}_\delta$ generated by the $\delta$-rays in the infinitesimal element $dx$ is given by [83]

$$\{dL(x)\}_\delta = a e r_e^2 \wedge' _{Ae} N_A \Im(x) n_0(x) \tag{3.47}$$

The above equation when integrated over the range of the particle $R(E_0)$ or by changing the variable over an energy $E_0$ leads to the light output expression

$$\begin{aligned}
L = a_G &\left[ \int_0^{E\delta} \{(1)/(a_R\eta_{s0}(E))\} \right] \\
&\times \left[ \ln\left\{ 1 - ((a_R\eta_{se}(E))/((1 + a_n\eta_{sn}(E) + a_R\eta_{se}(E))^{-1} \right\} \right] \\
&\times [\{(dE)/(1 + \eta_{sn}(E))/(\eta_s e(E))\}] \\
&+ \int_{E\delta}^{E0} [\{1 - \Im(E)\}/\{y(E)a_R\eta_{se}(E)\}] \\
&\times \left[ \ln\{1 - (((y(E)a_R\eta_{se}(E))/((1 + an\eta_{sn}(E) + a_R\eta_{sn}(E) + a_R\eta_{se}(E))\}^{-1} \right. \\
&\times (dE/(1 + \{(\eta_{sn}(E)/\eta_s e(E)\}) + \int_{E\delta}^{E0} [\{\Im(E)dE\}]/[\{1 + ((\eta_{sn}(E)/(\eta_s e(E))\}]
\end{aligned} \tag{3.48}$$

## 3.8 Digital Radiation Detector

**Fig. 3.14** The columnar structure of CsI grown by the author and the team at RMD by hot-wall epitaxy (Courtesy: Siemens AG)

The term $(dE/dx)$ in Eq. 3.44 is generally defined as the *energy loss per unit length*. It is also called the *stopping power* and depends linearly on the electron density of the absorber and quadratically on the charge of the particle being stopped.

The first term of the above Eq. 3.46 refers to the last part of the particle range, where the energy has decreased below the δ-ray production threshold $E \leq Axe_\delta$, and thus, $\Im(E) = 0$. The second term denotes the light output originating inside the primary column, while the third term is connected to the light produced by the knock-on electrons outside the primary column. For low incident energy, only the first term will remain.

Compared to the indirect conversion of light into energy, the real advantage of the direct conversion method is the minimal spreading inside the detector. This results in high spatial resolution and sharp images. However, lateral spreading in indirect radiation detector (scintillator) can be reduced (in some specific cases, e.g., CsI:Tl), and a comparatively larger light output can be produced by columnar growth of the polycrystalline photodetector material. Figure 3.14 shows a columnar structure of CsI:Tl grown by hot-wall epitaxy.

In photon-counting applications (nuclear medicine), indirect systems rely on photomultiplier tubes (PMTs) to count the visible photons emitted by the scintillator. The direct method, however, offers much better performance advantages to such systems [84]. Readout circuitry for photon (ph)-counting system is more complicated than for charge integration (int) system. As a result the contrast measured by these two methods can vary very widely which can be modeled by the following equations as

$$C_{int} = \left[ \int_0^{E\,max} EI_0\{(1 - e^{-\mu 1 d 1})(1 - e^{-\mu 2 d 2})\mathrm{d}E\} \right] \Bigg/ \left[ \int_0^{E\,max} EI_0\{(1 + e^{-\mu 1 d 1})(1 - e^{-\mu 2 d 2})\mathrm{d}E\} \right] \tag{3.49}$$

$$C_{ph} = \left[ \int_0^{E\,max} I_0\{(1 - e^{-\mu 1 d 1})(1 - e^{-\mu 2 d 2})\mathrm{d}E\} \right] \Bigg/ \left[ \int_0^{E\,max} I_0\{(1 + e^{-\mu 1 d 1})(1 - e^{-\mu 2 d 2})\mathrm{d}E\} \right] \tag{3.50}$$

where $\mu_1$, and $d_1$ are the X-ray attenuation coefficient and thickness of the imaging object, and $\mu_2$ and $d_2$ are the attenuation coefficient and thickness of the detector, respectively.

## 3.9 Direct Conversion Efficiency

Mercuric iodide ($HgI_2$) belongs to a family of layer-structured compound halide of mercury (Hg). The $\alpha$-phase red tetragonal structure of the single crystal is of interest for radiation detector and for medical imaging. The material is stable up to about 130 °C. Therefore, the single crystal growth of the material limits the growth temperature $T_g$ to less than 125 °C. The relatively high vapor pressure of $HgI_2$ is $12.5 \times 10^{-2}$ $T$ (at temperature ~120 °C), provides a satisfactory sublimation rate, and has been the key point for vacuum sublimation for both purification and growth process. Prolonging sublimation may result in loss of iodine from $HgI_{2-x}$ (where $|x| \ll 1$, $x > 0$) and ultimate degradation of the detector's properties and performance.

The dark electrical resistivity measurements of $HgI_2$ along $c$-axis following the relation $\{\rho = [(V)/(I \cdot d)]\}$, where $V$ is the applied voltage, $I$ is the dark current, and $d$ is the thickness of the sample, have been reported between $10^{12}$ and $10^{13}$ $\Omega$-cm, which is many orders of magnitude greater than that of other semiconductors such as silicon (Si), germanium (Ge), and II–V and IV–VI materials. Because of low-leakage current, the devices fabricated from $HgI_2$ show high-energy resolution and *FWHM* (full width at half maximum) which is related to

$$\text{FWHM (in eV)} = (2.355 E_{mean} e)(I_d \tau / 4q)^{1/2} \tag{3.51}$$

## 3.9 Direct Conversion Efficiency

where $I_d$ is the leakage current in amperes, $\tau$ is the time constant in seconds, $E_{mean}$ is the mean energy in eV required to create electron–hole pair, $e = 2.718$, and $q$ is the electronic charge. Dark current (along 001) of $HgI_2$ film has been reported as low as 0.3 pA for electric fields ($\varepsilon$) below 0.4 V/µm [85, 86].

*FWHM* is an important parameter, which indicates the resolution of the radiation detector and is related to the charge collection efficiency ($\eta$). The charge collection efficiency of a detector is defined as the actual charge collected ($Q$) by the electrodes under an applied bias to the charge that would have been collected had all the carrier trapping and recombination process been avoided ($Q_0$). It can be expressed mathematically as

$$\eta = (Q/Q_0) = (1/Q_0) \int_0^T dt \int_\Omega n \cdot \mu_n \cdot \nabla\varphi\nabla\psi d\Omega \qquad (3.52)$$

where $n$ is the concentration of electrons, $\varphi$ is the applied field, $\psi$ is weighting function, $\nabla\psi$ is the weighting potential, $T$ is the time interval, and $\mu_n$ is the mobility of electrons. For a good detector the material should be homogeneous, and the value of $\eta$ should be 1.

Again the FWHM can be expressed mathematically as a function of charge collection efficiency ($\eta$) [87]:

$$\text{FWHM (eV)} = (1 - \eta)\alpha(E_{ph})^{1/2} \qquad (3.53)$$

where $\alpha = 4.7 \times 10^5$ (eV)$^{1/2}$, $E_{ph}$ is the energy of the incoming photon in eV, and the value of $\eta$ is given by Schieber et al. [88] as

$$\eta = (\lambda_e/d)[1 - \exp\{(-d - x)/\lambda_e)\}] + (\lambda_h/d)[1 - \exp\{(-x)/\lambda_h)\}] \qquad (3.54)$$

where $x$ is the distance from the negative electrode to the charge sheet generated by the incident energetic photon, $\lambda_e$ and $\lambda_h$ are the mean drift lengths of electrons and holes, respectively, and $\lambda_e = \{(\mu_e\tau_e\varepsilon)/d\}$ and $\lambda_h = \{(\mu_h\tau_h\varepsilon)/d\}$ where $\mu_e$, $\tau_e$, $\mu_h$, and $\tau_h$ are the mobility ($\mu$) and lifetime ($\tau$) of electrons (sub e) and holes (sub h), respectively.

For a constant electric field, the charge collection efficiency ($\eta$) increases with both mobility and trapping time. Thus, the product $\mu\tau$ serves as *a figure of merit* for detector quality. Experimentally the *product* $\mu\tau$ may be extracted from the measured field of half-maximum charge collection $E_{hmc}$ in the collected charge vs. electric field plot. Some of the measuring methods are based on Hecht equation for charge collection as a function of bias [89]. Simulation of the $\mu\tau$ product to fit the experimental spectra width of $^{137}$Cs emission spectrum (660 keV gamma ray) has been used and reported by Baciak et al. [90].

The *product* $\mu\tau$ is a global measure of the material and electric charge transport properties in a material, such as *defects* and *impurity concentration, trapping*, and *de-trapping* centers, their behavior and cross section and consequently of detection quality as well as *collection* and *detection efficiency* and *resolution*. In general, semiconductor materials that show promise as good radiation detectors are having very high resistivity. As a result, ordinary Hall method cannot be applied to measure the resistivity. Therefore, a *dynamic method* such as *time of flight* is generally applied [91].

The measurement consists of deducing the drift velocity ($v_d$) of the charge carriers at one side of the planar detector. Carriers can be generated by different means such as $\alpha$-particles of $^{241}$Am, at 5.5 MeV, pulsed electron beam ($\approx$40 keV) [92, 93], or low-energy $\gamma$-rays from neutral sources. However, the pulse electron beam is considered to be more suitable and flexible method for this measurement. When the irradiated light is focused on the negative side of the electrode, the charge carriers are electrons, and when it is focused on the positive side, the charge carriers are holes.

The charge collection time in current mode is measured with a sensitive fast oscilloscope. The mobility $\mu$ of the carriers is related to ($v_d/\varepsilon$), where $v_d = d/t_R$, $d$ is the detector thickness, $t_R$ is the potential rise time, $\varepsilon$ is the electric field and is equal to ($V/d$), and $V$ is the applied voltage. The carrier lifetime $\tau_c$ before trapping can be deduced from collection time and efficiency and Hecht relation ($\mu\tau$) [94].

On the other hand, transit charge transport (TCT) in radiation detectors can simultaneously determine the mobility and trapping time [95]. The polarity of the irradiated electrode can be made positive or negative to allow separate analysis of hole or electron transport kinetics. Assuming that the de-trapping time is very large compared to the transit time $t_t = d/\mu V$, $d$ is the thickness of the detector, $V$ is the applied voltage, and $\mu$ is the mobility, we can express the voltage signal $\delta V(t)$ as

$$\delta V(t) = [\{(Q_0 d)/(\epsilon_0 \epsilon_r A)\}\{(1)/(1+s_r/\mu V)\}\{(\tau/t_t)(1-\exp(-t/\tau))\}] \quad \begin{matrix} 0 & t<0 \\ & 0<t<t_t \end{matrix} \quad (3.55)$$

or

$$[\{(Q_0 d)/(\epsilon_0 \epsilon_r A)\}\{(1)/(1+s_r/\mu V)\}\{(\tau/t_t)(1-\exp(-t_t/\tau))\}] \qquad t>t_t \quad (3.56)$$

where $Q_0$ is the charge without de-trapping and can be written as

$$Q_{0\,xk} = q \int_0^t dt' \int_\Omega d^3 r x v_x \cdot \nabla \psi_k \qquad (3.57)$$

$q$ is the charge on a single carrier, $v_x$ is the drift velocity, $x$ can be p (holes) or n (electrons), $\psi_k$ is the weighting potential for circuit $k$ and is given by the Poisson equation $\{(\nabla \cdot (\varepsilon \nabla \psi) = \rho(np)\}$, $\rho$ is the charge density per unit volume, $d$ is the thickness of the detector, $\epsilon_0$ is the permittivity of free space, $\epsilon_r$ is the relative dielectric constant of the detector material, $A$ is the area, $s_{rv}$ is the surface recombination velocity of the corresponding charge carrier, $\mu$ is the mobility of the carrier, $V$ is the applied voltage, $\tau$ is the lifetime of the carrier, and $t_t$ is the transit time.

## 3.9 Direct Conversion Efficiency

The *voltage signal* is proportional to the amount of charge collected. The inclusion of surface recombination velocity will allow us to determine the value of $s_{rv}$, and the trapping time can be obtained independently from the analysis of the shape of the waveform. Experimental measurements show that the total charge near the electrode consists of the charge on the electrode plus the charge generated in the shallow traps that are de-trapped in course of time $t$. Thus, the actual voltage signal (in Eqs. 3.54 and 3.55) will be modified and can be written as follows:

$$\delta V_{\text{actual}} = f \cdot \delta V(t) + \{(1-f)/(\tau_{\text{em}})\} \left\{ \int \delta V(t-t')[\exp(t'/\tau_{\text{em}})]dt' \right\} \quad (3.58)$$

where $f$ is the fraction of promptly generated free carriers and $\tau_{\text{em}}$ is emission time of the free carriers [96].

When the mean drift lengths of both types of carriers are much longer than detector thickness, the gamma rays ($\gamma$-rays) penetrate deep into the radiation detector, and the charge collection efficiency $\eta$ becomes approximately equal to 1. In the case of X-ray, which is mostly a shallow penetrating radiation, most of the carriers generated are close to one electrode, and only one type of carrier is seen to traverse in the detector. Thus, the detector thickness $d$ should be smaller than the drift length of the relevant carrier to achieve charge collection efficiency close to unity.

In $HgI_2$ crystals, electrons have much higher mobility than holes, and although electrons' lifetime is almost that of holes, the entrance electrodes of the X-ray detectors are always biased negatively so that electrons traverse the detector. The transit time for each charge carriers is approximately given by

$$T_r = (d^2/\mu V) \quad (3.59)$$

The transit time $t_r$ is larger than the variable travel time ($t$) of the charge carriers through the crystal ($t_r > t$), and the induced charge can be calculated as the charge carriers will travel through the crystal using the following equation [97]:

$$Q(t) = [\{(Q_0 T_m)\}]/[\{T_r(t/T_d) + (T_m/T_t)(1 - \exp(-t/T_m)\}] \quad (3.60)$$

where $Q_0$ is the number of electron–hole pair produced and is given by Eq. 3.22, $t_m = (t_t \, t_d)/(t_t + t_d)$, $t_r$ is the transit time, $t_d$ is de-trapping time, and $t_t$ is the trapping time.

As a matter of fact, the thickness of the $HgI_2$ detectors is limited between 0.4 and 5 mm primarily because of the short mean drift length of the holes. The $\mu_e\tau_e$ product for $HgI_2$ varies between $2 \times 10^{-4}$ and $5 \times 10^{-6}$ cm$^2$/V and $\mu_h\tau_h$ between $1 \times 10^{-5}$ and $5 \times 10^{-8}$ cm$^2$/V. The electric field ($\varepsilon$) for Schottky current ($I_S$) can be written as [98]

$$\varepsilon = (V/d)\gamma \quad (3.61)$$

where $V$ is applied bias, $d$ is the thickness of the sample, and $\gamma$ is the characteristic of the contact. For injecting contact, $\gamma < 1$, and for blocking contact, $\gamma > 1$.

Now we can establish a numerical model estimating all of the pulses that can be produced by a semiconductor radiation detector [99]. Since the excess carrier continuity equation involves linear operators, an adjoint continuity equation can be constructed and can be written as [100, 101]

$$(\partial n^+/dt) = \mu_n \nabla \varphi \cdot \nabla n^+ + \nabla \cdot (D_n \nabla n^+) - n^+/\tau_n + G_n^+ (\text{for electrons}) \qquad (3.62)$$

$$(\partial p^+/dt) = \mu p \nabla \varphi \cdot \nabla p^+ + \nabla \cdot (D_p \nabla p^+) - p^+/\tau_p + G p^+ (\text{for holes}) \qquad (3.63)$$

where $n^+$ and $p^+$ are electron and hole concentrations, $D_n$ and $D_p$ are the diffusivities of electrons and holes, respectively, $\tau_n$ and $\tau_p$ are the lifetime of electron and hole, and $G_n^+ (\mu_n \nabla \varphi \cdot \nabla \varphi_k)$ and $G p^+ (\mu_n \nabla \varphi \cdot \nabla \varphi_k)$ are the generation terms for electron and hole for circuit $k$, respectively. Now de-trapping of carriers causes an additional dispersion of charge, delays in charge collection, and enhances ballistic deficit. When de-trapping is included to Eqs. 3.62 and 3.63, additional terms will appear, and the continuity equation will be modified to

$$(\partial n_t/\partial t) = \{(n/\tau_n) - (n_t/\tau_n*)\} \qquad (3.64)$$

where $n_t$ is the concentration of trapped, excess carriers and $\tau_n*$ is the lifetime of the trapped carriers. These trapping and de-trapping of charge carriers are believed to be the source of *polarization* in the detector that degrades the performance of the detector.

## 3.10 Measurement of Alpha, Beta, and Gamma Radiation

The measurement of alpha ($\alpha$), beta ($\beta$), and gamma ($\gamma$) radiation from a mixed signal that allows discrimination has been reported by White and Miller [102]. The phoswich detector is a sandwich of three different phosphors. Pulses from the detector are digitized with a GaGe Compuscope 1,012, 12-bit 10-MHz digital oscilloscope card inside a PC, and the radiation types are separated by pulse discrimination, with cross-correlation analysis. For alpha ($\alpha$) detection, a 10-mg/cm$^2$ coating of zinc sulfide (ZnS) doped with silver (Ag) on a polymer backing has been used, followed by a beta ($\beta$) detection at a 0.1-in. layer of calcium fluoride (CaF$_2$) doped with europium (Eu) and gamma ($\gamma$) detection at a 1-in. thick sodium iodide doped with thallium NaI (Tl) crystal.

The beta ($\beta$) source used in the test is $^{90}$Sr/$^{90}$Y which gives an accuracy of 99.1 % with no particles identified as alpha ($\alpha$). A beta source of $^{204}$Tl gives 94.8 % accuracy with 5.2 % classified as gamma ($\gamma$). The $^{210}$Po alpha ($\alpha$) source has been found to give 99.7 % accuracy, with 0.3 % classified as gamma ($\gamma$). Gamma sources of $^{60}$Co and $^{137}$Cs give 97.2 and 99.5 % accuracy, respectively, with the remaining

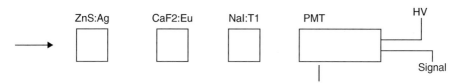

**Fig. 3.15** Schematic of the scintillators' arrangements coupled to a PMT (Courtesy: Google image)

2.8 and 0.5 % identified as beta (β). The schematic of the typical arrangement of the scintillators is shown in Fig. 3.15.

As the complexity and sophistication of medical imaging system have increased dramatically over the past several decades, improved understanding of the higher order of relationship between input and output, including the statistical fluctuations, or noise, and the reproduction of the fine spatial detail, has been a priority for a long time. The fundamental relationships that govern image quality in the medical system, many of which can be traced directly to the pioneering work of Albert Rose [103]. The stochastic nature of image quanta imposes a fundamental limitation on the performance of photon-based imaging systems and is embodied in the Rose model by Poisson statistics. So the signal-to-noise ratio (SNR) can be described in terms of statistics as

$$\text{SNR} = \{A(q_b - q_0)\}/(Aq_b)^{1/2} = C(Aq_b)^{1/2} \quad (3.65)$$

where $A$ is the area, $q_b$ and $q_0$ are quanta per unit area and mean quanta per unit area, respectively, and $C$ is the contrast. However, the Rose model, which described noise in terms of the statistical variance, became inadequate when noise is estimated from real image data. As a result, noise in a uniform image described by the stationary ergodic random process can be described by noise power spectrum (NPS) or Wiener spectrum. Therefore, the variance of $d(x, y)$ equals to that of the number of background quanta $N_b$ in a rectangular region. Ultimately, the convolution theorem can be used to show [104]

$$\sigma_b^2 = \int_{-\infty}^{\infty} \int_{-\infty}^{\infty} \text{NPS}(u,v) du\, dv = a_x^2 a_y^2 q_b$$
$$\times \int_{-\infty}^{\infty} \int_{-\infty}^{\infty} \sin c^2(\pi a_x u) \sin c^2(\pi a_y v) du\, dv = a_x a_y q_b = A q_b \quad (3.66)$$

where $\sigma_b$ is the variance, $a_x$ and $a_y$ are the object dimensions along $x$ and $y$ directions, and $u$ and $v$ are the spatial frequencies along $x$ and $y$ directions, respectively. The *detective quantum efficiency* (DQE) of a system can be derived from the measured MTF and NPS and can be expressed mathematically as [105]

$$\text{DQE}(f) = \{Gx(\text{MTF})^2(f)\}/\{qX\,\text{NPS}\} \quad (3.67)$$

where $G$ is the gain factor, and it can be 1 for data linearity; $X$ is exposure. The energy weighted $q$ for the DQE system for direct and indirect system varies less than 1 %, depending on the exposure. The modulation transfer function (MTF) and the noise power spectrum (NPS) are widely recognized as the most relevant metrics of resolution and noise performance in radiographic imaging [106].

## 3.11 Noise in a Radiation Detector

The output of a radiation detector always contains a certain amount of noise. A portion of this noise arises from the measurement procedure, and the remainder is the noise inherent to the physical properties of the detector known as the *fundamental noise*. The output signal of the detector can be expressed as the root mean square (rms) value $<S_{out}>$ and is defined as

$$< S_{out} > = (1/n) \sum_{i=1}^{n} < S_{out} >_I \tag{3.68}$$

and the noise can be expressed as the rms noise

$$( < S_{out} > )^{1/2} = [(1/n) \sum_{i=1}^{n} < S_{out} > i - < S_{out} > ]^{1/2} \tag{3.69}$$

Photon-based medical imaging systems impose a fundamental limitation due to the stochastic nature of the image quanta. Quality of the image improves, as more quanta are collected during imaging. Now if $q_b$ and $q$ are the mean quanta per unit area and mean number of quanta in a uniform object area of $A$, we can express the signal-to-noise ratio (SNR) as [107, 108]

$$SNR = \{A(\bar{q}_b - \bar{q}_o)\}/\{(A\bar{q}_b)^{1/2}\} = C(A\bar{q}_b) \tag{3.70}$$

where $C$ is the contrast $= \{(\bar{q}_b - \bar{q}_o)\}/(\bar{q}_b)$. Adopting Fourier-based wave notation, the variance $(\sigma_b)$ of the noise in a uniform image described by the stationary ergodic random process $d(x, y)$ can be expressed in terms of noise power spectrum (NPS) as

$$\sigma_b^2 = \int_{-\infty}^{\infty} \int_{-\infty}^{\infty} NPS_d(u, v) du \, dv = A\bar{q}_b \tag{3.71}$$

which is the Rose noise described in Eq. 3.71. However, Rose method must be used with care as misleading results are obtained when image quanta are statistically correlated or when sampling function that would normally be point spread function (PSF) of the system and does not correspond exactly with the size and shape of the object A [103]. In order to bridge Rose quanta-model with the Fourier-based wave approach, the $NPS_d(u, v)$ has been correlated with noise equivalent quanta (NEQ) as

## 3.11 Noise in a Radiation Detector

$$\text{NPS}_\text{d}(u) = \{\bar{q}^2\bar{G}^2\,\text{MTF}^2(u)\}/\text{NEQ}(\bar{q}, \bar{u}) \tag{3.72}$$

where $\bar{q}$, is the (uniform) average number of input quanta per unit area, $G$ is scaling factor relating $\bar{q}$, to average output $d$, $\text{NPS}_\text{d}(u)$ is the output NPS, and MTF is the modulation transfer function. The modulation transfer function (MTF) and the detective quantum efficiency (DQE) are the measures of the medical image quality and have been defined elsewhere in this book. We can correlate NPS with the MTF and DQE for a film screen as

$$\text{DQE}(\bar{q}, u) = \{(\log_{10}e)^2\gamma^2\text{MTF}^2(u)\}/\{\bar{q}\,\text{NPS}(u)\} \tag{3.73}$$

The Fourier-based quantum accounting diagram (QAD) analysis provides a theoretical estimate of the DQE based only on the mean gain, gain variance, and scattering MTF of each stage—parameters that can generally be estimated or measured from an analysis of each stage independently. The process is used routinely to determine the frequency-dependent DQE of a linear cascaded system. The process is applicable to determine the DQE digital imaging system. In case of a digital imaging system, cascaded linear system can be used with caution. The N-stage Zwieg-type cascaded model can be written in terms of DQE as

$$\text{DQE} = \{1\}/\{(1 + (1/\bar{g}_1) + (1/\bar{g}_1\bar{g}_2) + \cdots + (1/\bar{g}_1\bar{g}_2\ldots\bar{g}_n)\} \tag{3.74}$$

where $\bar{g}_n$ is the mean quantum gain of the $n$th amplification stage. The product $\bar{g}_1\bar{g}_2,\ldots,\bar{g}_n$ gives the normalized number of quanta at the $n$th stage.

On the other hand, the signal-to-noise ratio (SNR) can be defined mathematically as

$$\text{SNR} = [\{<S_\text{out}>\}/\{\Delta<S_\text{out}>\}^{1/2}] \tag{3.75}$$

The detectivity ($D$) of the detector is the inverse of noise equivalent power (NEP), and when normalized, the detectivity becomes a function of area ($A$) and the bandwidth ($\Delta f$) and can be written as

$$D_\text{norm} = D(A\Delta f)^{1/2} \tag{3.76}$$

### 3.11.1 Shot Noise

The most basic statistics of *shot noise*, namely, the mean and variance, were the framework of Campbell [109], and later on in 1918, Schottky studied the fluctuations of current in electric conductor and came out with a novel theory [110]. Hero [111] approximated the likely hood ratio of the shot noise process in the presence of additive white *Gaussian noise* which closely resembled the

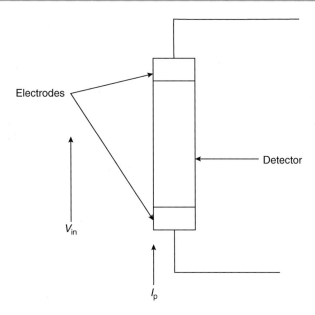

**Fig. 3.16** Schematic of the detector with two electrodes attached to the ends

underlying *Poisson process*. The *process* can be defined as the mean squared fluctuation of the number of emission events that are equal to the average count. From all the studies, the term *shot noise* draws an analogy between electrons and the small pellets of lead that hunters use for a single charge of gun [112].

In case of a granular structured film, especially when the detector is polycrystalline in nature, one can expect a *shot noise* out of the radiation detector, and it affects both the incident radiation and the electrical output of the detector. The average current due to the transport of electrons per unit time can be expressed mathematically as

$$<i> = \{<n_t>q\}/\{(\Delta t)\} \qquad (3.77)$$

where $n$ is the number of photons in the incident radiation flux striking the detector per unit time, $q$ is the electronic charge, and $n_t$ is the number of discrete events at an interval of time $\Delta t$. The probability of occurrence of these incidents can be represented by Poisson distribution, and the fluctuation in current $<\Delta i^2>$ is given by

$$<\Delta i^2> = <(i_t - <i>^2> \qquad (3.78)$$

When the above equation is expressed in terms of bandwidth ($\Delta f$), it transforms to

$$<\Delta i^2> = 2q<i>(\Delta f) \qquad (3.79)$$

## 3.11 Noise in a Radiation Detector

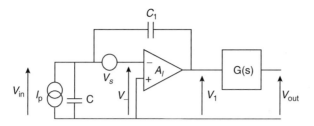

**Fig. 3.17** The schematic is an equivalent circuit diagram of a radiation detector coupled to a charge sensitive amplifier ($A_1$). $G(s)$ is the transfer function. $I_p$ represents the effects of parallel noise sources with a capacitor C parallel to $I_p$. $V_s$ represents the effects of the serial noise caused by the channel noise of the input amplifier ($A_1$)

Figure 3.16 shows the schematic of the detector with two electrodes attached to the ends. The bias is shown as $V_{in}$, and the flow of current through the detector has been shown as $I_p$.

Figure 3.17 is the schematic of an equivalent circuit diagram of a radiation detector (shown in Fig. 3.16 with electrodes) coupled to a charge sensitive amplifier ($A_1$). $G(s)$ is the transfer function. $I_p$ represents the effects of parallel noise sources with a capacitor $C$ parallel to $I_p$. $V_s$ represents the effects of the serial noise caused by the channel noise of the input amplifier ($A_1$).

When the current is turning off and on for a short interval of time such that the current pulses are independent and uncorrelated, we can approximate the current $I_p$ flowing into the circuit as a spike of current or delta function:

$$I_p(t) \Sigma_j q \delta(t - t_j) \tag{3.80}$$

where $t_j$ are the random events from a homogeneous Poisson point process of the arrival of electron. What we know about $t_j$ is that on average there should be $(\bar{I}/q)$ of them per unit time following the definition of current.

Now let us evaluate the noise spectrum due to the current pulse. For a well-behaved stationary random process, the power spectrum is equal to the Fourier transform of the autocorrelation function and can be presented in the mathematical form as

$$S_x(f) = (\bar{i_p^2}/\Delta f) = 2 \int_{-\infty}^{\infty} Rx(t') \exp(-2\pi f t') dt' \tag{3.81}$$

Now if we apply Eqs. 3.80 and 3.81 to find the autocorrelation of delta function current pulses, assuming that the delta function of the Fourier transform $F\{\delta(t)\} \leftrightarrow 1$, then the one-sided power spectral density (PSD) can be written as

$$S_x(f) = 2q\bar{I} \tag{3.82}$$

And the *Fano factor* which measures the unit of transferred charge is equal to $\{S_x(f)/2q\bar{I}\}$.

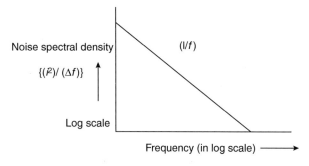

**Fig. 3.18** Flicker noise spectral density vs. frequency (log scale)

Factor 2 has appeared because positive and negative frequencies contribute identically. At zero temperature, the noise is related to the transmission probabilities and can be expressed mathematically as [113]

$$S_x(f) = 2q(2q^2/h)V\sum_{n=1}^{N} T_n(1-T_n) \quad (3.83)$$

The factor $(1 - T_n)$ describes the reduction of noise due to Pauli principle, $h$ is Planck constant, and $T_n$ is transmission probability. If we neglect it, we would have $S_x = S_{\text{poisson}} = (2q\bar{I})$. This is valid for the electrodes attached to the detector, which can be viewed as a parallel circuit of $N$-independent transmission channels. For metal, $N(A/\lambda_F^2)$ is very high ($\sim 10^7$) due to small Fermi wavelength ($\lambda_F \approx 1$ Å). In the detector, $T_n$ are defined as the eigenvalues of the product $(t - t_j)$. For the detector which is highly resistive, $q \neq e$, and the shot noise spectrum is denoted by Eq. 3.83. For the metal electrodes, Eq. 3.83 is valid, and it predicts that the shot noise should vanish when the conductance is quantized [114].

In Fig. 3.16, the bias resistor $R_b$ parallel to capacitors $C$ or $C_f$ will give rise to thermal noise due to the random thermal motion of the electrons. It is unaffected by the presence or absence of direct current, since typical electron drift velocities in a conductor are much less than electron thermal velocities. The thermal noise by a series voltage generator $\bar{v}^2$ is given by the following equation as

$$\bar{v}^2 = 4kTR_b\Delta f \quad (3.84)$$

When it is a shunt current generator $(\bar{i}^2)$, we can write:

$$(\bar{i}^2) = 4kT(1/R_b)\Delta f \quad (3.85)$$

where $\Delta f$ is the bandwidth at frequency $f$, $k$ is Boltzmann constant, $R_b$ is the bias resistance, and $T$ is the temperature in Kelvin. Experimental observation shows that the Johnson noise $(\bar{i}^2)$ shows in Eq. 3.85 produced by a 10 GΩ resistor is approximately about 1.26 f. [115].

## 3.11.2 Flicker Noise

There is another noise that can be expected in a discrete passive element-like resistor and is called as *flicker noise* (1/*f* noise). *Flicker noise* is always associated with direct current and displays a spectral density of the form as [116]

$$(\bar{i^2}) = K_1(I^a/f^b)\Delta f \tag{3.86}$$

The noise generated by the input of an FET amplifier ($A_1$) (Fig. 3.18) is represented by an infinite sum of power-law impulse response function. The power spectral density ($S_x(f)$) of the shot noise process (Eq. 3.82) can be written as [117, 118]

$$S_x(f) = <I>^2\delta(f) + \wp|\int_{-\infty}^{\propto} R_x(t)\exp(-j\omega t)\mathrm{d}t]^2|^2 \tag{3.87}$$

where autocorrelation function $R_x(t) = \lim_{T\to\infty(1/T)} \int_{-T/2}^{T/2} I(t)I(t+t')\mathrm{d}t$ [Wiener–Khintchine theorem]. The $\delta(f)$ is the Dirac delta function, $\wp$ is homogeneous Poisson point process rate, $\omega = 2\pi f$, and $f$ is the frequency. When the frequency changes from 0 to infinity ($\propto$), we can write the power spectral density as

$$S_x(f) \to [\wp\int_{-\infty}^{\propto}\delta(t)\mathrm{d}t]^2 = \wp K^2\{(B^{-\alpha})/(\alpha/2)^2\} \quad f \to 0 \tag{3.88}$$

and

$$S_x(f) \to \wp K^2\Gamma^2\{(\alpha/2)(\omega)^{-\alpha}\} \quad f \to \propto \tag{3.89}$$

where $\Gamma$ is the incomplete gamma function, and $B$ and $K$ are deterministic and fixed. Thus, at high frequencies ($f$) the power spectral density does indeed behave as $(1/f)^\alpha$.

## 3.11.3 Burst Noise

There is another type of low-frequency noise found in some thin-film transistors (TFTs), amplifier ($A_1$), field-effect transistors (FET), and charge-coupled devices (CCD), which is known as *burst noise*. The spectral density of burst noise (Fig. 3.18) can be shown to be of the form as [119, 120]

$$\bar{i^2} = K_2[\{(I_c)\}/\{(1+(f/f_c)^2\}\Delta f\}] \tag{3.90}$$

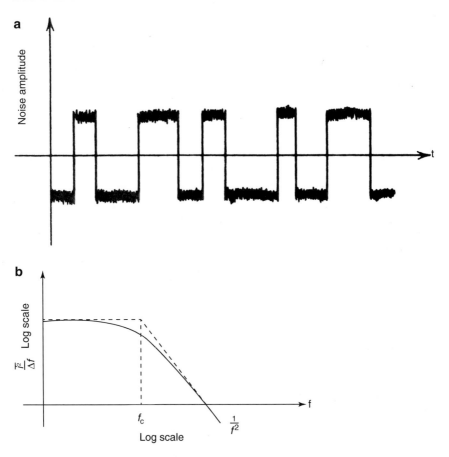

**Fig. 3.19** (a) Typical burst noise waveform. (b) Burst noise spectral density vs. frequency (Reproduced with permission, Wiley, Gray and Mayer)

where $K_2$ is a constant for a particular device, $i$ is the direct current, $c$ is a constant whose value is in the range of 0.5–2, and $f_c$ is a particular frequency for a given noise process.

Figure 3.19a shows a typical burst noise waveform, whereas Fig. 3.19b shows a plot of a burst noise spectral density vs. frequency. In addition to the effects described above, there is an additional noise that might come from the samples' *electromagnetic environment* (contacts, leads, amplifier, etc.). The *environment* emits noise, inducing fluctuations of the voltage across the sample, which in turn modifies the probability distribution $P$. Moreover, due to finite impedance of the *environment*, the noise emitted by the sample is responsible for the *Coulomb blockade* in case of high resistive materials (e.g., $HgI_2$, $PbI_2$, CdZnTe, CdTe), where it introduces a modification of $I(V)$ characteristics [121].

Now if we consider a resistor $R$ across $C$ and a parallel resistor due to the input impedance of the amplifier $A_1$ (Fig. 3.16), as $R_0$, it will give rise to an effective resistance $R_{\mathrm{eff}} = RR_0/(R + R_0)$. The effective resistance ($R_{\mathrm{eff}}$) will give rise to the spectral density of the second moment of the voltage fluctuations equal to $S_{v2} = R_{\mathrm{eff}}^2 (S_{12} + S_{i2})$, where $S_{12}$ is the noise power density (in $A^2/$Hz) and $S_{i2}$ is the noise spectral density of the current generator $i_0$. Thus, the only effect of the environment on the second moment of the noise is to rescale (by $R_{\mathrm{eff}}^2$) and add a bias-dependent contribution [122, 123].

Experimental observations with traveling heater method (THM)-grown CdTe show that $1/f$ noise lower at frequency $<100$ Hz, compared to high-pressure Bridgman method (HPBM)-grown CdZnTe (CZT). At the same time it has also been noticed that shot noise at frequencies ranging from 100 Hz to 10 kHz is lower in HPBM-grown CdZnTe (CZT) compared to THM-grown CdTe. Further studies reveal that in HPBM-grown CZT detectors, generation–recombination (G-R) is responsible for $1/f$ noise. However, below 100 Hz, $1/f$ noise is attributed to deep-level impurities or from trapped impurities. Moreover, the inhomogeneity of the metal contacts is also the source of $1/f$ noise [115]. The current noise crossing the detector has shown fluctuations and is thought to be the source of *blurring* in medical image.

## 3.12 Noise and Its Effect on Medical Imaging

It is desirable that the image acquired during medical imaging should have uniform brightness. Unfortunately, almost all the images acquired during medical imaging contain some sort of random variations in image brightness due to presence of noise. In medical images, noise suppression is particularly a difficult task, and it requires a delicate balance between the noise reduction and the preservation of the actual image. The presence of noise gives an image a *mottled, grainy texture* or snowy appearance.

In case of *ultrasound images*, it has been noticed that the images are frequently corrupted by *speckle noise* [124, 125], and the conventional method of reducing the noise is speckle filtering: the homographic Wiener filter [126, 127]. In general, nuclear images are more noisy than the images procured during magnetic resonance imaging (MRI) system or the images procured during computer tomography (CT) or during ultrasound (US) imaging. In MRI, the practical limits of the acquisition time impose a trade-off between signal-to-noise ratio and the image resolution, and wavelet domain noise filtration techniques are used [128]. Figure 3.20 shows the effect of noise in a medical image.

In all imaging procedures (using X-ray or $\gamma$-ray), it has been detected that the photons impinging the surface of the detector (image receptor) were in random pattern. The randomness of the photons gives rise to *quantum noise*. The amount of noise that will prevail in the system is determined by the variation of photon concentration from point to point within a small image area. Thus, it is possible to reduce *quantum noise* by increasing the concentration of photons (radiation

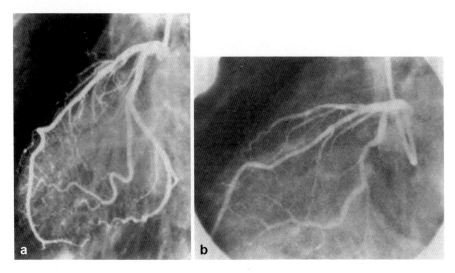

**Fig. 3.20** The image on the *right* (**b**) has more noise than image on the *left* (**a**) (Courtesy P. Sprawls, Ph.D., Image noise, in *The Physical Principles of Medical Imaging*)

exposure, meaning the dose, $D$) used to form an image. Unfortunately, when the receptor is human being, the total *dose* ($D$) applied is reduced to minimize the risk of too much radiation, which is considered to be harmful for a patient. Therefore, there is always some kind of compromization between the sharpness of the image and the total dose. Most X-ray imaging procedures are conducted at a point of reasonable compromise between these two important factors.

Quantum noise is usually a factor that limits the use of highly sensitive film in radiography. Other than the noises mentioned above, there are some other noises that can be present in an image. These noises are (a) *electronic noise* or snow, and (b) contrast noise that contributes to *blurring* of the images.

*Electronic noise* has a substantial effect on *energy resolution* of a detector, which ultimately affects the sharpness or contrast of the medical image. This is especially true when the dose is lowered to remain in the safe side. The lowering of the patient's dose ultimately produces a variation in the detector signal that is a percentage of the signal amplitude. In coplanar-grid geometry, the relatively high surface leakage current occurs when the grid bias is applied across the two grid electrodes and becomes a source of high noise level. However, it is argued that careful processing of the detector can minimize the surface leakage noise. It is actually the bulk leakage current which has been attributed to the source of noise in a detector besides $1/f$ noise [129, 130].

**Fig. 3.21** The differences in dead time ($\tau$) between two models

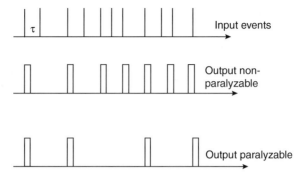

## 3.13 Dead Time

*Dead time* is defined as the time required for a *counting system* to process individual events. As for example, when we want to operate a radiation detector or detector system in pulse mode, the arrival of the two consecutive pulses must be separated by a finite interval of time, which is the *dead time* between the two pulses. Therefore, *dead time* can be defined as the time required for a *counting system* to process an individual event (it is also called pulse-resolving time $\tau$). Dead time is important in the sense that if two pulses arrive in a very short interval of time, one can notice distortion or complete loss of the second pulse [131]. Several components (detector, pulse amplifiers, pulse height analyzer, scalar, and computer interface) could contribute to the dead time.

*Medical imaging* is an important part of the medical physics, and its achievement depends on the performance of a radiation detector and the associated system. During *positron emission tomography* (PET) imaging, photon counting after ionization requires measurement of pulses produced by a radiation detector. Depending on the system, one or both particle arrivals will go unrecorded if the dead time between the pulses is very short [132].

The dead time of inorganic scintillator like NaI (Tl) or solid-state semiconductor detector is mainly caused by pulse amplifiers (pulse piling up and base line shift typical 0.5–5 μs), whereas in Geiger Mueller (GM) tube, it is caused by detector's pulse overlay (typical 50–200 μs).

### 3.13.1 Paralyzable and Non-paralyzable Models

*Counting systems* are often characterized as either *paralyzable* or *non-paralyzable* type. In *paralyzable* system, each event introduces a dead time $\tau$ whether or not that event actually was counted. As a matter of fact, most *nuclear medicine systems* are *paralyzable* type [133]. A *non-paralyzable* system, on the other hand, is one for which if an event occurs during the dead time $\tau$ of a preceding event, then the second event is simply ignored.

In Fig. 3.21 we can see that the difference in output signals between *paralyzable* and *non-paralyzable* systems. The dead time ($\tau$) is different in two models. In *non-paralyzable* system, events are lost if they occur within a time $\tau$ of a preceding recorded event, whereas in a *paralyzable* system, events are lost if they occur within a time $\tau$ of any preceding event, whether or not that event has been recorded. However, the detailed behavior of a specific counting system may depend on the physical processes taking place in the detector itself or on delays introduced by the pulse processing and recording electronics.

In the spirit of the definition of the two models as described above, let us set up two analytical models, one for the *non-paralyzable* system and the other for the *paralyzable* system. The relationship between observed count rate $R_0$ and true count rate $R_t$ for the *non-paralyzable* system can be written as

$$R_0 = \{(R_t)/(1 + R_t\tau)\} \tag{3.91}$$

or,

$$R_t = \{(R_0)/(1 - R_0\tau)\} \tag{3.92}$$

The observed counting rate $R_0$ will attain a maximum value $R_0^{max}$ and can be represented by

$$R_0^{max} = (1/\tau) \tag{3.93}$$

For *paralyzable* system the relationship between observed count rate and true count rate can be shown mathematically as

$$R_0 = \{R_t \exp(-R_t\tau)\} \tag{3.94}$$

and,

$$R_0^{max} = \{(1/\exp(-\tau)\} \tag{3.95}$$

The dead time losses are defined as $(R_t - R_0)$, and the percentage losses can be expressed mathematically as

$$\text{Percentage losses } (\%) = \{(R_t - R_0)/R_t\} \times 100 \tag{3.96}$$

When $R_t\tau \leqslant 0.1$, the above equation will transform into $\{(R_t\tau) \times 100\}$ percent.

Figure 3.22 shows the observed count rate ($R_0$) vs. true count rate ($R_t$) curves for paralyzable and non-paralyzable systems having same dead time $\tau$. Again dead time loss depends on total counting spectrum and is defined as $(R_t - R_0)$, and the percentage losses can be expressed mathematically as

## 3.13 Dead Time

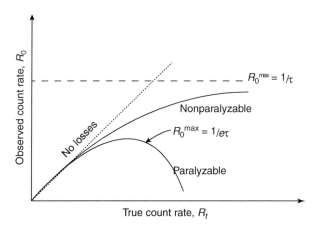

**Fig. 3.22** Observed count rate ($R_0$) vs. true count rate ($R_t$) curves for paralyzable and non-paralyzable systems having same dead time $\tau$

$$\text{Percentage losses }(\%) = \{(R_t - R_0)/R_t\} \times 100 \quad (3.97)$$

When $R_t\tau \leqslant 0.1$, the above equation will transform into $\{(R_t\tau) \times 100\}$ percent. The mathematical analysis of the two models for the calculation of the dead time $\tau$ is based on some assumptions that the detector is irradiated by a steady-state source of constant intensity. In reality, the X-ray is operated generally to produce pulses of a few microsecond widths with a repetition frequency of several kilohertz. Several conditions (radiation intensity is constant through the time period $T$, and the pulses are applied at a constant frequency $f$) have to be imposed: (i) If $\tau \ll T$, the pulse will have little effect, and the mathematical deductions that we have performed can hold. (ii) If $\tau$ is smaller but not too small compared to $T$, a small amount of counts may be registered and the situation is too complicated to arrive to a solution of the mathematical model [134]. (iii) When $\tau > T$, but less than the off-time between pulses (given by $\{(1/f) - T\}$), we will have only one detector count per source pulse, and the detector will fully recover at the start of each pulse.

Recently a hybrid model has been introduced by Lee et al. [135] using two different dead timings (paralyzable time $\tau_p$, and non-paralyzable time $\tau_{np}$). The mathematical hybrid model gives a relation between observed count rate $R_0$, true count rate $R_t$, and two dead times $\tau_p$ and $\tau_n$ as [136]

$$R_0 = \{(R_t \exp\cdot(-R_t\,\tau_p)\}/\{(1 + R_t\tau_{np})\} \quad (3.98)$$

In their model [135, 137], they also suggested a two-split numerical iterative method to determine the dead times for the hybrid model. The Monte Carlo G-M counter simulation (GMSIM) shows that the results obtained by simulation support the measured values of the dead times of the paralyzable and non-paralyzable systems.

In practice, detectors are usually operated at detection rates much lower than the inverse of the dead time to avoid high *dead time fractions* (DTF) and the associated large dead time corrections. The dead time fraction is defined as the ratio of missed to incident events or the fraction of the time detector spends in its recovery state (in case of Poissonian CW source). A DTF of 10 % is often a reasonable limit for detector operation. Castelletto et al. have forwarded a multiplexed detector array scheme for a photon-counting detection that can be operated at incident photon rates higher than otherwise possible by suppressing the effects of detector dead time [138, 139].

To quantify the advantages of multiplexed system of operations relative to non-multiplexed system, Castelletto et al. used approximate analytical and numerical Monte Carlo models, both for the cases of a CW Poisson distributed and a pulsed source [140]. A recursive expression for the effective final dead time obtained by analytical method from the iterative formula for the subsequent incident photon rate for pools of up to 12 detectors is $\tau_{d(i)} = \tau_{d(i-1)} - [1 - \{\exp(-\lambda \tau_{d(i-1)})\}/\{2\lambda\}]$. It is equal to the iterative formula for the subsequent incident photon rate for pools of up to 12 detectors obtaining a recursive expression for the effective final dead time by Monte Carlo modeling which is equal to $N_{d(i)} = N_{d(i-1)} - [1 - \{(1-p) N_{d(i-1)}^{+1}\}]/\{(2-p)p\}]$, where $\lambda$ is the mean photon rate, and $p$ is the probability that a live detector produces a count (event) for an pulse.

# References

1. Spyropoulou V et al (2007) Modeling the imaging performance and low contrast detectability in digital mammography. In: 4th international conference on imaging technologies in biomedical sciences, Milos Island, 2007
2. Efthimiou N (2007) Investigation of the effect scintillator material on the overall detection system performance by application of analytical model. Nucl Instrum Methods A 571 (1–2):270
3. Picone M (2002) Contribution a la simulation tridimensionnelle de detecteurs semiconductors en spectrometrie gamma. M.S. thesis, University of Joseph Fourier, Grenoble
4. Picone M, Gliere AA, Masse P (2003) A three dimensional model of CdZnTe gamma ray spectrometer. Nucl Instrum Methods A 504:313
5. Montemont G (2000) Optimization des performance de detecteurs CdTe et CdZnTe en spectrometrie gamma. M.S. thesis, University of Joseph Fourier, Grenoble
6. Hugonnard P Gliere A (1999) X-ray simulation and applications. In: Computerized tomography for industrial applications and image processing in radiology, Berlin. 15–17 Mar 1999
7. Petasecca M, Moscatelli F, Passeri D. Pignatel GU, Scarpello C (2005) Numerical simulation of radiation damage effects in p-type silicon detectors. Nucl Instrum Methods A 563(1):192
8. Iwata K, Hasagawa BH (1999) Numerical simulation of pixellated CdZnTe detector for medical radionuclide imaging application. IEEE Trans Nucl Sci 46(3):385
9. Del Sordo S (2009) Progress in the development of CdTe and CdZnTe semiconductor radiation detectors for astrophysical and medical applications. Sensors 9:3491
10. Niraula M, Nakamura A, Aoki T, Tomita Y, Hatanaka Y (2002) Stability issues of high energy resolution diode type CdTe nuclear radiation detectors in a long term operation. Nucl Instrum Methods Phys Res A 491:168
11. Mott NF, Gurney RW (1940) Electronic process in ionic crystal. Clarendon Press, Oxford

# References

12. Bao XJ, Schlesinger TE, James RB (1995) Electrical properties of mercuric iodide. In: Schlesinger TE, James RB (eds) Semiconductor for room temperature nuclear detector applications, vol 43. Academic, San Diego, p 124
13. Gerrish VM (1995) Characterization and quantification of detector performance. In: Schlesinger TE, James RB (eds) Semiconductor for room temperature nuclear detector applications, vol 43. Academic, San Diego, p 516
14. Derenzo SE, Weber MJ (1999) Prospects for first principle calculations of scintillator properties. Nucl Instrum Methods Phys Res A 422:111
15. Kabir MZ Yunus M Kasap SO (2005) The effects of large signals on charge collection in photoconductive X-ray detectors. In: Proceedings of the IEEE Canadian conference on electrical and computer engineering, Vancouver, 2005, p 197
16. Levi A, Schieber M, Burshtein Z (1983) Carriers surface recombination in $HgI_2$ photon detectors. J Appl Phys 54:2472
17. Chu M, Terterian S, Ting D, James RB, Szawlowski M, Visser GJ. Effects of p/n inhomogeneity on CdZnTe radiation detectors, Brook Heaven National Lab, NY. BNL-69312, DTRA funded (partial) program, DTRA01-01-C-0071
18. Barret HH, Eskin JD, Barber HB (1995) Charge transport in arrays of semiconductor gamma ray detectors. Phys Rev Lett 75:156
19. Bao XJ, Schlesinger TE, James RB (1995) Electrical properties of mercuric iodide. In: Semiconductors for room temperature nuclear detectors. Academic, San Diego, p 159
20. Layni S, Dikant J, Ruzicka M (1978) Acta Phys Slov 28(3):210] or *space charge formation* Mohammed T, Friant A, Mellet J (1984) IEEE Trans Nucl Sci NS-32(1):581
21. Mohammed TB, Friant A, Mellet J (1984) Structure MIS effects on polarization of $HgI_2$ crystals used for gamma ray detection. IEEE Trans Nucl Sci NS-32(1):581
22. Malm HL, Matini M (1974) Polarization phenomena in CdTe nuclear detectors. IEEE Trans Nucl Sci 21:322
23. Camarda GS, Bolotnikov AE, Cui Y, Hossain A, James RB (2007) Polarization studies of CdZnTe detectors using synchrotron X-ray radiation. In: NSS/MIC/RTSD symposium, Honolulu, 2007
24. Malm HL, Martini M (1974) Polarization phenomena in CdTe nuclear radiation detectors. IEEE Trans Nucl Sci 21(1):322 and Gerrish VM (1995) Characterization and quantification of detector performance. In: Schlesinger T, James RB (eds) Semiconductors for nuclear detection applications, vol 43. Academic, San Diego
25. Siffert P, Berger J, Cornet A, Stuck R, Bell RO, Serreze HB, Wald FV (1976) Polarization in cadmium telluride nuclear radiation detector. IEEE Trans Nucl Sci 23(1):159 and Camada GS, Bolotnikov AE, Cui Y, Hossian A, Awadalla SA, Mackenzie J, Chen H, James RB (2008) Polarization studies of CdZnTe detectors using synchrotron X-ray radiation. IEEE Trans Nucl Sci 55(6):3725
26. Mead CA, Spitzer WG (1964) A. Mead Conduction band minima Ga ($As_{1-x} P_x$). Phys Rev 134A:713
27. Gupta TK (2003) Contact resistance in hand book of thick and thin film hybrid microelectronics. Wiley, Hoboken, pp 193 and Blech J, Sello H, Greger LV (1983) Thin film integrated circuits, Chapter 23. In: Maissel LJ Glang R (eds) Thin film technology. McGraw Hill, New York
28. Cox CE, Lowe BG, Screen RA (1988) Small area high purity Germanium detectors for the use in energy range 100eV to 100 keV. IEEE Trans Nucl Sci 35(1):28
29. Darken LS (1993) Role of disordered regions in fast-neutron damage of HPGe detectors. Nucl Instrum Methods Phys Res B-74:523
30. Wald FW, Bell RO (1974) Semiconductors for room temperature nuclear detector application. Tyco Report US AEC AT (11–1), 3, 545
31. Franc J (2004) Defect structure of high resistive CdTe. IEEE Trans Nucl Sci 51(3):1176
32. Hau ID, Cindall C, Luke PN (2003) New contact development for Si (Li) orthogonal strip detectors. Nucl Intrum Methods Phys Res A 505:148

33. Owens A, Peacock A (2004) A compound semiconductor radiation detectors. Nucl Instrum Methods Phys Res A 531:18
34. McGregor DS, He Z, Siefert HA, Rojeski RA, Wehe DK (1998) CdZnTe semiconductor parallel strip Frisch grid radiation detectors. Trans Nucl Sci NS-45:443
35. McGregor DS, He Z, Siefert HA, Rojeski RA, Wehe DK (1997) Space charge carrier type sensing with parallel strip pseudo-Frisch grid CdZnTe semiconductor radiation detector. Appl Phys Lett 72:792
36. Barret HH, Eskin JD, Barber HB (1995) Charge transport arrays of semiconductor gamma ray detectors. Phys Rev Lett 75:56 and Frisch O (1944). British Atomic Energy Report BR-49
37. McGregor DS, Rojeski RA, He Z, Wehe DK, Driver M, Blakely M (1999) Geometrically weighted semiconductor Frisch grid radiation spectrometer. Nucl Instrum Methods A 422:164
38. Gregor DS, He Z, Siefert A, Wehe DK, Rojeski RA (1998) Performance characteristics of Frisch ring CdZnTe detectors. Appl Phys Lett 72:792
39. Butler J (1997) Nucl Instrum Method A 396:427 and Barrett HH, Eskin JD, Barber HB (1995) Charge transport in arrays of semiconductor gamma ray detectors. Phys Rev Lett 75:156
40. McGregor DS, Rojeski RA (1999) Performance geometrically weighted semiconductor Frisch grid radiation spectrometer. IEEE Trans Nucl Sci 46:250
41. McGragor DS, Rojeski RA (2001) US Patent No. 6175120
42. McNeil WJ, McGregor DS, Bolotnikov AE, Wright GW, James RB (1998) Project Report DOE, DE-FG07-031D14498
43. Luke PN (1995) Unipolar charge sensing with coplanar electrodes application to semiconductor detectors. IEEE Trans Nucl Sci 42:207
44. He Z et al (1995) Position sensitive single carrier semiconductor detectors. Nucl Instrum Methods A 380:228
45. He Z, Sturm BW (2005) Characteristics of depth sensing coplanar grid CdZeTe detectors. Nucl Instrum Methods A 554:291
46. Kozorezov AG, Wigmore JK (2007) Theory of the dynamic response of a coplanar grid semiconductor detector. Appl Phys Lett 91:023504
47. Montemont G, Arques M, Verger L, Rustique J (2001) A capacitive Frisch grid structure for CdZnTe detectors. IEEE Trans Nucl Sci 48:278
48. Owens A et al (2006) Hard X and $\gamma$-ray measurements with large volume coplanar grid CdZnTe detector. Nucl Instrum Methods Phys A 563:242
49. Luke PN, Amman M, Yaver H (1998) Coplanar grid CdZnTe detector with three dimensional position sensitivity. In: 8th symposium on semiconductor detectors, Schloss Elmau, 1998
50. Ramo S (1939) Currents induced by electron motion. Proc IRE 27:584
51. Mathy F, Gliere A, d' Allion EG, Masse P, Picone M, Tabary J, Verger L (2004) A three dimensional model of CdZnTe gamma-ray detector and its experimental validation. IEEE Trans Nucl Sci 51(5):2410
52. Soldner SA, Narvett AJ, Covalt DE, Szeles C (2004) Characterization of the charge transport uniformity of CdZnTe crystals for large volume nuclear detector applications. IEEE Trans Nucl Sci 51(5):2443
53. Amman M, Lee JS, Luke PN (2002) Electron trapping non-uniformity in high pressure Bridgman grown CdZnTe. J Appl Phys 92:3198
54. Perez JM, He Z, Wehe DK (2001) Stability and characteristics of large CZT coplanar electrode detectors. IEEE Trans Nucl Sci 48(3):272
55. He Z, Knoll GK, Wehe DK, Du YF (1998) Co-planar grid pattern and their effect on energy resolution of CdZnTe detectors. Nucl Instrum Methods A 411:107
56. Visvikis D et al (2006) Monte Carlo based performance assessment of different animal PET architectures using pixellated CZT detectors. Nucl Instrum Methods Phys Res A 569:225
57. Perie I, Armbruster T, Koch M, Kreidl C, Fischer P (2010) DCD – the multichannel current mode ADC chip for readout by DEPFET pixel detector. IEEE Trans Nucl Sci 57(2):743
58. Baciak JE, He Z, Devito RP (2002) Electron trapping variation in single crystal pixelated $HgI_2$ gamma ray spectrometer. IEEE Trans Nucl Sci 49(3):1264

# References

59. Shockley W (1938) Currents to conductors induced by moving point charge. J Appl Phys 9:635 and Ramo S (1939) Current induced by electron motion. Proc IRE 27:584
60. Kabir MZ, Kasap SO (2004) Charge collection and absorption – limited X-ray sensitivity of pixellated X-ray detectors. J Vac Sci Technol A 22:975
61. Baciak JE (2004) Development of pixelated $HgI_2$ radiation detectors for room temperature gamma ray spectroscopy. Ph.D. thesis, University of Michigan, Ann Arbor
62. Kabir MZ, Kasap SO (2003) Modulation transfer function of photoconductive X-ray image detectors: effect of charge carrier trapping. J Phys D: Appl Phys 36:2352
63. Day RB, Dearnaley G, Palms JM (1987) Determining the effective number bits of high resolution digitors. IEEE Trans Nucl Sci NS-14:487
64. Knoll GF, McGregor Proc DS (1993) Fundamentals of semiconductor detectors for ionizing radiation. Mater Res Soc 302:3
65. Shor A, Mardar YY, Soreq I (2003) IEEE Trans Nucl Sci Symp 5:3342
66. Kim JC, Kaye W, Zhang F, He Z (2010) Analysis of system-dependent factors affecting pixelated CdZnTe detector performance through simulation. In: IEEE NSS/MMC/RTSD, symposium, Knoxville, 2010
67. Shor A, Eisen Y, Mardor I (2004) Edge effects in pixelated CdZnTe gamma ray detectors. IEEE Trans Nucl Sci 51(5):2412
68. Neyts K, Beeckman J, Beunis F (2007) Quasi-stationary current contributions in electronic devices. Opto-Electron Rev 15(1):41
69. Knoll GF, McGregor DS (1993) Fundamentals of semiconductor detectors for ionizing radiation. Proc MRS 302:3
70. Moehrs S (2008) Ph.D. thesis, Dept. di Fisica Enrico Fermi, Università Degli Studi Di Pisa, Pisa
71. Iriarte A, Sorzano COS, Rubio JM, Marabini R (2009) A theoretical model for EM-ML reconstruction algorithms applied to rotating PET scanners. Phys Med Biol 54(1909)
72. Fano U (1947) Ionization yield of radiations. II Fluctuations of the number of ions. Phys Rev 72:26
73. Iwanczyk JS, Patt BE (1995) Electronics for x-ray and gamma ray spectrometers, Chapter 14. In: Schlesinger T, James RB (eds) Semiconductors for room temperature nuclear detector application. vol 43. Academic, San Diego, p 548
74. Swank RK (1974) Measurement of absorption and noise in X-ray image intensifier. J Appl Phys 45:3673
75. Hajdok G, Yao J, Battista JJ, Cunningham IA (2004) Signal and noise transfer properties of photoelectric interactions in diagnostic X-ray imaging detectors. Med Phys 33 (10):3601–3620
76. Metz CE, Vyborny CJ (1983) Wiener spectral effects of spatial correlation between sites of characteristic X-ray emission and reabsorption in radiographic screen film systems. Phys Med Biol 28:547
77. Zhao W, Ji WG, Rowlands JA (2001) Effects of characteristic X-rays on the noise power spectra and detective quantum efficiency of photoconductive X-ray detectors. Med Phys 28:2039
78. Rababani M, Shaw R, van Metter R (1987) Detective quantum efficiency of imaging systems with amplifying and scattering mechanisms. J Opt Soc Am A4:1156
79. Papoulis A (1991) Probability, random variable stochastic process, 3rd edn. McGraw Hill, New York
80. Quinon AR, Anderson CE and Knox WJ (1959) Fluorescent response of cesium iodide crystals to haevy ions. Phys Rev 115:886
81. Birks JB (1964) The theory and practice of scintillators. Pergamon Press, London, p 439
82. Hoek H (1992) Ph.D. thesis, Chalmers University of Technology, Gothenburg
83. Parlog M (2002) Response of CsI scintillators over a large range in energy and atomic number of ions. Nucl Instrum Methods Phys Res A 482(3):674
84. Scheiber C (2000) CdTe and CdZnTe detectors in nuclear medicine. Nucl Instrum Methods Phys Res A 448(3):513–524

85. Fornaro L et al. (2004) Growth of bismuth tri-iodide platelets for room temperature X-ray detection. IEEE nuclear science symposium conference, Rome, Italy, 16–22 Oct 2004, vol 7, p 4560
86. Gupta T et al (2000) X-ray security system, NIST funded joint program, Varian Medical Syst., Palo Alto, RMD, Water Town and Xerox Research, Palo Alto. Final Report NIST
87. Slapa M, Huth GC, Seibet W, Scieberand MM, Randtka PT (1976) Capabilities of mercuric iodide as a room temperature X-ray detector. IEEE Trans Nucl Sci NS-23:102
88. Schieber MM, Beinglass I, Dishon G, Holzer A, Yaron G (1978) Bulk performance of improved mercuric iodide nuclear detectors. IEEE Trans Nucl Sci NS-25:644
89. Bao XJ, Schlesinger TE, James RB (1995) Electrical properties of mercuric iodide, Chapter 4. In: Schlesinger TE, James RB (eds) Semiconductor and semimetals, vol 43. Academic, San Diego
90. Baciak JE, He Z, Devito RP (2002) Electron trapping variations in single crystal pixelated $HgI_2$ gamma ray spectrometers. IEEE Nucl Sci Symp 4:2335
91. Ali MH, Siffert P (1995) Characterization of CdTe nuclear detector materials. In: Schlesinger T, James RB (eds) Semiconductor and semimetals. Academic, San Diego
92. Quaranta A, Caali C, Ottaviani G (1970) A 40 keV pulsed electron accellator. Phys Rev Sci Instrum 41:1205
93. Stuck R (1975) Ph.D. thesis, Universite' Louis Pasteur, Strasbourg
94. Hecht K (1932) Zum mechanismus des lichtelektrischen primingless in isolierenden kristallen. Zeitchr Phys 77:235
95. Martini J, Mayer W, Zanio KR (1972) Drift velocity and trapping in semiconductors – transit charge technique. Appl Solid State Phys 3:181
96. Zuck A, Schieber MM, Khakhan O, Burshtein Z (2002) Delayer emission of surface generated trapped carriers in transient charge transport of single crystal and polycrystalline $HgI_2$. In: Proceedings of SPIE, in hard X-ray gamma ray detector physics IV, conference, Seattle, 2002
97. Dardenne YX, Wang TF, Lavietes AD, Mauger GJ, Ruhter WD, Kreek SA (1999) Cadmium zinc telluride spectral modeling. Nucl Instrum Methods Phys Res A 422:159
98. Yeargan JR, Taylor H (1968) The Polle-Frankel effect with compensation present. J Appl Phys 39:5600
99. Prettyman TH (1999) Method of mapping charge pulses in semiconductor radiation detector. Nucl Instrum Methods Phys Res A 422:232
100. Bell GI, Glasstone S (1970) Nuclear reactor theory. Van Nostrand Reinhold, New York
101. Arfken G (1966) Mathematical methods for physicists. Academic, New York
102. White TL, Miller WH (1999) A triple-crystal phoswich detector with digital pulse shape discrimination for alpha/beta/gamma spectroscopy. Nucl Instrum Methods A 422:144
103. Rose A (1953) Quantum and noise limitations of the visual process. J Opt Soc Am 43:715
104. Cunningham IA (1997) Analyzing system performance. In: Frey GD, Sprawls P (eds) Expanding role of medical physics in diagnostic imaging. Advanced Medical Publishing for American Association of Physicists in Medicine, Madison, p 231
105. Samei E, Flynn MJ (2003) An experimental comparison of detector performance for direct and indirect digital radiography systems. Med Phys 30(4):608
106. Samei E, Ranger NT, Dobbins JT III, Chen Y (2006) Intercomparison of methods for imaging quality characterization. I. Modulation transfer function. Med Phys 33:1454
107. Rose A (1953) Quantum and limitations of the visual process. J Opt Soc Am 43(715)
108. Cunningham IA, Shaw R (1999) Comparison of observer performance at 15 and 30 fps for reducing rates. J Opt Soc Am A 16(3):621
109. Campbell N (1909) The study of discontinuous phenomena. Proc Camb Philos Soc 15:117
110. Schottky W (1918) Uber spontane stromschwankungen in verschiedenen electrizitatsleitern. Ann Phys 57:541
111. Hero AO (1991) Timing estimation for a filtered Poisson process in Gaussian noise. IEEE Trans Inform Theory 37(1):92

# References

112. Beenakker C, Schonenberger C (2003) Quantum shot noise. Phys Today 56:37
113. Buttiker M (1990) Scattering approach to thermal and excess noise in open multipore conductors. Phys Rev Lett 65:2901
114. Houten HV, Beenakker C (1996) Phys Today, 22
115. Imad A, Orsal B, Alabedra R (2001) Experimental study of current noise spectral density versus dark current in CdTe:Cl and CdZnTe detectors. In: Bosman G (ed) Proceedings of the 16th international conference on noise in physical systems and 1/f fluctuations, ICNF, Gainesville, 2001
116. Gupta TK (2003) Handbook of thick and thin film hybrid microelectronics. Wiley, Hoboken, p 14, 48, 147
117. Spieler H (2004) Low noise electronics. Rev Part Phys Lett B 592:1
118. Lowen SB, Teich MC (1989) Generated 1/f shot noise. Electron Lett 25(16):1072
119. Gray PR, Meyer RG (1993) Analysis and design of integrated circuits, 3rd edn. Wiley, New York, Chapter 11
120. Jaeger RC, Broderson AJ (1970) Low frequency noise sources in bipolar transistors. IEEE Trans Electron Dev ED-17:128
121. Ingold G-L, Nazarov YuV (1992) Coulomb Blockade phenomena in nano-structure. In: Grabert H, Devoret MH (eds) Single charge tunneling. Plenum Press, New York
122. Blanter YM, Buttiker M (2000) Shot noise in mesocospic conductors. Phys Rep 336:1
123. Reulet B, Senzier J, Prober DE (2003) Noise thermal impedance of a diffusive wire. Phys Rev Lett 91:196601
124. Pizurica A, Philips W, Lemahieu I, Acheroy M (2003) A versatile wavelet domain noise filtration technique for medical imaging. IEEE Trans Med Imaging 22(3):323
125. Wagner RF, Smith SW, Sandrik JM, Lopez L (1983) Statistics of speckle in ultrasound B-scans. IEEE Trans Sonics Ultrason 30(3):136
126. Achim A, Bezerianos A, Tsakalides P (2001) Novel Bayesian multiscale method for speckle removable in medical ultrasound images. IEEE Trans Med Imaging 20(8):772
127. Jain AK (1989) Fundamental of digital image processing. Prentice-Hall, Englewood Cliffs
128. Nowak RD (1999) Wavelet based rician noise removable for magnetic resonance imaging. IEEE Trans Image Proc 8:1408
129. Luke PN, Amman M, Lee JS (2001) Factors affecting energy resolution of coplanar grid CdZnTe detectors. IEEE Trans Nucl Sci 48:282
130. Luke PN, Amman M, Lee JS (2004) Factors affecting energy resolution of coplanar grid CdZnTe detectors. IEEE Trans Nucl Sci 51(3):1199
131. Bushberg JT, Seibert JA, Leidholdt EM Jr, Boone JM (2002) The essential physics of medical imaging. Lippincott Williams and Wilkins, Philadelphia
132. Yu DF, Fessler JA (2002) Mean variance of coincidence counting with dead time. Nucl Instrum Methods A 488:362
133. Lee SH, Gardner RP, Jae M (2004) Determination of dead times in the recently introduced hybrid G-M counter dead time model. J Nucl Sci Technol Suppl 4:156
134. Knoll GF (2000) Radiation detection and measurements, 3rd edn. Wiley, New York
135. Lee SH, Gardner RP, Jae M (2004) Determination of dead times in the recently introduced hybrid G-M counter dead time model. Nuclr Suppl 4:156–159
136. Muller JW (1973) Dead time problems. Nucl Instrum Methods 112:47
137. Lee SH, Gardner RP (2000) A new G-M counter dead time model. Appl Radiat Isot 53:731
138. Castelletto SA, Degiovanni IP, Schettini V, Migdall AL (2007) Reduced dead time and higher rate photon counting detection using a multiplexed detector array. J Mod Opt 54:337–352
139. Rochase A, Besse P, Popovic R (2002) Actively recharged single photon counting avalanche CMOS photodiode with less than 9 ns dead time. In: The 16th European conference on solid state transducers, Prague, 2002
140. Gisin N, Ribordy G, Tittel W, Zbinden H (2002) Quantum cryptography. Rev Mod Phys 74:145

# Medical Imaging

# 4

## Contents

| | | |
|---|---|---|
| 4.1 | Introduction | 188 |
| | 4.1.1 Molecular Biology | 188 |
| | 4.1.2 Optical Imaging | 188 |
| | 4.1.3 Radionuclide Imaging | 189 |
| 4.2 | Radiation and Carcinogen | 189 |
| 4.3 | Molecular (Medical) Imaging | 191 |
| | 4.3.1 Radiology and Radiological Science | 192 |
| | 4.3.2 Thermography | 192 |
| | 4.3.3 Endoscopy | 193 |
| | 4.3.4 Microscopy or Medical Photography | 194 |
| 4.4 | More Advanced Technology for Medical Imaging | 195 |
| 4.5 | Advanced Tools (Instruments) for Medical (Molecular) Imaging | 197 |
| | 4.5.1 Fluoroscopy | 198 |
| | 4.5.2 Positron Emission Tomography (PET) | 203 |
| | 4.5.3 Single-Photon Emission Computed Tomography (SPECT) | 205 |
| | 4.5.4 Magnetic Resonance Imaging (MRI) | 205 |
| 4.6 | X-Ray Computed Tomography (CT) | 209 |
| 4.7 | Nuclear Medicine Imaging | 213 |
| 4.8 | Image Acquisition | 215 |
| | 4.8.1 Slice Thickness | 215 |
| | 4.8.2 Number of Projections and Number of Rays Used in CT | 215 |
| | 4.8.3 Simple Back Projection | 216 |
| | 4.8.4 Scattered Radiation | 220 |
| | 4.8.5 Spatial Resolution | 225 |
| | 4.8.6 Optical Density (OD) | 225 |
| 4.9 | Imaging Technology | 227 |
| | 4.9.1 Screen Film Technology | 227 |
| | 4.9.2 Digital Imaging Technology | 228 |
| | 4.9.3 Working Principle of Digital Imaging | 234 |
| 4.10 | Dependency of the Quality of the Medical Imaging System | 236 |
| | 4.10.1 Noise | 236 |
| | 4.10.2 Contrast | 238 |

T.K. Gupta, *Radiation, Ionization, and Detection in Nuclear Medicine*,
DOI 10.1007/978-3-642-34076-5_4, © Springer-Verlag Berlin Heidelberg 2013

| 4.11 | Energy Resolution | 242 |
| | 4.11.1 Full Width at Half Maximum (FWHM) | 243 |
| 4.12 | Digital Image Acquisition System | 244 |
| 4.13 | Summary | 246 |
| References | | 247 |

## 4.1 Introduction

*Medical imaging* refers to the techniques and processes used to create images of the human body or animals (or parts thereof) for clinical purposes or medical therapy. As a discipline in its widest sense or as a subdiscipline, *medical imaging* is the perturbation of the *cellular molecules* to identify the true identity of the cells or tissues of a living organism. As a matter of fact, the *biomedical engineering*, *medical physics*, or *medicine* unites *molecular biology* and in vivo *imaging* [1].

The multiple and numerous potentialities of this field of engineering coupled with *molecular biology, chemistry, optical imaging, structural engineering, radionuclide imaging,* and *instrumentation/tools* like MRI, CT, and PET are applicable to the diagnosis of cancer and neurological and cardiovascular diseases [1, 2]. Therefore, it will not be unrealistic to represent medical imaging by Venn diagrams or a set of diagrams that show hypothetical possible logic relations between a finite collection of sets (groups of things as mentioned). As a matter of fact, Venn diagram was conceived around 1880 by John Venn and comprises overlapping circles. The interior of the circle symbolically represents the elements of the set, while the exterior represents elements which are not numbers of the set. Figure 4.1 shows overlapping circles representing the Venn diagram with separate sets of different disciplines of *molecular (medical) imaging* system [3]. According to an expert in the field, *molecular imaging reflects a shift in emphasis and a shift in attitude, moving from the undeniably useful, but largely nonspecific, diagnostic imaging approaches that are currently employed in clinic* [1]. Thus, *molecular imaging* has become a part and parcel of *medical imaging*.

### 4.1.1 Molecular Biology

It is the study of biology at the molecular level. The field overlaps with other areas of biology and chemistry, particularly with genetics and biochemistry. It mostly deals with various systems of a cell, including interaction between DNA, RNA, and protein biosynthesis [3].

### 4.1.2 Optical Imaging

It falls under the categories of in vivo fluorescence and bioluminescence imaging. It plays an important role in interrogating biological systems, particularly, rodent models for human diseases [3].

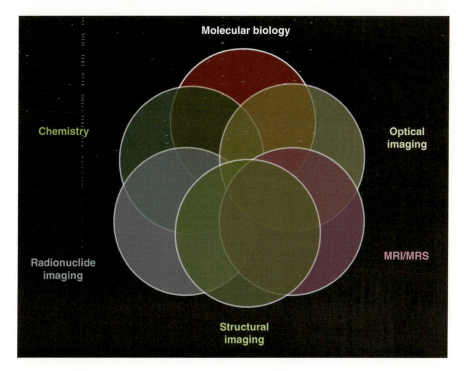

**Fig. 4.1** Venn diagram with different sets of molecular biology

### 4.1.3 Radionuclide Imaging

It is a technique that uses the introduction of tiny amounts of radioactive materials into a patient's body in order to obtain images. Common tests include *computed tomography* (CT), *positron emission tomography* (PET), *single-photon emission computed tomography* (SPECT), and scans. Figure 4.2 shows the scanned image of a human body captured by *computed tomography* (CT). X-ray-based CT scans are increasingly used for providing three-dimensional (3-D) view of a human body or parts thereof. It is estimated that by the year 2011, more than 100 million CT scans will be performed in the USA compared to three million in 1980 [4].

## 4.2 Radiation and Carcinogen

The *ionization radiation* can damage the biological tissues by breaking the water molecule—one of the constituents of the living tissue. The interaction of *ionizing radiation* with a molecule of water creating ion pair consisting of the ejected electron and a water molecule with an unpaired outer orbital electron is referred to as free radicals ($H_2O \rightarrow H^+ + HO^-$) (Fig. 4.3).

Two types of radiation are especially dangerous because they can modify DNA: ultraviolet radiation and ionizing radiation. Ionizing radiation mainly causes breaks

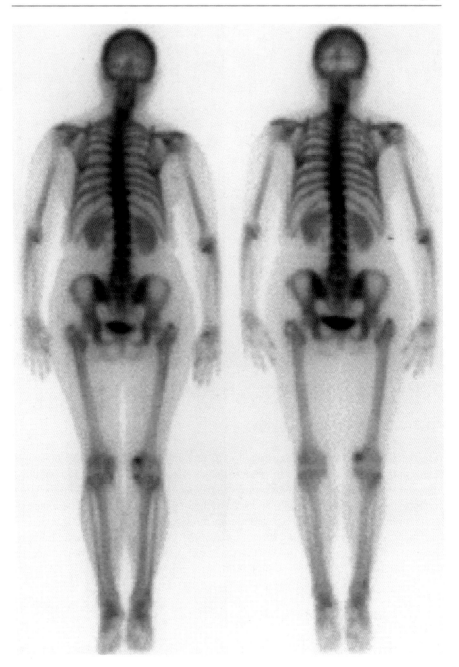

**Fig. 4.2** Scanning image of a human body obtained by CT scanning

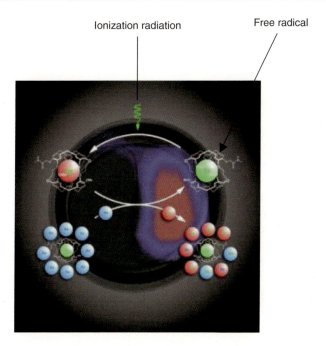

**Fig. 4.3** An artist's impression of the dissociation of a molecule when it is subjected to electromagnetic radiation which is ultimately imaged to detect the abnormalities inside a biological tissue or cell

in DNA chains. Both types of radiation can cause cancer in animals and can transform cells in culture [5].

Energy deposited by the *ionizing radiation* damages the chemicals of life within each cell. When the energy of ionization is deposited directly into the chemicals of life—such as DNA, RNA, proteins, and enzymes—damage is referred as direct effect. DNA is known to be critically important molecule located at the nucleus of the cell and is called the irreplaceable *master* molecule. If the *master* molecule is severely damaged, the cell can die or one can notice the abnormality of the cell/tissue [6]. Molecular imaging enables visualization of the cellular function and the follow-up of the molecular process in living organism without perturbing them. As a result, *molecular (medical) imaging system* has multiple and numerous potentialities in diagnosing diseases such as cancer and neurological and cardiovascular diseases [7].

## 4.3 Molecular (Medical) Imaging

*Medical imaging* in its widest sense is a part of *biological imaging* and incorporates *radiology, radiological sciences, thermography, endoscopy, medical photography,* and *microscopy* besides X-ray *fluoroscopy, computer tomography* (CT), *positron emission tomography* (PET), *single-photon emission tomography* (SPECT), *magnetic resonance imaging* (MRI), and *ultrasound*. As a matter of fact, *medical imaging* constitutes a subdiscipline of *biomedical engineering, medical physics*, or *medicine* depending upon the context [8, 9].

### 4.3.1 Radiology and Radiological Science

*Radiology* or the entire field of *radiological science* is the part of the medical science dealing with the use of electromagnetic (EM) energy emitted by X-ray generators or other such devices for the purpose of obtaining visual information as a part of medical imaging.

It is more than a century when physics made a dramatic entry into the field of nuclear medicine with the discoveries of X-rays and natural radioactivity, which are considered to have potential for *medical imaging* and *therapy* [10]. By and by, *X-ray radiography and therapy* have made impressive progress over recent decades as medical innovation has taken full advantage of the results of the fundamental research in physics, chemistry, and molecular biology [11]. As a matter of fact, X-rays are just one of the many (e.g., thermography, endoscopy, microscopy) technological tools that make up *medical imaging* technology.

### 4.3.2 Thermography

*Thermography, thermal imaging, thermographic imaging,* or *thermal video* all fall in one discipline of *infrared* (IR) *imaging,* and it displays the amount of IR energy that is emitted, transmitted, and reflected by an object of interest. As such, *thermographic cameras* detect radiation in the infrared range of the EM spectrum (roughly 900–14,000 nm) and produce images of that radiation, called *thermogram.*

The incident IR energy comprises of *emitted* energy, *transmitted* energy, and the *reflected* energy, and the energy profile is viewed through a *thermal imaging* device. In general, the emitted energy from the object is of interest, which is being measured. The thermographic camera employs a series of mathematical algorithms and builds a picture in the viewer, which is ultimately recorded [12]. Figure 4.4 shows the *IR picture* of the *female breast*. Different colors indicate temperature of the tissues on different parts of the breast. The scale on the right-hand side indicates the temperature of each color associated with the picture.

The principle of thermography is applied in *digital IR imaging* of a breast to diagnose any existence of cancerous tumor. It is based on the principle that metabolic activity and vascular circulation in both precancerous tissue and the area surrounding a developing breast cancer are almost always higher than in normal breast tissue. In an ever-increasing need for nutrients, cancerous cells increase blood circulation by holding open existing blood vessels, opening dormant vessels, and creating new ones. The activities naturally result in an increase in regional surface temperatures of the breast. The ultrasensitive thermoscopy with

## 4.3 Molecular (Medical) Imaging

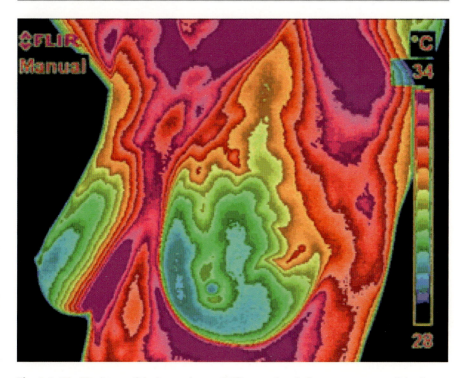

**Fig. 4.4** The IR picture of the human breast. Different colors indicate temperature of the tissues on different parts of the breast (Photo courtesy of Pacific Chiropractic and Research Center)

IR camera and sophisticated computer can detect, analyze, and produce high-resolution image [12, 13].

The difference between infrared (IR) film and thermography is that the former one is sensitive to IR radiation in the range between 250C and 500C, while the range of thermography is approximately −50C to over 2,000C. Thermography has different advantages over IR and other imaging systems such as the following: (1) it is capable of catching moving targets in real time, (2) it can be used to measure or observe an area inaccessible or hazardous for other methods, and (3) it can be used to find defects in shafts and metals.

### 4.3.3 Endoscopy

It is a procedure which enables a physician to look inside human body without an operation (Fig. 4.5a). It uses an instrument called an endoscope which has a tiny camera attached to a long thin tube (Fig. 4.5b). The tip of the instrument is inserted inside the body through a passage or opening to see some abnormalities of an organ. Sometimes the instrument is used to perform surgery, such as removal of polyps from a colon.

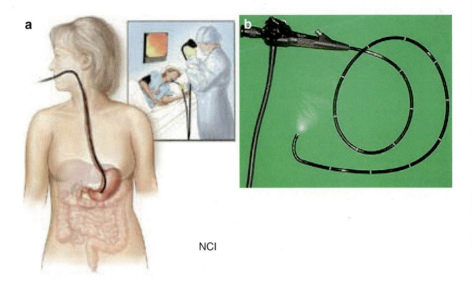

**Fig. 4.5** (**a**) Endoscope inside a human body to monitor any suspicious material inside the stomach. (**b**) Endoscope with an attached camera at the end of the long-necked tube (Courtesy: National Cancer Institute, US)

### 4.3.4 Microscopy or Medical Photography

Recently, the researchers at the University of California, Berkeley, have unveiled a new gadget that could be used by the physician to diagnose diseases like tuberculosis and malaria. The device has programmable mini mobile phone with a camera attached to the tip. It can be used as a microscope to examine blood samples in the field and help spot some of the world's deadliest diseases. The *CellScope microscope* as it is called has attachment clips on to the mobile phone that uses the built-in camera to process the images. As a result, physician can perform complex high-resolution microscopy on a blood or sputum sample placed on the slide. Sample evaluation could potentially be performed in real time while a patient is still in the presence of a health care worker. Figure 4.6 shows the picture of a mobile phone microscope (celloscope with attached camera).

The ability to make accurate, permanent, and objective images for the use in medical diagnosis, treatment and further prevention of the disease, medical photography are being used extensively in the medical Community. In this field, images are made to document all of the prominent features of a subject without exaggeration, enhancement, distortion, deliberate obliteration, or addition of details that might lead to misinterpretation by the radiologists. Figure 4.7 shows an exhaustive effort by a medical team taking photographs with a medical microscopic camera after surgical procedures.

**Fig. 4.6** Picture of a mobile phone microscope (CellScope with attached camera) (Photo courtesy of Wiki)

## 4.4 More Advanced Technology for Medical Imaging

The degree of importance of *imaging* to physicians and their ability to provide care was summed up by the New England Journal of Medicine in 2000. In a retrospective on the turn of the millennium, the editors of the journal called *medical imaging* one of the most important medical developments of the past 1,000 years—ranking with such milestones as the *discovery of anesthesia and discovery of antibiotics* [14]. Indeed, better *imaging* permitted the development of new treatments for *cancer* and *vascular* and *cardiac diseases*.

Over the past two decades, the *advances* in *medical imaging* have dramatically improved cancer diagnoses and its treatment. In fact, the death rates in America due to breast cancer have declined [15] because of minimally invasive imaging procedures. As for example, *X-ray fluoroscopy*—which allows biopsies of breast, bone, and other tissue without surgery—is dramatically reducing infections and recovery time [16].

Indeed, *CT scans* and *X-ray fluoroscopy* have enabled precise diagnosis of cancer and provided the physician to locate the affected area more precisely than ever. No wonder, *3-D imaging technology* has allowed surgeons to remove cancerous tissue while sparing neighboring healthy tissue. Use of imaging technologies like *CT*, *X-ray fluoroscopy*, *positron emission tomography* (PET), *single-photon emission computed tomography* (SPECT), *magnetic resonance imaging* (MRI), *ultrasound*, and *nuclear medicine* has contributed to the significant improvements in treating vascular and cardiac diseases and has made significant improvements in mortality and morbidity.

According to a survey in 2003 by a group of physicians, *X-ray fluoroscopy* is the therapy of choice for coronary angioplasty [17]. More broadly, studies revealed that

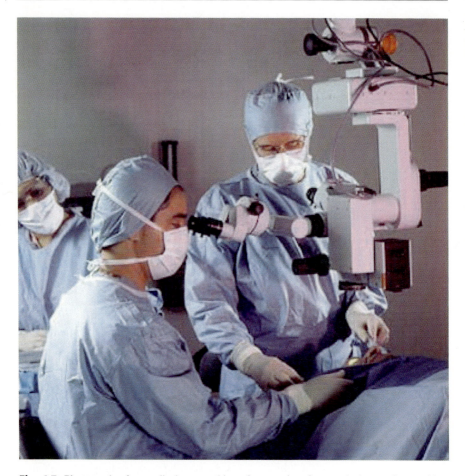

**Fig. 4.7** Photograph of a medical team taking photographs after surgical procedures with a medical microscopic camera

as much as 70 % of the survival improvement in heart attacks over the past 30 years is the result of changes in medical technology, including cardiac catheterization, coronary angioplasty, and cardiac stents—all made possible by X-ray-based or other *imaging technology* [18, 19].

The *computer tomography* (CT) scan, on the other hand, can identify stroke early, and the physicians can take early decision about which treatment will be better for the patient [20], including clot-busting drugs known as *thrombolytic therapy*. On the other hand, if the stroke is caused by hemorrhage, *X-ray fluoroscopy* guides delicate surgical procedures to close ruptured arteries [21].

There is no doubt that the *medical imaging* technology has empowered doctors and medical professionals to do a better job in diagnosing and treating patient and providing them with more medical care. As a result, there is a significant improvement in mortality and morbidity from vascular- and cancer-related diseases.

## 4.5 Advanced Tools (Instruments) for Medical (Molecular) Imaging

**Table 4.1** Innovations in imaging technologies and ultimate dose reduction

| Features | Amount of dose reduction |
| --- | --- |
| Automatic tube current modulation | Between 20 and 60 % |
| Pre-patient filtration (Cu/Al) | Up to 50 % |
| Pre-patient beam filtration and shaping | Up to 50 % |
| Use of pulse instead of continuous radiation | ~50–75 % |
| Ceramic CT detector material | ~50 % |
| Use of advanced hardware and software for X-ray beam collimation | Reduce multi-slice over-beaming by 35 % |
| Use of noise reduction filter | ~50 % |
| Tube current reduction in ECG | ~50 % (in cardiac CT) |
| Optimization of X-ray beam filtration | 15–25 % |
| Step and shoot acquisition and volumetric reconstruction | ~70 % (in cardiac CT) |

However, radiation exposure during imaging is a critical issue with regard to imaging children. However, National Cancer Institute points out that the benefit of the imaging technology outweighs potential risk. As a matter of fact, discussions of radiation risk are often highly complex. At the same time, the potential risk of radiation exposure from imaging technologies must also be understood in relationship to other potential sources of radiation. It goes without saying that the innovations in product design (automatic tube current modulation in X-ray radiation system, automatic exposure control, improvements in electronics, beam collimation, beam filtration, adaptive software filtration, and use of short radiation pulses) and operation have enabled *dose reduction* and ultimately reduced the potential risk of radiation (Table 4.1).

## 4.5 Advanced Tools (Instruments) for Medical (Molecular) Imaging

The area of advanced instrumentation for medical imaging covers image acquisition, modeling and quantification, image correction, and interpretation of medical images. In other words, *medical imaging* is often perceived to designate the set of techniques that produce images of the internal aspect of the body (without having to open it). The *advanced imaging system* may use any one of the following or the combination of any two of them [22–25]:

1. *Fluoroscopy* and *computer tomography* (CT)
2. *Positron emission tomography* (PET)
3. *Single-photon emission tomography* (SPECT)
4. *Gamma (γ) rays*
5. *High-frequency sound waves* (ultrasound)
6. *Magnetic fields* (MRI) to produce images of organs and other internal structures of the body

## 4.5.1 Fluoroscopy

It is a dynamic *X-ray imaging system* that is used in radiology mainly for real-time imaging and for aligning the patient with respect to the imaging system for filming. Fluoroscopy and radiography both use X-ray radiation to *image* the anatomy of the body of a patient [22, 23].

Figure 4.8 shows the schematic of a typical *fluoroscopy imaging system*. It consists of an X-ray system with a collimator, image intensifier (Fig. 4.9), couple of lenses for focusing the images, and a video camera to record the picture of the anatomy of the body of interest. The collimator limits the size of the X-ray beam and is automatically adjusted to the proper field of view. During panning over anatomical regions, the thickness of the anatomical region changes and accordingly the source-to-detector distance is also changed. The collimator on the other hand opens or closes to accommodate the differing height of the *image intensifier*.

The *image intensifier* is the most important part of the fluoroscopy system. It is composed of four principal components, namely, *the glass* or *metal input window, a fluorescent input screen, a high-voltage focusing electrode $E_1$,* and *other two electrodes $E_2$ and $E_3$* leading to anode and output fluorescent screen coupled to a *cine camera* (Fig. 4.10) and/or a video camera. Figure 4.9 shows the diagram of an image intensifier.

As the name suggests, the purpose of the image intensifier is to amplify the light signal produced by a phosphor to generate a bright X-ray image which can be viewed easily or recorded with a photographic or cine/video camera. The most important aspect of the image intensifier is that it can produce a live image in which the physicians and the radiologists can watch the dynamic changes such as ventricular contraction of the heart. Figure 4.10 shows the picture of a cine camera.

The input layer of the image intensifier consists of four layers, namely, input window, substrate, a phosphor material, and a photocathode. The photocathode is generally a combination of a thin layer of antimony (Sb) metal and alkali that releases electrons when stuck by visible light. The phosphor is, generally, a thick film with *columnar* cesium iodide doped with thallium (CsI:Tl) grown by vacuum evaporation on to the substrate. The *columnar structure* of CsI:Tl (Fig. 4.11) minimizes the internal reflection of light and thereby intensifies the number of photons received by the recording instrument [26]. The physical and chemical characteristics of the phosphors and the deposition techniques will be dealt in a separate chapter. Theoretical calculation shows that approximately 3,000 light photons (at about 420-nm wavelength) will be produced when the phosphor (CsI: Tl) is exposed to 60-keV X-ray.

During diagnosis of the anatomy of a patient, dynamic studies require a real-time imaging device capable of displaying a sequence of events with enough *temporal resolution* (images per second). *Temporal resolution* can be improved by acquiring images at a faster rate. Indeed, the live display of the X-ray image by fluoroscopic instrument made it possible to provide excellent *temporal resolution* [27, 28]. Dynamic studies are usually recorded on video tape recorders for playback and interpretation. In *cardiology*, *cine camera* with high-resolution film

## 4.5 Advanced Tools (Instruments) for Medical (Molecular) Imaging

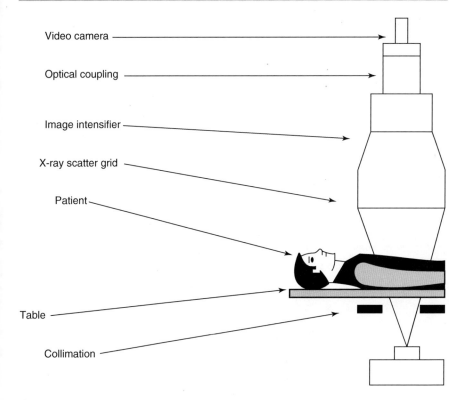

**Fig. 4.8** Schematic of a fluoroscopic imaging system

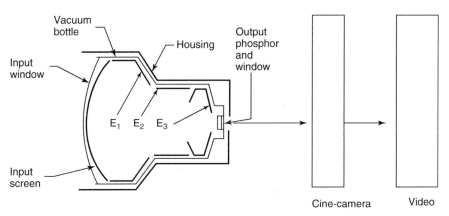

**Fig. 4.9** Diagram of image intensifier used in fluoroscopy imaging system

is attached to the image intensifier and record what is essentially high exposure rate onto film.

The *gastrointestinal* (GI) series and the barium enema are the most common fluoroscopic studies. In GI and angiographic studies, fluoroscopy is used to guide

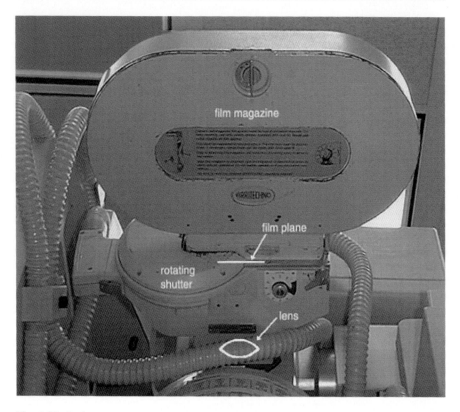

**Fig. 4.10** A cine camera

**Fig. 4.11** Columnar structure of CsI:Tl grown by hot-wall epitaxy [26]

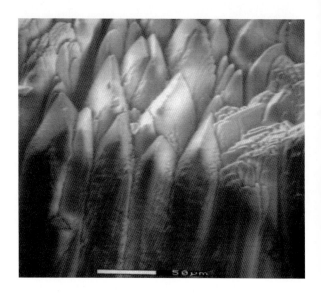

## 4.5 Advanced Tools (Instruments) for Medical (Molecular) Imaging

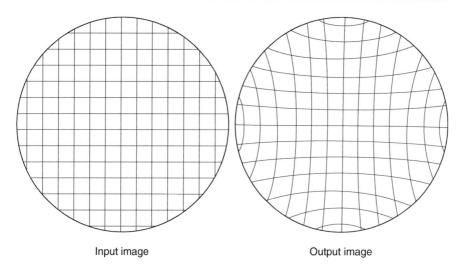

**Fig. 4.12** The pincushion distortion effects

barium through the GI tract or catheters through the vascular system. In both cases, there is the intention of identifying diagnostic views for radiographic film acquisition.

The anode, a window, and a phosphor, which is small relative to the input phosphor, are placed on the output side. To maintain an equal path length between all points on the input and the output phosphors, the input phosphor/photocathode is curved, and the amount of curvature is a function of the *field of view* (FOV) and length of the image intensifier. In order to maintain the length of the image intensifier, minimum curvature is maximized. Unfortunately, this results in unavoidable *pincushion distortion* of the image. Figure 4.12 shows how the straight grid lines of the input are curved inward on the output image.

The fluoroscopic imaging system plays an important role in radiologic imaging system. Image is generally created by a series of imaging chain, which allows the adjustment of the exposure rate used in producing images, by manipulation of the aperture in the optical stage between image intensifier and TV camera.

Continuous low current in the fluoroscopic systems can acquire images 30 frames per second in 33 ms. There are three major radiography/fluoroscopy designs available in the market, namely, (a) the over-table, (b) the under-table, and (c) the single or biplanar cine fluoroscopy.

Figure 4.13a and b show the diagrams of an over-table (OT) and under-table (UT) fluoroscopy systems, respectively. These instruments are versatile in designs with remote controls and a variety of imaging systems. Beside the positions of the X-ray tubes and the source to image distances (115–150 cm in *OT* system and 68–98 cm in UT system), most of the facilities offered by the over-table (*OT*) design are available in the under-table (*UT*) fluoroscopy system.

**Fig. 4.13** (a) An over-table fluoroscopy system and (b) an under-table fluoroscopy system (Courtesy: WIKI)

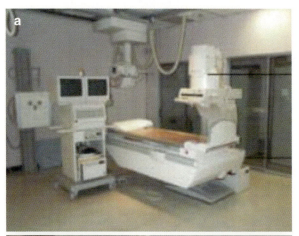

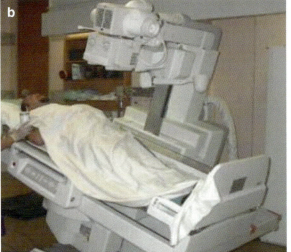

Recently, a movable C-arm design finds applications in surgery including orthopedics, bone fracture inspection, foreign body localization, and pacemaker implantation. Further, modification of the instrument which is known as double C-arm cine fluoroscopy system with horizontal and vertical image intensifiers has facilitated simultaneous viewing in two opposite planes (biplanar unit) (Fig. 4.14). The unit is used for investigating rapidly moving organs.

The maximum dose of the patient is controlled by controlling X-ray radiation and by the use of modern image intensifiers. The minimum focal spot to skin distance is also reduced to 30 cm. On the other hand, the use of large field image intensifier and high-definition video chain in the form of a plumbicon tube or CCD camera has given digital imaging technology considerable advantages to both the patient and the operator.

## 4.5 Advanced Tools (Instruments) for Medical (Molecular) Imaging

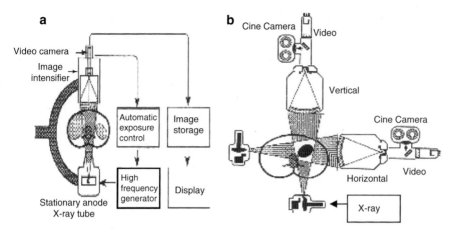

**Fig. 4.14** (a) Mobile C-arm fluoroscopy system and (b) C-arm biplanar cine fluoroscopy system (Courtesy: Thomson Scientific, G.B.)

### 4.5.2 Positron Emission Tomography (PET)

PET is a nuclear medicine imaging technique, which produces a 3-D image or picture of functional processes of a body or a part of the body [29, 30]. The history of PET can be traced to the early 1950s when a group of research workers in Massachusetts first discovered the medical imaging possibilities of a class of radioactive isotopes that decay by release of a positron. By introducing a positron emission radionuclide, a biologically active molecule, into the body, image of the radionuclide (tracer) in 3-D space within the body can be reconstructed. When the biologically active molecule is fluorodeoxyglucose (FDG), concentration of the radionuclide is imaged to explore the metabolic activities of the tissue. FDG-PET scanning and imaging is widely used in clinical oncology [31]. According to some experts, PET has a millionfold sensitivity advantage over other techniques used to study tissue metabolism and neuroreceptor activity in the brain.

Figure 4.15 shows an experimental setup for *positron emission tomography* (PET). The picture shows how the instrument works, the scanning procedure, and how the tracer is introduced to get an image of a particular area like kidneys, lungs, heart, chest, or abdomen of a living animal. *PET* is a nuclear imaging tool (instrument) which produces a 3-D map of the body or a part of the body.

Figure 4.16 shows the medical image of the kidneys with renal arteries and renal veins. The picture has been captured by scanning positron emission tomography (PET system).

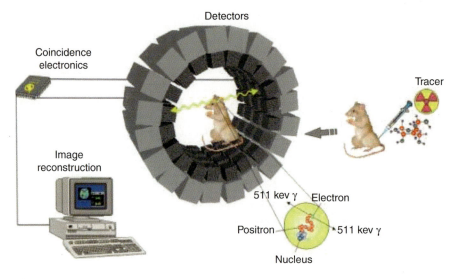

**Fig. 4.15** Experimental setup for positron emission tomography (PET) (Courtesy of Google image from *The Mesothelioma Library*)

**Fig. 4.16** PET imaging of human lungs (Photo courtesy of Canadian Association of Radiologists)

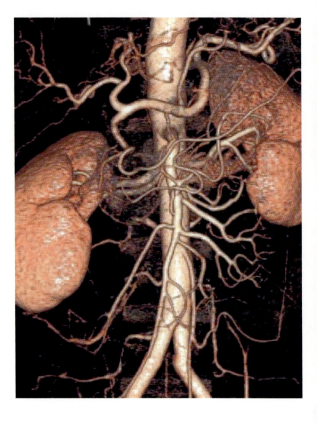

## 4.5.3 Single-Photon Emission Computed Tomography (SPECT)

The history of SPECT goes back to the 1960s when the idea of a single-photon emission imaging technology was introduced by D. E. Kuhl and R. Q. Edwards. Like PET, SPECT acquires information on the concentration of the radionuclide introduced inside the patient's body. However, SPECT imaging involves the rotation of a photon detector array around the body to acquire data from multiple angles [32–34].

In SPECT, only a single photon is detected from the emission of the radionuclide. As a result, a special lens known as a collimator is used to acquire the image data from multiple views of the body, which caused a tremendous decrease in detection efficiency as compared to PET. In X-ray CT, scan attenuation is measured, not the transmissions source. To compensate for the attenuation experienced by emission photons from the injected tracers in the body, contemporary SPECT machines use mathematical reconstruction algorithms to increase resolution.

To acquire SPECT images, the gamma camera is rotated around the patient. Projections are acquired at defined points during rotation typically 3–6°. In most cases, a full 360° rotation is used to obtain an optimal reconstruction. The time taken to obtain each projection is variable, but 15–20 s is typical.

Although the resolution of SPECT-scanned image is not as good as PET-scanned image, the availability of new pharmaceuticals for brain and head and economic aspects has made SPECT more appealing to the medical professionals and the radiologists [35, 36].

Figure 4.17 shows the picture of a SPECT scanning machine from Siemens. The model is Symbia T-SPECT CT scanner, the first of its kind with a combination of the functional sensitivity of a SPECT system which can provide detailed anatomical information for diagnostic multi-slice CT scan.

## 4.5.4 Magnetic Resonance Imaging (MRI)

The discovery of *magnetic resonance imaging* (MRI) and its development and application in modern medical imaging is one of the spectacular and successful events in medical imaging science and technology. Nuclei of the same type have different resonance frequencies depending on the chemical composition caused by typical field screening effect by the electrons in the molecule [37, 38].

In the center of the magnet, a typical symmetric field inhomogeneity is created that has a sort of three-dimensionally *saddle-shaped* field distribution. Then, only in the vicinity of the *saddle point* is the field sufficiently homogeneous to raise a measurable resonance signal. The system which is known as *FONAR* (field-focusing NMR) can not only perform localized measurement of resonance signals and relaxation parameters but it can also produce an image in a point by point scan when the specimen is moved [39].

The MRI is actually a map of the very weak magnetization, originated from some of the atomic nuclei in the body tissue, in the presence of an external magnetic

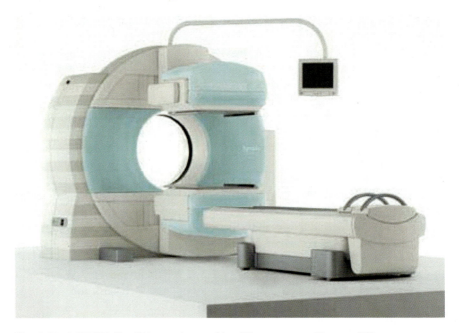

**Fig. 4.17** A SPECT (Symbia) scanning machine (Photo-courtesy: Siemens, US)

**Table 4.2** $T_1$ and $T_2$ values for various tissues

| Tissue | $T_1$ (0.5 T) (ms) | $T_1$ (1.5 T) (ms) | $T_2$ (ms) |
|---|---|---|---|
| Fat | 210 | 260 | 80 |
| Liver | 350 | 500 | 40 |
| Muscle | 550 | 870 | 45 |
| White matter | 500 | 780 | 90 |
| Gray matter | 650 | 900 | 100 |
| CSF | 1,800 | 2,400 | 160 |

field. Since this magnetization is proportional to the density of the nuclei, the MR image shows the distribution of the selected atoms. The magnetic resonance imaging has the capability of not only mapping the tissue density but also it can it sense alterations in the chemical structure which actually differentiate MR from other imaging techniques. The ability resides in the fact that different tissues have values of $T_1$ and $T_2$ significantly distinctly, especially between normal and cancerous tissue. Table 4.2 shows the characteristics of $T_1$ and $T_2$ times with respect to molecular motion, size, and interactions.

The relaxation processes that $T_1$ and $T_2$ characterize depend on the size of the molecules and the tissue type constituting the spins. Emphasizing the differences of spin density (the local hydrogen concentration) as well as $T_1$ and $T_2$ relaxation, time constants of different tissues are the key to exquisite contrast sensitivity of the MRIs [40, 41].

## 4.5 Advanced Tools (Instruments) for Medical (Molecular) Imaging

**Fig. 4.18** A magnetic resonance imaging machine (Courtesy: Siemens Med., US)

The *time* to *acquire* (AT) an image in MRI system is equal to (time of repetition (TR)) × (phase encode steps) × (number of signal averages), and the total number of slices is equal to {(TR)}/{(TE, time of echo + $K$, a constant)}. The value of $K$ is dependent on the capabilities of the instrument. *Acquisition time* (AT) can be shortened either by *synthesizing* data or by fast *spin echo* (FSE) technique. The first one takes advantage of the symmetry and redundancy characteristics of the frequency domain signals in k-space. The second one uses multiple phase encode steps per TR interval to speed up the data collection rate and fill k-space much faster.

Acquisition of 3-D images requires the use of broadband, nonselective radio frequency (RF) pulse to excite large volume spins simultaneously. Two-phase encode gradients are discretely applied in the slice encode and phase encode directions, before the frequency encode gradient. A 3-D FT (Fourier transform) is independently applied for each column, row, and depth axis in the image matrix cube. Volume orientation may be either isotropic or anisotropic. Isotropic orientation is preferred to anisotropic because of better resolution.

Figure 4.18 shows a *magnetic resonance instrument* or *imaging* (MRI) with a long arm for positioning the patient. MRI is an indispensable diagnostic tool because of its noninvasive, nonionizing, real-time portable and moderate-cost nature. Diagnostic usage for MRI has greatly expanded over the past couple of decades because it offers many advantages as an imaging modality.

Figure 4.19 shows the *magnetic resonance imaging* (MRI) of knee showing different parts. MRI uses a powerful magnetic field, radio frequency pulses, and a computer to produce detailed pictures of organs like the knee bones as shown in the figure below. The images can be examined on a computer monitor, printed or copied to hardware of the computer and disk. The most important criterion of the MRI is that it does not use any *ionizing radiation* (X-ray) [42, 43].

(a) *Spatial Resolution*: The spatial resolution of MR image depends on the *field of view (FOV), receiver coil characteristics, gradient field strength, sampling*

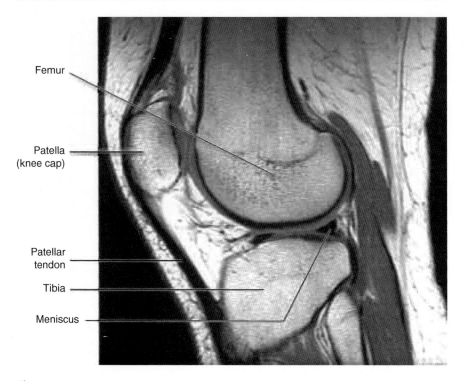

**Fig. 4.19** Magnetic resonance imaging of knee showing different parts (Courtesy: WIKI)

*bandwidth,* and *image matrix*. In general, MR provides almost same resolution of CT images when the contrast of image is high and the field of view is reasonably large. As a general rule, the higher the field, the better will be the resolution. Higher field strength generates a larger *signal to noise ratio* (SNR). At the same time, longer $T_1$ relaxation timing and reduced depth of RF penetration can limit the advantages of the higher field strengths in terms of *spatial resolution* and *contrast sensitivity*.

As a matter of fact, *contrast sensitivity* is a major attribute of MR for the discrimination of soft tissues and contrast effects due to blood flow. Higher field strength of the magnet increases the $T_1$ relaxation timing and decreases sensitivity.

(b) *Signal to Noise Ratio*: The *signal to noise ratio* (SNR) of a MR image depends on the FOV, magnetic field strength, quality factor of the coil, slice thickness, image matrix, and intrinsic signal intensity. Another important parameter that affects the quality of the image is the magnetic field inhomogeneity, which ultimately causes a more rapid decay of the signal from $T_2$ relaxation due to dephasing effects. However, modern MRI machines are provided with self-shielded magnets, and automatic shimming has been able to minimize the field inhomogeneity artifacts.

## 4.6 X-Ray Computed Tomography (CT)

Prior to the end of the nineteenth century, doctors lacked the ability to definitively diagnose many internal medical problems without having to cut open the patient. The invention of X-ray and its uses in medical imaging have brought obvious benefits to health care, but as with all new technologies, it also required some changes in behavior and processes. Thus, each time as the demand continues to push, the medical image processing is changing to accommodate better technology and instrumentation. The issues include cost and productivity, the need to acquire skills, radiation doses, overuse, and image quality. Moreover, some of the ethical and legal issues surrounding teleradiology remain unclear [44].

## 4.6 X-Ray Computed Tomography (CT)

Needless to say that medical imaging continues to empower doctors and medical professionals to view the human body with ever-increasing clarity and accuracy [44]. For example, computed tomography (CT) has all but eliminated the practice of exploratory surgery with its associated invasive risks and lengthy recovery periods. Indeed, as a result of the use of radiation, medical imaging and nuclear medicinal therapy has brought tremendous improvement in health science, saving lives of millions, especially, in the field of CT, X-ray, and nuclear medicine. The introduction of medical imaging technology, new product, and system innovations during the past 20 years has reduced radiation dose for many imaging procedures by 20–75 % while preserving the ability of imaging technologies to aid physicians in diagnosing and treating disease. These dose reductions have been achieved through innovations in product operation and design, software applications, operating practices, and procedure algorithms. Figure 4.20 shows a three-dimensional view of a human heart scanned by computed tomographic (CT) technique. Dr. D. J. Brenner and E. J. Hall from the Center for Radiological Research at Columbia University, New York, argue that the potential carcinogenic effects from using CT scans may be underestimated or overlooked. It is estimated that more than 62 million CT scans per year are currently given in the USA, compared to three million in 1980. Because CT scans result in a far larger radiation exposure compared with conventional plain X-ray, this resulted in a marked increase in the average personal exposure in the USA, which has about doubled since 1980, largely because of the increased CT usage [45]. Figure 4.20 shows a scanning image of a heart by computed tomographic (CT) technique.

In spite of criticism from different corners of the medical professionals, CT has become the forefront technology in medical imaging system because of so many innovations during its early clinical use. As a result of the invention of the CT scanner by Godfrey Hounsfield and Allan Cormack, many possibilities of X-ray imaging system have been opened up not only in medicine but also in many other industrial applications such as nondestructive testing and soil core analysis.

The CT scanner acquires a large number of CT projections around the patient that can be 800 or more. Each data point that is acquired by the detector array is a

**Fig. 4.20** A scanning image of a heart by computed tomographic (CT) technique (Photo courtesy of D. J. Brenner and E. J. Hall, Columbia University Medical Center, New York)

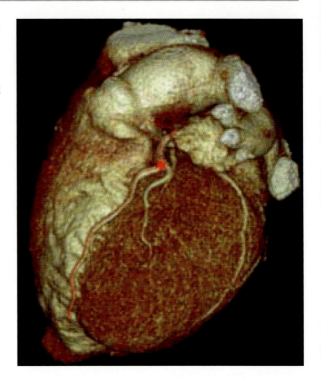

transmission measurement through the patient along a given line. During acquisition, the detector integrates attenuation information along a known path of the narrow X-ray beam. The CT projection data ($P_x$) can be expressed mathematically as

$$P_x = \text{Log}\{I_0/I_x\} = \mu x \tag{4.1}$$

where $I_0$ and $I_x$ are the intensities of X-ray without and with the object, $\mu$ is the attenuation coefficient, and $x$ is the distance the ray has traveled.

The processing of the data before image reconstruction is independent of the intensity of the X-ray beam, but during actual processing step, it normalizes variations in the beam intensity and provides attenuation measurements. However, during *back projection* reconstruction, data from a number of different angular profiles are smeared back onto the image matrix (back-projected). As a result, areas of high and low attenuation tend to reinforce with each other, building up the image in the computer.

CT images are two-dimensional matrix numbers with different possibilities for the value of each pixel and each number corresponding to a spatial location of the image and in the patient. The numerical value of each pixel in the computer represents a certain level of gray on the CT image. Other names for the numbers that correspond to the brightness of each pixel in a CT image are *Hounsfield* units or *CT numbers*. The *CT numbers* for a certain pixel correlate *linear attenuation*

## 4.6 X-Ray Computed Tomography (CT)

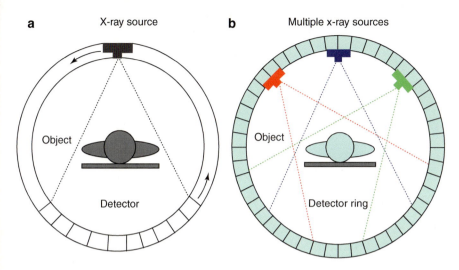

**Fig. 4.21** (**a**) Conventional third-generation CT scanner and (**b**) the multiplexing CT scanner (courtesy: J. Zang, UNC)

*coefficients*, which are ultimately related to the *composition* and *density* of a material under study and can be represented mathematically as

$$\text{CT number} = 1{,}000 \times (\mu_{\text{pixel}} - \mu_{\text{water}})/\mu_{\text{water}} \qquad (4.2)$$

where $\mu_{\text{pixel}}$ is the linear attenuation coefficient for a given pixel and $\mu_{\text{water}}$ is the attenuation coefficient in water.

*Compton scattering* is the primary type of interaction that contributes to the subject contrast in the images and depends both on the mass density (g/cm$^3$) of the material under study as well as the number of electrons per unit mass (electrons/g and is equal to $N$ ($Z/A$), where $N$ = Avogadro's number, $Z$ = atomic number, and $A$ = atomic mass). As a result, bone tissue has a higher CT number than lung tissue which is highly aerated.

The medical imaging systems and the processes have come a long way, and it is now matured. As a matter of fact, the art and the understanding of an image modality involve the appreciation of maturation of that modality. As for example, the *first generation* of CT scanner, which was a rotate/translate pencil beam system, had been modified in the *second generation* with a linear array of 30 detectors to increase the utilization of the X-ray beam by 30 times over the single detector used per slice. But soon, the difficulties like pragmatic problems with cable wear and in perfect balancing were felt with too many detectors. As a result, to ease the difficulties, the linear detector array was designed to form an arc wide enough to allow the X-ray beam to interrogate the entire patient in the *third generation* (Fig. 4.21).

But soon, further modifications in the *fourth-generation* CT scanners were performed. Here each individual detector is responsible for acquiring a fan of

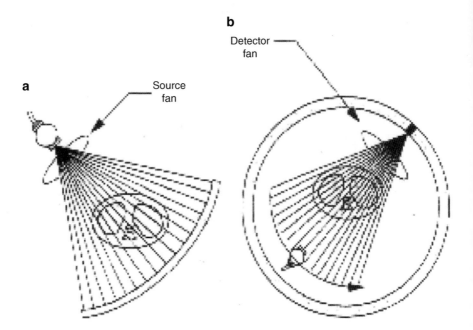

**Fig. 4.22** (a) Schematic of the fan beam in third-generation scanner and (b) schematic of fourth-generation scanner (Courtesy: Bushberg et al., with permission from LWW pub.)

transmission data, $I(x)$, and its own reference measurement of $I_0$, instead of measuring them separately as in third-generation system. The CT projection value $P$ is then calculated from the relation below:

$$P = \text{Log}\{(g_1 I_0)/(g_2 I(x))\} \qquad (4.3)$$

where the gains $g_1$ and $g_2$ are taken to be equal, since $I_0$ and $I(x)$ are acquired by the same detector. Therefore, the value of $P$ is not sensitive to detector drift. However, an additional motion of detector array during CT acquisition introduces constraint to rotate the X-ray tube inside the detector ring that surrounds the patient, and this results very often magnified the focal spot.

Figure 4.22 shows the schematic of third- and fourth-generation scanners. Third-generation scanners require excellent detector calibration and stability, whereas the fourth generation requires excellent X-ray tube output stability. The scanners generally use solid state scintillator materials, which emit visible light when struck by X-ray. Details of the scintillator materials will be dealt in a separate chapter.

By and by, the search for a better-equipped CT scanner for medical imaging system has been developed. The novel fifth-generation CT scanner produces electron beam in a cone-like structure behind the gantry that strikes the tungsten and is electronically steered around the patient. Figure 4.23 shows the schematic of a fifth-generation CT scanner system.

## 4.7 Nuclear Medicine Imaging

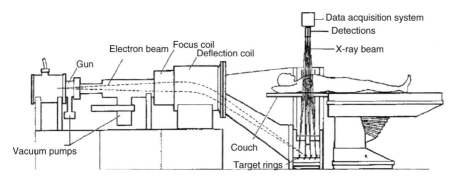

**Fig. 4.23** Schematic of a fifth-generation CT scanner system that uses no moving parts during the scan (Courtesy: Bushberg et al., with permission from LWW Pub.)

Figure 4.23 shows the schematic of a fifth-generation CT scanner, which has faster in scanning speed than the previous scanners. However, the use of unconventional X-ray generation system has imposed some limitations (such as lower number of generation of X-ray photons compared to third- and fourth-generation scanners) with respect to *X-ray tube* output when it is used for shorter time scan. A recent innovation has been made possible to redesign the X-ray tube and the use of helical scanning in third- and fourth-generation scanning system. The advent of *helical scanning* has provided a full 360° set of projections during the rotation of the gantry. At the same time, helical scanning allows the use of less contrast agent, since the total scan time is shorter, and increases the throughput of patients through the scanner.

### 4.7 Nuclear Medicine Imaging

Nuclear medicine planar images are taken to study the volume activity from overlying tissue. Circulating activities like blood, liver, lung, DMSA (kidney function), and myocardial activities are part of routine work of nuclear medicine images. The planar images can be grouped as static planar and dynamic planar [46].

Static planar imaging is a conventional procedure for following radiopharmaceutical distribution. Image acquisition generally provides anterior, posterior, and several lateral views with oblique views. Low-, medium-, and high-energy collimators are used for planar clinical studies. High-sensitivity collimators with low energy are used for planar lungs, planar myocardium, dynamic renography, and MIBG adrenals. But for SPECT of brain and heart, DMSA of kidney requires high-resolution low-energy collimator. On the other hand, thyroid metastases and PET need high-energy collimator.

The gamma camera image is generally represented as matrix, and 256 × 256 is presently used to display static images. It is recommended that for visual acuity, the pixel size should be smaller than the data pixel size, and 64 *gray levels* seem to be optimum for a good picture. Both contrast and resolution in a nuclear medicine image depend whether the lesion shows as a negative or positive intake. It has been noticed that a color image gives a higher contrast than a gray scale [47]. The number of *gray scale* (G) depends on the number of bits m used to store the gray level values and can be expressed mathematically as

$$G = 2^m \tag{4.4}$$

Suppose that we have 256 gray levels, with 8 bits, then for $512 \times 512$ image with 256 gray levels, the storage required is $512 \times 512 \times 8 = 262$ kB (kilobyte).

Dynamic studies (DS) are used for multiphase studies where the fast vascular phase of the bolus through the kidney is captured at 0.5–1.0 s intervals for 30 frames. In DS, rapidly changing distributions of the radionuclide or moving organs can be captured at a fast rate either by (a) fast pass bolus study or (b) gating the cardiac cycle. As a matter of fact, in DS, very rapid sequences can be captured in list mode where the individual counts are stored in a memory disk and are reconstructed into matrices.

The main problem with the planar views is to separate the overlapping interference, which reduces the contrast of the image. Single-photon emission computed tomography (SPECT) which uses single-head rotating gamma camera tomography system has improved the contrast and resolution of the image considerably. There are two methods followed to collect data from the rotating gamma camera: (a) step and shoot and (b) continuous rotation. During data collection, the camera takes a series of images at an equal angular spacing called projection with the detector stopping at each projection. Recently, use of multiple detectors has been developed for head scanning and provides superior resolution to the rotating gamma camera. Here the entire array rotates through $180°$ giving 40 projections in 5 s.

Cyclotron produced $\beta^+$ isotopes are used in positron emission tomography (PET) during nuclear medicine imaging that takes advantage of the unique coincident gamma radiation. PET provides the highest sensitivity of all diagnostic imaging techniques. The last imaging system that is used in nuclear medicine without ionizing radiation is magnetic resonance imaging (MRI) [48]. However, the application of tracer techniques in MRI is severely limited by low signal to noise ratio (SNR) which requires high concentrations of stable NMR isotope to give a good image resolution. In spite of all the pros and cons, the SPECT, the PET, and the MRI have reached a degree of maturity where it is possible to assess where, if at all, any overlap occurs in their clinical usefulness [49]. All of these technologies have been discussed in details in Chap. 1. A comparative analysis of computer axial tomography (CAT), positron emission tomography (PET), single-photon emission tomography (SPECT), and magnetic resonance imaging (MRI) is tabulated in Table 4.3.

**Table 4.3** Comparative analysis of different imaging technology

| Technology | Advantages/disadvantages |
| --- | --- |
| CAT | Good signal strength, no attenuation problems, multiple axes with fixed source-to-detector position |
| PET | Good signal strength, no attenuation problems, multiple axes, higher sensitivity than SPECT |
| SPECT | Good signal strength, multiple axes, attenuation problems, lower sensitivity than PET |
| MRI | Multiple axes, but poor signal (radio)strength with attenuation problems |

## 4.8 Image Acquisition

### 4.8.1 Slice Thickness

Two lead jaws that collimate the incident X-ray beam control the thickness of the slice. However, there is trade-off with respect to the *slice thickness*. As for example, wider slice increases the number of X-ray photons and yields better contrast resolution with minimum noise. However, it degrades the spatial resolution and thereby reduces the partial volume averaging of small objects. At the same time, wider slice will require more tube current to maintain the same *statistical integrity* (signal to noise ratio, *SNR*). Moreover, as the slice thickness is not planar, the uneven slice sensitivity profile will be the result of the relatively large focal spots combined with the edge effects of the collimator and X-ray scatter.

The main factor that limits image quality in nuclear medicine studies is noise caused by statistical fluctuations in the number of detected gamma rays. Thus, with SPECT detector design, a high detection efficiency is needed ($>85$ %), which implies that the detector depth should be at least two attenuation lengths thick preferably three. While this criterion can be met by any scintillator, a short attenuation length is necessary to minimize the degradation of spatial resolution.

### 4.8.2 Number of Projections and Number of Rays Used in CT

The image quality of a CT scanner depends on the number of rays and the number of projections used. As for example, for *third-generation system*, the number of detectors used determines the number of rays used in each projection. For *fourth generation,* however, it is determined by how frequently the individual detectors are sampled as the X-ray tube sweeps around the fan angle. The increased number of projections used in each scanning has a profound effect on the image quality due to the increased number of X-ray photon availability.

Figure 4.24 shows a CT scanner with quarter detector shift, which shows that the center of the gantry is aligned to one-fourth of the way of the edge of the center detector. By shifting the detector array by one-half of the detector width, the ray sampling is doubled and it reduces the possibility of *aliasing*.

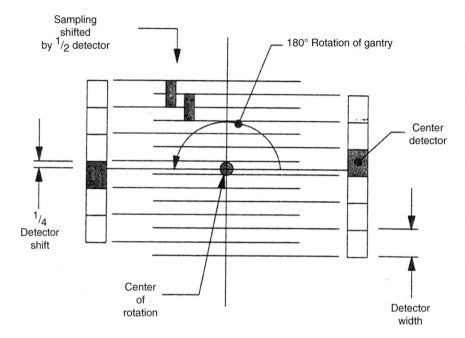

**Fig. 4.24** The incorporation of a quarter detector shift in a CT scanner

In CT scanners, planar detectors are used with the planar receptor, which is often provided with large FOV (field of view) image intensifier system to yield high resolution. The system is an inexpensive alternate to CT acquisition in radiotherapy treatment [50, 51].

The raw data consist of the digitized transmission measurements for each ray for each projection and preprocessed before CT reconstruction. They are also used for tomographic reconstruction in specialized applications.

### 4.8.3 Simple Back Projection

In simple back projection, the image matrix is initially set to zero, and the transmission value of the projected data is added to each pixel in the image that corresponds to the linear path through the object corresponding to the transmission measurement. However, the simple back projection data are smeared, preprocessed data and lack of compensating blurring effect of the image, which is due to overlaying of the data. In order to compensate for the blurring, the image is filtered by some mathematical process known as *convolution*. Filtering of a simple one-dimensional projection data before back projection is faster than filtering a two-dimensional back-projected image.

## 4.8 Image Acquisition

Intrinsic blurring in simple back projection appears to be convoluted with a *convolution kernel* that has a $(1/r)$ shape to it, which means that the intensity that should be assigned to a pixel at a certain point in the image is spread around into the neighboring pixels. In filtered back projection, a *convolution kernel* is designed to undo the $(1/r)$ blurring that is endemic to simple back projection. In essence, the mathematical convolution operates on neighboring projection data points, takes some of the intensity it finds there, and puts it back into the parent data point, thereby correcting for the $(1/r)$ blur. Now, if $P(r)$ represents the projection data and the convolution kernel is denoted as $k(r)$, the corrected projection data $P_{\text{corr}}(r)$ is computed as

$$P_{\text{corr}}(r) = P(r) \times k(r) \qquad (4.5)$$

CT scanners during filtering may use *convolution* in the *spatial domain* or *frequency domain,* and from the mathematical point of view, both the cases are similar. In the frequency domain, the kernel $k(r)$ can be transformed into Fourier terms as $k(f) = \text{FT} \{k(r)\}$, with frequency domain convolution $P(f) = \text{FT} \{P(r)\}$. On the other hand, the convolution in the spatial domain when applied FT is written as

$$
\begin{aligned}
P_{\text{corr}}(f) &= P(f) \times k(f) = (1/\text{FT})\{P_{\text{corr}}(f)\} \\
&= (1/\text{FT}) \left[ \text{FT} \{P(r) \ \text{FT} \{k(r)\}\} \right]
\end{aligned}
\qquad (4.6)
$$

Different CT manufacturers supply numerous filters often designed on different theme, with each filter having a slightly different shape. The polyenergetic X-ray beam does not create problem in radiographic imaging, but it affects linear attenuation coefficient. As a result, lower energy photons are attenuated more rapidly through the slice than the higher energy photons. Thus, when the slice is thicker, the energy of the X-ray beam is increased which known as *beam hardening,* which causes violation of the linear superposition of attenuation values that is required by back projection reconstruction [52]. The phenomenon that a polychromatic X-ray beam becomes more penetrating or harder as it traverses through matter. The X-ray beam used in medical imaging is polychromatic with broad energy spectrum.

A *cupping artifact* is a classic example of *beam hardening,* and it occurs in the soft tissue regions between large bony structures and results in decreased image intensity, which is not a true representative of the tissues. However, b*eam hardening* is reduced when the X-ray beam is pre-hardened by the insertion of copper or other attenuator in the beam near the X-ray tube. Figure 4.25 shows the dependence of the attenuation coefficient ($\mu$) on the thickness of the barrier.

Beam hardening can cause two types of artifacts, namely, (1) *volume artifact* and (2) *motion artifact.* The former one results when a variety of different tissue types are contained within a single *voxel.* Unfortunately, when bone and soft tissue are contained in a single voxel, partial volume averaging results in incorrect diagnosis.

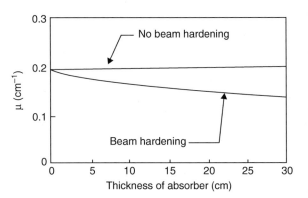

**Fig. 4.25** Beam hardening and its effect on $\mu$ as a function of the thickness of the absorber (Courtesy: Bushberg et al., with permission from LWW Pub.)

However, a solid knowledge of cross-sectional anatomy is the best defense against partial volume artifacts. On the other hand, the motion artifact arises due to the motion of the patient, even due to the motion of the cardiac and respiratory systems. Likewise, during CT acquisition, the first half of the projection data may back project the anatomy in one orientation, while the second half of the data may back project the anatomy in a different orientation. Even the slight motion of soft tissues during scanning can give rise to image *ghosting*. Rapid scan time and immobilization of the patient during CT acquisition can reduce the motion artifacts.

CT, a digital imaging process, produces separates axial sectional images with no intersection interference. There are two general classes of CT available in the market, namely, the linear or conventional CT and the computed axial CT (*CAT*). Linear CT is simple but suffers from two major limitations like swing angle ($\theta$) and (*b*) slice thickness.

The *CAT* on the other hand produces transaxial radiological images free from intersectional interference or blurring. The rotating X-ray source surrounded by a group of 800 or more detectors (Fig. 4.26) is collimated as a fan beam and rotated through 360° through 1° or even 1/2° per projection.

In practice, the attenuated beam is measured using a detector array opposite to the X-ray beam and displayed as a two-dimensional (2-D) image. The matrix of the absorption coefficients (dependent on the applied voltage in kV) is obtained from the scanning pattern, and the individual values in matrix can be calculated by using either (1) iterative technique or (2) back projection. As the former one takes a considerable amount of time of the computer, the second method has been very popular. In order to store the volume slice element or voxel in computer memory, a reference number has been established and the number is called CT number (CT number = $[1{,}000 \times \{(\mu_{\text{tissue}} - \mu_{\text{water}}) / (\mu_{\text{water}})\}]$).

The raw signal data undergoes processing before taking part in image reconstruction, and then the reconstructed image itself can be processed to enhance features or to remove artifacts. *Fast Fourier transform* (FFT) is the preferred method for reconstruction algorithm for filtered back projection. The data in each

## 4.8 Image Acquisition

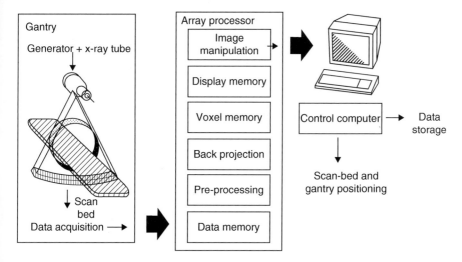

**Fig. 4.26** The rotation of the fan beam around the patient. The X-ray tube is on the *top*, and detector array is shown *below* (Courtesy: Bushberg et al., with permission from LWW Pub.)

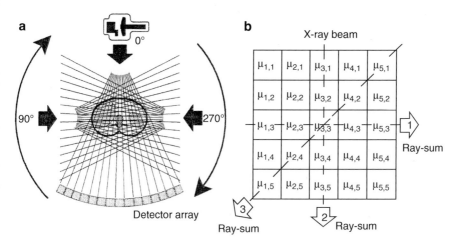

**Fig. 4.27** A block diagram of a typical CT system data collection and control showing pipeline processing in the array processor (DJ Powset et al., Thomson Sci. GB)

profile are treated as a mixed frequency, and the entire image reconstruction then takes place as a series of amplitudes in the frequency domain. A block diagram of a typical CT system data collection and control showing pipeline processing in the array processor is shown in Fig. 4.27. From Fig. 4.26, we can see that the gantry with the X-ray generator tube and the detector form the main components of the array connected to a computer. The array processor reconstructs the section images and displays the final results at the console.

## 4.8.4 Scattered Radiation

*Scattered* X-ray radiation has profound effects on the values of $P(x)$, $I(x)$, and attenuation coefficients ($\mu$). Correct design of the CT scanner and the use of post-patient collimation can reduce the multiple scattering reaching the detector that comprises the CT image.

In mammography, due to thick or dense breasts 40–80 % rays are expected to scatter which may reduce the image contrast by 30–40 %. New mammography units with a moving grid system reduce the scattering of radiation at the cost of increased radiation dose. Another problem that is frequently been observed during the use of slit radiography is the reduction of the background optical density. As a matter of fact, elimination of most of the scattered radiation produces a range of radiation exposures that may exceed the dynamic range of the film.

More recently, a technique called *magnification* (*or* divergence of the X-ray beam), which introduces an air gap between the patient and the image receptor, has become an accepted way to improve diagnostic accuracy in mammography examination. The technique also provides an escape of scattered radiation from the beam as it traverses the air gap so that less scattered radiation reaches the film. The focal spot of the X-ray tube is maintained within 0.2 mm for a smaller magnification factor (higher than 1.5). Figure 4.28 shows the grid, the emulsion film with the single screen, and the arrangements of the X-ray tube with a moly filter to radiate the breast during imaging process.

Scattered radiation can play havoc in screen film radiography, and scattering event can effectively cause redirection of X-rays that travel in straight line. Scattered radiation contributes a constant background fog level to the image. In digital images, where contrast can be arbitrarily increased by using window and leveling techniques, the effect of scattered radiation increases the noise in the image. For quantitative imaging applications, such as digital subtraction angiography and digital radiography, scattered radiation causes inaccuracies in transmission measurements and therefore reduces the performance of the techniques. The amount of scattered radiation energy can be expressed relative to the amount of primary radiation energy deposited at the same point in the image. The ratio is termed as *scatter to primary ratio* (SPR), and it increases with thicker patient anatomy.

Figure 4.29a shows that the light beams ($A$ and $B$) coming out of the detector are unaffected by scattering. In Fig. 4.29b, however, the effect of scattering $S$ appeared in the denominator. As a result, the contrast of the image will be affected by the amount of scattering ($S$).

To minimize the *scattered radiation*, three procedures are generally adopted: (1) The use of air gap between the patient and the detector system. Unfortunately, use of air gap imposes additional magnification of the object and ultimately degrades the resolution. Thus, the use of air gaps has certain limitations. (2) The use of radiographic grid, which is composed of a series of very small parallel bars of highly attenuating materials like lead (Pb) or tantalum (Ta). (3) The grid channels formed between the grid bars are made tall and thin in cross section, and the long axis is pointed toward the X-ray source. As the grid spaces are aligned

## 4.8 Image Acquisition

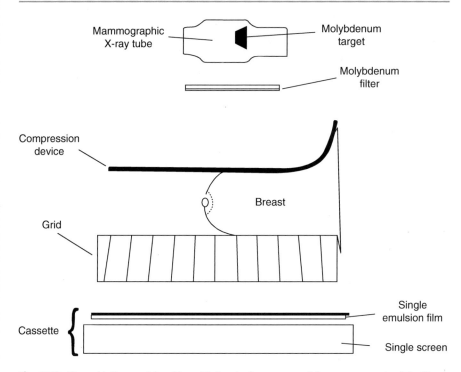

**Fig. 4.28** The grid, the emulsion film with the single screen, and the arrangements of the X-ray tube with filter to radiate the breast during mammographic imaging process

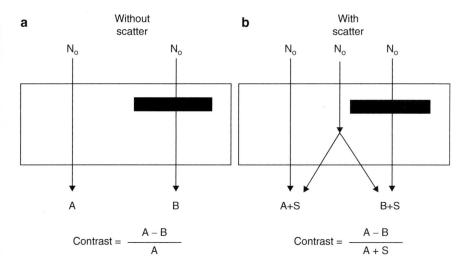

**Fig. 4.29** Scattering inside scintillator can influence the contrast of the image. (**a**) Without scattering and (**b**) with scattering

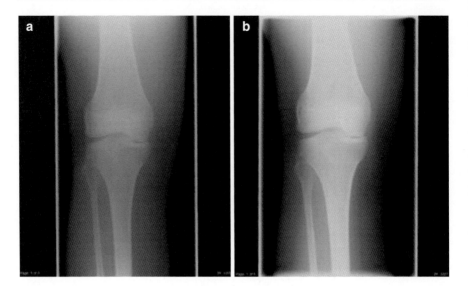

**Fig. 4.30** Two images of the AP projection of the knee phantom were obtained at 60 kV at the table top (*left*) and using the scatter removal grid (Bucky) (*right*) (Photo courtesy of RSNA; Courtesy: WIKI)

with the focal spot, the scattered radiation will not exit the patient but rather shows a broad angular distribution. However, a grid is only effective if it lets a higher percentage of X-rays through than scattered X-rays. The ratio of the primary transmission of a grid is called the *selectivity*. Beside selectivity, there are other factors that relate the properties of an effective anti-scatter grid. These are *primary transmission factor*, the *Bucky factor*, the *grid ratio*, the *grid strip density*, the *lead content*, and the *contrast improvement factor*. The Bucky factor $B$ ($=1/T_t$, where $T_t$ is the total transmission radiation) gives the factor by which the entrance dose with grid must be increased in order to compensate the reduction in dose at the image receptor caused by lower amount of scattered radiation and the small absorption of the primary radiation by the interspacing material [53].

In Fig. 4.30, two images of the AP projection of the knee phantom were obtained at 60 kV at the table top (left) and using the scatter removal grid (Bucky) (right). The final $S$ numbers of both images were ~350, indicating an air kerma incident on the computed radiography imaging plate of ~6 μ (0.6 mR) in both cases. The table top image on the left, however, required a technique of 3 mAs, whereas the one on the right required 10 mAs, since the scatter removal grid removes most of the scattered photons that emerge from the phantom. The Bucky factor is thus 3.3 (i.e., 10 mAs/3 mAs), and this is a quantitative measure of the increase in patient dose resulting from the use of the scatter removal grid. Note the improvement in image quality achieved by removal of most of the scatter radiation.

Grids are chiefly characterized by the grid ratio, grid frequency, and focal distance. The grid ratio is a measure of the height of the lead strip to the interspace distance and is a good measure of the selectivity of primary to scatter transmission. In general, a grid with a higher grid ratio will reject scatter better than a lower grid ratio, due to the limited angle that is allowed by the grid structure. However, a higher ratio grid typically has a higher dose penalty for its use (for screen film imaging). This is known as the "Bucky factor" which represents the increased dose to the patient when using a grid compared to not using a grid when the film optical density is matched. With digital imaging, there is also a dose penalty when a grid is used, and the benchmark is the signal to noise ratio (as opposed to film optical density).

The grid frequency is a measure of the number of grid lines per unit distance (inches or centimeters) and is in the range of 40–50 lines/cm (100–120 lines/in.) for low-frequency grids, 50–60 lines/cm (120–150 lines/in.) for medium-frequency grids, and 60–70+ lines/cm (150–170+ lines/in.). Low-frequency grids are used with systems having a moving grid assembly (known as a Bucky device) that oscillates during the exposure to blur the grid lines. Medium- and high-frequency grids are typically used with stationary grid holders (e.g., portable radiography and many digital radiography systems). High-frequency grid use is particularly important for digital radiography systems to avoid aliasing artifacts (see section on radiography artifacts) that arise from insufficient sampling of high-frequency patterns that are interpreted in the output signal as low-frequency (aliased) signals. The grid focal distance is determined by the angle of the lead strip geometry that is progressively increased from the center of the grid to the periphery, to account for the diverging primary X-ray beam emanating from the focal spot.

Typical focal distances are 100 cm (40 in.) and 180 cm (72 in.), although there are many specialized grid focal distances. Focal range is an indicator of the flexibility of grid positioning distance from the focal spot and is a function of the grid ratio and frequency. General-purpose grids for portable radiography have a fairly large range (e.g., 80–130 cm), while special purpose grids have a much narrower focal range. Grid artifacts arise from improper positioning of the grid device, such as tilting the grid at a non-perpendicular direction to the incident X-ray beam, not centering the grid to the X-ray beam central axis, using a focused grid outside the specified focal range, and placing the grid upside down (converging geometry is directed opposite of the focal spot).

Figure 4.31 shows the cross section of an anti-scatter grid, and the grid ratio is expressed as the ratio of the height ($H$) to the width ($W$) of the grid. The interspaces are aimed at the focal spot to make sure that the maximum number of the primary rays passes through the grid and strikes the detector.

In general radiographic studies, stationary grids do not pose any threat. However, when radiographs are digitized at sufficient resolution for the high-frequency line pattern from stationary grids, the high-frequency lines give rise to *moiré pattern*, which is the effect of aliasing between the digital sampling and the image of the grid bars. The scattered radiation emitted by the patient is dependent upon several other parameters like field of view (FOV) and the use of X-ray spectrum, besides the thickness of the patient. As a result, the selectivity and the performance of the grid depend on the measurement conditions. The third parameter that is suggested to reduce scattered radiation is aggressive field collimation.

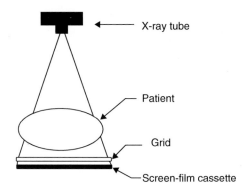
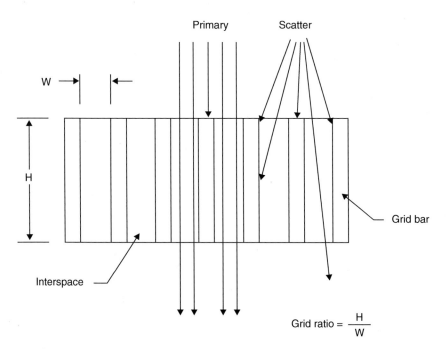

**Fig. 4.31** The schematic of an anti-scatter grid

Unfortunately, aggressive collimated X-ray has the risk of missing diagnostically important information of the image.

Three-dimensional (3-D) *single-photon emission computed tomography* (SPECT) images suffer from poor spatial resolution mainly due to 3-D scattering of the emitted photons. As a result, the interpretation of the images becomes very difficult for the clinician. However, image blind deconvolution technique has been reported to improve both the spatial and inter-slice resolution of the SPECT volumes [54]. On the other hand, Compton collimated imaging has been reported to improve the detection of $\gamma$-rays emitted by radioisotopes used in SPECT. The study reports the coincident Compton scatter events in both modules scatterer (silicon detector) and absorber (sodium iodide detectors are observed for photons emitted by $^{57}$Co source with principal gamma ray ($\gamma$-ray) energies of 122 and 136 keV) [55].

### 4.8.5 Spatial Resolution

Many of the assortment parameters that affect the spatial resolution in radiographic images also affect the resolution in CT. In radiography, the resolution characteristics in the image are approximately the same over the entire extent of the image. In CT, however, because of the way in which the image is reconstructed, the resolution characteristics change depending upon the location in the image. The resolution limitations, however, do not run strictly in the horizontal and vertical directions, but rather have radial components and circumferential components.

In CT, the photographic film is the most widely used detection medium in X-ray imaging applications. But the principal disadvantage of the X-ray film is its low sensitivity due to poor absorption (~1 % of the radiation is absorbed). Recently, much of the screen–film technology for acquiring images have been modified and improved by the use of a single intensifying screen positioned behind the X-ray film to reduce the spread of light before it reaches the film. A single screen eliminates image blurring caused by parallax and crossover that occurs with dual film configuration.

### 4.8.6 Optical Density (OD)

Automatic control of radiographic exposure, better design of detectors and better detector materials, and their placement between the breast and image receptor provide high sensitivity to variations in breast thickness and composition. These improvements have yielded excellent uniformity and consistency in the optical density of mammograms. However, image noise arises from the sources like quantum mottle, screen structure, X-ray to light conversion, film grains, and film processing artifacts. However, developments in film chemicals, speed, developer replenishment rate, and development time have reduced image noise considerably.

The subtle differences in X-ray attenuation of various tissues in the breast create special challenges for mammographic imaging. Dedicated units of X-ray mammography developed over the past couple of years provide smaller X-ray focal spots for better magnification, longer target receptor distances to reduce geometric unsharpness, more effective compression devices to eliminate motion and to separate tissue structures, moving grids to reduce scattered radiation, and automatic exposure control, and improved screen-film combinations have greatly improved the modern mammographic imaging system.

Film *radiography* was the first imaging modality used in radiology and became the most popular and widely used diagnostic tool, which is used even today. Recent advancement in screen, film, and the chemicals used in film photography has

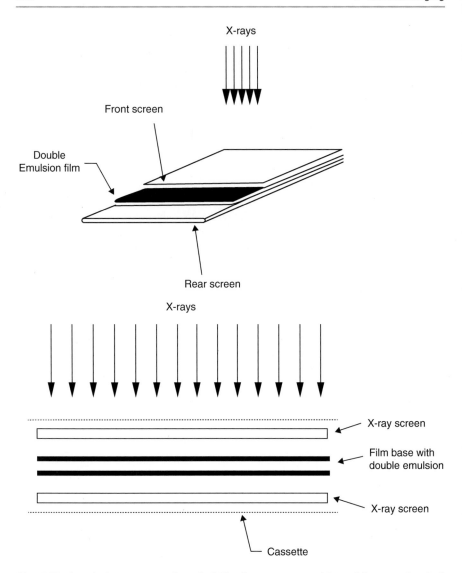

**Fig. 4.32** A typical arrangement for a dual film detector system with emulsion coated on both sides. The film is sandwiched by two intensifying screens

resulted in a marked improvement in radiographic image quality. *Film radiograph* is a *projection image*, meaning that the X-ray attenuation information concerning the *three-dimensional (3-D) anatomy* of the patient is projected onto the *two dimensions (2-D)* of the radiograph. For general radiography, the film emulsion on both the front and the back is tightly sandwiched by two X-ray intensifying screens (Fig. 4.32). The screen–film–screen sandwich is housed by the cassette, which provides a dark environment which transport and expose the film.

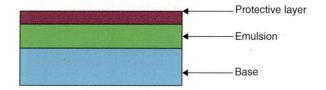

**Fig. 4.33** Schematic of a radiographic film structure

As the *optical density* (OD) at a point on the film is related to the thickness and composition of the patient along a line between the focal spot to that point on the film, low values of the optical density appear white, and higher OD values are darker. As a result, air-filled lungs being a low-density projection of tissue appear dark on the radiograph. It has been found difficult to determine the depth of a particular abnormal structure, and additional radiograph at different angles is necessary to locate unambiguously a particular lesion in all three dimensions.

Radiographic image contains a great deal of redundant information. Therefore, instead of storing the total value for each pixel, the difference value from its neighbor is stored, reducing the image's storage requirements. Pixels in image areas that contain little detail have neighboring pixels with the same value. These can be compressed by storing these values along with the number of pixel locations having this value as *run length*. Images can also be compressed by replacing small pixel regions with single pixels that are average of all values in the region.

## 4.9 Imaging Technology

Image processing is a rapidly growing area of computer science and has been growing fast for the last two decades. The technological advances of the imaging technology, computer processors, and mass storage devices have fueled the growth of the image processing. Fields that were using traditionally analog imaging system are now switching to digital imaging system for their flexibility and affordability. Important examples are medicine, film and video production, photography, remote sensing, and security monitoring.

### 4.9.1 Screen Film Technology

The film consists of a photographic emulsion made up of a mixture of *silver halide* grains and gelatin. The emulsion is deposited on a transparent base and protected by a thin protective layer. Figure 4.33 shows the schematic of a radiographic film structure.

The silver halide (generally bromide) varies in grain size depending on the application. On exposure to radiation, the halide emulsion interact with the photons, and after electron release, migration, and recombination with silver ions, the latent image is eventually built up with sensitivity specks of few silver atoms. Latent image centers act as a catalyzer in the chemical reaction by absorbing the developer solution, which contains electron transfer agents (reduction process) and also agents to regenerate the electron transfer (amplification). Development of the latent image to grains of viewable size results in an amplification factor of about $10^9$. Undeveloped crystals are dissolved in the fixing solution. Finally, washing and drying complete the film processes.

Sometimes a double layer of the emulsion is also recommended to increase the interaction of the radiation light. However, in double layer, some light crosses the film base and exposes the emulsion on the opposite side, which causes blurring of the image. For mammography, single-layer emulsion is mostly recommended where X-ray photons are primarily absorbed near the surface. The quality of the image besides the emulsion, the exposure energy, and the interactions of the light photons with the emulsion also depends on other variables like temperature and processing cycles. These variables are optimized to maximize the reduction of exposed crystals and to minimize reduction of nonexposed ones. The optical density ($D$) of the film is dependent on the ratio of the transmitted light ($I$) to the incident light ($I_0$). Mathematically, we can write $D$ as

$$D = \{-\log\,(I/I_0)\} \tag{4.7}$$

Again, the radiographic signal ($\Delta D = 0.434\ \Gamma\ (\Delta E/E)$), where $\Gamma$ is the film contrast and ($\Delta E/E$) is the exposure contrast. The value of $\Gamma$ is in between 2 and 3. On the other hand, the efficiency of the film screen is dependent on conversion efficiency, absorption efficiency, and transmission efficiency of the film.

The development of *screen film* systems has reached the stage where further advances within tightly bound constraints of quantum efficiency, speed, and image quality are not far from those set by bounds of physics, as opposed to the bounds of technology. Indeed, incremental but important advances in performance now come mainly from the insight cast by fundamental analysis and evaluation afforded by modern imaging theory. A further important factor is due to the parallel development of competing technologies, especially in the *digital domain* [56].

Indeed, recent development in computer technology has enhanced analysis of all radiological diagnosis more accurately than ever. As a result, the *digital X-ray imaging systems* that are having higher sensitivity due to increased absorption compared to photographic film have been introduced on the market. These systems are used in *medicine*, *dentistry*, *material inspection* and *analysis*, *security applications*, and *research in high-energy physics*.

### 4.9.2  Digital Imaging Technology

The digital image processing is composed primarily with extracting useful information from images. Ideally, it is done by computers with little or no human intervention. Image processing algorithms may be placed at three levels. At lowest level are those techniques that deal directly with the raw, possibly noise pixel values, with denoising and edge detection being good examples. In the middle are algorithms that utilize low-level results for further means, such as segmentation and edge linking. At the highest level are those methods that attempt to extract semantic meaning from the information provided by the lower levels, for example, handwriting recognition [57].

## 4.9 Imaging Technology

The advantage of digital image processing like *spatial filtering* to enhance the edges and structure of the images to reduce noise is its ability to manipulate raw image data using computer techniques. There are two types of digital spatial filtering applied: (1) *low pass*, to reduce noise, and (2) *high pass*, to improve the edge or contrast of the image. In the field of fluoroscopy and nuclear medicine, noise reduction of the images is particularly important, and it is achieved by averaging a number of images in sequence. It has been observed that *image noise* decreases as the *pixel* count increases. However, increasing pixel count does not improve *resolution*.

In *digital X-ray imaging technology,* which has revolutionized our modern X-ray imaging system, the size and spacing of the detector will influence the resolution. Smaller detectors show higher resolution. The thickness of the detector as well as the thickness of the slice plays important role in determining the resolution of the image. Now we can collect, store, analyze, and use more and more information at a faster and faster space. The forces behind the X-ray imaging are no exception. The many medical modalities, such as *computer tomography* (CT), *position emission tomography* (PET), *single-photon emission tomography* (SPECT), *magnetic resonance imaging* (MRI), and *ultrasound,* are inherently digital. However, standard X-ray radiography and fluoroscopy are still primarily based on analog technologies, specifically, screen–film and the image intensifier [57]. However, flat-panel detectors have emerged as the next generation digital technology. Since X-rays are not focused, the imager is necessarily on the scale of the object being imaged, which requires an enormous integrated circuit. Fortunately, a large-area IC technology based on amorphous silicon TFTs is available both on glass and flexible plastic substrates.

Active-matrix liquid crystal displays (AMLCDs) have grown tremendously for the past few years. The introduction of thin-film transistors (TFTs) has drastically improved the resolution and performance of the liquid crystal display because each pixel of the TFTs is driven by an individual TFT which has superior characteristics, such as the absence of ghosts-shadow image, a gray scale capability, and a large viewing angle, compared to conventional (passive matrix) LCD [58]. Plasma-enhanced chemical vapor deposition (PECVD) of the thin films of amorphous hydrogenated silicon (a-Si:H), silicon nitride ($SiN_x$), and heavily phosphorus-doped silicon followed by proper etching of the films are the two critical processes followed in the fabrication of TFTs in active-matrix liquid crystal display. $SiN_x$ is used as the gate dielectric, the channel, or the final passivation layer. The heavily phosphorus-doped silicon film between the a-Si:H and the metallic interconnection lines formed the ohmic contact. During plasma etching process, the etching rate, selectivity, and profile control are monitored and controlled for better performance of the devices. Large-area arrays (tens of centimeters in square) are required in medicine and security applications. Figure 4.34 shows a typical thin-film transistor array.

(a) *Application of TFT Array*: Direct and indirect active-matrix flat-panel imagers (AMFPIs) have been adopted for many X-ray image procure arrangements involving projection imaging (such as general radiography, cardiology,

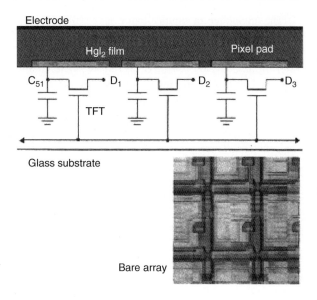

**Fig. 4.34** Thin-film transistor array

mammography, and radiotherapy), tomosynthesis (including breast and chest imaging), and cone beam computed tomography (e.g., radiotherapy, ENT diagnosis, and surgery). AMFPIs are based on a two-dimensional array of imaging pixels—each of which contains a thin-film transistor (TFT) which typically works as a switch and an indirect detection device, typically a discrete photodiode [59]. When a large positive bias voltage is applied to one of the gate lines, the TFT switches in the selected row are closed, causing them to conduct electricity. With the TFTs energized, each pixel in the selected row discharges the stored signal electrons on to the data line. At the end of each data line is a charge-integrating amplifier which converts the charge packet to a voltage.

(b) *Technical Details of TFT Array*: The electronics can be mounted either to the side of the array, out of the beam, or can be mounted behind the array and protected by a thin layer of lead for diagnostic and interventional procedures. The amorphous silicon has properties sufficient for detection of electronics, but it is not suited to subsequent signal processing. For this reason, every column and row of the array is brought to the glass, where it is connected to a standard crystalline silicon chip by means of a tape-automated bonding (TAB). The glass side of the TAB package may have 128 channels at a 50–100-μm pitch.

The amorphous silicon photodiodes are typically the n-i-p type. The layers in the photodiode consist of an electron-rich layer at the bottom, an intrinsic or undoped layer in the middle, and a hole-rich layer at the top. This type of amorphous silicon photodiode has the advantages of low dark current and a capacitance that is independent of the accumulated signal. Experimental observation shows that compared to crystalline silicon photodiodes like those used in CMOS imagers, the dark current in amorphous silicon photodiodes is orders of magnitude less.

**Fig. 4.35** Pixels covered by scintillator material (HgI$_2$) (Photo courtesy of author)

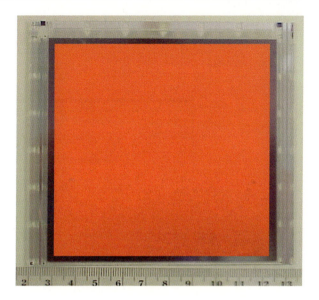

Recently, the AMFPIs are fabricated on flexible special type of plastic substrates [60]. Considerable efforts have been put to create TFTs on flexible substrate of a quality very similar to that conventional a-Si:H and small-area matrix devices. Such efforts are largely driven by the desire to produce light, robust, flexible, and inexpensive displays that also have potential to be scaled up to very large areas for applications in the context of consumer electronics, automobiles, signage, and advertising.

Figure 4.35 shows a TFT array covered by HgI$_2$ material deposited by adopting screen-printing technology. The screen-printing technology will be described in details in a separate chapter.

Figure 4.36 shows the schematic of the geometry used in Monte Carlo simulations. The illumination of the AMFPI can be performed either on the front or on the back surface. Monte Carlo simulations with CsI (Tl) as scintillating material on AMFPI have been applied for front and back of the glass substrates. Simulations do not show any improvements in MTF measurements when glass is used as substrates. However, the DQE values show much improvement when the AMFPI is illuminated on the front of the glass substrates. On the other hand, the plastic substrates do not show any change in both DQE and MTF values when illuminated either from back or front side of the AMFPIs [61].

Recently, there has been a lot of interest to introduce flexible and plastic substrates. The substrates are rugged, lightweight, thinner, amenable to roll-to-roll manufacturing, not prone to breakage, and most importantly cost-effective when compared to glass substrates. *Active-matrix organic light-emitting diodes* (AM OLEDs) display media offers the advantage of being a solid state and rugged structure for flexible displays in addition to the many potential advantages of an

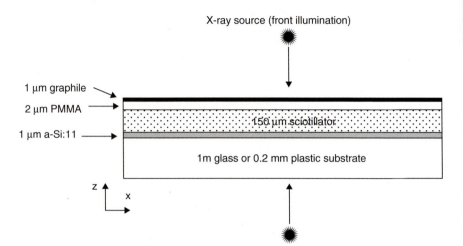

**Fig. 4.36** Schematic of the geometry used in Monte Carlo simulations

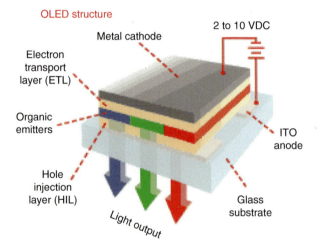

**Fig. 4.37** Schematic of an OLED structure

AM *OLED* over currently dominant *active-matrix liquid crystal display* (AM LCD) [62]. Figure 4.37 shows the schematic of an OLED structure.

Amorphous silicon (Si) thin-film transistors (TFTs) are the workhorses of the well-established AM LCD technology. However, the processing temperature (~300 °C) used in the conventional a-Si TFTs used in AM LCD could not be used with plastic substrates. In recent years, significant advances have been made in the process temperature reduction (from 300 to 150 °C), and a-Si TFTs have been fabricated on plastic substrates (e.g., polyethylene naphthalate, PEN) with a performance comparable to 300 °C process with respect to mobility, threshold voltage,

## 4.9 Imaging Technology

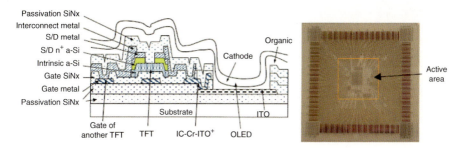

**Fig. 4.38** Cross section of integrated TFT/PLED (*left*) and TFT array on plastic substrate (*right*) (Reproduced with permission from SPIE, SPIE Symp. Orlando, FL)

and leakage current. However, a-Si TFTs are available only in the n-channel mode which restricts the choice of pixel addressing circuitry that could be utilized. Figure 4.38 shows the cross section of integrated TFT/PLED on the left and TFT array on plastic substrate on the right.

The fabrication procedure starts with the deposition of ITO followed by chromium (Cr) layer deposited by thermal evaporation and patterned as the bottom electrodes of the TFTs. This process is followed by the deposition of silicon nitride (SiN$_x$), intrinsic a-Si and n$^+$ a-Si layers by plasma-enhanced chemical vapor deposition (PECVD) method. Then a second layer of Cr is deposited as the source/drain metal contacts. After patterning of the n$^+$ a-Si layer to define the channel and patterning of the intrinsic a-Si to define the TFT island, contact holes are etched through SiN$_x$ for contacts between interconnect metal and gate metal. In order to protect the active electronics, a passivation layer of SiN$_x$ is deposited and a window is etched to expose ITO for the PLED.

An AM OLED pixel driver uses a PMOS transistor using a bottom emission with the light emission through indium tin oxide (ITO) anode at the bottom. Connecting the OLED to the drain side of the TFT ensures that the gate–source ($V_{DD}$ bus) bias across the drive TFT is held constant to achieve a constant current drive. However, when using the n-channel a-Si TFT, the power supply ($V_{DD}$) bus will be connected to the TFT drain, and the OLED will be on the source side the TFT. As a result, the TFT gate voltage will be divided between the TFT gate–source voltage, $V_{gs}$, and the OLED ($V_{OLED}$). Any variations in the OLED devices across the display will result in variation in $V_{gs}$ and thus variations in the drive current and the luminance. Thus, an n-channel TFT cannot provide a constant current when the OLED properties vary across the display surface.

Figure 4.39 shows two TFTs ($T_1$ and $T_2$) attached to a photodiode (PLED). Photons striking the photodiode are converted into two carriers of electrical charge, called electron–hole pairs. The generated charge migrates to the pixel electrode in accordance with the polarity of the bias being applied to the X-ray photoconductor and is stored in a storage capacitor ($C$) within the TFT array. By subsequently scanning the TFTs line by line, the charge information stored in the storage

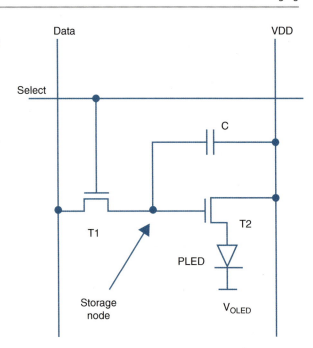

**Fig. 4.39** One of the pixels of the 64 × 64 pixel array of the test panel

capacitor can be read out from the data bus lines. The data bus line termination connects to charge sensitive amplifiers (CSAs) and A/D converters, and the scanned charge information is converted to digital image signals and output sequentially.

### 4.9.3 Working Principle of Digital Imaging

In digital X-ray imaging process, a discrete image representation is produced ($D_{ir}$), where $D_{ir} = D\ (i\Delta x,\ r\Delta y)$, $m$ and $n$ are integers, and $\Delta x$ and $\Delta y$ are the sample distances in the $x$ and $y$ directions, respectively. The discrete Fourier transform of the array ($D_{ir}$) is given by

$$D_d(v_x, v_y) = \sum_{i,r=-\infty}^{\infty} D_{ir}\{\exp\left[-j\,2\pi\,(v_x\,i\Delta x + v_y\,r\Delta y)\right]\Delta x \Delta y \qquad (4.8)$$

The array $D_{ir}$ can be recovered from $D_d\ (\Delta x,\ \Delta y)$ by the inverse Fourier transform. However, there is an important relationship between the discrete Fourier transform $D_d$ of the array ($D_{ir}$) and the ordinary, continuous transform of the function $D\ (x,y)$ that was sampled to produce ($D_{ir}$). In the frequency domain ($v_x$, $v_y$), $D_d\ (v_x, v_y)$ equals the sum of $D\ (v_x, v_y)$ evaluated at integer multiples of ($v_x, v_y$):

## 4.9 Imaging Technology

$$D_d(v_x, v_y) = \sum_{k,l=-\infty}^{\infty} D(v_x + k/\Delta x, v_y + 1/\Delta y) \qquad (4.9)$$

The frequency pairs $(v_x + k/\Delta x, v_y + 1/\Delta y)$ are termed *aliases* of the pair $(v_x, v_y)$. In any sampled data, the highest frequency which can be obtained in the sampling process is called *Nyquist* frequency $v_N$, where $v_N = \frac{1}{2} \Delta x$ or $\frac{1}{2} \Delta y$ for $x$ and $y$ directions, respectively. It is seen that the frequencies in the sampled image that are greater than $v_N$ reappear as lower spatial frequencies. Here, the spatial frequency values higher than $v_N$ are indistinguishable from the appropriate submultiple frequency with lower value.

On the other hand, spatial resolution gives the smallest size of the object, which can be resolved, whereas the noise sets limit on the smallest contrast differences that can be detected. The quality of the data transfer in imaging system in terms of *detective quantum efficiency* (DQE) [63] as has been introduced by Rose and Clark Jones [64] is a measure of the detector efficiency as a measure of detector's performance, and it can be defined as

$$(DQE) = \{(SNR)_{out}^2 / (SNR)_{in}^2\} = 1 \ (1 + B) \qquad (4.10)$$

where the relative variance $B$ is defined in terms of gain $m$ and relative variance $b$ as [65] and can be expressed mathematically as

$$B = \{b_1 + (b_2/m_1) + (b_e/m_1.m_2) + (b_3/(m_1, m_2, m_e)) \cdots$$
$$+ (b_n/(m_1, m_2, \ldots m_{n-1}, m_e))\} \qquad (4.11)$$

Equation (4.4) requires the gain and true variance of the following stage, $m_3$ and $(b_3.m_3^2)$, respectively, to be proportional to its input signal. The gain $m_{e, s}$ and variance $b_{e, s}$ of each sublayer can be estimated as the probability of primary photon interaction $(m_2, s/m_2)$ in each sublayer so that the total gain of the escape stage is

$$m_e = \sum [(m_{2,s} \cdot m_{e,s})/m_2] \quad \text{for} \quad s = 1 \text{ to } n, \qquad (4.12)$$

where $n$ is the number of sublayers present in the film and the total noise $N$ is given by

$$N^2 = \sum \left\{ m_{2,s}.m_{e,s}^2.b_{e,s} \right\}/m_2 \Big] \qquad (4.13)$$

One of the design criteria for X-ray imaging systems is to ensure whether the quantum noise predominates. When it does, one should ensure that the overall quantum gain whether light photons or electrons, at intermediate conversion stages, must be greater than any losses [66]. It can be shown that [67] if I is the intensifier X-ray exposure rate, in mR/s, $G$ is the conversion factor of the intensifier in $cd/m^2$

per mR/s, $V$ is the optical coupling transmittance, and f is the lens number, then we can write

$$(1/10) < IGV/f^2 < 5 \tag{4.14}$$

As the X-ray input exposure is increased, the light falling on the detector must be adjusted to maintain the inequality. This is undertaken by adjusting $f$ in Eq. (4.14) by means of diaphragms.

The probability of generating a characteristic K$\alpha$/$\beta$ photon from a phosphor (scintillator) will depend upon the probability of backward ($p_B$) and forward ($p_F$) escape probabilities, where the total escape probability $p$ can be expressed as $p = (p_F + p_B)$. The probability of photon escape is found to occur at different photon energies according to the constituents of the phosphor material.

## 4.10 Dependency of the Quality of the Medical Imaging System

### 4.10.1 Noise

The random variation in image brightness is designated as *noise*. Almost all medical images contain some sort of visual noise that tends to produce variation in brightness, and the image appears *snowy*. It has been observed that the *radiography images* are less noisy than *CT* or images obtained by *fluoroscopy*. The most significant feature of noisy pictures is the reduced visibility of certain features within the images. Though the noise in a picture is undesirable, yet at some point, one has to compromise certain other factors of the medical imaging. As for example, in X-ray imaging, the primary compromise is with patient exposure and dose. Likewise, in MRI and nuclear imaging, the primary concern is with imaging time [68].

X-ray and gamma ($\gamma$) ray photons are received in a random fashion on a receptor, and the random distribution of the photons on the receptor surface causes noise, which is designated as *quantum noise*. The quantum noise level is determined by the concentration of photons actually absorbed by the receptor on a particular surface area rather than the number of photons delivered to the receptor. The photon concentration or the exposure time is determined by the *sensitivity* of the receptor, which varies over a considerable range in X-ray projection imaging (radiography and fluoroscopy) [69].

The sensitivity of a radiographic receptor (film cassette) is determined by the characteristics of the film and the screen and the way they are matched. However, increasing the receptor sensitivity by changing any factors that decrease the number of photons absorbed by the receptor will increase quantum noise. As a matter of fact, quantum noise is usually the factor that limits the use of highly sensitive film in radiography. Thus, one can say that the most efficient receptor will be able to convert all the absorbed incident photons into more visible light photons without increasing quantum noise at the same time reducing the exposure dose. Absorption efficiency,

# 4.10 Dependency of the Quality of the Medical Imaging System 237

however, reduces *noise* but at the cost of *blurring* of the image, which tends to blend each image point with its surrounding area. However, sometimes image blurring is admitted to reduce the visibility of noise as seen in digital image processing [70].

There is a distinct difference between film–screen and digital radiographic receptors with respect to quantum noise. Digital radiographic receptors do not have a fixed sensitivity like film–screen receptor. One of the most valuable characteristics of a digital receptor is a wide exposure dynamic range, meaning that errors do not result in image with loss of contrast of the film. Another advantage of the digital radiography is its ability to capture full range of exposure coming from patient's body where there are a large number of variations in body density and penetration, such as chest.

In digital radiography, it is important that the appropriate exposure and proper techniques be applied for each procedure. The optimum exposure for digital imaging is described as *an acceptable noise level without unnecessary exposure to the patient*. The digital radiographic community uses an arbitrary factor $S$ to quantify the image and display the exposure information related to the quality of the image. The high $S$ (like 1,000) indicates that the image is formed with a low exposure, and a noise is expected. Similarly, an $S$ value of 50 will indicate a high exposure, and consequently, the noise will be less with a better picture.

In conventional fluoroscopy imaging system, the receptor sensitivity is on the order of 1–10 μR. The relatively low exposure rate is certainly a plus, but the image becomes noisy. Moreover, in normal fluoroscope viewing, we do not see one image frame at a time but an average of several frames. Recently, several research laboratories are engaged in developing a better receptor and image processing technologies to decrease the radiation dose without compromising the quality of the image (less noise). Unfortunately, there is no known way to overcome the fundamental limitation of quantum noise level.

In TV, the video images often contain noise, known as *snow*, which is in the form of random electric currents often produced by the thermal fluctuations within the device of the electronic circuitry. This is particularly noticeable when the input signal is weak and is seen very much in fluoroscopy imaging.

Three fundamental concepts, *resolution, contrast* (Fig. 4.40), and *noise*, help to clarify the presence or absence of an anatomical object inside body. The *resolution* for a conventional radiographic film is defined as the number of line pair per mm it can resolve. Mammography demands a higher resolving power film. As the X-ray photon energy increases, it penetrates further into the phosphor material, causing diffusion of deeper light event, so there is a significant geometrical broadening when it reaches the film emulsion at the phosphor surface. Using thinner phosphor screen can control light diffusion, but sensitivity of a film is greatly reduced, since only a fraction of the beam is absorbed. On the other hand, low keV photons react at the surface of the plate; the photon undergoes minimal diffusion, giving rise to a sharp image (10–15 Lp/min).

*Geometrical unsharpness, movement unsharpness, and radiographic unsharpness influence spatial resolution in radiographic image.* The geometrical unsharpness is influenced by distance between X-ray tube, patient, and image

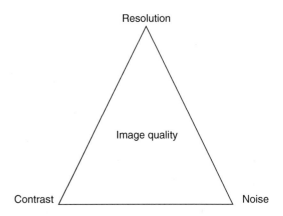

**Fig. 4.40** Three main factors that control the image quality

surface as well as X-ray focal spot. The second one is due to movement of the organ and patient. The radiographic unsharpness is due to poor contact of film and deep diffusion of light, and as a result, image blurring is often noticed.

*Point spread function* (PSF) is the simplest test of spatial resolution. A PSF has limitations as an image test tool since it only represents discrete points on the image surface. A line source, in the form of metal (10 μm) slit, provides a better indication regarding image quality and is defined as line-spread function (LSF). It is easier to make a line source using two knife edges with 10-μm gap in between.

### 4.10.2 Contrast

*Contrast* means *difference*, which can be in the form of different shades of gray light intensities or colors. Image contrast, on the other hand, is a measure of difference between adjacent regions of the image, and in medical science, it is used to assess differences between adjacent tissues. The contrast of an image depends on the variations in the flux of photons arriving at the receptor and is thus related to the yield of absorbed photons per incident photons. On the other hand, the physical contrast of an object must represent a difference in one or more tissue characteristics. For example, in radiography, object can be imaged relative to their surrounding tissue if there is adequate difference in either density or atomic number and if the object is sufficiently thick. For X-ray imaging, contrast between fat and muscle is low, whereas contrast between bone and soft tissue is high.

*Noise* in the images is more noticeable if the overall *contrast* transfer of the image system is increased. This is particularly important when using image displays with adjustable contrast, such as some video monitors used in fluoroscopy and the viewing window in CT, MRI, and other forms of digital images.

It is difficult to compare the contrast sensitivity of various imaging methods. However, certain methods do have higher sensitivity, for example, CT has contrast sensitivity than conventional radiography (because CT has the ability to image soft

**Fig. 4.41** The concept of contrast

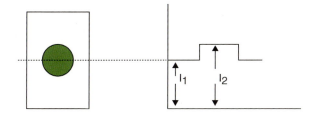

tissues). Thus, one can say that a system with low-contrast sensitivity will allow one to visualize only objects with relatively low inherent physical contrast.

The contrast sensitivity of human viewer changes with viewing conditions (like background brightness, object size, viewing distance, glare, and background structure). For general image viewing, view boxes should have a luminescence of at least 1,500 nits, whereas 3,500 nits is recommended for mammography. The detectability of an object is more closely related to the angle it forms in the visual field, though background brightness plays an important role especially when the viewing object is small. On the other hand, the distance of the object from the field of view generally peaks at a distance of approximately 2 ft. The glare or the bright areas in the field of view creates some undesirable effects on the viewing of the object. Last but not least is the background structure or texture of an object's background, because a smooth background always produces maximum visibility.

In Fig. 4.41, $I_1$ is the transmitted light intensity for the background and $I_2$ is that for the disk. The contrast of the disk relative to the background is then can be expressed mathematically as

$$\text{Contrast} = (I_2 - I_1)/I_1 \qquad (4.15)$$

It is difficult to quantify the performance of an image only in terms of factors like *noise, contrast,* and *resolution*. No one can say that a *screen–film* combination has a better *spatial resolution* than a *CT scanner*, and therefore, it is better for all medical imaging applications. At the same time, it is hard to say that a CT scanner is better than a film–screen system because it provides higher contrast or that MRI is better than CT because of its high soft tissue contrast. For these several reasons, there are some other parameters to be defined in order to quantify an image. As for example, the contrast with which an object is represented decreases as the object size approaches *full width at half maximum* (FWHM) of the system point spread function (even though its thickness is unchanged). In order to clarify the concept of *FWHM*, an intermediate term like *modulation transfer function* (MTF) is introduced. MTF can be defined as the contrast produced by an imaging system as a function of spatial frequency of the object.

Another linking concept is *Wiener spectrum* (WS), which measured level on noise as a function of spatial frequency and equals the *Fourier transform* (FT) of the

240 4 Medical Imaging

autocorrelation function in a uniformly exposed radiographic image. The WS can be represented mathematically in terms of MTF as

$$WS = (G^2/n) \, (MTF)^2 \tag{4.16}$$

where $n$ is the number of photons absorbed and $G$ is the gamma factor of the film. For radiology, 0.2–1 Lp/mm frequency range has been accepted since these frequencies are visible and their effects are noticed [50].

The final concept, which is needed to evaluate an image on the basis of spatial resolution, noise, and contrast, is known as *Rose model* [71]. Rose model plays an important role in establishing the fact that image quality is ultimately limited by the statistical nature of the quanta. However, this model has some practical limitations in analyzing the most modern medical imaging systems.

The model provides a mathematical relation between object area ($A$), signal to noise ratio ($s$), and number of photons use to image the object ($n$), which is equal to ($\varphi/A$, $\varphi$, is the photon fluence), and contrast ($C$) can be written as

$$C = \{s/(n)^{1/2}\} \tag{4.17}$$

For a special case of uncorrected background quanta, noise is described by Poisson statistics, and the Rose signal to noise ratio (SNR) is proportional to the contrast and to the square root of the background quanta multiplied by area ($A$). This result led to the general expectation that the lesion detectability should be proportional to the contrast of the object and to the square root of object area and radiation dose.

The output signal to noise ratio ($S_{out}$) is defined in terms of the quantum efficiency of a radiation detector, and the output noise includes the statistical noise in $S_{out}$ as well as readout noise of the system, that is,

$$\text{Out put noise} = N_{out} = \left[\sigma_{stat} + \sigma_{rout}^2\right]^{1/2}$$
$$= \left[S_{out} + \sigma_{rout}^2\right]^{1/2} = \left[(QE) \left(S_{in} + \sigma_{rout}^2\right)\right]^{1/2} \tag{4.18}$$

where $S_{out} = QE \, (S_{in})$.

The Rose model can identify a low-contrast object in a noisy image. It also helps to correlate the concept of *receiver operating characteristic curve* (ROC) which is considered to be the ultimate test of an imaging system. It enables display *sensitivity*, *specificity*, and overall accuracy of a particular display system to be compared with another system so that decisions can be made regarding the success or improvement that the new system may give and a measure of its diagnostic *sensitivity* and *specificity*.

The complexity and the sophistication of medical imaging systems have increased dramatically over the past several decades. The stochastic nature of *image quanta* imposes a fundamental limitation on the performance of photon-based imaging

# 4.10 Dependency of the Quality of the Medical Imaging System

system. The relationship between mean number of image quanta ($n$) is related to the contrast C and the *signal to noise ratio* (SNR) as shown in Eq. (4.17). On the other hand, *Fourier transform* (FT) has been used to express spatially varying signals (images) in medical systems in terms of spatial frequencies (cycles/mm). The sinusoidal image patterns are transferred with only a scalar change in amplitude, and Rossmann and his contemporaries expressed these factors as spatial frequency-dependent *modulation transfer function* (MTF) [72, 73]. The theory of *linear imaging systems* to describe many fundamental principles and characteristics has been used in radiography, computed tomography (CT), nuclear medicine, ultrasound, and other areas. However, use of *MTF* is less established in magnetic resonance imaging (MRI) system, but may have significant role to play [74].

Rose-based concept of detective quantum efficiency (DQE) of a linear imaging system can be correlated with noise equivalent number of quanta (NEQ). The NEQ is a metric describing image quality where as DQE describes the ability of a particular imaging system to effectively use all available quanta.

$$\text{DQE} = \text{NEQ} \, (fs)/(\bar{n}) = \left\{ (\text{SNR})^2_{\text{out}} (fs/\bar{n})/(\text{SNR})^2_{\text{in}} (fs/\bar{n}) \right\}$$

$$= \left\{ (\Delta M)^2 / \sigma_m^2 \right\} / \left\{ (\Delta N)^2 / \sigma_n^2 \right\} \tag{4.19}$$

where $fs$ is the spatial frequency (cycles/mm) and $\bar{n}$ is the average number of quanta per unit area.

The accuracy of detecting a display system is its clinical success for detecting diseases. Detecting a positive lesion (true positive) or reporting a negative finding (true negative) becomes more difficult, and the possibility of making mistakes will depend on the quality of the image and on the viewing conditions and the skill of the observer. A test with high specificity is good for detecting the absence of the disease, preventing unnecessary treatment. In order to minimize the probability of false-positive detection, about 100 images are taken in series and are given to a panel of observers for judging.

A typical ROC curve is shown in Fig. 4.42. The TPF along the $Y$-axis indicates the percentage of patients correctly identified as positive findings, and along $X$-axis, the percentage of patients that are falsely called positive but are really normal. The upper right corner of the curve A shows ~100 % sensitivity and ~0 % specificity, and we can call the interpretations of the results are perfect or ideal. The curve B gives a purely random result (50:50), and curve C is a practical display of the imaging interpretations (average of 100 images).

Figure 4.43 illustrates imaging concepts used to model imaging system performance. The most basic concepts of noise, contrast, and spatial resolution are placed outside, while the intermediate linking concepts of signal to noise ratio (SNR), Wiener spectrum, and modulation transfer function (MTF) are placed inside the diagram, bridging pairs of basic concepts. Finally, the Rose model, the ROC analysis, and the related contrast detail analysis are shown inside the diagram.

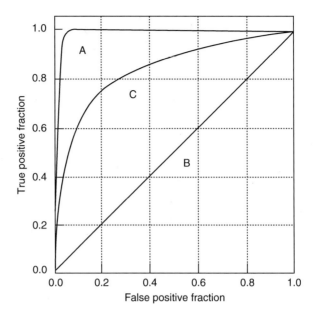

**Fig. 4.42** A series of ROC curves. The plots are the true positive fraction (*TFF*) of an observer response against the false-positive fraction (*FPF*)

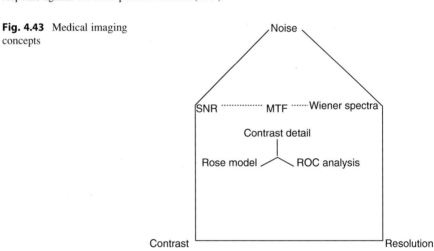

**Fig. 4.43** Medical imaging concepts

## 4.11 Energy Resolution

In many applications, the energy distribution of the incident radiation of a detector is of prime interest. This part of the measurement comes under radiation spectroscopy, and one important property that it measures is the response of the detector to a monoenergetic incident of radiation. The figure illustrates the differential pulse

## 4.11 Energy Resolution

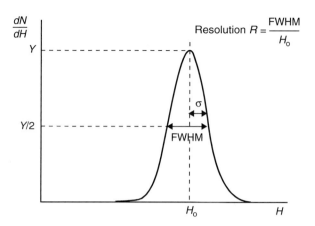

**Fig. 4.44** A Gaussian-shaped peak and the standard deviation ($\sigma$) and FWHM

height distribution of the detector, which can be defined as the response function of the detector for the energy used in this distribution. The most common method of displaying pulse amplitude information is to plot the ratio of the differential number of pulses to differential amplitude increment (d$N$/d$H$) against the vertical scale, which represents pulse amplitude ($H$) in volts. The sharper the peak, the smaller will be the fluctuations within the pulse. On the contrary, a wider pulse with more fluctuations gives a smaller peak. The full width at half maximum (FWHM) is defined as the width of the distribution of the pulse at a level which is just half the maximum ordinate of the peak.

### 4.11.1 Full Width at Half Maximum (FWHM)

The figure below (Fig. 4.44) shows the plot of differential number of pulses observed with differential amplitude increment of pulses d$H$ and represented by (d$N$/d$H$) along $Y$-axis. Along $X$-axis, linear pulse amplitude ($H$) has been plotted. The FWHM is illustrated on the figure, which is defined as the width at the distribution at a level, which is just half the maximum ordinate peak.

The resolution $R$ can be expressed mathematically as

$$= (\text{FWHM}/H_0) = \{(2.35 K \, (N)^{1/2})/(KN)\}, \tag{4.20}$$

where $K(N)^{1/2} = \sigma$ and $N$ is the number of charges. From the above equation, one can see that the resolution ($R$) is dependent on $N$. On the other hand, the energy resolution of a detector is expressed in terms of a new factor, which has been named as *Fano factor* as

$$R = 2.35 \, (F/N)^{1/2} \tag{4.21}$$

**Table 4.4** Some factors that influence *contrast* of an image

| Radiographic/subject contrast | Image contrast |
|---|---|
| Tissue density | The radiographic contrast |
| Tissue thickness | X-ray film characteristic |
| Electron density of tissues | Screen characteristic |
| Effective atomic number ($Z$) | Windowing level of CT and DSA |
| Energy of the X-ray (keV)[a] | |
| X-ray spectrum (after filtering) | |
| Scatter rejection | |

[a]Higher energy, greater penetration depth, and both the bone and soft tissue become transparent. On the other hand, lower kVp can enhance image contrast; however, higher kVp is required when the contrast is very high (chest). The current in milliampere (*mA*) controls the quantity of X-ray (intensity), and when the film is too white, *mA* control can bring up the intensity and optical density

where $F \equiv$ (observed variance in the number of charge carriers/Poisson predicted variance). Table 4.4 shows some factors that influence *contrast* of an image.

## 4.12 Digital Image Acquisition System

The digital image acquisition system can be operated in two modes, fluoroscopic and radiographic. In fluoroscopic mode, image is observed in two monitors. The image-hold feature of the digitizer average at least 10 video frames is displayed on the second monitor and stored on hard disk.

In digital radiographic mode, the digital images are generated by digitizing the output of a progressive scan video circuit and stored on a hard disk of the video digitizer. The processed image then can be displayed and digitized for storage. In digital radiographic mode, the collimation system automatically limits exposure to correspond to the area displayed fluoroscopically. An automatic time exposure is used in radiographic mode. To ensure that the digital film images are similar to the video display terminal, a calibrated light meter is used to measure the optical densities of each step. The brightness, contrast, and gamma slope adjustments of the display are adjusted until the optical densities of the film and the video display are same. Table 4.5 shows the entrance exposure rates in the fluoroscopic mode, and Table 4.6 shows the entrance exposure values in the conventional and digital radiographic modes.

Another important research field is *hadron therapy*, which uses detectable beams of heavy particles (protons, neutrons, pions, and ions) which are highly effective when nearby critical organs do not allow another therapy.

*Image Reconstruction*: *Reconstructed* images typically have $64 \times 64$ or $128 \times 128$ pixels, with the pixel sizes ranging from 2.8 to 7 mm. The number of projections acquired is chosen to be approximately equal to the width of the resulting images. However, it is to be remembered that during scanning there should not be any movement of the patient. Movement can cause significant

# 4.12 Digital Image Acquisition System

**Table 4.5** Entrance exposure rates in the fluoroscopic mode

| Voltage $\times (10^3)$ | Current $(10^{-3})$ | Entrance exposure rate (R/min)[a] |
|---|---|---|
| 73 | 1.8 | 1.4 (0.059) |
| 82 | 1.8 | 1.7 (0.073) |
| 90 | 2.8 | 2.5 (0.107)[b] |
| 105 | 2.8 | 5.1 (0.219)[c] |
| 105 | 2.8 | 5.2 (0.223)[d] |

[a]Values in the parenthesis $\times 10^{-4}$ (C/kg)/s are equivalent SI units
[b]Average child
[c]Average adult
[d]Maximum

**Table 4.6** Entrance exposure values in the conventional and digital radiographic modes

| Voltage $\times (10^3)$ | Digital/conventional | Ratio (digital/conventional) |
|---|---|---|
| 123 | 57.1 (0.147)/49.8 (0.128) | 1.4 (0.059) |
| 103 | 56.5 (0.146)/72.3 (0.186) | 1.7 (0.073) |
| 86 | 66.8 (0.172)/109.0 (0.281) | 2.5 (0.107)[a] |
| 76 | 97.5 (0,252)/294 (0.758) | 5.1 (0.219)[b] |
| 62 | 216 (0.556)/515 (1.329) | 5.2 (0.223)[c] |

Cox et al. [75]. Courtesy of RSNA 1990 from RSNA annual meeting 1988
Note: Exposures were measured for 33-cm field of view. Values in parenthesis $\times 10^{-4}$ C/kg are equivalent SI units
[a]Maximum
[b]Average adult
[c]Average child

degradation of the reconstructed image quality, although movement compensation reconstruction techniques can help with this. At the same time, uneven distribution of the radioisotope has the potential to cause artifact. Attenuation of the gamma rays within patient can lead to significant underestimation of activity in deep tissues, compared to superficial tissues. However, modern SPECT/CT has been tagged as a modality solution for this type of problem.

In SPECT, the gamma rays should travel normal to the surface of the scintillator. Fan beam and cone beam collimators are no good because the impinging photons on the detector will not be parallel and will introduce artifact. The photons that impinge on the detector ring at an oblique angle can penetrate into scintillator crystal at a significant distance before they interact and are detected. The magnitude of these errors is increased with the crystal thickness and the angle of incidence. Therefore, attenuation lengths are desired to minimize crystal thickness while maintaining high efficiency [76].

Medical CT systems rely on two basic imaging modalities: transmission tomography and emission tomography. In the former case, beams of X-ray are passed through the body being imaged from various positions and at various angles. Each beam is detected on the side of the body opposite from the beam, and the detected beam intensity is compared with the intensity of the source. If we define the logarithm of the ratio of the intensity of the detected beam ($I_D$) to the intensity of the transmittance beam as ($I_T$), we can write the following equation as:

$$I' = \int_L f(x, y) \, du \qquad (4.22)$$

where $f(x,y)$ is the absorption coefficient of the object at point $(x,y)$ and $L$ is the line along which the beam travels. The CT image scanner will obtain different $I'$ values for various lines $L$ and will use the information to compute an approximation to $f(x,y)$ throughout the object.

Emission tomography, on the other hand, measures radiation emitted by a radioactive substance by a movable array of detectors. In absence of attenuation, and with ideal detectors, the detected intensity will be the sum of the linear intensity and can be expressed as

$$I = \int_L f(x, y) \, du \qquad (4.23)$$

where $f(x,y)$ represents the concentration of the radioactive substances at $(x,y)$ and $L$ represents the line passing through the detector perpendicular to the detector arrays.

SPECT images are constructed from the observation data by gamma ($\gamma$)-camera following an orbit around the patient body at regular spaced angles. At each position, the registered photon counts at the ($\gamma$)-camera are conveniently processed and stored as discrete 2-D images. Bayesian reconstruction methods have been extensively used to reconstruct medical images since they can improve the reconstruction with respect to the classical, nonstatistical methods, such as filtered back projection (FBP) and algebraic reconstruction technique (ART) [77, 78]. Given an estimate of the line process, $L$ and observation of $Y$, the image estimation $X$ is performed by using the deterministic method and it is extended for the use on tomographic images and modified to take account only neighbors not separated by an active line element [79].

## 4.13 Summary

The emerging and rapidly growing field of medical imaging has provided new opportunities to directly visualize the biology of living organism without surgery. The medical imaging technology, which is a part of molecular engineering, is not a new technology at all. In fact, fluorescently and radioactively labeled contrast agents that bind to specific proteins, particularly cell surface receptors, have been employed for decades in cell culture, tissue slice, and autoradiographic studies. As a whole, these techniques have been transferred into the nuclear imaging modalities such as positron emission tomography (PET) and single-photon emission computed tomography (SPECT), into humans, with an early emphasis devoted to studying spatially localized neuroreceptor systems in the human brain [22, 80, 81]. With the advancement of science, especially the materials science and the engineering, all major imaging modalities are contributing to the new field, each with its unique mechanisms for generating contrast and trade-off in spatial resolution, temporal resolution, and sensitivity with respect to biological process of interest.

The critical step of medical imaging, which is within the field of molecular and genomic imaging, is the approach to define the gene or protein (generally known as target) that is to be imaged. Once the target is chosen, the next step is to develop techniques to image that target in a living object. Above all, at the imaging system level, there is a need to improve imaging performance, either by increases in spatial or temporal resolution, sensitivity, or commonly both. The ability to accurately quantify the medical imaging studies is of great importance, because characterization of the quantitative accuracy or reproducibility of the medical imaging methods will build a solid foundation required for the future anticipated growth of medical imaging science for the betterment of human health and diseases.

# References

1. Semmler W, Schwaiger M (2008) Molecular imaging, vol II. Springer/LLC, New York and also Cherry SR (2004) In vivo molecular and genomic imaging. Phys Med Biol 49:R-13 and Ventor JC, Levy S, Stockwell T, Remington K, Halpern A (2003) Massive parallelism, randomness, and genomic advances. Nat Genet 33:219
2. Bushberg JT (2001) The essential physics of medical imaging. Lippincott Williams and Wilkins, Pittsburgh, PA and also Weissleder R (2002) Scaling down imaging: molecular mapping of cancer in mice. Nat Rev Cancer 2:11
3. Butel JS (1999) Viral carcinogenesis: revolution of molecular mechanisms and etiology of human disease, carcinogenesis. Oxford J 21(3):205
4. Cormen TH, Leiserson CE, Rivest RL, Stein C (2009) Introduction to algorithms, 3rd edn. 3rd ed. The MIT Press, Cambridge. Chap. 5, p 79 and also Welch MJ, Redvantly CS (eds) (2002) Handbook of radiopharmaceuticals. Wiley, New York
5. Darnell J, Lodish H, Baltimore D (1986) Molecular cell biology. Scientific American Books Inc. Pub., New York. Chap. 23, p 1035 and also Greer and LF, Szalay AA (2002) Imaging of light emission from the expression of luciferases in living cells and organisms: a review. Luminescence 17:43
6. Halperin EC, Freeman C, Prosnitz LR (eds) (2007) Principles and practice of radiation oncology. Lippincott Williams & Wilkins, Philadelphia, p 76, Chap. 2
7. Bogdanov AA Jr, Licha K (eds) (2004) Molecular imaging. Springer, New York and also Schwaiger M (2008) Molecular imaging. Springer, New York
8. Brown BH, Smallwood RH, Barber DC, Lawford PV, Hose DR (1999) Medical physics and biomedical engineering. Francis and Taylor, New York
9. Cherry S, Sorenson J, Phelps M (2003) Physics and nuclear medicine, 3rd edn. Saunders, Philadelphia
10. Ziessman HA, O'Malley JP, Thall JH (2005) Nuclear medicine. Mosby Pub., St Louis, MO and also Cherry SR, Sorensen JA, Phelps ME (2006) Physics in nuclear medicine. Saunders, Philadelphia
11. Mayles P, Nahum A, Rosenwald JC (2002) Handbook of radiotherapy. Francis and Taylor, New York
12. Nyirjesy I et al (1986) Clinical evaluation, mammography and thermography in the diagnosis of *breast carcinoma*. Thermology 1:170
13. Belliveau N et al (1998) Infrared imaging of the breast. Breast J 4(4):9

14. Drazen JM (ed) (2000) Looking back on the millennium in medicine. N Engl J Med 342(1):42
15. Berry DA et al (2005) Effect of screening and adjuvant therapy on mortality from breast cancer. N Engl J Med 353:1784
16. Jelinek JS et al (2002) Diagnosis of primary bone tumors with image guided percutaneous biopsy. Radiology 223:731
17. Andersen HR et al (2003) A comparison of coronary angioplasty with fibrinolytic therapy in acute myocardial infection. N Engl J Med 349(8):733
18. Cutler DM, McClellan M (2001) Is technology change in medicine worth it? Health Affairs, Millwood
19. Heidenreich P, McClellan M (2001) Trends in heart attack treatment and outcomes. In: Cutler DM, Berndt ER (eds) Medical care output and productivity. University of Chicago Press, Chicago, IL
20. Vo DK, Lin W, Jin-Moo L (2003) Evidence based neuro-imaging in acute ischemic stroke. Neuroimaging Clin N Am 13:167
21. Johnston SC, Dudley RA, Gress DR, Ono L (1999) Endovascular and surgical treatment of unraptured cerebral aneurysms. Neurology 52:1799
22. Cherry S (2004) In vivo molecular and genomic imaging: new challenging for imaging physics. Phys Med Biol 49:R-13
23. Ritman EL (2002) Molecular imaging in small animals, role for micro CT. J Biomed Opt 6:432 and also Li G et al (2009) New fluoroscopic imaging technique for investigation of GDOF knee kinematics during trade mill gait. J Orthop Surg Res 4:6
24. Benvensite H, Blackband S (2002) MR microscopy and high resolution small animal MRI: applications in neuroscience research. Prog Neurobiol 67:393
25. Dayton PA, Ferrara KW (2002) The targeted imaging using ultrasound. J Magn Reson Imaging 16:362 and also Lanza GM and Wickline SA (2001) Targeted ultrasonic contrast agents for molecular imaging and therapy. Prog Cardiovasc Dis 44:13
26. Gupta TK et al (1999) High speed X-ray imaging camera using a structured CsI (Tl) scintillator. IEEE Trans Nucl Sci NS-46:232
27. Su Z et al (2005) Systematic investigation of the signal properties of polycrystalline HgI2 detectors. Phys Med Biol 50:2907
28. Siewerdsen JH, Jaffray DA (1999) A ghost story: spatio-temporal response characteristics of an indirect-detection flat panel imager. Med Phys 26:1624
29. Cherry SR et al (1997) A high resolution PET scanner for imaging small animals. IEEE Trans Nucl Sci 44:1161
30. Del Guerra A, Di Domenico G, Scandola M, Zavattini G (1998) High speed spatial resolution small animal YAP-PET. Nucl Instrum Method 409:508
31. Aurek KK, Machac J (1999) Improvement of fluorine-18 fluorodeoxyglucose images in simultaneous F-18 FDG/Tc-99m collimated SPECT imaging. Med Phys 26(6):917
32. Tainter KH, Lokitz S, Vascose C (2009) Image quality in multimodality PET/SPECT preclinical imaging. IEEE Nuclear Science Symposium, Orlando, 25–31 Oct 2009
33. Meng LJ, Tan JW (2009) An ultrahigh resolution SPECT system based on novel energy resolved photon counting CdTe detector. IEEE Nuclear Science Symposium, Orlando, 25–31 Oct 2009
34. Ter-Antonyan R, Browsher JE, Greer KL, Metzler SD, Jasczak RJ (2008) High sensitivity converging collimators for brain SPECT. IEEE Nuclear Science Symposium, Dresden, 19–25 Oct 2008
35. Bushberg JT (2001) The essential physics of medical imaging. Lippincott Williams and Wilkins, Pittsburgh
36. Zeintl J, Vija AH, Yahil A, Hornegger J, Kuwert T (2009) Quantitative accuracy of slow rotating dynamic SPECT imaging. IEEE Nuclear Science Symposium, Orlando, 25–31 Oct 2009
37. Vlaardingerbroek MT, den Boer JA, Luiten A (2008) Magnetic resonance imaging. Springer, New York
38. Johnson GA et al (2002) Magnetic resonance histology for morphologic phenotyping. J Magn Reson Imaging 16:413

# References

39. Kang JH et al (2009) Characterization of cross compatibility of PET components and MRI. IEEE Nuclear Science Symposium, Orlando, 25–31 Oct 2009 and also Moller HE et al (1999) Sensitivity and resolution in 3-D NMR microscopy of the lung with hyperpolarized noble gas. Magn Reson Med 41:800
40. Benveniste H, Blackband SJ (2002) MR microscopy and high resolution small animal MRI. Prog Neurobiol 67:393 and also Pautler RG, Silva AC, Koretsky AP (1998) In vivo neuronal tract tracing using manganese-enhanced magnetic resonance imaging. Magn Reson Med 40:740
41. Beck BL, Blackband SJ (2002) Phased array imaging on a 4.7T/33 cm animal research system. Rev Sci Instrum 72:4292
42. Keereman V et al (2009) Scatter effects of MR components in PET-MR inserts. IEEE Nuclear Science Symposium, Orlando, 25–31 Oct 2009 and also Louie AY, Meade TJ (2000) Recent advances in MRI. Biochem Sci (Special issue):7
43. White P (2002) Legal issues in tele-radiology – distant thoughts. Br J Radiol 75:201
44. Paulus MJ et al (2000) High resolution X-ray computed tomography. Neoplasia 2:62 and also Paulus MJ et al (1999) A new X-ray computed tomography system for laboratory mouse imaging. IEEE Trans Nucl Sci 46:558
45. Brenner DJ, Hall EJ (2007) Growth of CT scan use increases risk of higher radiation exposures, may lead to significant public health problem. Bostan, MA, N Engl J Med 45 (357):2277
46. Charon Y, Lanie P, Tricoire H (1998) Radio-imaging for quantitative autoradiography in biology. Nucl Med Biol 25:699
47. Seemann T (2002) Digital image processing using local segmentation. Ph.D. Thesis, Monash University, Australia, April 2002 and also Otsu N (1979) A thresholding selection method from gray-level histogram. IEEE Trans Syst Man Cybern 9:62 and also Muller H, Muller W, Squire DM, Maillel SM, Pun T (2001) Performance evaluation in content-based image retrieval. Pattern Recognit Letts 5:22
48. Haacke EM, Brown RW, Thompson MR, Venkatesan R (1999) Magnetic resonance imaging. Wiley-Liss, New York
49. Berstein MA, King KF, Zhou XJ (2004) Hand book of MRI pulse sequences. Academic, San Diego
50. Chen M, Pope T, Ott D (2004) Basic radiology. McGraw Hill, New York
51. Nosher J, Bonder L (2001) Interventional radiology. Wiley-Blackwell, New York
52. de Casteele V, van Dyck E, Sijbers J, Raman E (2003) A biomedical energy model for correcting beam hardening artifacts in X-ray tomography. IEEE 29th Annual Proceedings, 22–23 March 2003, pp 57–58
53. Aichinger H, Dierkey J, Barfuk SJ, Sabel M (2004) Radiation exposure and image quality in X-ray diagnostic radiology. Springer, New York
54. Mignotte M, Meunier J (2000) Three-dimensional blind deconvolution of SPECT images. IEEE Trans Biomed Eng 47(2):274
55. Studen A et al (2004) First coincidence in pre-clinical Compton camera prototype for medical imaging. Nuc Ins Meth A 531(1–2):258–264
56. Cunningham IA, Shaw R (1999) Signal to noise optimization of medical imaging system. Opt Soc Am A 16(3):621
57. Houston JD, Davis M (2001) Fundamentals of fluoroscopy. WB Saunders Co, Philadelphia, PA
58. Kuo Y, Okajima K, Takeichi M (1999) Plasma processing in the fabrication of amorphous silicon thin-film transistor arrays. IBM J Res Dev 43(1/2):73–88
59. Antonuk LE (2004) a-Si:H TFT-based active matrix flat panel imagers for medical X-ray applications. In: Kuo Y (ed) Thin film transistors, materials and processes, vol 1, Vol-1: amorphous silicon thin film transistors. Kluwer Academic Pub, Boston
60. Wong WS, Daniel JH, Chabynic ML, Arias AC, Ready SE (2006) Thin film transistor fabricated by digital lithography. In: Klauk H (ed) Organic electronics: materials, manufacturing, and applications. Wiley-VCH, Germany

61. Antonuk L, Wang Y, Behravan M, E-Mohri Y, Zhao Q, Du H (2007) Quantitative exploration of performance enhancements offered by active matrix X-ray imagers fabricated on plastic substrates. Proc SPIE 6510:65100

62. Meng Z, Wong M (2002) Active matrix light emitting diode displays realized using metal induced unilaterally crystallized polycrystalline silicon thin film transistor. IEEE Trans Biomed Eng 49(6):991

63. Danty JC, Shaw R (1974) Image science: principles, analysis and evaluation of photographic-type image processes. Academic, London

64. Rose A (1948) Noise and signal response in lead sulphide photoconductivefilms. Adv Electron Phys 1:131 and also Jones RC (1959) Adv Electron Phys 11:87

65. Breitenberger E (1955) Scintillation spectrometer statistics. Prog Nucl Phys 4:56

66. Ter-Pogossian NR (1967) The physical aspects of diagnostic radiology. Harper Row, New York

67. Kuhl W (1969) X-ray image intensifier today and tomorrow. Medica Mundi 14:57

68. Marry RA et al (2006) MRI vs. conventional fluoroscopy. Radiology 238(2):489

69. La Riviere PJ (2009) The price of tomography: SNR comparison of acquisition strategies for X-ray fluorescence imaging. IEEE Nuclear Science Symposium, Orlando, 25–31st Oct 2009

70. Brendel BJ, Roessl E, Schlomka J-P, Thran A, Proska R (2009) A novel CT perfusion protocol for quantitative contrast material mapping. IEEE Nuclear Science Symposium, Orlando, 25–31st Oct 2009

71. Bental J, Kim Y, Kundel HL, Horiz SC, van Metter RL (2000) Hand book of medical imaging. SPIE Press, WA

72. Cunningham IA, Shaw R (1999) Signal to noise optimization of medical imaging system. J Opt Soc Am A 16(3):621

73. Barrett HH, Swindell W (1981) Radiological imaging – the theory of image formation, detection, and processing. Academic, New York

74. Steckner MC, Drost DJ, Prato FS (1994) Computing the modulation transfer function of a magnetic resonance imager. Med Phys 21:483

75. Cox GG et al (1990) Digital fluoroscopy and radiographic system, RSNA annual meeting, 1990, Chicago, IL

76. Tavemier S (2006) Radiation detectors for medical applications. Springer, New York

77. Lopez A, Molina R, Katsaggelos AK, Mateos J (2001) SPECT image reconstruction using compound models. Rept Comision National de Ciencia y Tecnologia, Universiad de Granda, Spain

78. Oskoui-Fard P, Stark H (1988) Tomographic image reconstruction using the theory of convex projections. IEEE Trans Biomed Eng 7:45

79. Leahy R (2000) Recent developments in iterative image reconstruction for PET and SPECT. IEEE Trans Biomed Eng 19:257

80. Wagner HN et al (1989) Imaging dopamine receptors in the human brain by positron tomography. Science 22:1264

81. Garnett ES et al (1983) Dopamine visualized in the basal ganglia of living man. Nature 305:137

# Basic Principles of Radiation Detectors

**5**

## Contents

5.1 Introduction .................................................................. 251
5.2 Working Principle of the Detectors Used in Nuclear Medicine ...................... 253
5.3 Organic Scintillators ........................................................ 255
5.4 Light Output in an Organic Scintillator ......................................... 257
    5.4.1 Fluorescence and Phosphorescence .......................................... 259
5.5 Kinetics of Quenching in Organic Scintillators ................................... 259
    5.5.1 Mean Lifetime of the Organic Scintillator ................................... 261
5.6 Scintillation Efficiency of an Organic Scintillator ............................... 261
5.7 Structural and Electronic Properties of Scintillators ............................. 262
5.8 Detector Counting Efficiency ($\eta_C$) ....................................... 265
5.9 Time Resolution of an Inorganic Scintillator .................................... 266
5.10 Interaction of Ionizing Radiation with Scintillators ............................. 266
    5.10.1 Photoelectric Effect ...................................................... 267
    5.10.2 Compton Effect .......................................................... 267
    5.10.3 Pair Production .......................................................... 269
5.11 Ionization Losses ........................................................... 269
5.12 Inorganic Scintillators ...................................................... 272
    5.12.1 Light Yield of an Inorganic Scintillator ................................... 275
5.13 Defect Formation by Ionizing Radiation ........................................ 277
5.14 Solid-State Detector ........................................................ 278
    5.14.1 Uniform Excitation of the Solid-State Detectors ............................ 280
References ...................................................................... 282

## 5.1 Introduction

The radiation detectors are mainly divided into two large groups: (1) *scintillation detectors* and (2) *semiconductor diodes*. The first one can detect high-energy radiation through generation of light that can subsequently be registered by a *photodetector*, either a *photomultiplier* tube (PMT) or an *avalanche photodiode* (APD) [1–7]. The *PMT* or *APD* converts the light signal into electrical signal,

T.K. Gupta, *Radiation, Ionization, and Detection in Nuclear Medicine*,
DOI 10.1007/978-3-642-34076-5_5, © Springer-Verlag Berlin Heidelberg 2013

which is ultimately measured. The main advantage of these detectors is their large detection volume [8, 9].

The key requirement for best efficiency of a scintillator is that it should be transparent to its own radiation. Therefore, in order to optimize the performance of a scintillator, it is important that nearly all light that is created within the scintillator be received by the subsequent photodetector. A good scintillator has emission band gaps that do not overlap with the optical absorption band gaps because an overlap will cause excessive self-absorption of light in the scintillator. Therefore, the material is chosen in such a way that the radiative component of recombination dominates over non-radiative components [10].

The optical properties of a scintillator do not depend on the *lattice orientation*, but the presence of significant impurities around *grain boundaries* severely affects the optical properties. In particular, the image quality of a gamma camera is seen to degrade when light output across a *grain boundary* of the scintillator is affected.

The incoming photons received by the photocathode are influenced by the refractive index (RI) of the materials of the imaging system. It is to be noted that the RI of the materials of the imaging system should match each other. In a case where they do not match with the light coming out of the surface of the scintillator and the entering surface material of the photocathode, one can notice that one component of light is being reflected and the other component is being refracted at the interface. From our experimental observations, we have found that the RI of most of the scintillating materials is between 1.44 and 2.20, and the RIs of air, glass, and quartz that exist in the pathway of the outgoing light are ~1, 1.5–1.7, and ~1.47, respectively. In our present medical imaging system, the light that comes out of the scintillator passes through air before entering into the photocathode, with an envelope made out of either glass or quartz. So we should expect some loss of light before it is received by the photocathode. Additional light losses can occur inside the scintillator itself (internal reflection, and in case of a single crystal, it depends on the crystal structure too). However, polished scintillator surface can minimize the loss of output light.

Total internal reflection of light occurs in a material at a critical angle ($\theta_c$) or Brewster's angle and can be expressed mathematically as

$$\theta_c = \text{Sin}^{-1}\left(\frac{n_2}{n_1}\right) \tag{5.1}$$

where $n_1$ and $n_2$ are the refractive indices of the material in which light originates (scintillator) and the material across the interface (air), respectively. It would be a good idea to use a good diffuse reflector to return the light into the crystal when the angle of incidence is less than the critical angle.

The second category of detector is *semiconductor PN junction* reverse-biased diodes where the absorbed radiation creates electron–hole (e–h) pairs. By applying electric field across the PN junction, the e–h pairs are separated and a direct electrical signal is produced [11]. Recently, silicon *PIN* (*I stands for insulator*) diodes are used as nuclear radiation detectors and have been found to be more

## 5.2 Working Principle of the Detectors Used in Nuclear Medicine

efficient than *PN* junction diodes. The *PIN* diode detector, where *I* stands for insulating layer in between the *P* and *N* junctions, detects radiation by generating a transient photocurrent when the incoming radiation is absorbed in the insulating (*I*) layer. In practice, *I* region of the *PIN* device is kept large to maximize the photocurrent.

When the *PIN* device is exposed to the nuclear radiation, the energy quanta, which are absorbed by the elements of the lattice structure, produce electron–hole pairs provided the energy quanta are larger or equal to the band gap energy of the material. Now when the device is reverse biased, *I* layer is depleted of free carriers and electrons are swept to *P* layer and the holes are swept to the *N* layer, respectively. The most important criterion of *PIN* detector is that they are not restricted to use at low temperature like *PN* junction devices.

Semiconductor radiation detectors (SRD) play an important role in radiation instrumentation. Being small in size, SRD have many advantages as nuclear detectors such as (1) they have outstanding energy resolution in nuclear spectroscopy, (2) they can be pixelated easily for high spatial resolution, and (3) they can be integrated easily with readout electronics. Advancement in the semiconductor industry has made silicon as one of the most popular material for radiation detectors in medicine with applications in diagnostic and cancer treatment [12].

Extensive studies of both groups of detectors revealed their pros and cons [13]. The diodes typically suffer from inadequate electron–hole (e–h) production and loss of e–h pairs during collection due to the imperfection of the device. The most common semiconductor materials used for radiation detection are lithium-doped silicon (SiLi) and germanium (Ge). Both of them require high voltage typically on the order of kilovolts to maximize the charge collection and to increase the drift velocity. In case of scintillators, the efficiency of converting the high-energy radiation into light is typically on the order of 10–12 %.

Recently some workers are working on some direct band gap semiconductors like CdS and ZnO doped with donors in a reducing atmosphere [14]. The discovery of epitaxy and its application in device manufacturing has made possible the existence of lattice-matched multilayer devices with two alternating materials like $InP/Ga_{0.47}In_{0.53}$ or $Al_x\,Ga_{1-x}\,As/\,GaAs$. Upon interaction with the ionizing radiation, the created electrons and holes will quickly diffuse into InGaAs well and will recombine. The difference in band gap energies will ensure that all the light absorbed in InGaAs and InP will remain transparent to emitted photons [15, 16]. It is believed that these devices might be the most promising and efficient semiconductor detectors in medical science and will replace silicon-based devices in the near future.

## 5.2 Working Principle of the Detectors Used in Nuclear Medicine

The detection of ionizing radiation by *scintillation light* produced by certain materials is one of the oldest and useful methods adopted in nuclear medicine even today. The material in the form of single crystal or in the form of a polycrystalline film deposited either by thin (vacuum evaporation, hot-wall epitaxy, chemical vapor deposition process, or sputtering) or thick film (screen printing) [17] that produces light on being absorbed by ionizing radiation is called a *scintillator*.

The *scintillation* or the *flashlight* effect persists as long as the radiation energy is being absorbed. According to some experts, scintillation phenomenon can be described as the *phenomenon of luminescence in transparent solids, fluids, or gases originating at the propagation of ionizing radiation through them* [18, 19]. However, some critics have distinguished the phenomenon of *luminescence* and *scintillation*. According to their conceptions, *scintillation* is the energy loss of ionizing radiation through matter, whereas *luminescence* is the results from the radiative relaxation of an active ion of the material after it is being excited between its fundamental state and excited energy levels by an electrostatic discharge or a pulse of light.

As a matter of fact, luminescence can be subgrouped into four categories, namely, (1) *excitation luminescence*, where the ionization/excitation by radiation can create e–h pairs or bound e–h pairs called excitons, a vivid example of this type of luminescence is bismuth germanate (BGO, $Bi_4Ge_3O_{12}$); (2) the *dopant luminescence*, where radiative recombination of self-trapped excitations occurs at dopant ions, as we found in gadolinium oxysulfide (GOS:Ce,$Gd_2O_2S$); (3) *charge transfer luminescence*, where excitation due to charge transfer is different for initial and final states and the selection rules for electromagnetic transition are loosened, as in the case of yttrium aluminum garnet (YAG:Yb,$Y_3Al_5O_{12}$); or (4) *core valence luminescence*, where after excitation of the core-valence electron, an electron in the valence band recombines with the resultant hole radiatively, as we found in the case of barium fluoride ($BaF_2$), calcium fluoride (CsF), and lithium fluoride (LiF).

However, according to some observers, *scintillators* are *energy converters* that convert the incident energetic photon (UV, X-ray, or gamma ray) into a number of photons of much lower energy having longer wavelengths in the visible or near-visible range. According to them, *scintillators* are *transparent dielectric media* where *luminescence is induced by ionizing radiation*. Some of these luminescence centers are *cations* or *anionic complexes* of the lattice or doping ions such as $Ce^{3+}$.

There are wide varieties of materials that have shown scintillation properties when properly excited with radiation light [20]. These materials can be grouped into three main categories: (a) *organic*, (b) *inorganic*, and (c) *solid-state* scintillators. The ideal scintillating materials should possess the following properties:

(a) Conversion of the kinetic energy of the charged particles into highly *efficient scintillation light*.
(b) The medium should be *transparent to the wavelength of the scintillation light* so that minimum amount of light will be absorbed or lost during transition.
(c) The conversion should be *linear* and the material should have *good optical quality* with refractive index comparable to glass (~1.5).
(d) The *decay time* of the induced luminescence should be *short*.

The general idea behind the *scintillation process* is the conversion of energy of an incident gamma quantum or particle into great many low-energy photons. The ionization event put the scintillator in a state of nonequilibrium situation. As time goes on, the scintillator begins to relax toward a new equilibrium situation. During relaxation, a multitude of elementary processes such as creation of primary electronic excitation occur. As a result of this excitation, one can expect an avalanche of secondary excitations including electrons, holes, photons, and plasmons. In addition to those events, electronic excitations will also produce thermalized electron–hole (e–h) pairs and low-energy excitations which ultimately transform into light photons, that is, scintillation [20–22].

Electrons and holes produced by ionizing radiation have several ways to be involved in scintillation process after thermalization. The simplest one is the direct radiative recombination of free thermalized electrons in the conduction band with holes from the valence band or from outer electronic shells. The excitation luminescence is efficiently *quenched* causing a sensitization of luminescence of the activating ions in presence of impurity centers or activating ions. In case of heavily doped or self-activated scintillators like $CeF_3$, the direct excitation of the activation centers by ionizing radiation provides an important contribution to the scintillation.

## 5.3 Organic Scintillators

Organic scintillators are aromatic hydrocarbon compounds containing a benzene cycle. In organic scintillators, the mechanism of light emission is a molecular effect. The most distinguishing feature of an organic scintillator is their very rapid decay time. It proceeds through excitation of molecular levels in a primary fluorescent material, which emits bands of ultraviolet (UV) light during de-excitation. The UV light is absorbed in most organic materials with an absorption length of few mm.

The organic scintillators have low Z value; as a result, the organic scintillators do not show photo-peak and give rise to a Compton continuum in their $\gamma$-ray pulse height spectrum. However, recent developments in organic scintillators have made to add high Z-elements. Unfortunately, the addition of high Z-elements decreases the light output. A typical energy band diagram of an organic scintillator is shown in Fig. 5.1 with the distinguishing features of spin singlet state with that of a triplet state. The states above the singlet ground state $S_0$ are excited states. In case of triplet state, the ground state is $T_0$ and above that all the remaining states are excited states. Being hit by radiation light, the scintillator material emits light, which arises from the transitions made by the free valence electrons of the molecules. These delocalized electrons are not associated with any particular atom in a molecule and they occupy the $\pi$-molecular orbitals [23, 24].

From our experimental observations we know that stilbene has lower scintillation property than anthracene, but it is preferred in those situations where pulse shape discrimination is to be used to distinguish among scintillations induced by charged particles and electrons. On the other hand, naphthalene in ethyl alcohol has also been used as organic scintillators [25].

The *plastic scintillators* like NE102A and NE10 have the advantage that they can be produced in a variety of shapes according to the application needs. Most frequently they are used as rectangular plates from 0.1 up to 30.0 mm and area from

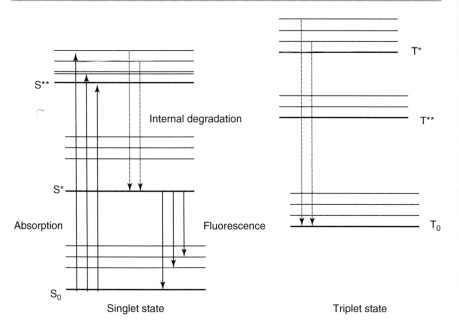

**Fig. 5.1** The energy level diagram of an organic scintillator molecule

a few mm$^2$ to several mm$^2$. Some of these materials are used in the detection of β particles. Unfortunately, the devices made out of these materials are fragile and difficult to obtain in large sizes.

There are some other organic materials that have been used as scintillators such as (1) PPF2, 7-diphenyl-9,9-dipropyl fluorine; (2) d-amyl-PPF, 2,7-bis(4-t-amyl phenyl)-9,9-dipropyl fluorine; and (3) d-CH3O-PPF 2,7-bis(4-methoxy phenyl)-9,9-dipropyl fluorine [26]. Synthesized methyl and/or methoxy substituted 1,3-diphenyl-2-pyrazolines in benzene [27].

Sometime organic materials are dissolved and sealed in a tube for better handling as a detector. These liquid scintillators are used for measuring β emitters; particularly they are classified as weak β emitters. The sample is dissolved or suspended in an organic chemical, which fluoresce when acted by ionizing radiation [28, 29]. The radiation is then converted into pulses and detected by photomultiplier tube. Organic scintillators are generally used for direct detection of β particles (fast electrons) or alphas (positive ions), fast neutrons through proton recoil process. Table 5.1 shows some of the basic properties of organic scintillators.

Liquid scintillator counting (LSC) may be used to quantize alpha particles for a wide range of applications, such as environmental monitoring, nuclear fuel processing, and high-level waste management. Alpha-emitting radionuclides when quantitated by LSC generally produce a symmetrical peak around the energy maximum of the alpha particle, whereas in beta-emitting radionuclides, a beta emitter produces a continuous spectra from zero to energy maximum of the beta particle. As a matter of fact, LSC reduced the time required to analyze radioactive

## 5.4 Light Output in an Organic Scintillator

**Table 5.1** Properties of some organic scintillators

| Properties | Anthracene | Polysyrene | Stilbene | NE 102 | NE 10 | NE224 liquid | NE316 Sn-loaded liquid |
|---|---|---|---|---|---|---|---|
| Density g/cm$^3$ | 1.25 | 1.06 | 1.22 | 1.032 | 1.032 | 0.877 | 0.96 |
| H to C ratio | 0.714 | 1.0 | 0.857 | 1.105 | 1.105 | | |
| Emission spectrum (nm) | 448 | 430 | 400 | 425 | 437 | 425 | 425 |
| Decay time constant (ns) | 31 | 2–3 | 3.5 | 2.5 | 3.3 | 2.6 | 4.0 |
| Scintillation amplitude | 100 | 56 | 11 | 65 | 60 | 80 | 35 |
| Main application | α, β, γ, and n | α, β, γ, and n | γ and n | α, β, γ, and n | γ and n | γ only | γ and X-rays |

samples from hours to minute. For low-energy (soft) β emitters, LSC offers unmatched convenience and sensitivity. LSC detects radioactivity via the same type of light emission events, which are used in solid scintillators. The solvent portion of an LSC is about 60–90 % [30].

During the isolation of alpha-emitting radionuclides in nuclear fuel processing and other related processes, organic extraction techniques are employed to separate alpha-emitting radionuclides from one another, as well as from contaminating beta nuclides [31, 32]. Quenching agents are generally used to prepare samples that affect the resolution.

For organic scintillators such as anthracene, stilbene, and many of the commercially available liquid and plastic scintillators, the response of electrons is linear for particle energies above about 125 keV. The response of organic scintillators to charge particles can be expressed in terms of the fluorescent energy emitted per unit path length ($dL/dx$) and the specific energy loss for the charged particles ($dE/dx$). In absence of quenching, $(dL/dx) = S (dE/dx)$, where $S$ is the normal quenching efficiency, and the light output $L = SE$, where $E$ is the energy output. The light output vs. photon energy of a typical liquid scintillator is shown in Fig. 5.2.

## 5.4 Light Output in an Organic Scintillator

The organic scintillator is doped with a second fluorescent material, which is called *wavelength shifter*, to extract a light signal due to conversion of UV light into visible light. The second fluorescent substance is chosen in such a way that its absorption spectrum matched to the emission spectrum of the primary fluor, and its emission spectrum, adapted to the spectral dependence of the photocathode quantum efficiency [33].

These two active fluorescent materials are either dissolved in a suitable organic solvent (s) or mixed with a monomer of a material capable of polymerization. For organic scintillators, the differential light output ($dL/dx$) due the ionizing radiated

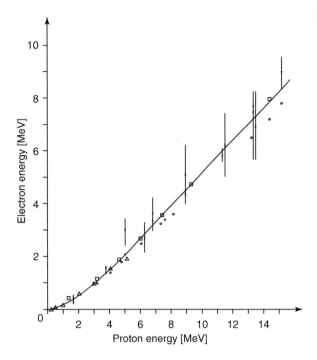

**Fig. 5.2** The light output vs. photon energy for a liquid scintillator (NE-213) (Reprinted with permission, Maier and Nitschke)

light is related to the stopping power of the ionizing particle (d$E$/d$x$) which is known as Birks' law [34], and mathematically we can express it as

$$(dL/dx) = \{(dE/dx)\text{Eff}_{\text{sci}}\}\{[1]/[1 + qB\,(dE/dx)]\} \quad (5.2)$$

where $\text{Eff}_{\text{sci}}$ is the efficiency of the scintillator, $B$ is *Birks' parameter*, and $q$ is the *quenching* parameter. The $qB$ in Eq. 5.2 has saturation constant value between 0.01 and 0.02 g/cm$^2$ MeV. For fast electrons (d$E$/d$x$ is small), the above equation transforms to

$$(dL/dx) \approx \{(dE/dx)\text{Eff}_{\text{sci}}\} \quad (5.3)$$

However, for alpha ($\alpha$) particle, where (d$E$/d$x$) is large, the Eq. 4.2 can be written as

$$(dL/dx) \approx \{\text{Eff}_{\text{sci}}\}/\{qB\} \quad (5.4)$$

Now by integrating the Eq. 5.4, we get

$$L(E) \approx \{\text{Eff}_{\text{sci}}\}/\{(qB)\,R(E)\} \quad (5.5)$$

where $R$ is the range of the particle.

5.5 Kinetics of Quenching in Organic Scintillators 259

It has been found from the experimental studies that liquid and plastic scintillators show *luminescence quenching* which is wholly responsible for the interaction between the excited solvent and the quencher, which competes with the energy transfer process from the solvent to the fluor (fluorescent solute). The advantage of studying energy transfer and quenching in plastic scintillators is that the rigidity of the medium hinders molecular diffusion and collision.

### 5.4.1 Fluorescence and Phosphorescence

Both the fluorescence and phosphorescence phenomena occur due to de-excitation of electrons. When a fluorescence material is exposed to thermal or radiation energy, it produces high-energy photons that collide with electrons in the material causing them to be excited and jump to a higher energy level. The electrons in the excited states have the same spin as they had in the ground state. However, the electrons in the excited states are not stable and they release energy in the form of light and are de-excited. When they are de-excited, they come back to the ground state. Phosphorescence is similar to fluorescence except that the light emission continues even after the radiation energy or the thermal energy is removed.

The other way to distinguish *fluorescence* and *phosphorescence* is primarily on the basis of the duration of the *afterglow*—a slow decay of luminescence being called *phosphorescence*. However, in *scattering (Raman and Rayleigh)* of light, there is no absorption of light. Rayleigh and Raman scattered light following oscillation of an induced dipole moment consists of two polarization components that are aligned perpendicular and parallel to the scattering plane (depolarization) [35–37].

*Afterglow* can be attributed due to the following reasons: (1) *natural*, (2) *metastable states or traps* (impurities, defects, etc.), and (3) *speed of energy transfer and number of luminescence centers*. Besides pulse shape discrimination techniques, the knowledge of radioluminescence decay curves is, in itself, the fundamental importance in radiation physics and chemistry, and it enables to get information on the nature of the basic processes induced by nuclear particles in organic media [37].

## 5.5 Kinetics of Quenching in Organic Scintillators

The organic scintillator under study consists of solute (polystyrene), fluor (F = diphenylanthracene, DPA), solvent (S), and a dopant which is a quencher (Q = chloro- or bromo-substituted organic compounds). In the ternary matrix, the concentration of the fluor is kept fixed, while the concentration of the dopant is a variable quantity. Let the ternary matrix be perturbed by a flux of beta ($\beta$) ray from a strontium ($Sr^{90}$) source. The luminescence quenching in the liquid scintillator is attributed to the interaction between the excited solvent and the quencher, and it competes with the energy transfer process from the solvent to the fluor (fluorescent solute). One of the advantages of studying quenching in plastic scintillators is that the rigidity of the medium hinders molecular diffusion and collision.

The theoretical values for the solvent quenching constant ($\alpha$) and the solute quenching constant ($\beta$) can be expressed mathematically as [38]

$$\alpha = k_Q/(k_e + k_i + k_s + k_F[F]) \tag{5.6}$$

$$\beta = k'_Q/(k'_e + k'_i + k'_s[F]) \tag{5.7}$$

And the quantum efficiency of energy transfer from S* to F is $\varphi_{ET}$ and can be presented mathematically as

$$\varphi_{ET} = k_F[F]/(k_e + k_i + k_s + k_{ET}[F] + k_Q[Q]) \tag{5.8}$$

where the quantities $k_e$, $k_i$, $k_s$, $k_{ET}$, and $k_Q$ are rate constants for emission, internal quenching, self-quenching, energy transfer to the fluor, and quenching by the quencher for excited solvent $S^*$, respectively. Similarly, the quantities with prime are the rate constants for the excited fluor $F^*$. The square brackets indicate the concentration in mole/liter (L).

Again the solvent and the solute quenching constants can be deducted mathematically as

$$\alpha[Q] = \{(\tau_{S_0}/\tau_S) - 1)\} \tag{5.9}$$

$$\beta[Q] = \{(\tau_{F_0}/\tau_F) - 1)\} \tag{5.10}$$

where $\tau_S$ and $\tau_F$ are the mean lifetime of the excited state of the solvent (S*) and that of the fluor (F*) when the quenching reagent is present. $\tau_{S0}$ and $\tau_{F0}$ are the mean lifetime of the excited state of the solvent (S*) and that of the fluor (F*) when the quenching reagent is absent. Now we can find out the values of the $\alpha[Q]$ and $\beta[Q]$ presented in Eqs. 5.9 and 5.10 from the slope of the straight line fitted to the least-squares method to the plots of $(\tau_{S_0}/\tau_S)$ and $(\tau_{F_0}/\tau_F)$.

Experimental studies show that *fluorescence quenching* is dependent on the solvent, the quencher, and the nature of the substituent. The heavy atom quencher increases fluorescence, which has been attributed to internal or external spin–orbit coupling effects. These effects are seen to increase the probability of singlet–triplet absorption and decrease the lifetime and quantum yield of phophorescence [39].

Further studies on fluorescence of the organic scintillators show higher degree of solvent quenching in ionizing radiation compared to UV exposure. The most possible cause is thought to be due to the larger amount of energy transfer and highly excited states of the solvent (such as polystyrene). It is also inferred that the energy transfer from the excited state of fluor to the triplet state of quencher is responsible for solute quenching.

### 5.5.1 Mean Lifetime of the Organic Scintillator

New interest for *organic scintillators* arose when it was recognized that the emission decay involves a slow component, which depends on the ionizing power of the nuclear radiation. The phenomenon was soon applied to particle discrimination. The origin of the slow component results from triplet interaction leading to a singlet state. However, a part of these states is produced very rapidly and gives rise to prompt emission intensity, which decays exponentially with a time constant identical to the mean lifetime $\tau$ [40]. Mathematically, the mean lifetime ($\tau$) of an excited state can be written following Stern–Volmer kinetic relation as

$$(\tau_0/\tau) = 1 + \beta\tau_0[\text{Q or S}] \tag{5.11}$$

where $\tau_0$ is the lifetime of the excited state in absence of quencher Q and S stands for solute.

And the triplet lifetime $\tau_T$ is defined as

$$\tau_T = 1/\beta \tag{5.12}$$

Again the intensity ($I$) of radiation light at time $t$ and the initial intensity ($I_0$) at time $t = 0$ can be correlated as

$$I = I_0 \exp(-t/\tau) \tag{5.13}$$

## 5.6 Scintillation Efficiency of an Organic Scintillator

The *quantum efficiency* (QE) of fluorescence is defined as the fraction of all incident particle energy, which is converted into visible light. Unfortunately, all the incident particles are not converted into visible light but ended into heat due to *quenching*. On the other hand, the *quantum efficiency* of an activated scintillator is very much dependent on the ratio of the band gap of the material to the energy of the activator in the radiating state. It is also related to the position of its ground and excited states to the valence and conduction bands, respectively [41, 42]. The change in photon energy causes a shift of the fluorescence spectrum to longer wavelength, relative to the absorption spectrum, which is referred to as *Stokes shift*.

From the definition of the *quantum efficiency* of an organic scintillator (solute + solvent + quencher (if there is any)), we can write the number of photons absorbed per volume per unit time ($\varphi_{pa}$) as [43]

$$\varphi_{pa} = \varphi_{pi}\alpha\exp(-\alpha s) \tag{5.14}$$

where $\varphi_{pi}$ is the number of photons incident on the cell containing the liquid scintillator, $\alpha$ is the optical absorption coefficient of the solvent, and $s$ is distance within the solution from the front window of the cell.

Let us consider that the liquid scintillator within the cell is perturbed by a radiation light and the output signal is received by a photomultiplier tube (PMT). As a result of the incoming energy flux, we will notice a current $(I_s)$ in the PMT when viewing solvent emission under optical excitation and a PMT current $(I_c)$ when viewing solvent emission under $C^{14}$ excitation conditions. The ratio between the two currents can be expressed mathematically as:

$$(I_s/I_c) = (\phi_f/\Gamma\eta_s) \tag{5.15}$$

where $\phi_f$ is the fluorescence quantum yield of the solvent, $\eta_s$ is the scintillation efficiency, and $\Gamma$ is equal to

$$\Gamma = \left( \int F_p(r,0) r \, dr \right) / \left( \pi R^2 \int F_p(r,s) dr \, ds \right) \tag{5.16}$$

where $F_p$ is the fraction of the photons received by the photomultiplier tube (PMT), $r$ is the radial distance away from the axis of the cylinder, and $R$ is the radius of the exciting radiation beam. Now, if we assume that there is no emission from the back of the cell, we can write the scintillation efficiency as

$$\eta_s = \left\{ [\{(I_c\phi_f)/I_s\}I_0\alpha]/\pi R^2 \right\}$$
$$\times \left\{ \int (\exp(-\alpha s) F_p(r,s) r \, dr \, ds) / \int (F_p(r,s) r \, dr \, ds) \right\} \tag{5.17}$$

where the integral extends from $r = 0$ to $r = R$. The mathematical expression of Eq. 5.17 is established on the assumption that the radiation beam is homogeneous over $R$. For extreme case when the absorption coefficient $\alpha$ is very large and the absorption is confined to the immediate vicinity of the front window, the above expression (5.17) transforms to

$$\eta_s = \{(Ic\phi_f)/I_s\}N_p\Gamma \tag{5.18}$$

where $N_p$ is the total number of photons absorbed per unit time.

## 5.7 Structural and Electronic Properties of Scintillators

The structural and electronic properties of *organic, inorganic,* and *ionic materials* are important for the selection of a detector material for medical imaging [44–47]. A scintillator can be considered as essentially a photon conversion device. The most important similarity between ordinary luminescence and scintillation is in the role played by so-called luminescence centers that is generally created by an

## 5.7 Structural and Electronic Properties of Scintillators

appropriately chosen *dopant ion* or *activator*. The ideal materials used for scintillators in medical imaging system should have *high density, high effective number, high luminous efficiency, chemical stability, short decay time, no (or minimum) afterglow, good spectral match to photodetectors*, and *cost-effective*. But in reality, it is hard to find any single material, which will possess all these properties. Thus, there is a need to compromise choosing from available materials that will be most suitable for a particular application such as planar X-ray, X-ray CT, SPECT, or PET.

Modeling [48, 49] a scintillator material in terms of structural and electronic properties means studies of its atomic structure as a function of *density* [50] and *electron excitations* vs. *structure* [51].

Scintillators are used as radiation detectors for X-rays, gamma ($\gamma$) rays, electrons, neutrons, alpha ($\alpha$) and beta ($\beta$) particles, and neutrinos. Efficient detection of these particles (through radiation) depends on how well and how quickly they lose their energy in the detector. One of the most important properties of a scintillating material is its *stopping power*, which depends not only on the *atomic weight* (Z-factor) of the material but also on the *density* of the material [52]. As for example, cadmium tungstate ($CdWO_4$, CWO, density 7.9 g/mL) produces higher light output than bismuth germanate (BGO, $Bi_4Ge_3O_{12}$ density 7.19 g/mL) because of its higher density and atomic weight than BGO. Density and atomic weight are therefore important factors that are used to assess the effectiveness of a scintillator (scintillating material).

*High density* in scintillating materials is an advantage in applications where a compact high-efficiency counter is required. In other words, we can say that the detection efficiency (or $\gamma$-ray stopping power) of a scintillator depends on the density of the material [53]. Bismuth germanate (BGO, $Bi_4Ge_3O_{12}$) and lutetium oxyorthosilicate (LSO, $Lu_2SiO_5$:Ce) have dominated the field of positron emission tomography (PET) because their high density and high atomic number (Z) imply a short attenuation length and consequently high spatial resolution. On the other hand, low-density organic scintillators have made them inefficient for detecting photons [54, 55].

Structural properties of a scintillator (radiation detector) are important because they affect a material's absorption and emission characteristics and its response to heat, *pressure*, and other external and internal forces [56]. A systematic study of several hundred inorganic crystal structures on the basis of electronic structure calculations reveals that subgroup of more than 3,000 entries of crystals that contain one or more of the elements like mercury (Hg), thallium (Tl), lead (Pb), or bismuth (Bi,) have a Bravais lattice that is either cubic ($\Gamma_c$), orthorhombic ($\Gamma_o$), tetragonal primitive ($\Gamma_q$), or hexagonal ($\Gamma_h$).

The size of the energy gap between the empty and filled electron states greatly influences the frequency of light emitted (a wider gap corresponds to a higher frequency of light) by a particular semiconductor. Recently, there are several technologies to alter the band gap of a semiconductor. These are known as *substrate engineering*, or *strain engineering*, where the host semiconductor is mixed with another semiconductor having different band gap energy (e.g., $Si_{1-x}$ $Ge_x$, implanting germanium into silicon). The *strain engineering* increases the mobility of charge carriers of an engineered semiconductor [57, 58].

Small band gap materials are not efficient scintillators, because when the emission energy is greater than the band gap of the material, the emitted photons are absorbed by the crystal (host lattice) itself and never exit the crystal. Moreover, due to photoionization the electron in the excited state is injected back into conduction band by thermal activation. For example, in $Lu_2O_3$:Ce all the $Ce^{+3}5d^1$ states are located within the conduction band of the solid and the cerium luminescence is totally quenched [59].

The crystal structure (e.g., cubic, monoclinic, or orthorhombic) determines the internal loss of light, which ultimately affects the light output of the scintillator. Most scintillators are made in such a way that they produce light in the visible region of the electromagnetic (EM) spectrum. Since the photons produced in the scintillator have to be guided to the photodetector, they must travel through the material without significant attenuation. Attenuation of photons causes loss of output light and information, besides induction of nonlinearity in the detector response. Unfortunately, most of the efficient scintillators produce ultraviolet photons with short attenuation length and therefore quickly get absorbed without producing sufficient information.

In order to get rid of this type of problem, the host scintillating material is mixed homogeneously with another scintillating material (second scintillator which is called *wavelength shifter*) that will absorb these photons and will emit visible light photons. Since the visible photons have longer attenuation lengths, their lifetime in the material is longer and consequently they will increase the light efficiency of the mixed material.

Traps in the scintillating materials not only produce prompt scintillation but also delay the emitted light, which ultimately provides thermal energy to the material known as *thermoluminescence* (TL). However, TL is not of much concern in the scintillators, but the saturation of the metastable traps causes undesirable broadening of the pulse shape. On the other hand, scintillating materials having impurity levels will trap electrons. As a result, there will be slow transition of the electrons and a longer decay time and more afterglow.

Experimental observations have shown that cubic barium fluoride ($BaF_3$) is a fast scintillator, whereas cubic lead fluoride ($PbF_3$) fails to show the scintillation property. However, there is some controversy about the orthorhombic crystals such as lead iodide ($PbI_2$). Indeed, the calculations for orthorhombic phase indicate that it should scintillate as a fast scintillator. On the other hand, the orthorhombic and cubic phases of $PbF_3$ indicate that the energy bands for both are similar. Therefore, the luminescence characteristics of both phases will be similar. The other interesting property of an orthorhombic crystal is that they may not scintillate all the time.

Applying pressure to a scintillator influences volume compression, distortion in crystal structure, density, and microscopic arrangement of atoms and phase diagrams especially in $ABX_4$ materials ($CdWO_4$, $PbWO_4$, etc.) [60]. However, pressure-driven phase transition has not been observed in $CdWO_4$ [61]. The crystalline structure of these $ABX_4$ materials is scheelite, and they are highly ionic with

## 5.8 Detector Counting Efficiency ($\eta_C$)

$A^{+2}$ cations and tetrahedral $BX_4^{-2}$ anions. They have short B–X bond lengths of approximately 1.78 Å that are quite rigid under compression [62].

## 5.8 Detector Counting Efficiency ($\eta_C$)

The isotropic materials propagate light efficiently, but their *counting efficiency* depends on the crystal structure and on the presence of the defects. In fact, all the components (*conversion efficiency, transport of charge carriers, luminescence, yield*, and *light collection*) are to some extent dependent on the *structural orientation of the lattice*. Experimental observations show that alloying the lattice ions by isovalent ions, one can alter the conversion efficiency of the scintillation materials. However, the activator concentration and its homogeneous distribution in the crystal lattice determine their *optical* and *electrical behavior* [63].

The *counting efficiency* ($\eta_C$) of a detector is related to the amount of radiation emitted by a radioactive source to the amount measured in a detector. The $\eta_C$ of a detector can be expressed mathematically as

$$\eta_C = (P_c/P_e) \tag{5.19}$$

where $P_c$ is the number of photons counted in the detector and $P_e$ is the number of photons emitted by the source and is equal to ($B_F \times B_R \times T \times A$). Here $B_F$ and $B_R$ are the branching fraction for that mode of decay and the branching ratio for that of photon energy, respectively, $T$ is the total counting time interval in seconds, and $A$ is the activity. The source activity $A$ is related to the *elapsed time* ($\tau$), and it varies exponentially as

$$A = A_0 \exp[(-t/\tau)] \tag{5.20}$$

where $A_0$ is the initial activity, $t$ is time interval since the source strength is calibrated, and $\tau$ is the lifetime (=half-life × 1.4427).

The value of $\eta_C$ is affected by three factors, namely, $S_D$ (the fraction of all space that the detector subtends), $T_p$ (the fraction of photons transmitted), and $F_p$ (the fraction of the photons absorbed). Again $T_p$ is calculated on the basis of attenuation coefficient and thickness of the medium it traverses and can be expressed mathematically as

$$T_p = \exp[-(\mu_1 \times d_1)\exp[-(\mu_2 d_2)] \tag{5.21}$$

where $\mu_1$ is the attenuation coefficient of air ($1.0 \times 10^{-4}$ cm$^{-1}$), $d_1$ is the distance traveled by the photons in air, and $\mu_2$ and $d_2$ are the attenuation coefficient (0.20 cm$^{-1}$) and thickness of aluminum. Now the fraction of the photons absorbed ($F_p$) by the detector is calculated from the following equation:

$$F_p = 1 - \exp[-(\mu \times d)] \tag{5.22}$$

Therefore, the detector efficiency ($\eta_C$) can be written as

$$\eta_C = S_D \times T_P \times F_P \tag{5.23}$$

Now we can combine these factors together to get an expression for the number of counts in the photo-peak ($N_P$) as

$$N_P = \eta \times R \times B_F \times B_R \times T \times A \tag{5.24}$$

## 5.9 Time Resolution of an Inorganic Scintillator

A good scintillator should show a good timing resolution. It is required for the determination of coincident gamma quanta in *positron emission tomography* (PET), for *time of flight measurement in high-energy physics* (HEP), or to perform *positron lifetime measurements* in material science. The best time resolution of an absorption event is obtained when the scintillator's pulse is short with a very fast rise and decay times. The time resolution is usually measured by using a start–stop spectrometer with a time-to-amplitude converter and an analyzer in the output of the two scintillation detectors. True resolution $\Delta\tau$ can be estimated as [64–66]

$$\Delta\tau \propto \left(\tau / N_{ph}^{\frac{1}{2}}\right) \tag{5.25}$$

where $\tau$ is the duration of scintillation and $N_{ph}$ is the total number of scintillating photon.

## 5.10 Interaction of Ionizing Radiation with Scintillators

The *photons* of the collimated X-ray or gamma ($\gamma$) rays incident on a scintillator (absorber) of thickness $\times$ (in cm) are first absorbed, and then a part or whole of it comes out of the scintillator as light pulse. The transmitted light, which is attenuated, is reduced in intensity, and the reduced intensity I can be expressed mathematically as

$$I = I_0\{\exp(-\mu x)\} \tag{5.26}$$

where $I_0$ is the initial intensity of the incident light and $\mu$ is the absorption or attenuation coefficient of the scintillator (absorber) and can be represented by

$$\mu = (dI/Idx) = \{(n_e\sigma_e)/Z\} \tag{5.27}$$

## 5.10 Interaction of Ionizing Radiation with Scintillators

where $n_e$ is the electron density, $\sigma_e$ is the absorption cross section of an electron, and $Z$ is the atomic number of the scintillating material. The interaction of the incident radiation with the scintillator can be classified into three categories (according to the energy of the interacting photons in the incident radiation), namely, (a) *photoelectric effect*, (b) *Compton effect*, and (c) *electron positron pair production* (Fig. 5.4).

### 5.10.1 Photoelectric Effect

It is a low-energy event, which extends the concept of quantization to the very nature of the radiation light itself. In this the electromagnetic quantum energy $h\nu$ ejects a bound electron from one of the shell due to absorbed photon and acquires an energy $E_e$ which is equal to the difference of photon energy $E_p$ and the binding energy of electron $E_b$. The acquired energy $E_e$ can be expressed mathematically as

$$E_e = E_p - E_b \tag{5.28}$$

At high energy, $(E_p/E_b) \geq 1$, the cross section of the photoelectric effect $\sigma_{ph}$ depends on $Z$ and $E_p$ as

$$\sigma_{ph} \propto (Z^5)/(E_p) \tag{5.29}$$

Thus, *photoelectric effect* plays a leading part for heavy elements with large $Z$, and the absorption coefficient $\mu$ is approximately proportional to $Z^4$.

The rate $(n/t)$ at which photoelectrons (with charge $e$) are emitted can be written as

$$(n/t) = (I/e) \tag{5.30}$$

where $I$ is the photocurrent and $t$ is the time. The maximum energy of the photoelectrons is $KE_{max}$ and is equal to

$$KE_{max} = (E_{photon} - \varphi) = eV, \tag{5.31}$$

where $E_{photon} = h\nu$, $\nu$ is the frequency, $h$ is Planck constant, $\varphi$ is the work function of the metal surface, $V$ is the stopping potential, and $e$ is the electronic charge.

### 5.10.2 Compton Effect

The effect is known after Compton who in 1923 discovered that a well-defined photon from a radiation beam having wavelength $\lambda_0$ (energy $E_p$) interacts with an electron of a metallic foil and is scattered through an angle $\theta$ with a well-defined wavelength $\lambda_1$ (energy $E_p'$).

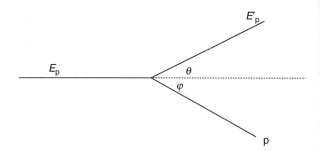

**Fig. 5.3** The schematic diagram of Compton scattering of a γ-ray by an electron

Figure 5.3 depicted the scattering event of a photon with initial energy $E_p$. $E'_p$ is its energy after being scattered through an angle $\theta$. The original energy of the rest electron (511 keV) is $E_0$. It gains a momentum of magnitude $p$ after being struck by the incident gamma ray and projected at an angle $\varphi$ with respect to the incident direction. From the conservation of energy and momentum, we can set up a mathematical relation of $E'_p$ as follows:

$$E'_p = [\{(E_p)\}]/[1 + \{E_p/(m_e c^2)\}(1 - \cos\theta)\}] \tag{5.32}$$

where $\theta$ is the angle between the direction of the original photon path and the scattered one, $m_e$ is free electron mass, and $c$ is the velocity of light. The scattered electron is transferred to the scintillation event and adversely affects the light output of the scintillator. The *Compton scattering* per electron is independent of Z (atomic number), and therefore, the cross section of *Compton scattering* $\sigma_C$ per atom is proportional to Z and can be related to the energy $E_p$ and number of electrons in an atom (Z) as

$$\sigma_C \propto (Z/E_p) \tag{5.33}$$

The scattering and absorption processes contribute to the attenuation of a beam of radiation beam passing through matter. The flux of the attenuated beam decreases exponentially with increasing thickness ($t$) of the barrier. The quantity ($\sigma_a N$), where $\sigma_a$ is the total cross section of the atom and $N$ is the number of atoms, is called the attenuation coefficient. It has the dimension cm$^{-1}$ and is reciprocal of the thickness of the barrier (e.g., aluminum foil) required to attenuate the beam by a factor $e$. This thickness is called the *attenuation length* ($1/\sigma_a N$) [67].

The *linear attenuation coefficient* ($\mu$) is limited by the fact that it varies with density of the absorber material, even though the absorber material is the same. Therefore, a more appropriate term called *mass attenuation coefficient* ($\mu/\rho$, where $\rho$ is the density of the material) is used for all practical purposes. The m*ass attenuation coefficient* is really of more fundamental value than are the linear coefficients because all mass attenuation coefficients are independent of the actual density and physical state of the absorber [68]. All commercial and promising scintillators can be divided into four main categories, namely, oxides, halides,

## 5.11 Ionization Losses

chalcogenides, and glasses. Except glasses, almost all the other inorganic (almost there are more than 400) are crystalline in structure. These *crystalline scintillators* are usually compounds having two or more elements, and the mass attenuation coefficient $\mu_m$ is given by

$$\mu_m = W_A\mu_{mA} + W_B\mu_{mB} + W_C\mu_{mC} + \cdots \qquad (5.34)$$

where $W_A = \{(xM_A)/(xM_A + yM_B + zM_C)\}$ (having three elements, A, B, and C). Similarly, $W_B = \{(yM_B)/(xM_A + yM_B + zM_C)\}$, $W_C = \{(zM_C)/(xM_A + yM_B + zM_C)\}$, and $(\mu_{mA})$, $(\mu_{mB})$, and $(\mu_{mC})$ are mass attenuation coefficients of the elements A, B, and C, respectively. The effective $Z_{eff}$ is given by

$$Z_{eff} = \left(W_Z Z_A^4 + W_B Z_B^4 + W_C Z_C^4\right) \qquad (5.35)$$

Absorption processes due to *Compton effect* are undesirable for position-sensitive (scintillators) $\gamma$-ray detectors, because two separated light sources degrade the position resolution of the detector. Figure 5.4 gives theoretical values of the mass attenuation coefficient for NaI for $hv = 0.01$–$100$ Mev, plotted from theoretical calculations of G. R. White, National Bureau of Standards.

### 5.10.3 Pair Production

The third mechanism by which the gamma ($\gamma$) ray can be absorbed is the *pair production*. Pair production can only occur when the energy of the gamma ray exceeds twice the rest-mass energy of an electron (1.02 MeV). Pair production is always followed by annihilation of the positron, usually with simultaneous emission of two 0.51-MeV photons. The probability of pair production increases with the nuclear charge and the energy is approximately as

$$\sigma_{pr} \propto Z^2 \ln 2E_p \qquad (5.36)$$

Figure 5.5 shows three interactions described above and the relative importance of the three major types of gamma ray interaction, namely, (a) *photoelectric effect*, (b) *Compton effect*, and (c) *electron positron pair production*. The y-axis shows the Z values of the absorber and the x-axis denotes the energy $hv$ in MeV.

### 5.11 Ionization Losses

Heavy charged particles such as alpha particle interact with matter primarily through Coulomb forces. Depending on the energy of the incoming particle, the impulse may be sufficient enough to remove completely the electron of the absorber. The energy in this particular case is defined as the *ionization energy*, and the differential energy loss (d$E$/d$x$) of the particle within the absorber due to the impact is called the *linear stopping power* of the absorber. The mean ionization

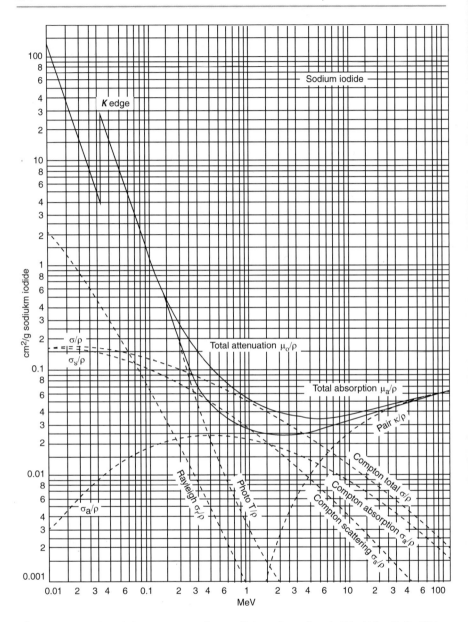

**Fig. 5.4** Illustration of the mass attenuation coefficients for sodium iodide (After G. R. White, NBS)

losses of a charged particle are determined by the well-known Bethe–Bloch formula, which can be written as

$$-(dE_{ion}/dx) = \left(4\pi N_A z^2 e^4/m_e V^2\right)(Z/A)\left[\{\ln(2mv^2/\bar{I}(1-\beta^2)\} - \beta^2\right] \quad (5.37)$$

## 5.11 Ionization Losses

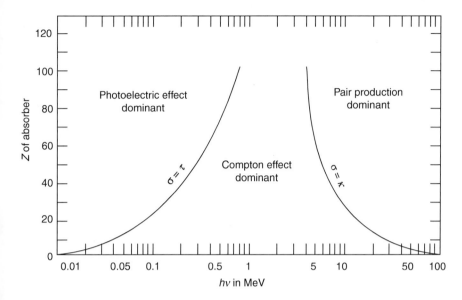

**Fig. 5.5** Illustration of the relative importance of the three major types of gamma ray (Reproduced with permission: John Wiley)

where $x$ = penetration depth in g/cm², $N_A$ = Avogadro's number, $Z_e$ = charge of the primary particle, $e$ = electronic charge, $m_e$ = electron rest mass, and $Z$ and $A$ = atomic and mass numbers of the absorber, respectively, $\bar{I}$ = mean excitation and ionization potential, and $\beta = v/c$, $v$ and $c$ are the velocities of the charged particle and light, respectively.

From the above Eq. 5.37, we can see that for nonrelativistic particle differential energy varies inversely with velocity $v$, or in other words with the energy of the particle. At the same time ionization power of a charged particle (proton, α particle, ionized atom, etc.) is $z^2$ times greater than the ionization power of an electron of the same velocity. The energy losses due to ionization also depend on the property of the solids due to constancy of the ratio $Z/A$. For low-energy electrons, protons, α particles, and heavy ions, small scintillator thickness (~0.5 mm) is enough for detection of these particles. But when the penetrating particle has high energy, it can lose energy preferentially by *bremsstrahlung*, that is, by radiation of a photon due to deceleration in the Coulomb field of the nucleus. The energy loses due to radiation can be approximated as

$$-(dE_{rad}/dx) \approx Z_{eff}^2 n\, E_k \tag{5.38}$$

where $n$ is the number of atoms and $E_k$ is the kinetic energy of the electron.

By *bremsstrahlung* an electron (positron) loses energy at a rate proportional to its energy. The stopping power of the *bremsstrahlung* varies asymptotically and approaches $X_0 E$ where $X_0$ is the *radiation length* and $E$ is the energy of the particle. On the other hand, collision losses (ionization or excitation) vary only slowly.

272                                                    5 Basic Principles of Radiation Detectors

When the collision loss rate equals the *bremsstrahlung* rate at that point, we defined the energy as *critical energy* $E_c$ and can be approximated as

$$E_c(\text{MeV}) \approx (800/Z_{\text{eff}}) \tag{5.39}$$

The *radiation length* (in cm) $X_0$ in material is approximated as

$$(1/X_0) = \left[\{4\alpha \cdot Z(Z+1)(r_e^2 N_A)\}/\{(A)\}\right] \ln(183/Z^{1/3}) \tag{5.40}$$

where $A$ is the atomic number, $\alpha = (1/137)$ (fine structure constant), and $r_e = (e^2/m_e c^2)$ is the classical radius of electron with mass $m_e$ and $c =$ velocity of light. When the scintillator material is a mixture of different compounds $(A_x, B_y, C_z)$ with $Z >> 1$, the $X_0$ of Eq. 5.40 transform to

$$(1/X_0) = \left[\{4\alpha \cdot r_e^2 N_0)\}/\{(A)\}\right]$$
$$\times \left[xZ_A^2 \ln(183/Z^{1/3}) + yZ_B^2 \ln(183/Z^{1/3}) + zZ_C^2 \ln(183/Z^{1/3})\right] \tag{5.41}$$

---

## 5.12 Inorganic Scintillators

The use of scintillators for radiation detection is a century-old event. The earliest scintillators $CaWO_4$ and $ZnS$ go back to the time of Rontgen and Crookes and Rutherford [69, 70]. The basic principle of scintillating properties of the inorganic materials is the conversion of the energy of the incident radiation or particle into great many low-energy photons. The ionization event creates an inner shell hole and an energetic primary electron, followed by radioactive decay (secondary X-rays), non-radiative decay (Auger process), and inelastic electron–electron scattering.

After ionization event the scintillator is in a nonequilibrium position and tries to come to an equilibrium position by relaxation through a multitude of elementary processes such as creation of primary electronic excitation, which will produce an avalanche of secondary excitations including electrons, holes, photons, and plasmons. The scintillation process is followed through different sequences such as absorption of the ionizing radiation, generation of electrons and holes (e–h), relaxation of the system and production of secondary events, transfer of energy from the e–h pair to the luminescence centers, and ultimate illumination from the luminescence centers. Thus, the scintillation process can be called the migration stage due to the transfer of energy to the luminescent centers. For example, thallium (Tl) in halide materials (NaI, LiI, CsI, $CaF_2$, $BaF_2$,) are mainly involved in the electronic recombination such as

$$\text{either,} \quad Tl + h \rightarrow Tl^{2+}, \quad Tl + e \rightarrow hv \tag{5.42}$$

## 5.12 Inorganic Scintillators

$$\text{or,} \quad Tl + e \rightarrow Tl^0, \quad Tl^0 + h \rightarrow h\nu \tag{5.43}$$

The excitonic mechanism of energy transfer to luminescence centers occurs less often than recombination luminescence in inorganic crystals like sodium iodide (NaI) and cesium iodide (CsI). When NaI and CsI crystals contain more than 0.02 % thallium (Tl), the complex activator centers (aggregates) are formed. These centers also represent deep traps in the crystals, and as a result, NaI and CsI with 0.02–0.03 % $Tl^+$ activator give rise to higher light yield. It has been detected that the hole migration to $Tl^0$ centers plays an important role in the scintillation process in CsI:Tl [71]. The scintillation spectrum of CsI:Tl extends from 300 to 800 nm in the visible spectral region with a peak of about 560 nm at room temperature. However, at liquid helium temperature, CsI:Tl shows six absorption band peaks at 228, 238, 246, 257, 278, and 290 nm. In NaI:Tl, the $Tl^0$ and Tl2+ centers produce absorption bands at 620 and 312 nm, respectively [72].

Trivalent cerium ($Ce^{3+}$) has also been extensively used as an activator in many solid-state phosphor and scintillator materials. $Ce^{3+}$ may be excited to a 5d state by ionizing radiation either directly by intraionic process within $Ce^{3+}$ or indirectly. When $Ce^{3+}$ loses its 4f electron, it forms $Ce^{4+}$ by direct ionization or by capturing a hole:

$$Ce^{3+} + h \rightarrow Ce^{4+}, \quad Ce^{4+} + e \rightarrow (Ce^{3+*}) \tag{5.44}$$

$Ce^{3+*}$ will emit a 5d $\rightarrow$ 4f photon via

$$(Ce^{3+*}) \rightarrow Ce^{3+} \; h\nu \tag{5.45}$$

In alkaline earth fluorides, $Ce^{3+}$ enters the lattice substitutionally at a divalent alkaline earth site. If there is no nearby charge compensating ion, $Ce^{3+}$ acts as an electron trap and gives rise to

$$Ce^{3+} + e \rightarrow Ce^{2++} \tag{5.46}$$

On the other hand, if it captures a hole, it will result in $Ce^2 + h \rightarrow Ce^{3+*}$.

The luminescence properties of cerium-activated ($Ce^{3+}$) materials are well understood for most compounds used as phosphors [73, 74]. The free cerium atom has a single electron in ground $4f(^2f_j)$ state and the excited states are 5d ($^2D_j$) and $6s(^2S_{1/2})$. The emission of $Ce^{3+}$ ions usually occurs from the lowest 5d level to the ($^2F_{7/2}$) and ($^2F_{5/2}$) levels of the 4f ground state. Depending upon the host crystal and the temperature, transitions to the ($^2F_{7/2}$) and ($^2F_{5/2}$) levels may be resolved.

A wide variety of rare earth (RE) ions mainly cerium ($Ce^{3+}$) and europium ($Eu^{2+}$) are usually used for scintillators with short decay time and high efficiency at room temperature (RT). The energy gap between ground 4f and excited 5d states of $RE^{3+}$ ions is typically of several electron volts, and 4f $\rightarrow$ 5d transitions can usually be

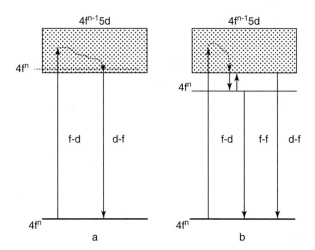

**Fig. 5.6** Two possible locations of a $4f^n$ excited level and $4f^{n-1}5d$ configuration of a rare earth ion

observed in wide band gap host materials. The energy of $4f \rightarrow 5d$ absorption and $5d \rightarrow 4f$ emission transitions is smaller for cerium ($Ce^{3+}$), praseodymium ($Pr^{3+}$), neodymium ($Nb^{3+}$), and terbium ($Tb^{3+}$) ions [75]. Figure 5.6 shows two possible locations of a $4f^n$ excited level and $4f^{n-1}5d$ configuration of a rare earth ion. Among the divalent $Eu^{2+}$, samarium ($Sm^{2+}$), ytterbium ($Yb^{2+}$), and thulium ($Th^{2+}$) are stable in the crystals without an irradiation.

$Eu^{2+}$ is the most stable $RE^{2+}$ ion in crystal. The ground-state electronic configuration of $Eu^{2+}$ is that of a half-filled 4f shell, that is, 4f7 ($^8S_{7/2}$). $CaF_2$:Eu, LiI:Eu, CaI:Eu, and $SrF_2$:Eu materials are used as scintillators [76]. The scintillating materials are formed upon isomeric substitutions of cations in the host structure by $Eu^{2+}$ ions. It has been observed that the behavior of $Ce^{3+}$ ion in solids is similar to those of $Eu^{2+}$ ion. In relatively narrow band gap crystals, the excited levels of $Ce^{3+}$ ions can fall into the conduction band and the $Ce^{3+}$ luminescence is quenched like it is in $BaF_2$:$Eu^{2+}$ [77]. Crystals with wide band gaps like $LaF_3$ ($E_g \sim 10$ eV), $LuPO_4$ ($E_g \sim 8.27$ eV), and $LuAlO_3$ ($E_g \sim 7.65$ eV) show high scintillation efficiency when activated with $Ce^{3+}$, while $Y_2O_3$:Ce ($E_g \sim 5.6$ eV) does not emit at room temperature at all.

In $Nd^{3+}$ and $Pr^{3+}$ the allowed $5d \rightarrow 4f$ radiative transitions can be used in scintillation materials like $LaF_3$, $BaF_2$, ($Nd^{3+}$), and $Y_2SiO_2$ ($Pr^{3+}$). The ionic radii of $Y^{3+}$ (0.106 nm), $Lu^{3+}$ (0.009 nm), and activators $Ce^{3+}$ (0.118 nm) and $Pr^{3+}$ (0.116 nm) play a role in the efficiency of the energy transfer to luminescence centers. Activator $Ce^{3+}$ ion shows higher light output in the host oxides compared to $Pr^{3+}$ ion.

According to the ionization mechanism of the activator, the inorganic scintillation crystals can be classified into three categories [78, 79]. (a) In NaI:Tl, CsI:Na, $CaF_2$:Eu, and $Lu_2SiO_5$:Ce, the ionizing energy diffuses through the host crystal and produces an excited state at or near an activator atom that is present in low ($\approx 0.1$ %) concentration of the activator. (b) In CsI, NaI, and $Bi_4Ge_3O_{12}$, the scintillators are called self-activated due to the involvement of the excitonic process. (c) In $BaF_2$, CaF, and $RbCaF_3$, the ionizing radiation produces an electron vacancy in an upper

## 5.12 Inorganic Scintillators

**Table 5.2** Classification of main electronic transitions in inorganic scintillators

| Electronic transitions | Luminescence centers | Typical compound | Decay time at 300 K |
|---|---|---|---|
| Mercury like $^3P_1(sp) \rightarrow {}^1S_0(s)^2$ | Tl+, In+, Pb+, Bi+ | NaI:Tl, $Bi_4Ge_3O_{12}$ | $\sim 3 \times 10^{-7}$ |
| Interconfiguration 5d $\rightarrow$ 4f | $Ce^{3+}$, $Nd^{3+}$, $Eu^{2+}$ | $CeF_3$ $Gd_2SiO_5$:Ce, | $(3\text{–}7) \times 10^{-8}$ |
| Interconfiguration $4f^65d \rightarrow 4f$ | $Eu^{2+}$ | $CaF_2$:Eu, LiF:Eu | $\sim 1 \times 10^{-6}$ |
| $np^6 \rightarrow nd$ | $WO_4^{2-}$, $MoO_4^{2-}$ | $CaWO_4$, $PbWO_4$ | $\sim 6 \times 10^{-6}$ |
| Excitonic $(3\Sigma_u^+ \rightarrow {}^1\Sigma_g^+)$ | $V_k$ | $CaF_2$, $SrF_2$, $BaF_2$ | $\leq 10^{-6}$ |
| Core to valence $np^6 \rightarrow np^6(X)$ | $Ba^{3+}$, $Cs^{2+}$, $Rb^{2+}$ | $BaF_2$, CsF, $BaLiF_3$ | $\sim 1 \times 10^{-9}$ |
| Excitonic like | – | CsI, $CdI_2$ | $\leq 1 \times 10^{-8}$ |
| Point defect | $[V_{zn}Te_0Zn_{2-}]$ | ZnSe:Te | $5 \times 10^{-6}$ |

level of one atom and the vacancy is promptly ($\approx 1$ ns) filled with an electron from the valence band of another atom to produce scintillation.

The scintillation process in activated scintillating materials undergoes various stages [79], such as (Table 5.2):

(a) Ionization, where the average energy needed to produce electron–hole (e–h) pairs is almost three times the energy gap of the material.

(b) Relaxation of electrons and holes: In ionic crystals it is ~1 eV, and in alkali crystals the self-trapped hole forms $V_k$ center, where the two halogen atoms share the hole forming covalent bond [80].

(c) Carrier diffusion: When the band gap of the material is high (ionic crystals), the valence bands are filled and an excess electron is spatially diffuse in an essentially empty conduction band. On the other hand, when the self-trapped hole forms a $V_k$ center, the breaking of the covalent bond is an energy barrier for hole transport that retards their diffusion. As a result, the activated scintillating materials whose holes form $V_k$ centers (NaI:Tl, CsI:Tl, and CsI: Na) generally have slow rise time.

(d) Sequential capture.

(e) Radiative emission [81].

### 5.12.1 Light Yield of an Inorganic Scintillator

One of the most important things of a scintillator is its light output, which depends not only on the energy losses but on the density of the ionization as well. The effect of density ionization of a scintillator is expressed in terms of the $(\alpha/\beta)$ ratio, and the energy efficiency $(\eta)$ is expressed as

$$\eta = \{(\alpha/\beta)(h\nu_m/E_g)\} = (h\nu_r/E\gamma)N_{ph} \tag{5.47}$$

where $\alpha = (N_{ph}/N_{eh})$, $\beta$ = numerical coefficient, $h$ = Planck constant, $v_m$ = scintillator minimum emission frequency, $hv_r$ = mean energy emitted by photon, $E\gamma$ =gamma quantum energy, $E_g$ = band gap of the crystal, and $N_{ph}$ and $N_{eh}$ = number of photons and electron–hole pairs. The mean energy of emitted photons $(hv_r)$ at frequency $v_r$ is proportional to the intensity of scintillator emission at frequency $v_r$ and can be expressed as

$$(hv_r) = \left[ \int_{v_{min}}^{v_{max}} \{hv_r J(v_r) dv_r\} \right] \Big/ \left[ \int_{v_{min}}^{v_{max}} \{I(v_r) dv_r\} \right] \qquad (5.48)$$

where $v_{min}$ and $v_{max}$ are the minimal and maximal frequencies of the emission spectrum, respectively. The intensity of the emission $I(t)$ (in photons/s) of the scintillator with one type of luminescent center decays exponentially with time constant $\tau$ and can be expressed mathematically as

$$I(t) = I(0) \exp(-t/\tau) \qquad (5.49)$$

where $I(0)$ is the initial intensity, that is, photons/s at time $t = 0$, and $\tau$ is decay time constant which is equal to the inverse of radiative rate (wr) and is given by

$$\tau = [\{(cm_e/8\pi q^2)\}].[\{(\lambda^2/fn)\}\{(3)/(n^2 + 2)^2)\}] \qquad (5.50)$$

where $f$ is the oscillator strength, $n$ is the refractive index of the crystal, $\lambda$ is the wavelength of the transition, and $q$ is the electronic charge. From the above equation, one can presume that ultraviolet emitting crystal will give faster decay time. The impurity or thermal quenching may decrease decay time, but this is accompanied by the decrease of the *light yield*.

The decay time of the scintillator (detector) should be as short as possible when the counting rate of the detector is high. To obtain short decay time without quenching, the *core to valence* (C–V) transition can be used. The C–V transition, which is also known as *cross luminescence* or *Auger luminescence*, is a new type of luminescence first observed in barium fluoride ($BaF_2$). The *intrinsic luminescence* due to C–V has a *short decay time*, a *high thermal stability* in all parameters, and relatively *high light yield*. Therefore, the crystals with *radiative C–V transitions* are promising candidates for a new class of scintillator.

The *radiative C–V transitions* occur provided that holes are formed in the upper core band when we have the condition that the energy of the incident quanta $(hv_i)$ is $\geq E_{cc}$ where $E_{cc}$ is the energy gap between the upper core band and the conduction band. The quantum efficiency of the C–V luminescence at $(hv_i) \geq E_{cc}$ for some crystals is high, while the energy efficiency at $(hv_i) \leq E_{cc}$ does not exceed 1.5 % [79]. C–V scintillation efficiency is very low when it is excited with $\alpha$ particles because high-density crystals produce low-energy secondaries when they are excited with $\alpha$ particles.

## 5.13 Defect Formation by Ionizing Radiation

*The light yield* of the inorganic scintillator having band gap energy $E_g$ can be presented in the mathematical form as

$$L \approx \left\{ (0.5/E_g) \times 10^6 (\text{photons/MeV}) \right\} \ldots = (I_0 \tau / E\gamma) \tag{5.51}$$

However, for practical measurements, there are some corrections to be made of the mathematical formulation given above, namely, (1) *absorption correction*, (2) *rate correction*, and (3) *target absorption*.

1. *Absorption correction*: The absorption correction is done because of the fact that the incident radiation that reaches the monitor is not detected by the scintillator. The correction factor can be written as $C_{abs}$ = (total number of photons coming out of the radiation source $(P_T)$/(total number of photons received by the scintillator $(S_T)$) = $(P_T/S_T)$. Again, the true calibration factor $f_{cal}$ differs from the measured calibration $f'_{cal}$ by factor of $C_{abs}$ and can be written as $f_{cal} = C_{abs} f'_{cal}$. The true calibration factor can also be measured from the simulation provided all parameters of the flux monitor are known [82].

2. *Rate correction*: The photons that are coming out of the radiation source $(N_{pR})$ are not detected completely by the detector $(N_{pD})$. So we can write

$$(N_{pR}) = (N_{pD}) / (\eta_{PD}) \tag{5.52}$$

where $(\eta_{PD})$ is efficiency of the detector. If the experimental arrangement consists of any veto paddle [83], the Eq. 5.52 will transform to

$$(N_{pR}) = (N_{pD})/(\eta_{PD}) + \left\{ [(\eta_v (1 - \eta_{PD}) + \eta_{PD}/2)] / [(\eta_{PD}^2)] \right\} \left\{ \left( N_{pD}^2 / C \right) \right\} \tag{5.53}$$

where $\eta_v$ is the efficiency of the veto, $C$ is the accelerator per bunch, and $\eta_{PD}$ is the efficiency of the detector.

It is true that the light output will be high when $E_g$ is small and the crystal should be transparent for its own light emission, that is, $E_g > h\nu_m$. Crystals with small $E_g$ usually have covalent (homopolar) types of chemical bonds and it is difficult to obtain large light output ($A^{II} B^{VI}$ scintillators). However, with an ionic crystal with a narrow band gap and at $\alpha \approx 0$, the scintillator will yield maximum light output. The *light yield*, however, in inorganic scintillators is nonlinear when it is calculated on the basis of the microscopic spatial distributions of electron–hole pairs with an atomistic kinetic Monte Carlo (KMC) model of energy transfer [84].

## 5.13 Defect Formation by Ionizing Radiation

Ionization radiation can bring several damages to the crystal depending upon the energy and duration of the ionizing radiation. Due to the radiation damage, the intrinsic electronic excitation is transformed to *lattice defects*. Some of the primary produced defects are stable and the number of defect increases with the duration of the dose. In wide-gap crystals, these defects are called *color centers* that produce absorption bands. One can expect loss of light output if these bands are present in the emission regions of the scintillator. In addition to the lattice defects, ionizing radiation can produce several numbers of *shallow traps* due to inhomogeneities in the crystal lattice.

Electrons and holes trapped on inhomogeneous sites can liberate thermally at room temperature, and the electrons can undergo luminescent recombination with holes giving rise to *phosphorescence*. The resulting light emission after irradiation, if detected by the readout device, can cause additional noise and can degrade the overall energy resolution. During single crystal growth, the impurity distribution is not uniformly spread over the entire crystal. As a result, the light output along the length of the crystal will be different after irradiation.

High-energy radiation can produce F–H pair (Frenkel defects) through the capture of electrons as a result of anion vacancy in ionized halogen crystals [85, 86]. In ionic crystal, a pair of anion vacancy ($\alpha$-center) and interstitial ions (I center) or a pair of F–H centers is formed during the non-radiative annihilation or dissociation of a self-trapped exciton (STE) [87, 88].

When the distance between F and H center is small, one can expect an immediate recombination of the center without any thermal activation. Thus, we can expect two stages of defect formation in ionic crystals: the primary process of unstable F–H pair creation and dynamic motion of the II center and secondary processes of thermal migration and stabilization of the H center. In ionic crystals defects are formed mostly in the anion sublattice because STHs are localized on anions.

In case of ionizing radiation, the holes are supposed to be self-trapped within a very short time (ps) because of the absence of a potential barrier for self-trapping of the hole in contrast to the case of exciton, which is rather complex (because efficiency of the formation of F center depends on temperature and the amount of impurity content within the crystal). Radioactivation of the crystals having heavy elements causes serious problem like degradation in optical transmission when irradiated by hadrons (protons, neutrons, mesons, etc.).

## 5.14 Solid-State Detector

The ionizing radiation in materials often involves the generation of free electrical charges that can be separated under the influence of an electric field. In semiconductors, ionizing radiation created electron–hole (e–h) pairs, and an applied field gives them a motion which is the origin of the basic electrical signal from a semiconductor detector. The charge produced is equal to the total number of pairs ($N_0$) created multiplied by the unit of electronic charge $q$. A measurement of $qN_0$ is often taken as proportional to the initial energy of the interacting quantum of radiation, whether it is from incident photons or charged particles.

The semiconductor detectors that measure the ionizing radiation operate in different modes, like *current mode*, *charge integration mode*, *individual quantum*

## 5.14 Solid-State Detector

*pulse mode*, and *flash pulse mode*. In the *current mode operation*, detector output is measured as a direct current averaged over response time, which is typically a fraction of a second. The measured current reflects the product of the interaction rate multiplied by the average *induced charge* from a single interaction. The *charge integration mode* is applied when the direct current is too small to use current-induced mode. The charge across a capacitance for an extended period is measured which gives the total integrated charge. In individual quantum pulse mode, a charge-sensitive preamplifier with low noise can observe the voltage output pulse which will reflect the induced charge from the detector. The individual pulse can be stored and displayed as a differential pulse height spectrum in a multichannel analyzer. The flash pulse mode does not give any information about energy spectrum of the radiation and the incident radiation appears in form of burst radiation. The pulse is processed and the effect of the single pulse whose amplitude is proportional to the summed energy is being deposited in the detector volume from the burst.

When the detector is a semiconductor, charges created under the influence of X-ray radiation are collected by means of electrical contacts applied across the detector volume. The electrical contact is made by a highly conductive metal, which will influence the transport of carriers due to band bending in the semiconductor during alignment of the Fermi levels of the metal and the semiconductor.

A metal semiconductor contact has negligible contact resistance, and the voltage change across the contact should be negligible with respect to the applied voltage. Thus, the voltage applied across the contact will have negligible power, and thermal equilibrium will be established current will flow through the contact. The semiconductor material (fourth group of the periodic table, e.g., silicon or germanium) can be n-type (doped with elements from the fifth group of the periodic table) or p-type (doped with third group material of the periodic table). For n-type semiconductor, the work function $\varphi_n$ should be less than the work function of the metal $\varphi_m$, and for p-type semiconductor, it is just the opposite of the previous one. The problem is generally circumvented by degenerately doped semiconductor surface such that Schottky barrier is formed which is sufficiently thin to allow for current tunneling directly through the barrier.

The electrical signal produced from a semiconductor detector is due to the motion of the individual charge carriers within the active region of the device. The magnitude and time profile of the induced charge can be derived by simple arguments based on the conservation of energy [89]. As soon as the charge carriers (electrons and holes) are created, the two types of charges will drift in opposite directions. Once one set of carriers has been collected, the remaining set continues to drift for an additional time and is collected from the opposite end of the active region. At some point there is no drift of the charges and the signal voltage will drop to zero with time. Subsequent shaping of the pulse is carried out in a linear amplifier using shaping times that are longer than the collection times so that the information carried by the full amplitude of the pulse is preserved.

## 5.14.1 Uniform Excitation of the Solid-State Detectors

The number of carriers collected will depend on the recombination effects. The short time trapping and de-trapping of the charges will determine the shape of the pulse. If the hole extraction time is long compared to the hole lifetime, the hole may undergo multiple trapping and de-trapping with effective mobility ($\mu_h'$) as

$$\mu_h' = \mu_h\{(\tau_1/(\tau_1 + \tau_2)\} \tag{5.54}$$

where $\mu_h$ is the mobility without trapping and $\tau_1$ and $\tau_2$ are average trapping time and the time spent in the traps. The hole excitation time can be written as

$$t_h' = t_h\{((\tau_1 + \tau_2)/\tau_1)\} \tag{5.55}$$

where $\tau_h$ is the extraction time of holes in absence of trapping. When electrons and holes both undergo recombination either by direct or trap-assisted recombination, charge carriers are completely removed in transit and no longer contribute to the pulse height. However, the probability of recombination in the depleted region is negligible, but recombination can be significant in the vicinity of dense charge creation [90].

In an ideal planar detector (neglecting the effects of carrier loss due to recombination), the charge measured can be expressed as

$$Q = [\{(qN_0)/W\}] \\ \times [\{(\tau_e x_e/t_e(1 - \exp(-t_e/\tau_e^*)\} + \{(\tau_h x_h)/t_h)(1 - \exp(-t_h/\tau_h^*)\}] \tag{5.56}$$

where $q$ is the electronic charge, $N_0$ is the total number of carriers, $W$ is the width of active region, $\tau_e$ and $\tau_h$ are the lifetime of the electrons and holes, $x_e$ and $x_h$ are the distance the electrons and holes traveled before they recombine, $t_e^*$ and $t_h^*$ are the mean drift time of the electrons and holes, and $t_e$ and $t_h$ are the time required for the carriers to be collected.

In case one carrier is severely trapped, the maximum pulse height location may shift closer to the contact toward which the trapped carriers traveled. On the other hand, if one carrier travels at a higher velocity than the other, but retains the same value of $\tau^*$, the maximum pulse height location skews in the direction of the contact toward which the slower carriers travel. When the charge excitation is confined to a localized point and the total charge created is simply the gamma ray ($\gamma$-ray) energy divided by the necessary average excitation energy for an electron–hole pair, the average charge collected can be simply be expressed as

$$<Q> = \left[\int_0^w \{f(x)Q(x)\mathrm{d}x\}\right]/\left[\int_0^w (f(x)\mathrm{d}x\right] \tag{5.57}$$

# 5.14 Solid-State Detector

where $f(x)$ is the spatial distribution function for $\gamma$-ray interactions. When $f(x)$ is constant, the average charge collected is found to be

$$<Q> = N_0 q \left[ \delta_e + \delta_e^2 (e^{-1/\delta e} - 1) + \delta_h + \delta_h^2 (e^{-1/\delta e} - 1) \right] \qquad (5.58)$$

where $\delta$ is the carrier extraction factor and is equal to $(\tau^* v/W)$, h and e in the subscripts represent holes and electrons, and $v$ is the carrier velocity. The variance of the charge collected can be calculated from the following equation:

$$\sigma_Q^2 = <Q^2> - <Q>^2 \qquad (5.59)$$

If we assume that statistical deviations in the total drift length of the carriers and in the total number of e–h pairs created, we can write Eq. 5.59 as

$$\begin{aligned}
\sigma_Q^2 = N_0^2 \, q^2 [\delta_e^2 &+ 2\delta_e^3 (e^{-1/\delta e} - 1) + \delta_h^2 + \{(\delta e^3/2)(1 - e^{-2/\delta e})\}] \\
&+ 2\delta_h^3 (e^{-1/\delta h} - 1) + \{(\delta_h^3/2)(1 - e^{-2/\delta h})\} + 2\delta_e \delta_h \\
&+ \{2\delta_e^2 \delta_h (e^{-1/\delta e} - 1)\} + \{2\delta_e \delta_h^2 (e^{-1/\delta h} - 1)\} \\
&+ \{2(\delta_e \delta_h)^2/(\delta_e - \delta_h)\}\{(e^{-1/\delta e} - e^{-1/\delta h})\} - <Q>^2 \qquad (5.60)
\end{aligned}$$

When the effects of charge trapping is negligible, each of the $N_0$(e–h) pairs that are created by incident radiation contributes exactly the same increment to the total induced charge. The resulting pulse amplitude is therefore subjected only to statistical fluctuations in $N_0$. The corresponding fractional variance $\{\sigma_n^2 = (\sigma_N/N_0)\}$ is given by $F/N_0$, where $F$ is the *Fano factor*. In case of strong trapping, the contribution of one carrier (say electrons) is proportional to the total drift length traveled by all $N_0$ electrons. The total drift length $X = x_1 + x_2 + x_3 + \cdots + x_{N0}$ and

$$P(X)\mathrm{d}x = \left[ \{(X/\lambda)^{N_0-1} e^{-X/\lambda}\}/\{N_0 - 1)!\}(\mathrm{d}x/\lambda) \right] \qquad (5.61)$$

where $\lambda$ is the mean drift length. The expected value of $X$ is presented by $<X> = N_0\lambda$ and the standard deviation is $\sigma_x = \lambda\sqrt{N_0}$ [91]. The fractional standard deviation in the induced charge is

$$\sigma_d = (\sigma_D/<Q_D>) = \left[ (1/N_0)\{(\lambda_e^2 + \lambda_h^2)\}/\{(\lambda_e + \lambda_h)^2\} \right] \qquad (5.62)$$

If one carrier drift length is much greater than the other, then the above equation will transform into

$$\sigma_d = \sqrt{(1/N_0)\left(\{(\lambda_e^2 + \lambda_h^2)\}/\{(\lambda_e + \lambda_h)^2\}\right)} \qquad (5.63)$$

If one carrier length is much greater than the other $(\lambda_e \gg \lambda_h)$, the above equation will reduce to

$$\sigma_d = \sqrt{(1/N_0)} \qquad (5.64)$$

The total statistical fluctuation is then given by

$$\sigma_s = \sqrt{\sigma_s^2 + \sigma_d^2} = \sqrt{(F + 1)/N_0} \qquad (5.65)$$

Since the *Fano factor* for many semiconductors is on the order of 0.1, it is evident that the statistical contributions due to fluctuations in the total drift length $(\sigma_d)$ are generally much larger than those due to fluctuations in the number of carriers $(\sigma_n)$. If the drift lengths are identical $(\lambda_e = \lambda_h)$, $\sigma_d$ will be equal to $\sigma_d = \sqrt{(1/2N_0)}$.

In semiconductor materials where significant trapping occurs, energy resolution from $\alpha$ particles is always found to be better than that measured from $\gamma$ particles [92]. Medical imaging using ionizing radiation requires X-ray for transmission imaging and the use of radioactive tracers in emission. Position-sensitive detectors used in radiographic mode are characterized by high flux, short-duration exposure, and a detector system that can give high spatial resolution. It is usually operated in integrating mode.

## References

1. Gupta TK et al (1999) High speed X-ray imaging camera using a structured CsI (Tl) scintillator. IEEE Trans Nucl Sci NS-46:232 and also Kasap SO and Rowlands JA (2002) Direct conversion flat panel X-ray image sensors for digital radiography. Proc IEEE 90:591
2. Wilson J (1989) Optoelectronics. Chap. 7. Prentice Hall Int., Hemel Hempstead and also Zhang N, Grazioso RF, Doshi NK, Schmand MJ (2006) RF transformer coupled multiplexing circuits for APD PET detectors. IEEE Trans Nucl Sci 55(5):2570
3. Zappettini A et al (2009) Boron oxide encapsulated vertical Brodgman grown CdZnTe crystals as X-ray detectors. IEEE Trans Nucl Sci 56(4):1743
4. Knoll GF (1999) Radiation detection and measurements, 3rd ed. Chapters: 11, 12 and 13. Wiley, New York and also Spencer DF, Aryaeinejad R, Rebber EL (2002) Using the Cockroft-Walton voltage multiplier with small photomultiplier. IEEE Trans Nucl Sci 49(3):1152
5. Globus M (2006) Scintillation detectors for medical imaging. Springer, New York
6. Garlick GFJ (1966) Luminescence of inorganic solids. Chapter 12. Academic, New York and also McClish M, Farell R, Vandepuye K, Shah KS (2007) A reexamination of deep diffused silicon avalanche photodiode gain and quantum efficiency. IEEE Trans Nucl Sci 53(5):3049
7. Rodnyi PA (1977) Physical process in inorganic scintillators. CRC Press, Boca Raton and also Gramsch A, Avila RE, Bui P (2003) Measurement of the deep interaction of an LSO scintillator using a planar processed APD. IEEE Trans Nucl Sci 50(3):307
8. Degenhardt C et al (2009) The digital silicon photomultiplier-A novel sensor for detection of scintillation light. In: IEEE nuclear science symposium and medical imaging, Orlando, 25–31 Oct 2009 and also Piemonte C (2009) Production of large area silicon photomultipliers for a PET/MR scanner. In: IEEE nuclear science symposium and medical imaging, Orlando, 25–31 Oct 2009

# References

9. Kamada K et al (2009) Pr:Lu3Al5O12 scintillator read out using enhanced avalanche photodiode. In: IEEE nuclear science symposium and medical imaging, Orlando, 25–31 Oct 2009
10. Wernick MN, Aarsvold JN (2004) Emission tomography: the fundamentals of PET and SPECT. Elsevier Sci Pub, New York
11. Derenzo SE, Weber MJ, Bourret-Courchesne E, Klintenberg MK (2003) The quest for ideal semiconductor scintillator. Nucl Instrum Method A 505:111
12. Tavernier S, Getkin A, Grinvov B (2006) Radiation detectors for medical application. Springer, New York and also Darambara DG, Speller RD, Sellin P (2002) Development of a dual detector system based on Si:H arrays and multielement silicon detectors for diffractive enhanced breast imaging. IEEE Trans Nucl Sci 49(3):1012
13. Knoll G (2000) Radiation detection and measurements. Wiley, New York
14. Derenzo SE, Bourett-Courchesne E, Weber MJ, Klintenberg MK (2005) Scintillation studies of CdS (In): effects of coupling strategies. Nucl Instrum Method Phys Res A 537:261
15. Kastalski A, Luryi S, Spivak B (2006) Semiconductor high-energy radiation scintillation detector. Nucl Instrum Method A 565:650
16. Sze SM (1998) Modern semiconductor device physics. Wiley, New York and also Burstein E (1992) Anomalous optical absorption limit in InSb. Phys Rev 93:632
17. Gupta TK (2003) Hand book of thick and thin film hybrid microelectronics. Wiley InterScience, Hoboken and also Choi C, Kyon C, Kang C, Nam S (2007) Comparison of compound semiconductor radiation films deposited by screen printing method. Proc SPIE Med Imaging 6510:651042-1-8
18. Mao R, Zhang L, Zhu R-Y (2008) Optical & scintillation properties of inorganic scintillators in high energy physics. IEEE Trans Nucl Sci 55(4-part-2):2425
19. Lecog P, Annenkov A, Gektin A, Korzhik M, Pedrini C (2006) Inorganic scintillators for detector systems: physical principle and crystal engineering. Springer, New York
20. Brekenridge RG, Russel BR, Hahn EE (eds) (1956) Photoconductivity conference. Wiley, New York and also Overdick M et al (2009) Status of direct conversion detectors for medical imaging with X-rays. IEEE Trans Nucl Sci 56(4):1800
21. Jones W, March NH (1973) Theoretical solid state physics, vol 2. Wiley, New York and also Kastalsky A, Luryi S, Spivak B (2006) Semiconductor high energy radiation scintillator detector. Nucl Intrum Method A565:650
22. Ziman JM (1964) Principles of the theory of solids. Cambridge University Press, London and also Windish R et al (2002) Light extraction mechanisms in high efficiency surface textured light emitting diodes. IEEE J Select Top Quantum Electron 8:248
23. Leo WR (1994) Techniques for nuclear and particle physics. Springer, New York
24. Brannon E, Olde GL (1962) The existence of a neutron. Radiat Res 16:1 and also Craun RL, Smith DL (1970) Analysis of response data for several organic scintillators. Nucl Instrum Method 80:239
25. Christophorou LG, Carter JG (1966) Improved organic scintillators in 2-ethyl napthalene. Nature 212:816
26. Ross H, Noakes JE, Spraudling J (1991) Liquid scintillation counting and organic scintillators. CRC Press, Boca Raton
27. Gusten H, Seltz W (1978) Organic scintillators with unusually large stroke shifts. J Phys Chem 82(4):459
28. Funt BL, Neparko E (1956) Kinetics of luminescence quenching in liquid scintillators. J Phys Chem 60(3):267
29. Horrocks DL (1964) Alpha particle energy resolution in a liquid scintillator. Rev Sci Instrum 35(3):334
30. Verreze V, Hurtzen C (2000) A multiple window deconvolution techniques for measuring low energy beta activity in samples contaminated with high energy beta impurities used liquid scintillation. Appl Radiat Isot 53:289
31. Young D (1990) PhD thesis, Tsinghua University, Beijing
32. McKlveen JW (1974) PhD thesis. Nuclear Eng. Dept. University of Virginia

33. Leory C, Rancoita PG (2004) Principles of radiation interaction in matter and detection. World Scientific, New York
34. Hirschberg M et al (1992) IEEE Trans Nucl Sci 39(4):511 and also Glen GF (1999) Radiation detection and measurement, 3rd edn. Wiley, Hoboken, p 227
35. Tropea C, Yarin AL, Foss JF (2007) Handbook of experimental fluid mechanics. Springer, New York
36. Xu C, Shear JB, Web WW (1997) Hyper Rayleigh and hyper Raman scattering. Anal Chem 69(7):1285
37. Kneipp J, Kneipp H, Rice WL, Kniepp K (2005) Optical probes for biological applications based on surface enhanced Raman scattering from indocyanine green on gold nano-particles. Anal Chem 77:2381
38. Hirayama F, Basile LJ, Kikuchi C (1968) Energy transfer and quenching in plastic scintillators. Mol Cryst 4:83
39. Lakowicz JR (2006) Principles of fluorescence spectroscopy. Springer, New York, p 317 and also Kasha M (1952) Collisional perturbation of spin orbital coupling and the mechanism of fluorescence quenching. J Chem Phys 20:71
40. Laustriat G (1968) The luminescence decay of organic scintillators. Mol Cryst 4:127
41. Birks JB (1964) The theory and practice of scintillation counting. Pergamon Press, Oxford/London and also Yaffe MJ, Rowlands JA (1997) X-ray detectors for digital radiology. Med Biol 42:1
42. Weber MJ (ed) (1998) Selected papers on phosphors, LED's and Scintillators. SPIE Optical Engineering Press, Bellingham and also Kabir MZ, Kasap SO, Rowlands JA (2005) Photoconductors for X-ray image sensors. Springer, New York
43. Skarstad P, Ma R, Lipsky S (1968) The scintillator efficiency of benzene. Mol Cryst 4:3
44. Gupta TK et al (1999) Structural CsI (Tl) scintillators for X-ray imaging applications. IEEE Trans Nucl Sci 45(3):492 and also Laustriat G (1968) The luminescence decay of organic scintillators. Mol Crystal 4:127
45. Klepeis JE, Shirley EL, Surh MP (1994) Predicting the structural and electronic properties of scintillators. Energy and Technology Riview, Aug/Sept, p. 33
46. Ananenko A et al (2004) Structural dependence of CsI(Tl) film scintillation properties. Semicond Phys Quantum Electr Optoelectron 7(3):297
47. Gupta TK et al (2003) Polycrystalline lead iodide films for digital X-ray sensors. Nucl Instrum Method A505:269
48. L'Annuriziata MF, El Baradei MM, Burkart W (2003) Hand Book of Radioactivity Analysis. Academic Press, San Diego
49. Dorenzo SE, Weber MJ, Bourret-Courchesne E, Klintenberg MK (2003) The quest for idea scintillator. Nucl Instrum Method A505:111
50. Derenzo SE, Weber MJ (1999) Prospects of first principle calculations of scintillator properties. Nucl Instrum Method A 422:111
51. Klintenberg M, Derenzo S, Weber MJ (2002) Potential scintillators identified by electronic structure calculations. Nucl Instrum Method A 486:298
52. Ahmed SN (2007) Physics and engineering of radiation detectors. Academic, San Diego
53. Letant SE, Wang TF (2006) Semiconductor quantum dot scintillation under $\gamma$-ray irradiation. Nano Lett 6(12):2877
54. Hussein MA (1999) Handbook on radiation probing, imaging & analysis. Kluwer Pub, Boston
55. Dendyand PP, Heaton B (1999) Physics for diagnostic radiology, 2nd edn. CRC Press, Boca Raton, p 144
56. Valli P (2008) Lattice dynamics and electronic properties of the scintillator host material. Solid State Comm 147(1–2):1
57. Gupta TK (2009) Copper interconnect technology. Springer, New York, Chapter 1
58. Sze SM (1998) Modern semiconductor device physics. Wiley, Hoboken, NJ, p 161
59. Rond C (2007) Luminescence from theory to applications. Wiley, Hoboken, NJ

# References

60. Errandonea D, Manjon FJ (2008) Pressure effects on the structural and electronic properties of $ABX_4$ scintillating crystal. Prog Mater Sci 53(4):711
61. Jeitschko W, Sleight AW (1973) The crystal structure of $HgMoO_4$ and related compounds. Acta Cryst B-29:869
62. Angloher G et al (2002) Limits on WMP dark matter using sapphire cryogenic detectors. Astropart Phys 18:43
63. Lecog P, Annenkov A, Gektin A, Korzik M, Pedrini C (2006) Inorganic scintillators for detector system, physical principle and crystal engineering. Springer, New York
64. Rodnyi PA (1997) Physical processes in inorganic scintillators. CRC Press, Boca Raton, p 49
65. Gupta T, Antonuk L et al (2013) Development of active matrix flat panel imagers incorporating thin layers of polycrystalline HgI2 for mammographic X-ray imaging. Phys Med Biol 58:703–714
66. Gupta TK et al (2006) Phys Med Res 51:R117
67. Eisberg RM (1961) Fundamentals of modern physics. Wiley, New York
68. Evans RD (1955) The atomic nucleus. McGraw Hill Pub, New York, p 712, Chapter 25
69. Dorenzo SE, Weber MJ, Bourret-Courchesne E, Klintenberg MK (2002) The quest for the ideal inorganic scintillator. Nucl Instr Method
70. Rutherford E, Chadwich J, Ellis CD (1930) Radiation from radioactive substances. Cambridge University Press, London
71. Gutan VB, Shamovsky LM, Dunina AA, Gorobets BS (1974) Two types of luminescence centers in CsI:Tl. Opt Spectrosc 37:717 and also Gupta TK et al (1999) High speed X-ray imaging camera using a structured CsI (Tl) scintillator. IEEE Trans Nucl Sci NS-46:232
72. Ishikane H, Kwanashi M (1975) The scintillation process in NaI:Tl. Jpn J Appl Phys 14:64
73. Hoshina T (1980) 5d-4f radiative transition probabilities of Ce3+ and Eu2+ in crystals. J Phys Soc Jap 48:1261
74. Chernov SP et al (1985) 5d4fn-1-4fn absorption and luminescence of and laser materials of $(Ce^{3+})$, $(Pr^{3+})$, $(Nb^{3+})$ ions in $BaY2F8$ single crystals. Phys Status Solidi (A) 88:K169
75. Schlessiger M, Szczurek T (1973) 4f-5d transition studies of some rare earth ions in $CaF_2$. Phys Rev B 8:2367
76. Viktorov L, Shorikov V, Zhukov V, Shulgin B (1991) Inorganic scintillating materials. Izv Akad Nauk SSSR Neorg Mater 27:1699
77. McClureand DS, Pedrini C (1985) Excitations trapped in impurity centers in highly ionic crystals. Phys Rev B 32:8465
78. Derenzo SE, Weber MJ (1999) Prospectus for first principle calculations of scintillator properties. Nucl Instr Method A 422:111 and also Blasse G (1994) Scintillator materials. Chem Mater 6:1465
79. Rodnyi PA (1997) Physical process in inorganic scintillators. CRC Press, Boca Raton
80. Song KS, Williams RT (1993) Self trapped excitons. Springer, New York
81. Blasse G, Grabmaier BC (1994) Luminescent materials. Springer, New York
82. Mavrichi O (2006) Master thesis, Dept of Phys and Eng. Phys. Univ. of Saskatchewan, Saskatoon, Can
83. Komives A et al (2005) A backscatter suppressed electron detector for the measurement of a "a". J Res Natl Inst Stand Technol 110:431
84. Kerisit S, Rosso KM, Cannon BD, Gao F, Xie Y (2009) Computer simulation of the light yield nonlinearity of inorganic scintillators. J Appl Phys 105:114915
85. Peisl H, Balzer R, Waidelich W (1966) Scotty and Frenkel disorder in KCl with color centers. Phy Rev Lett 17:1129
86. Toriumi K, Itoh N (1981) Behavior of defects introduced by a nanosecond intense electron pulse at high temperatures in alkali halides. Phys State Solidi (B) 107:375
87. Pooley D (1966) F-center production in alkali halides by electron hole recombination and subsequent $<110>$ replacement sequence: a discussion of electron–hole recombination. Proc Phys Soc 87:245

88. Hersh HN (1966) Proposed excitonic mechanism of color center formation in alkali halides. Phys Rev 148:928
89. Knoll GF (1989) Radiation detection and measurement, 2nd edn. Wiley, New York, pp 401–407
90. Sah CT, Noyace RN, Shockley W (1957) Carrier generation and recombination in P-N junctions and P-N junction characteristics. Proc IRE 45:1228
91. Evans RD (1985) The atomic nucleus. Krieger Co., Malabar
92. Knoll G, McGregor DS (1993) Fundamentals of semiconductor detection for ionizing radiation. Mat Res Soc Symp Proc 302:3

# Theoretical Approach of Crystal and Film Growths of Materials Used in Medical Imaging System

**6**

## Contents

6.1 Introduction .................................................................... 287
6.2 Theory of Crystal Growth ...................................................... 289
    6.2.1 Factors Affecting Crystal Growth ........................................ 291
6.3 Theoretical Modeling of Growing Single Crystal/Polycrystal Used in Radiation
    Detection and Medical Imaging ................................................. 293
    6.3.1 Floating Zone (FZ) Method .............................................. 293
6.4 Growth of a Crystal on a Seed ................................................. 295
    6.4.1 Czochralski (CZ) Technique ............................................. 296
6.5 Physical Vapor Transport (PVT) and Bridgman–Stockbarger (BS) Processes ........ 299
    6.5.1 Physical Vapor Transport (PVT) Method .................................. 299
    6.5.2 Bridgman–Stockbarger (BS) Method ...................................... 301
    6.5.3 High-Pressure Vertical Bridgman Method (HPB) ........................... 302
6.6 Traveling Heater Method (THM) ................................................ 304
6.7 Metal Solution Growth ......................................................... 305
6.8 Purification of Crystal ........................................................ 307
    6.8.1 Distillation ........................................................... 307
    6.8.2 Zone Melting Technique (ZMT) ........................................... 308
6.9 Thin- and Thick-Film Technology ............................................... 308
6.10 Solgel Coating (SGC) .......................................................... 310
References ......................................................................... 311

## 6.1 Introduction

Mankind from the prehistoric age has admired *crystals*. However, the significance of that beauty for the engineers and the scientists is on the basis of structural symmetry, simplicity, and purity. These characteristics endow crystals with unique physical and chemical properties, which have been used to cause a major transformation of the electronic industry. The systematic growth of synthetic crystals might be viewed as an art rather than science and has been described by some experts in this field as *a new agriculture*. The list of the *synthetic crystals* grown in the laboratory is far from exhaustive, but it shows how several old substances develop unique properties when they have the form of crystals. Figure 6.1 shows the many

T.K. Gupta, *Radiation, Ionization, and Detection in Nuclear Medicine*,
DOI 10.1007/978-3-642-34076-5_6, © Springer-Verlag Berlin Heidelberg 2013

**Fig. 6.1** Picture of a many facets cut diamond crystal—*a symbol of bond* (*love?*)

facets cut in a *synthetic diamond crystal*—a highly refractive, colorless crystalline allotrope of carbon.

Diamonds feature more predominantly than any other gemstone in the history and cultural heritage of the human race. *They were prized for their scarcity for centuries and still remain a symbol of wealth and prestige to this day* [1].

Figure 6.2 shows a natural diamond approximately 35 × 29 mm in size embedded in the kimberlite host rock. From the figure, we can see numerous visible eyes in the kimberlite rock similar to olivine crystals called *chrysolite*.

Diamonds were first mined in India over 4,000 years ago, but the modern diamond era only began in 1866 [1]. But recently, a scientist named Tracy Hall of General Electric (GE) is the first person to report of *synthesizing diamond* in the laboratory. The crystals were grown in the GE laboratory, from a solution of graphite in metal solvents such as nickel at a very high pressure (between 50,000 and 70,000 AP) and temperature (about 1,700 K).

Recently, at the University of Florida, a pilot plant has been established to grow commercial grade diamond using *BARS* diamond growth equipment. The largest crystal examined to date is a 10-carat, half-inch thick single-crystal diamond manufactured by Carnegie Institution's geophysical laboratory.

From time immortal diamond has been a fascinating gemstone. It has been used as an expression of beauty, love, and wealth, which has led to enormous amount of scientific research. As a result, besides being available in the market as a natural single-crystal gemstone, polycrystalline thin films of diamond have been successfully grown by plasma-enhanced chemical vapor deposition (PECVD) method. These films are very much in use as *scintillator* in *high-energy physics* and in *high-radiation environments*, such as *hadron colliders* and *radiation-hard sensors* [2–4]. The possibility of fabricating diamond detectors has been enticing research

**Fig. 6.2** A natural diamond approximately 35 × 29 mm in size embedded in the kimberlite host rock (Photo courtesy: Gem and Mineral Miners Inc.)

for decades. It is *extremely rugged* and possesses highest thermal conductivity at room temperature. On the other hand, the *wide band gap* (~6.6 eV) of diamond provides a low dark current—an important criterion for *radiation detectors* that are used in *medical imaging* [5–7].

*Threat reduction*, *nuclear nonproliferation*, and *medical imaging* activities require improved radiation detectors. The performance of these detectors can be significantly enhanced if the materials currently entrusted with the detection of radiation are placed with optimized materials. In order to gain a unique expertise knowledge about the requisite optimization of the radiation detector materials, one has to study the *theory and nucleation of the growth* of the materials either in the form of *a crystal*, *a polycrystal*, or in the form of *a thin* or *a thick film*.

## 6.2  Theory of Crystal Growth

*Crystals* are said to be the ordered arrangements of atoms (or molecules). In crystalline solids, the atoms are arranged in a periodic fashion and when this periodicity is seen to exist throughout the entire solid, the solid material is defined

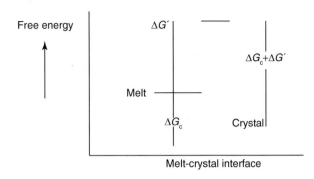

**Fig. 6.3** Schematic of free energy vs. crystal–melt interface during attachment

as being formed *a single crystal*. The *crystal growth process* starts with the nucleation stage. Several atoms or molecules in a supersaturated vapor or liquid start forming *cluster*. When the *cluster* is having a critical radius $r$, which is also defined as the *critical energy* barrier $\Delta G$, the radius starts to grow larger and eventually forms a crystal. From the standpoint of thermodynamics, we can express this *critical energy* barrier $\Delta G$ as

$$\Delta G = (4\pi r^2 \gamma + (4/3)\pi r^3) G_V \tag{6.1}$$

where $\gamma$ is the surface tension and $G_V$ is free energy change per unit volume forming the stable solidification from vapor or liquid.

Figure 6.3 shows the schematic of free energy vs. crystal–melt interface during attachment. Now, the free energy difference between the melt and the activated state $\Delta G'$ can be related to molecular attachment $M_a$ as

$$M_a = \exp(-\Delta G'/RT) v_a \tag{6.2}$$

where $R$ is the gas constant, $T$ is the temperature in degree Kelvin, and $v_a$ is the rate of molecular attachment. Similarly, we can write the rate of molecular detachment ($M_d$) as

$$M_d = \exp\{-(\Delta G_c + \Delta G')/RT\} v_a \tag{6.3}$$

where $\Delta G_c$ is the free energy difference between the melt and the crystal. Now, we can define the growth rate $G_r$ as the difference of these multiplied by the thickness per molecular layer $d_m$ and the fraction of the sites on the crystal surface available for attachment $f$ [8]. Thus, the *growth rate* $G_r$ can be expressed mathematically as

$$G_r = f d_m \exp(-\Delta G'/RT) v_a [1 - (-\Delta G_c)/RT] \tag{6.4}$$

# 6.2 Theory of Crystal Growth

Now, when the cooling rate is small, we can write $\Delta G_c$ as

$$\Delta G_c = \Delta H_c \Delta T / T_L << RT \tag{6.5}$$

where $\Delta H_c$ is the latent heat of crystallization and $T_L$ is the liquidus temperature.

Again, Stokes–Einstein relation for diffusion coefficient $D$ can be expressed mathematically as

$$D = kT / 3\pi \eta d_m = d_m^2 v \, \exp(-\Delta G' / RT) \tag{6.6}$$

where $\eta$ is the viscosity of the melt and $v$ is an attempt frequency. Finally, we can write the growth rate $G_r$ of the crystal as [9]

$$G_r \approx fk / 3\pi \eta \, d_m^2 \Delta H_c \, \Delta T / RT_L \tag{6.7}$$

There are two broad categories of crystal growth mechanisms, namely, the *lateral* and *continuous*. These two growth mechanisms have been identified as the main growth techniques. The *continuous* growth rate is determined by the nucleation rate and the rate of layer spreading [10, 11].

Detailed studies of the growth mechanisms of the crystals confirm that the crystal growth is a process coupled with *fluid dynamics, heat and mass transfer, surface kinetics and morphology, phase transitions,* and *chemical reactions.* Nevertheless, it is very important to know the thermal history of the growing crystal, in order to improve the crystal quality and process yield. It is also true that the temperature distribution in the growing crystal has a dominating influence on the formation of *thermal stress, which* can ultimately lead to *plastic deformations* and initiation of *defects of structures.*

## 6.2.1 Factors Affecting Crystal Growth

The rapid advance in information processing is one of the greatest achievements in modern technology as well as human history. It has been accomplished through the development of devices that are being in use in different fields, and they are built on single-crystal wafers. Therefore, the growth of single-crystal wafers is regarded as the key foundation underlying these innovations.

Almost all crystal growth experiments involve phase reactions. It is the objective to grow a single phase, namely, single crystal, from some other phase, a liquid or gas, which may have the same composition or may have a composition quite different from the final product. Therefore, successful growth from phase requires knowledge of the thermodynamic equilibria among the reacting vapor constituents, which may involve species other than the initial gaseous species. The kinetic of growth is critically dependent on the concentration or the diffusion of the desired reacting species. Vapor phase growth can proceed either through *reversible* or *irreversible* reactions. In reversible reaction, the solid material can be brought into vapor phase and then reformed under controlled conditions by reversing the direction of the reaction ($GaCl + As_2 \rightleftarrows = 2GaAs + 2HCl$). On the other hand, irreversible reaction will proceed only one way ($SiH_{4-x}Cl_x + (x-2)H_2 \rightarrow = Si + xHCl$).

During the reversible or irreversible reaction, there are three primary modes of heat transfer that are considered to be present during the growth of a crystal. These are *radiation, conduction,* and *convection*.

Crystal growth melts are usually heated by *radiation*. Heat losses also occur by radiation from the surface of the growing crystal and/or from the free melt surface itself when not completely contained by the growing crystal. The two most important factors in radiative processes are the temperatures of the emitting and the absorbing surfaces and their geometric positions relative to each other.

The conduction–radiation process mostly controls solidification of the crystal. Considering the thermal conductivity, the steady-state energy balance can be modeled following temperature distribution in the melt. The mathematical expression can be expressed as

$$\Delta(k\Delta T) \tag{6.8}$$

As the above process is coupled to radiation–conduction transport and accounting radiative heat flux in a particular medium, the above Eq. 6.8 will be modified to

$$\Delta(k\Delta T) + (1/3)\Delta J = 0 \tag{6.9}$$

Or,

$$\Delta J = 3a^2(J - 4n^2\sigma T^4) \tag{6.10}$$

where $J$ is the mean irradiance and $(1/3)\,\Delta J$ acting as an additional volumetric heat source due to radiation and being used in energy equation [12, 13].

Due to inherent complexities associated with *conduction–radiation transfer* calculations such as long-distance nature of radiation (solid-angle interaction) and functional dependence of conduction–radiation properties, conduction–radiation transfer calculations are very often simplified. Consequently, simplification or neglect of detailed conduction–radiation analyses may inaccurately predict the temperature field, interface shape/location, and thermal stress distribution within the growth furnace [14]. A nonlinear heat conduction process has been attributed for a crystal growth from liquid melt [15].

*Convection* or the fluid motion, on the other hand, is due to *natural* or *thermal* convection where fluid motions are caused by the *action of gravity field* on density gradients in the fluid. It may also be thought due to the orientation of the heating or cooling surfaces with respect to vertical direction of gravity which provides an important source of enhancing mass transfer and accounts for the growth rates in crystal. Therefore, the interrelations of convective flows with compositional segregation effects have been thought as the most important part of the crystal growth especially for the melt growth where solid compositional profiles can be most easily measured. Another basic criterion of *convection process* on compositional profiles in growing crystals is the distribution of the solute along the growth direction, which occurs during the unidirectional solidification of the stirred melts.

The important aspects of crystal growth configurations with regard to thermal convection are (i) the *confined nature of the fluid flows* and (ii) the presence of *vertical and non-vertical heat flows* that depend on aspect ratio of the boundary walls. In addition to these two, there may be some *forced convection* introduced by the rotational field of the growing crystal and *rotation of the fluid* or *melt around crystal*.

The other convection-related phenomenon is *surface tension* due to temperature and compositional gradients that are produced during growth process at the contact point of the melt between the crystal and the container. The float zone growth of silicon is considered to be the predominant mode of convection. Above all, convective process can produce a wide variety of spatial and temporal patterns of transport rate that can be the source of undesirable macro- and micro-segregation patterns in the growing crystal [16].

## 6.3 Theoretical Modeling of Growing Single Crystal/ Polycrystal Used in Radiation Detection and Medical Imaging

Most of the bulk single crystals are grown from solution or melt, where either solutal or thermal or both gradients are present. Although the solution growth has dominated in the field of single crystal growth, melt growth has also been very popular in semiconductor industry. The melt growth has been grouped mainly into three categories, namely, *floating zone* (FZ), *Czochralski* (CZ), and *Bridgman– Stockbarger* (BS) [17].

### 6.3.1 Floating Zone (FZ) Method

The method that is adopted for floating zone single-crystal growth is essentially a zone melting system in a vertical direction without a seed. The main advantage of the float zone process is that it is a containerless process, which can completely or partially eliminate contamination associated with the flux carrier. Another advantage is the reduction of thermal stresses that are caused by differential thermal expansion between the crystal and the crucible [18]. The FZ process is extensively used for the growth of silicon (Si) single crystal with a very low oxygen concentration. It is also used for high-purity crystals of metals. For refractory metals (with very high melting temperatures), this method is unique since virtually no suitable

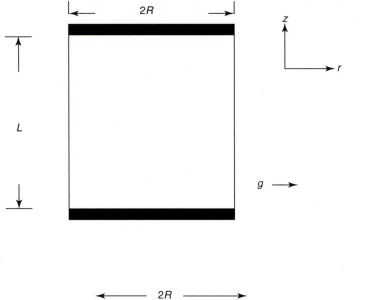

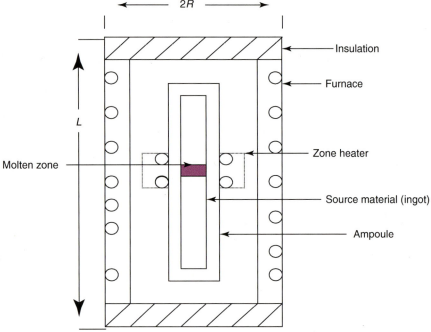

**Fig. 6.4** Schematic of a floating zone crystal growth procedure. $g$ gravity acceleration, $z$, $r$ vertical and radial coordinates

containers exist for these materials and the chance of contamination from the container does not exist. Figure 6.4 shows the schematic of a floating zone crystal growth apparatus. Floating zone's unique advantage is the elimination of the melt container, which is invariably a source of contamination. The quartz ampoule is loaded with the source material. Near the middle is an additional heater is called the zone heater to add additional heat to the source. Outside is the main heater, which consists of the furnace frame.

Figure 6.4 shows the schematic of a floating zone crystal growth process. In the case of differential pulling, the crystal is pulled from a molten zone of the feed material. The zone remains stationary while the feed material is advanced upward. Floating zone disadvantage is that it does not lend itself to the growth of the material with high vapor pressure or materials with volatile constituents. From an economic perspective, floating zone (FZ) growth is more expensive than Czochralski (CZ) method.

In a literature, the floating zone method has been modeled as a cylindrical liquid bridge of radius $R$ and length $L$ hold between two disks (Fig. 6.4). It is assumed that a laminar incompressible fluid is present in presence of thermal fluctuation forces. As a result one can expect different steady thermo-capillary flows due to the variation of different parameters like temperature gradient conditions and surface tension gradient [19].

## 6.4 Growth of a Crystal on a Seed

Crystal growth is a complex process and it is sometimes called a state of the art because of an unbridged gap between theory and experiment. The crystal growth is influenced by several factors particularly when controlled chemical composition and homogeneity and crystalline perfection, orientation, and size are considered. There is no unique method for growing a single crystal of a material. As such there are several methods that have evolved during the course of experimental research. As a rule of thumb, the materials that have low melting points ($<$1,000 °C) can be grown by less expansive methods like Bridgeman–Stockbarger (BS) rather than Czochralski (CZ) method.

The advantages of control over orientation, growth rate, perfection, and doping make growth on a seed an attractive improvement over growth by random nucleation. Growth on a seed in a cool part of a system while excess solute (often called nutrient) is in contact with the solvent in a hotter region of the system. The method is often called *growth in a thermal gradient* and is in some respects analogous to *hydrothermal gradient growth*. The seed is mounted at the end of a platinum-sheathed ceramic rod (which is capable of withstanding the central zone (highest zone temperature)), which projected just below the molten source material inside a platinum or rhodium crucible (sometimes graphite element can be used). The rod carrying the seed is rotated (~200 rpm) by gear-train motor (30 s clockwise and 30 s counterclockwise).

The nucleation rate will depend upon stirring speed since stirring shortens the diffusion path close to the growing face [20]. Since supersaturation increases with cooling rate, sometimes a greater critical stirring speed is required. Surprisingly, it is observed that the greater the critical stirring rate, the larger the mass of the melt. As a matter of fact, the seed is the principal sink on which excess solute deposits, so apparently the larger the melt mass the longer the diffusion path.

Czochralski (CZ) growth, named after the inventor, is the most common and versatile method ever known to grow single crystal with the aid of a seed material. The simplicity of the process and relatively high degree of crystal purity helped establish the CZ process as the dominant crystal growing technology in the early years of crystal growing industry. The development of dislocation-free ingot growth and automatic diameter control in the late 1960s led to the rapid growth of growing high melting temperature materials mainly silicon.

### 6.4.1 Czochralski (CZ) Technique

But when the melting temperature of the source material is high ($>$1,000 °C), neither the physical vapor transport (*PVT*) nor Bridgman–Stockbarger (*BS*) method is suitable. In that case, *Czochralski* method is mostly adopted to grow single crystal of the melts [21].

Jan Czochralski (CZ) invented the process in the year 1916. The process is similar to Verneuil process with a movable seed crystal. However, in CZ process, the seed is located above the liquid melt and not below the flame. The growth also takes place under a temperature gradient, which is maintained along the length of the growth of the crystal, and the mechanism is to produce a single nucleus from which a single crystal will propagate to grow.

The method lends itself to convenient chemical composition control by using the appropriate melt composition. At the same time, one can control the desired ambient atmosphere and temperature very accurately. However, the CZ method cannot be used readily to the growth of materials whose vapor pressure or that of their constituents is high at the melting point. Moreover, there are primary engineering problems in accommodating rotation and pulling of the crystal and with the thermal configuration requirements for maintaining thermodynamic equilibrium between the vapor and the melt. The height of the liquid crystallized per cm/s is given by

$$H_1 = \{(D^2/4)(dP/dt)\}/\{(D^2/4)(\rho_1/\rho_s) - (D^2/4)\} \tag{6.11}$$

Assuming the diameter of the crystal ($d_c/2$) is equal to the diameter of the crucible ($D/2$), and ($dP/dt$) is the pulling rate and $\rho_1/\rho_s$ is the ratio of the density of the source material in liquid and solid states conditions. A general rule of thumb is that the final crystal diameter should be between 50 and 60 % of the crucible diameter [22]. When this value is exceeded, growth becomes more difficult.

## 6.4 Growth of a Crystal on a Seed

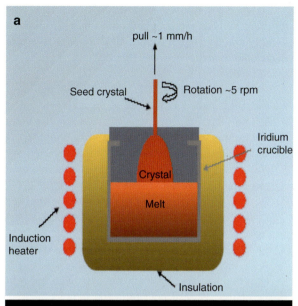

**Fig. 6.5** (a) Schematic of the single-crystal growth by Czochralski method and (b) picture of the seed inside the furnace during crystal pulling (Picture courtesy: Anwendung Kernphysikalischer Meßmethoden und Detektoren in Medizin und Industrie, R. Novotony, SS04, Teil 3C)

The systems used in *Czochralski* method to grow single crystal can be subdivided into four subsystems: (a) furnace, (b) crystal pulling mechanism, (c) ambient control, and (d) computer-based control system. The temperature inside the furnace can go over 2,500 °C. Accordingly, the crucible carrying the melt should be chosen in such a way that it should be able to withstand the temperature of the melt. During growth, the temperature of the melt is kept at few higher degrees higher than the melting temperature of the source material.

Figure 6.5 shows the pictures of Czochralski method. Figure 6.5a shows the different parts of the apparatus and Fig. 6.5b shows the inside of the hot furnace. Temperature can reach couple of thousand that depend on the melting temperature of the source material. A brief description of the accessories and their functions are given below:

1. Seed crystal is attached to one end of the pulling rod.
2. The rod is pulled slowly out of the melt.
3. A rigid vibration-free frame is used to mount the instrument.
4. The pulling rod is pulled slowly ~1 to 2 mm/h.
5. A stable and controllable power supply is used to heat the crucible.
6. Well-designed crucible with a given crucible height vs. crucible diameter ($D$) and height of the liquid ($H_1$).
7. The charge and the pulling rod with the seed are kept inside the vacuum chamber especially when the material to be crystallized evaporates or dissociates when molten.
8. Specially designed coils or induction rods can achieve a maximum temperature of the melt to 3,000 °C.

The charge material is loaded in a crucible and is heated above the melting point of the charge. A seed held by a rod is positioned above the crucible and dipped partially into the melt. The melt temperature is adjusted until the seed crystal can support a meniscus. After equilibration, the seed is pulled very slowly (1–2 mm/h). During crystal growth, provisions are there to rotate (rotation rate 10–30/min) the spindle carrying the seed that pulls the crystal from the melt and the crucible that holds the melt [23, 24]. The operation is performed with minimum vibration and sufficient precision. There is a provision to change the ambient inside the chamber for materials that are susceptible to change their compositions when open to the environment. The control systems, for example, pull rate, crucible rotation, seed rotation, melt temperature, and gas flow are controlled by a programmable computer.

In this method, the growth from a free surface accommodates the volume expansion associated with solidification of many materials that is opposed to solidification in a confined configuration. As a result, the method can grow very large single crystal of very large diameter.

In most *Czochralski* system, an induction generator heats the crucible. Therefore, energy to melt the charge and maintain the liquid state of the melt is supplied only from the crucible walls with no heating above or below the crucible. Therefore, positioning of the crucible within the work coil of the generator determines the coupling efficiency of the generator and therefore the temperature gradient through the crucible length. As mentioned above, the position of the crucible, furnace geometry, and the rotation of the crystal can strongly influence the temperature profile of the furnace and the liquid.

It has been observed that large crystal with homogeneity is difficult to grow by ordinary CZ method. This is because the melt composition changes during crystallization, which results in a compositional nonuniformity in the grown crystal [25]. Recently, a method called *double-crucible CZ* (DCCZ) method has been developed which is equipped with an automatic power supply and a radio-frequency (RF) generator. The platinum (Pt) crucible, which is used to carry the source material, has a double chamber structure, which divides the melt into two parts. The inner crucible is a Pt cylinder placed on the bottom of the outer crucible. The outer molten source material enters into the inner crucible through a Pt pipe welded through the hole at the bottom of the inner crucible.

## 6.5 Physical Vapor Transport (PVT) and Bridgman–Stockbarger (BS) Processes

Two frequently used crystal growth methods are *physical vapor transport* (PVT) and *Bridgman–Stockbarger* (*BS*) methods. PVT method is the crystal growth under vapor solid equilibrium conditions. The temperature of the source material (generally in the form of powder) is higher at the nucleation/crystal growth region. As a result, the imposed temperature gradient leads to mass flow resulting in a net mass transport of vapor species toward the crystal growth side.

The prevailing wisdom for conducting detached growth is to establish a liquid meniscus between the crystal and the crucible wall and maintain it by controlling the pressure difference between the gas space above the melt and the gap. The practice follows early theoretical analyses of the stability of the process that concentrated on meniscus and indicated that de-wetted growth depends on the wetting angles of the melt with the ampoule and the crystal and the pressure applied along the gas–liquid meniscus [26, 27].

### 6.5.1 Physical Vapor Transport (PVT) Method

Physical vapor transport method for single-crystal growth is the synonym of physical vapor deposition for thin-film deposition. PVT has gained much attention for the growth of single crystal because of (i) simplicity of the method, that is, sublimation–condensation in a closed ampoule which is nonreactive to the source at the melting temperature, (ii) a low-temperature operation, (iii) the existence of a low atomic roughness between the vapor solid interface, (iv) considerable morphological stability during growth, and (iv) its ability to grow a highly pure crystal when the process is properly executed. However, presence of numerical chemical species demands multicomponent concentration diffusion treatments. Not only that, the steep temperature and concentration gradients often employed lead readily to nonsteady oscillatory convective phenomena. These complexities are associated with thermal- and/or solute-driven convection generated by the interaction of gravity with density gradients. Experimental observations identify the convection phenomena as detrimental to the PVT growth.

In two-dimensional (2-D) model, the oscillatory convective phenomena related to Rayleigh numbers and the convective flow structure changes from multicellular to unicellular for the base parameter state [28, 29]. At steady state, the molar gas

**Fig. 6.6** Schematic of a PVT system

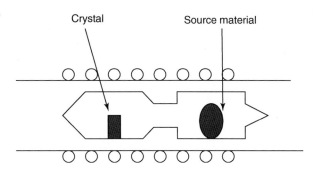

flux, $F_g$, dissolved into the melt and moves toward the freezing interface, which must be equal to the sum of the gas into the gap plus that being incorporated in the growing solid. Both numerical and material balance results in two solutions, which are beyond the scope of this book [30].

The volume ($V$) of the solid grown from the vapor phase of the source material can be equated mathematically as

$$V = (F/n) \tag{6.12}$$

where $F$ is the flux (number of atoms or molecules passing per unit are per unit time) and $n$ is the number of atoms from the source material. The equation (6.12) can be further extended and can be written in terms of gas phase mass transfer ($h_g$) and chemical reaction rate constant ($k_s$) as

$$V = \{(k_s h_g)/(k_s + h_g)\}\{(C_T/n)\, Y\} \tag{6.13}$$

where $C_T$ is total number of atoms or molecules per cubic centimeter in the gas phase of the source material, $Y$ is mole fraction of the reacting species, and $n$ is the number of source atoms incorporated into a unit volume of the film.

The stagnant film thickness which gives rise to the single crystal can be equated with the diffusivity of the active species in the gas ($D_g$) and can be presented mathematically as

$$\delta = (D_g/h_g) \tag{6.14}$$

The thickness of the crystal is an arbitrary quantity and has to be determined by experimentally for any given set of conditions.

Figure 6.6 shows the schematic of a PVT method to grow single crystal. The material under study (or the source material) is loaded in a quartz ampoule and is sealed under vacuum at a pressure between $10^{-5}$ and $10^{-6}$ Torr. The sealed sample is loaded inside a two-zone controlled furnace. The source is kept at a higher

## 6.5 Physical Vapor Transport (PVT) and Bridgman–Stockbarger (BS) Processes

temperature than the growth zone. A temperature gradient of 40–50 °C between the source temperature and the growth zone has been found to work well for the single-crystal growth.

The growth of the crystal is always terminated before the source material is fully sublimed. This is considered to be an important part of the process as the growth of the crystal in PVT process is dynamic. The deposition involves gaseous form of the source to the crystal competing against the sublimation of the growing crystal itself due to the elevated temperature. The yield is affected by the flow rate of the inert gas and the source temperature. Experimental studies show that platelet type of growth is very common at high-temperature end of the growth zone whereas lath-like and needle-shaped growth are more commonly found further down the stream [31–33]. PVT method has been used to grow single crystals of lead iodide ($PbI_2$) and mercuric iodide ($HgI_2$) for medical imaging purposes.

### 6.5.2 Bridgman–Stockbarger (BS) Method

*Bridgman–Stockbarger* (*BS*) method is very convenient for compounds and alloys that can be grown as single crystal in sealed silica tubes. The sealed silica tube with the charge is dropped vertically inside the furnace. Various methods have been suggested for supporting the crucible and lowering it through the temperature gradient. A temperature gradient of about 25 °C/cm is usually adequate and dropping rates can vary from 0.1 to 10 cm/h. In *BS* method, the crystal growth starts from the bottom of the melt. However, less constraint is placed on the crystal if a horizontal method is used. Two furnaces can be arranged horizontally and the tube carrying the source material can be drawn horizontally. For compounds with a relatively low vapor pressure, the crystal can be pulled directly at atmospheric pressure. Crystals of compounds with a high vapor pressure may also be prepared by pulling if it is done in a completely sealed system. Crystals of cadmium telluride (CdTe), cadmium zinc telluride (CZT), strontium iodide ($SrI_2$), and cesium lithium yttrium chloride (CLYC) have been grown by this method.

The *BS* method is basically a controlled freezing process, which takes place under liquid–solid (L–S) equilibrium conditions. The shape of L–S interface is often used as an indicator of the crystallographic imperfections such as the density dislocations. Numerical approaches to the solidification problem in *vertical and/or horizontal* Bridgman methods have appeared from time to time since 1970 [34–36].

Figure 6.7 shows the schematic of the vertical and horizontal Bridgman processes. The position, shape, and local interface morphology are tightly controlled by heat flow. The important local heat flow at the interface can be balanced as

$$K_s G_s = R \rho_s L + K_l G_l \tag{6.15}$$

where $K$ is thermal conductivity, $G$ is temperature gradient, $R$ is growth rate, $\rho$ is density, $L$ is latent heat of fusion, and subscripts s and l denote solid and liquid phase, respectively.

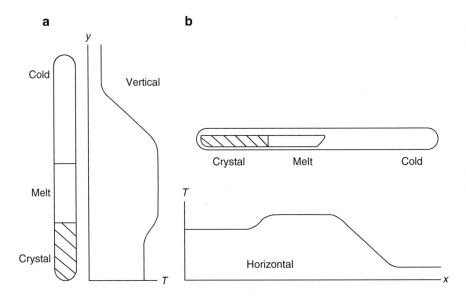

**Fig. 6.7** Schematic of Bridgman–Stockbarger process. (**a**) represents the vertical process and (**b**) represents the horizontal process

The interface shape has been calculated on the basis of the melting point isotherm which, in turn, is set by heat transfer. Further study of solid–liquid interface during the growth reveals that convective heat transfer through the molten phase is much more important than conduction heat transfer under typical growth conditions. In such system, the conditions affecting interface shape are a balance among latent heat, conduction in the solid, and convection in the melt [37].

In Bridgman method, ampoules' superheating and supercooling affect the stoichiometry of the compound when forming a single crystal. Moreover, the temperature gradients along the length of the furnace can introduce stress inside the crystal, which might change the structural symmetry of the crystal. Structural symmetry is very important when the crystal is used as a scintillator due to internal reflection. In order to minimize these effects, multiple-zone furnaces (up to 18 zones) with lower thermal gradients (within the range from 2 to 45 °C) have been fabricated to minimize under cooling solidification [38–40]. Figure 6.8 shows the schematic of a multizone vertical Bridgman furnace.

### 6.5.3 High-Pressure Vertical Bridgman Method (HPB)

The schematic diagram of an HPB is shown in Fig. 6.9. It is a variant of the conventional vertical Bridgman (VB) furnace housed inside a high-pressure chamber [41].

## 6.5 Physical Vapor Transport (PVT) and Bridgman–Stockbarger (BS) Processes

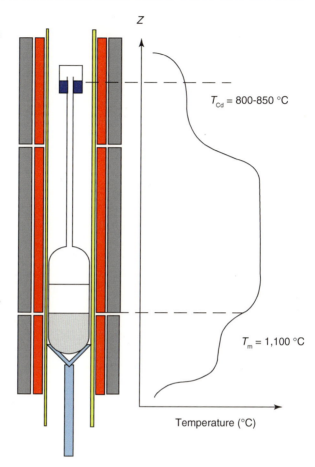

**Fig. 6.8** Schematic of a multizone vertical Bridgman furnace (Reprinted with permission from SPIE)

Figure 6.9 shows the sketch of a typical high-pressure Bridgman furnace (HPB). According to the experts, the technique used in HPB allows for a wide choice of crucible materials, including porous graphite. The porosity of the crucible helps to evacuate the crucible during the evacuation time of the chamber and helps to bake out the source material at high temperature [42].

The source material is sealed inside a proper ampoule (quartz or graphite-coated quartz) under high vacuum. During sealing the ampoule is heated. The vacuum-sealed ampoule is moved relative to multizone furnace. The central zone temperature is maintained at the melting temperature of the source material. A fine control of the stoichiometry is maintained during the growth process [41, 43].

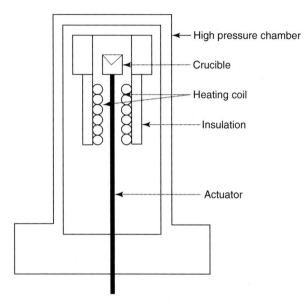

**Fig. 6.9** Sketch of a typical high-pressure Bridgman furnace (HPB)

## 6.6 Traveling Heater Method (THM)

THM falls into the category of solution growth, where the molten zone through a solid homogeneous source moves in slow motion [44]. The advantage of THM method is that it is a low-temperature growth process. In cadmium zinc telluride crystals, the phenomenal spectral performance and small size and low concentration of tellurium (Te) inclusion/precipitates indicate that THM is suitable for mass production of CZT radiation detectors. Due to the ampoule's movement, a temperature gradient is created and the material is seen to transport by convection and diffusion across the solvent zone. However, strong convective flows adversely affect the quality of the crystal growth. It has been found experimentally that the temporal evolution of the shape of the solid–liquid interface is of importance when a liquid is encapsulated in a crucible and has a motion [45, 46]. It has been suggested that proper mixing in the liquid solution may minimize the convective effect [47].

According to some experts, application of small rotating magnetic field (at 400 Hz) around 20-G magnitude under *microgravity* conditions can improve liquid solution mixing and can suppress residual *bouncy convection* in the solution. The buoyancy-driven convection changes the steady-state flow of the molten mass of the source. Figure 6.10 shows the schematic of a traveling heater method.

The induction of a magnetic field into the melt dampens the melt perturbations, especially when the magnetic field is in motion (traveling magnetic field, TMF) [48]. The effect of the induced TMF on the melt convection has been studied for various growth systems, for example, vertical gradient freeze (VGF), liquid-encapsulated CZ (LEC), and vapor pressure-controlled CZ (VCZ) [49].

## 6.7 Metal Solution Growth

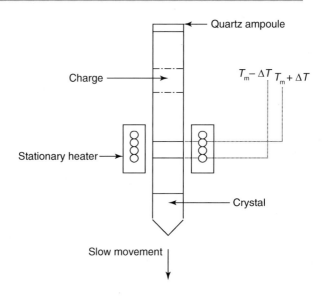

**Fig. 6.10** Schematic of a traveling heater method for growing single crystal

For binary and ternary alloys, a steady state is seen to reach at the point where the solvent zone dissolves a solid composition. In order to prevent constitutional supercooling inside the melt, the temperature profile should have a sharp temperature peak, producing a narrow molten zone (approximately the same dimension as the crystal diameter) with a high-temperature gradient at the growth interface [50].

When the thermal and the chemical equilibrium situation is achieved, the ampoule is moved very slowly (approximately on the order of 0.1 mm/h) relative to the heater. During growth, the asymmetrical growth temperature profile gives rise to a temperature difference between the upper (dissolution) and the lower (growth) liquid–solid interfaces. The source material at the top dissolves into solution, and the solute moves into liquid zone toward the seed and the crystallization starts at the lower interface, which is at a lower temperature than the dissolution interface [51].

## 6.7 Metal Solution Growth

*Metal solution growth* (MSG) is another method for growing single crystal when the melting temperature of the charge is high, for example, the growth of a single crystal of diamond. Diamond crystals are grown from solution of graphite in metal solvents such as nickel (Ni) at very high pressure (~70 kbar) and temperature (~1,700 K). Figure 6.11a shows pressure–temperature diagram for nickel–carbon eutectic melting line. The temperature in Kelvin along $X$-axis and the pressure in thousands of atmosphere along the $Y$-axis have been plotted. Figure 6.11b shows the phase diagram of diamond.

Discovery of *chemical agents* was one of the breakthroughs in the first successful work on the synthesis of diamond. The transitional metals like iron (Fe), nickel (Ni), cobalt (Co), manganese (Mn), chromium (Cr), platinum (Pt), and their alloys

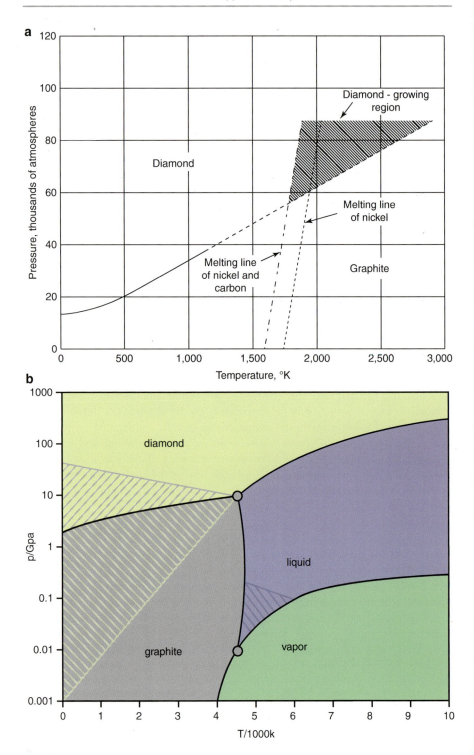

**Fig. 6.11** (a) Pressure–temperature diagram for Ni–C eutectic melting line and (b) shows the phase diagram of diamond (Courtesy: Wiki)

## 6.8 Purification of Crystal

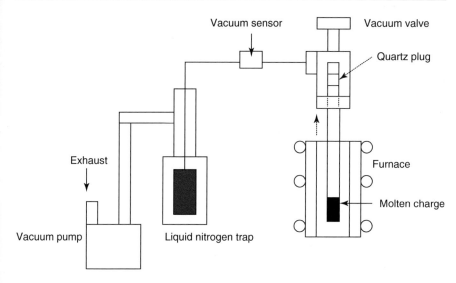

**Fig. 6.12** Schematic of a distillation process

are used as *agents*. The metal agent used is melted at high temperature and pressure in presence of carbon (the agents here are called the catalyst solvent), which converts graphite (carbon) into diamond. After solidification, carbon atoms are seen to organize in the same crystal lattice as their natural counter parts. As such, both the diamond crystal and the carbon show the same basic physical and chemical properties.

The role of catalyst is still not fully understood. It is infer that carbon-bearing molecules are catalytically decomposed on catalyst surface, resulting in incorporation of carbon atoms into the catalyst. Then once the super saturation occurs, carbon atoms precipitate from the catalyst leaving to the formation of crystal seed and later into a crystal of diamond.

## 6.8 Purification of Crystal

### 6.8.1 Distillation

Most of the time, the material procured from the vendors is not pure and it cannot be used as a charge material for crystal growth. In order to grow a highly purified crystal from the melt, the procured material from the vendor has to undergo a series of purification. Firstly, the material undergoes distillation inside a sealed quartz ampoule. During sealing of the source material, the quartz tube is heated and simultaneously evacuated by connecting to a mechanical pump (Fig. 6.12). The distillation process mainly eliminates any moisture or any occluded gases that may be present during manufacturing of the material.

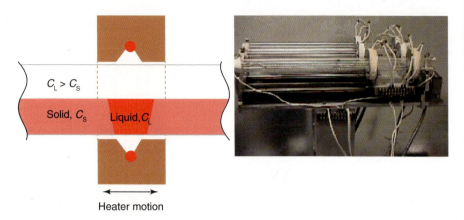

**Fig. 6.13** Schematic of the zone melting process (*left*) and picture of the apparatus used for zone melting process (Photo courtesy: RMD)

### 6.8.2 Zone Melting Technique (ZMT)

After distillation, the distilled material undergoes the second process, which is called zone refining. Zone refining has been a routine exercise for purification of the material before crystal growth. By using a zone-refined material, crystals of extremely high chemical purity have been obtained. In ZMT process, the material is confined in a quartz boat inside a horizontal quartz tube furnace. A temperature gradient is maintained along the quartz tube furnace.

In zone melting, a molten zone is established at one end of the charge (or between the seed and the charge, if a seed is used) and is advanced by moving either the container or the furnace. There is a distinct difference between normal freezing and the zone melting with regard to the segregation of impurities. In principle an effective distribution of segregation coefficient can be achieved in zone melting technique due to the relatively small volume of the melt. Directional solidification during zone refining has been known as one of the best methods for purification of materials and is characterized by their simplicity and versatility.

Figure 6.13 shows the schematic of the zone melting process (left) and the picture of the apparatus used for zone melting process in the laboratory. The ingot is passed through the melting zone for several times (multi-pass zone refining) to achieve the purest material from the grown ingot. A programmable computer controls the translatory motion of the multi-pass zone refining apparatus.

## 6.9 Thin- and Thick-Film Technology

The revolutionary growth of medical science especially in the branch of nuclear medicine for the detection, imaging, and treatment of cancerous cell, is geared towards quality improvement of radiation detectors. As a result, basic and applied

## 6.9 Thin- and Thick-Film Technology

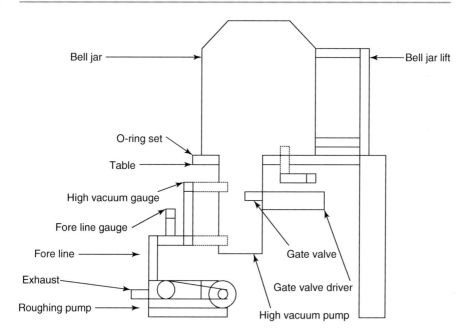

**Fig. 6.14** Schematic of a vapor deposition system with accessories

studies on crystal growth of semiconductors and other electronic materials have been carried out over a wide spectrum including nucleation, growth rate, segregation, growth interface composition morphology and stability, crystalline defects, compositional homogeneity, and thermodynamics of the source liquid. Although the technology is matured and ultimate preparation of single crystal has been reached, the procedure is complicated and expensive. Moreover, to grow a large crystal especially of the materials that are of potential interest in medical science has become a challenge.

As a result, a less expensive and a less complicated procedure in making a polycrystalline thin film of the scintillating materials has been developed successfully. The method is matured and is capable of producing large polycrystalline films inside a vacuum chamber, which is the ultimate goal of medical imaging system.

The simplest of all the thin-film deposition method is the *resistive vacuum evaporation*. A roughing pump system reduces pressure inside the chamber from atmospheric level to a level where *high vacuum pump* can work. A *high vacuum oil diffusion* or more expensive and sophisticated *turbo or cryopump* is used to attain the ultimate pressure ($>10^{-7}$ Torr) inside the vacuum chamber [52].

Figure 6.14 shows the schematic of a vapor deposition system with almost all the accessories. The draw back with the resistive vacuum deposition method is its inability to deposit stoichiometric film when the charge material is a compound. Besides, the method is unsuitable for the materials that have high meting temperatures. In order to deal with these problems, a more versatile and sophisticated method like *chemical vapor deposition* (CVD) and sputtering method have been used successfully.

CVD, as it implies, involves a gas-phase chemical reaction occurring above a solid surface which generally involves thermal (hot filament) or plasma (D.C. R.F. or microwave) activation or use of a combustion flame (oxyacetylene or plasma torches). Recently, there has been an interest in making devices using thin films of diamond grown by chemical vapor deposition (CVD) [53]. It is also true that simple CVD method has not been successful in some cases and a more sophisticated CVD method like *plasma-enhanced CVD* (PECVD) has been applied successfully.

Unfortunately, most of the films grown by these methods are thin (maximum 150–250 μm). In nuclear medicine, for medical imaging and therapy, physicians do need to work with the higher radiation dose to achieve better results (e.g., for better resolution in mammography, high rate dose (HRD) treatment for prostate cancer or brachytherapy). As a result, thin-film scintillating materials are not sufficient to block X-rays when the dose is high. In order to protect the patient from unwanted radiation dose, they need thick film (400–5,000 μm) of the scintillator. Experimental observations show that the vacuum evaporation or the CVD methods cannot provide thick film ~1,000 μm thickness or more.

To grow such thick films, a different technique called *screen printing method* has been adopted [54, 55]. The method is matured and was developed during late 1950s and early 1960s funded by the department of defense (DOD) at RCA laboratory. The method of making thick film is very easy and cost-effective but to make slurry (ink) of the scintillating material is tricky and requires very good knowledge of chemistry.

Figure 6.15 shows the schematic of a screen printing process. The screen can be made of silk or stainless steel according to the requirements. The slurry (or the ink) is made out of a mixture of a suitable binder and the material to be screen printed. The electrical resistivity of the binder is matched with the host material and the viscosity of the mixture is adjusted according to the mesh of screen used. The distance between the screen and the substrate is adjusted to get the required film thickness. The ink is pushed by the squeeze either manually or by an automatic or semiautomatic system so that the ink (slurry) passes through the opening region of screen mesh and deposited on the substrate. The deposited film is air dried and baked at a suitable temperature. The screen printing technology is by far the most inexpensive way to fabricate thick films (400–1,000 μm) [55].

## 6.10 Solgel Coating (SGC)

SGC is another inexpensive method, which does not involve any kind of heating during the growth of the film [50]. The process is old almost more than a century, but it has gained tremendous momentum in the last two decades after pioneering work of Dislich [56, 57].

The method involves the deposition of the desired material through a mixture in the form of suspended precursor from metallic alkoxides or organometallic materials. It is based on the phase transformation of a solution (the *sol*). The suspended particles in the *sol* are polymerized at low temperature to form a *wet gel*. The *wet gel* is densified through thermal annealing and finally dried over a

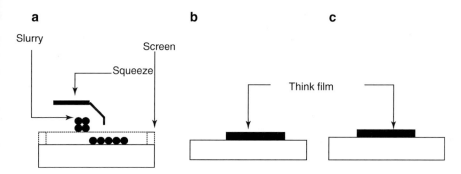

**Fig. 6.15** Schematic of a screen printing process. (**a**) The slurry is pushed through the screen mesh on the substrate. (**b**) Air-dried film on the substrate. (**c**) High-temperature baked film

substrate. Annealing changes the morphology of the film. The advantage of the method includes high purity and high degree of homogeneity over large surface area. The versatility, ease of making film, and its cost-effectiveness have made the procedure very attractive and unique [58].

The method has been applied to deposit scintillating materials for radiation detection [59, 60]. During the last two decades, numerous inorganic oxide materials have been identified as potential candidates for radiation detectors because of their high $\rho \cdot Z_{eff}$, where $\rho$ is the density of the materials and $Z_{eff}$ is their effective atomic number. These inorganic oxides generally have high melting points and single-crystal growth is expensive. Solgel chemistry has been initially developed as a cost-effective alternative route to glasses.

# References

1. May PW (2000) Diamond thin film – a 21st century material. Philos Trans R Soc Lond A 358:473
2. Hollingsworth MS (2010) IEEE NSS-MIC-RTSD joint conference, Knoxville, TN, 30 Oct–6 Nov 2010
3. Wermes N (2004) Trends in pixel detectors. IEEE Trans Nucl Sci 31(3):1006
4. Zao S et al (1993) Electrical properties in CVD diamond films, semiconductors for room temperature radiation detector applications. Material research symposium. Apr 1993, San Francisco, MRS Pub. Pittsburgh, p 257
5. Bol J et al (2007) Diamond thin film detectors for beam monitoring devices. Phys Stat Solidi 204:2997–3003
6. Adam W et al (2006) Radiation hard diamond sensors for future tracking applications. Nucl Instrum Methods A 565:278
7. BenMoussa A et al (2009) Recent developments of wide band gap semiconductor base UV sensors. Diamond Relat Mater 18(5–8):860
8. Kirkpatrick RJ (1975) Crystal growth from the melt: a review. Am Mineral 60:798–814
9. Jackson KA (1967) Current concepts in crystal growth from the melt. In: Progress in solid state chemistry, 4th edn. Pergamon Press, Oxford
10. Jackson KA, Uhlmann DR, Hunt JD (1967) On the nature of crystal growth from the melt. J Cryst Growth 1:1

11. Culvert PD, Ulhmann DR (1972) Surface nucleation growth theory for large and small crystal cases and the significance of transient nucleation. J Cryst Growth 12:291
12. Barvinschi F, Duffar T, Santailer JL (2000) Numerical simulation of heat transfer in transparent and semitransparent crystal growth process. J Opto Electron Adv Mater 2(4):327
13. Lapika P, Furmanski P (2010) Fixed grid simulation of radiation-conduction dominated solidification process. J Heat Transfer 132(2):023504
14. Naraghi MHN, Kassemi M (1988) Radiative transfer in rectangular enclosures. In: Proceeding of national heat transfer conference, vol 7, Kyongju, Korea, pp 457–462
15. Rudolph J, Winkler J, Woittenneck F (2008) Non-linear finite & infinite dimensional systems. In: Flatness based approach to heat conduction problem in a crystal growth. Springer, New York
16. Carruthers JR (1979) Dynamics of crystal growth. In: Bardsley W, Hurle DTJ, Mullin JB (eds) Crystal growth. North Holland, Amsterdam
17. Lan CW (2004) Recent progress of crystal growth and modeling and growth control. Chem Eng Sci 59:1437
18. Younsi R, Harkati A, Ouadjaout D (2007) Computational fluid dynamics applied to thermocapillary convection during crystal growth of silicon. Adv Model Optimiz 9(7):195
19. Younsi R, Harkati A, Ouadjaout D (2007) Computional fluid dynamics applied to thermocapillary convection during floating crystal growth of silicon. Adv Model Optimiz 9(2):195
20. Gilman JJ (ed) (1963) The art and science of growing crystals. Wiley, New York
21. Park JS, Seo M, Oh HJ, Junk JH (2008) Silicon ingot diameter modeling in CZ process and its dynamic simulation. Korean J Chem Eng 25(4):623–630
22. Brandle CD (1979) Czochralski growth of large oxide crystals. In: Bardsley W, Hurle DTJ, Mulin JB (eds) Crystal growth. North Holland, Amsterdam
23. Shimura N et al (2006) Zr doped GSO:Ce single crystal and their scintillation performance. IEEE Trans Nucl Sci 53(5):2519
24. Huber JS, Moses WW, Andreaco MS, Peterson O (2001) An LSO scintillator array for a PET detector module with depth interaction measurement. IEEE Trans Nucl Sci 48(6):664
25. Kuroda K (2002) Progress in photorefractive non-linear optics. CRC Press, Boca Raton, FL
26. Wilcox WR, Regel LL (1995) Detached solidification. Microgravity Sci Technol 8:56
27. Durby JJ, Lun L, Yeckel A (2007) Strategies for the coupling of global and local crystal growth models. J Cryst Growth 303:114
28. Kim GT, Duval WMB, Glicksman ME (1997) Effects of asymmetric temperature profile on thermal convection during physical vapor transport of $Hg_2Cl_2$. Chem Eng Commun 162(1):45
29. Markham BL, Greenwall DW, Rosenberger F (1983) Convective and morphological instability in vapor crystal growth. Report submitted to NSF under Grant DMR-7913183 and by the National Aeronautics and Space Administration under Grant NSG-1534, Department of Physics University Utah, Salt Lake City, 84112
30. Wang Y, Regal LL, Wilcox WR (2002) Approximate material balance solution to the moving meniscus model of detached solidification. J Cryst Growth 243:546
31. de Almeida VF, Carlos Rojo J (2002) Simulation of transport phenomena in aluminum nitride single crystal growth. Technical report ORNL/TM-2002/64, Oak Ridge National Laboratory, Oak Ridge, Mar 2002
32. Markham BL, Greenwell DW, Rosenberger F (1981) Numerical modeling of diffusive-convective physical vapor transport in cylindrical vertical ampoules. J Cryst Growth 51(3):426
33. Kim GT (2005) Convective-diffusive transport in mercurous chloride ($Hg_2Cl_2$) crystal growth. J Ceram Process Res 6(2):110
34. Sanghamitra S, Wilcox WR (1975) Influence of crucible on interface shape, position and sensitivity in vertical Bridgman Stockberger techniques. J Crystal Growth 28:36
35. Haung GE, Elwell D, Feigelson RS (1983) Influence of thermal conductivity on interface shape during Bridgman growth. J Cryst Growth 64: 441 and Lan CW, Young DT (1998) Dynamic simulation of the vertical zone melting crystal growth. Int J Heat Trans 41:4351

# References

36. Tiller WA (1963) Principles of solidification. In: Gilman JJ (ed) The art and science of growing crystals. Wiley, New York, p 276
37. Lun L, Yeckel A, Reed M, Szeles C, Daotidis P, Derby J (2006) On the effects of furnace gradients on interface shape during the growth of cadmium zinc telluride in EDG furnace. J Cryst Growth 35:290
38. Choi BW, Wadley HNG (2000) In situ studies of Cd1-x Znx Te nucleation and crystal growth. J Cryst Growth 208:219
39. Batur C, Duval WMB, Bennett RJ (1999) Performance of Bridgman furnace operating under projective control. In: IEEE proceedings American control conference, Vol. 6, San Diego, June 2–4 1999
40. Rudolph P, Koh HJ, Schafer N, Fakuda T (1995) The crystal imperfection depends on superheating.....semiconductor compounds. J Cryst Growth 1666(1–9):578
41. Szeles C, Driver MC (1998) Growth and properties of semi-insulating CdZnTe for radiation detector applications. In: SPIE conference on hard X-ray and gamma-ray detector physics and applications, San Diego, July 1998, vol 3446, p 1
42. Szeles C (2004) Advances in crystal growth and device fabrication technology of CdZnTe room temperature radiation detectors. IEEE Trans Nucl Sci 50(3):1242
43. Sen S et al (1996) Reduction of CdZnTe substrate defects and relation to epitaxial HgCdTe quality. J Electron Mater 25:1188
44. Bell RO, Hemmatand N, Wolf F (1970) Cadmium Telluride growth from tellurium solution as a material for nuclear radiation detectors. Phys Stat Sol 1:375
45. Seidh A et al (2001) 200 mm GaAs crystal growth by the temperature gradient controlled LEC Method. J Cryst Growth 225:561
46. Jurisch M et al (2005) LEC and VGF growth of SiGaAs single crystals – recent developments and current issues. J Cryst Growth 275(1–2):283
47. Dost S, Lent B (2007) Single crystal growth of semiconductors from metallic solutions. Elsevier, Netherlands
48. Rudolph P (2008) Travelling magnetic fields applied to bulk crystal growth from the melt. J Cryst Growth 310:1298
49. Kasjanow H et al (2008) 3d numerical modeling of asymmetry effects of a heater magnet module for VGF and LEC growth under travelling magnetic fields. J Cryst Growth 310:1540
50. Capper P (2003) Bulk crystal growth of electronic optical and optoelectronic materials. Wiley, Hoboken
51. Chen H, Luke PN et al (2008) Conduction of large crystals of cadmium zinc telluride grown by travelling heater method. J Appl Phys 103(1), American Institute of Physics
52. Gupta TK (2003) Handbook of thick and thin film hybrid microelectronics. Wiley Interscience, Hoboken, p 222
53. Kagan H et al (1993) Electrical properties in CVD diamond films. In: James RB, Siffert P, Schlesinger T, Franksed L (eds) Electrical properties in CVD Diamond Films, vol 302. Materials Society, Pittsburgh, PA, p 257
54. Gupta TK et al (2007) Novel X-ray security systems: fast, accurate and affordable. Varian Medical System, CA, Xerox (PARC), CA and RMD, MA, NIST final report
55. Gupta TK (2003) Handbook of thick and thin film hybrid microelectronics. Wiley Inter-Science, Hoboken
56. Dislich H (1971) Preparation of multicomponent glass without fluid metals. Glastechn Ber 44:1
57. Schmidt H (2006) Considerations about sol-gel process from classical sol-gel to advanced chemical nanotechnologies. J Sol-Gel Sci Technol 40(2–3):115
58. Brinker CJ, Scherner CW (1990) Sol-Gel science. Academic, San Diego
59. Pedroza G, de Azevedo WM, Khoury HJ, Silva EF Jr (2002) Gamma ray detection using sol-gel glass doped with lanthanide ions. Appl Radiat Isot 56(3):563
60. Wu YC, Parola S, Villanucva-Ibanez O, Mugnier J (2005) Structural characterization and wave guiding properties of YAG thin film obtained by sol-gel process. Opt Mat 27:1471

# Device Fabrication (Scintillators/ Radiation Detectors)

# 7

## Contents

| | | |
|---|---|---|
| 7.1 | Introduction | 316 |
| 7.2 | Compound Halides | 316 |
| | 7.2.1 Alkali Halides | 316 |
| | 7.2.2 Sodium Iodide (NaI) | 318 |
| | 7.2.3 Cesium Iodide (CsI) | 318 |
| 7.3 | Halides of Heavy Metals | 321 |
| | 7.3.1 Lead Iodide (PbI$_2$) | 322 |
| | 7.3.2 Mercuric Iodide (HgI$_2$) | 326 |
| | 7.3.3 Thallium Bromide (TlBr) | 331 |
| 7.4 | Lanthanide Halides | 333 |
| | 7.4.1 Cerium-Activated Lanthanum Chloride (LaCl$_3$:Ce) | 333 |
| | 7.4.2 Cerium-Activated Lanthanum Bromide (LaBr$_3$:Ce) | 334 |
| | 7.4.3 Rubidium Gadolinium Bromide (RGB) (Cerium Doped) RbGd$_2$Br$_7$:Ce | 335 |
| | 7.4.4 Cerium-Doped Lutetium Iodide (LuI$_3$:Ce) | 335 |
| | 7.4.5 Cerium-Activated Lutetium Oxyorthosilicate (LSO) | 336 |
| 7.5 | Complex Oxides with High Atomic Number | 342 |
| | 7.5.1 Cerium-Activated Gadolinium Silicate (GSO, Gd$_2$SiO$_5$) | 342 |
| | 7.5.2 Bismuth Germanate (BGO, Bi$_4$ Ge$_3$ O$_{12}$) | 343 |
| | 7.5.3 Cadmium Tungstate (CdWO$_4$ or CWO) | 347 |
| | 7.5.4 Lead Tungstate (PbWO$_4$ or PWO) | 348 |
| | 7.5.5 Cerium-Activated Lutetium Gadolinium Silicon Oxide (Lu$_{0.4}$Gd$_{1.6}$SiO$_5$:Ce) | 349 |
| | 7.5.6 Cerium-Activated Ortho-aluminates (Perovskites) | 350 |
| 7.6 | Compound Semiconductor | 351 |
| | 7.6.1 Cadmium Telluride (CdTe) and Cadmium Zinc Telluride (CZT) | 351 |
| | 7.6.2 CdTe Crystals | 351 |
| 7.7 | Cadmium Zinc Telluride (CZT) | 353 |
| | 7.7.1 Growth of Single of Cadmium Zinc Telluride (CZT) | 353 |
| 7.8 | Elemental Semiconductor | 354 |
| | 7.8.1 Diamond | 355 |
| | 7.8.2 Silicon (Si) | 356 |
| | 7.8.3 Germanium | 359 |
| References | | 360 |

T.K. Gupta, *Radiation, Ionization, and Detection in Nuclear Medicine,* DOI 10.1007/978-3-642-34076-5_7, © Springer-Verlag Berlin Heidelberg 2013

# 7.1  Introduction

The use of scintillators to detect radiation is a century old, and the discovery and the quest for new inorganic/organic scintillators are ongoing processes. The discovery of thallium-doped sodium iodide (NaI) followed by the earlier discovery of cadmium tungstate ($CdWO_4$) brought revolution in the luminescent physics. In a burst of exploration during the following few years, the scintillation properties of most pure and activated alkali halide crystals were investigated between 1940 and 1950 [1, 2]. Lithium-containing compounds to detect neutrons and the core valance barium fluorides ($BaF_2$) were discovered almost in the same decade. The search for new inorganic and organic scintillators is going on unabated. Indeed, the past two decades have witnessed a variable renaissance in research and development of new scintillator materials prompted in large part by the need for scintillators for precision calorimetry in high-energy physics and for high light output scintillators for medical imaging, geophysical exploration, nonproliferation and national security applications, and numerous other scientific and industrial applications [3–9].

The science and art of fabrication of these highly efficient scintillators are fascinating, and the engineering to build a radiation detector is a part and parcel of medical science, high-energy physics, and exploration geophysics. After the disaster of 9/11, these scintillators (radiation detectors) are also used in commercial sectors to reduce nuclear threats. Most national security applications are concerned with the detection of gamma rays, with energies in the range approximately 10 keV to 10 MeV, and neutrons arising from fissile materials. Modern detector systems consist of three steps: *careful selection of materials*, *device fabrication either from single crystal* or *from polycrystalline films*, and *instrumentation*. Therefore, we have devoted this chapter on the physics, chemistry, and engineering on the growths of single crystals and polycrystalline films to fabricate most efficient detectors. The instrumentation and the measurements will be dealt with in a separate chapter. The following chart shown in Fig. 7.1 is the chronology of the development of the materials to fabricate the scintillators or radiation detectors.

## 7.2  Compound Halides

### 7.2.1  Alkali Halides

Alkali halides (Li, Na, K, Rb, and Cs) are the family of ionic compounds with simple chemical formula XY, where X is the alkali metal and Y is the halogen (F, Cl, Br, and I). The internal crystalline structure at room temperature is center cubic usually face centered (fcc). The advantage of the cubic structured materials is that the internal reflection is less. As a result, these materials show high light output (which is one of the most desirable for a scintillator) compared to hexagonal or orthorhombic or monoclinic. Table 7.1 shows some of the alkali halide compounds.

## 7.2 Compound Halides

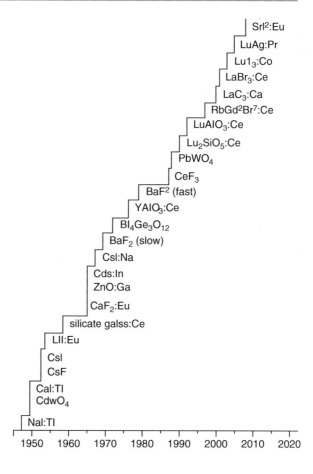

**Fig. 7.1** Chronology of the discovery of inorganic scintillators

**Table 7.1** Alkali halide compounds

| Halides | Alkali metals | | | | |
|---|---|---|---|---|---|
| Fluorine | LiF | NaF | KF | RbF | CsF |
| Chlorine | LiCl | NaCl | KCl | RbCl | CsCl |
| Bromine | LiBr | NaBr | KBr | RbBr | CsBr |
| Iodine | LiI | NaI | KI | RbI | CsI |

In 1940, with the development of photomultiplier tube (PMT), Hofstadter first developed thallium-activated sodium iodide (NaI) scintillator [1]. In a burst of exploration during 1950s, lithium-containing compounds showed many prospects as neutron detector. As these materials were identified, the procedure to purify and to fabricate detectors either in the form of single crystal or polycrystalline thin films became a routine work. As a result, several procedures were invented or discovered to make single crystals or polycrystalline films from these materials.

## 7.2.2 Sodium Iodide (NaI)

Sodium iodide (NaI) is one of the oldest members of the optical engineering family and is considered by many scientists as a standard optical device in the field of detection technology. In 1950, the discovery of thallium-activated sodium iodide attracted the attention of many investigators who thought that it will be used as a tool in modern scintillation spectroscopy of gamma ($\gamma$) rays. It is worth mentioning that in spite of four decades of research, NaI remains as a preeminent scintillator in the field of spectroscopy and nuclear medicine. The most remarkable property of thallium-activated NaI is its remarkable light yield.

### 7.2.2.1 Crystal Growth

As the melting point of NaI is low (~650 °C), the single crystal of NaI can be grown easily and inexpensively from the melt in an ambient of nitrogen. Large crystals of NaI *single crystal* have been grown by *directional solidification* of the material. A graphite mold is loaded with sodium iodide (NaI) with thallium (Tl) powder in appropriate proportion and heated at 700 °C, and the molten mass is cooled slowly from the bottom in air. On cooling, NaI:Tl solidified and formed a *single crystal* out of the molten mass.

Large crystals of thallium-doped (0.001 mol fraction) NaI has also been grown by *Bridgman–Stockbarger* (BS) method. However, *press-forged* recrystallization of NaI(Tl) crystal under heat and pressure has been reported to produce similar quality crystals as produced by BS method. The main disadvantage of NaI:Tl crystals is that they are very hygroscopic in nature and mechanically not very strong. Thallium-doped (Tl) sodium iodide (NaI) crystals are normally encapsulated in aluminum housings with a wall density of 170 mg/cm$^2$ and an end cap density of 117 mg/cm$^2$.

Figure 7.2 shows the picture of a NaI:Tl single crystal grown by BS method. In order to verify the structure of the single crystal, X-ray diffraction pattern was performed at Massachusetts Institute of Technology (MIT), Cambridge, MA, in a Rigaku X-ray diffractometer model Rotaflex RTP 500 with copper k$\alpha$ radiation. Table 7.2 shows some of the physical properties of NaI.

Thallium-activated sodium iodide crystals are susceptible to radiation damage and are not rugged. Recently, the single crystal ingots of thallium-activated NaI are recrystallized under heat and pressure. The resulting material is a polycrystalline material (Polyscin®) with randomly oriented crystal grains. But surprisingly, the polycrystalline thallium-activated NaI retains its optical character but with improved mechanical strength.

## 7.2.3 Cesium Iodide (CsI)

Cesium iodide (CsI) is another member of the alkali halide family with many desirable properties (with high atomic number, high density, fast response, and high radiation resistance). It has gained substantial popularity as a scintillating material and is available commercially either doped with sodium (Na) or thallium (Tl). Cesium iodide (CsI) has high resistance to thermal and mechanical shock

## 7.2 Compound Halides

**Fig. 7.2** Picture of a single crystal of tellurium-doped sodium iodide

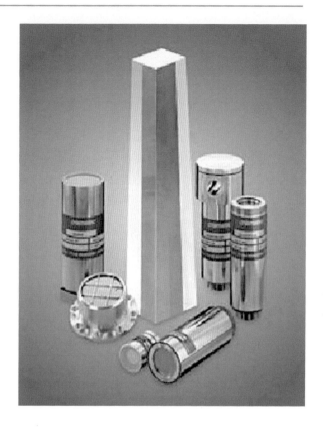

**Table 7.2** Characteristics of thallium-activated NaI

| Material | NaI |
|---|---|
| Density | 3.67 |
| Hygroscopic | Strong |
| Emission wavelength (nm) | 415 |
| Refractive index | 1.85 |
| Primary decay time (ns) | 250 |
| Light yield (photons/MeV) | $38 \times 10^3$ |
| Melting point (°C) | 651 |
| Light yield (photons/MeV) | 40,000 |
| Structure | bcc |

due to absence of cleavage plane. The emission spectrum is very well matched to the sensitivity characteristics of silicon PIN diode. The luminosity of thallium-activated cesium iodide CsI:Tl can be as high as the theoretical value. Figure 7.3 shows the picture of a CsI:Tl single crystal grown by BS method, and some of the properties of thallium-activated CsI is tabulated bellow in Table 7.3.

The thallium (Tl)-activated cesium iodide (CsI) crystals are being frequently used in medical imaging system (e.g., gamma-ray imaging, X-ray CT, digital

**Fig. 7.3** Shows the picture of a single crystal of CsI:Tl grown by BS method

**Table 7.3** Properties of cesium iodide scintillator (activator)

| Scintillator (activator) | CsI | CsI (Tl) | CsI (Na) |
|---|---|---|---|
| Density (g/mL) | 4.51 | 4.51 | 4.51 |
| Hygroscopic | Slightly | Slightly | Yes |
| Emission wavelength (nm) | 315 | 550 | 420 |
| Lower cutoff | 260 | 320 | 300 |
| Refractive index | 1.95 | 1.79 | 1.84 |
| Primary decay time (μ-s) | 0.016 | 1.0 | 0.63 |
| Light yield (photons/MeV) | $2 \times 10^3$ | $54 \times 10^3$ | $42 \times 10^3$ |
| Crystal structure | Cubic | Cubic | Cubic |

radiography, mammography, dentistry, fluoroscopy). The material has high stopping power and good energy resolution [10–12].

### 7.2.3.1 Crystal Growth

Typically cesium iodide (CsI) and thallium with 99.99 % purity are used for growing single crystal. *Bridgman–Stockbarger (BS)* method is very popular for growing single crystal of CsI:Tl. The weight percent of thallium within the matrix is maintained approximately about 0.1 mol% [13, 14]. The charge material is loaded in a quartz ampoule inside a glove box and vacuum encapsulated after degassing. The powder is then melted inside the quartz ampoule, and the quenched solid undergoes through zone refining for a couple of times. The refined material is reloaded in a quartz ampoule inside a glove box and inserted in a *BS* furnace. The vacuum-sealed quartz ampoule with the charge material is pulled slowly (~0.1–0.5 mm/h) through the furnace with a temperature gradient of 10°/cm. Inside the furnace, argon gas at 2 atm is introduced throughout the entire time of the crystal growth.

### 7.2.3.2 Vacuum-Deposited Film: Electronic Portal Imaging Devices

*Vacuum-Deposited Film*: *Electronic portal imaging devices* are developed as a digital X-ray imaging system because of the inherent difficulties associated with film-based image capturing system. In order to cover a 25 cm × 25 cm flat-panel imaging arrays with CsI:Tl material, a hot-wall deposition method had been followed at radiation monitoring devices (RMD). The high-vacuum hot-wall

**Fig. 7.4** Shows the scanning electron microscopic photograph of a columnar film of thallium-activated cesium iodide deposited by hot-wall epitaxy (Courtesy RMD Inc.)

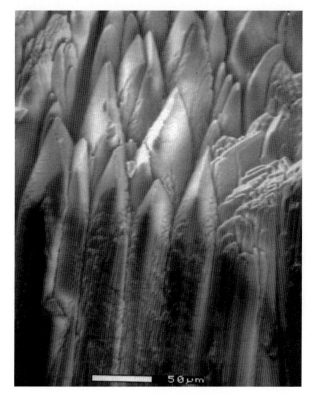

epitaxial growth system was capable of depositing future 40 cm × 40 cm flat panel. The process was capable of depositing *micro-columnar structure of CsI:Tl* (Fig. 7.4).

The CsI:Tl micro-columnar structures as shown in Fig. 7.4 are high-density fibers of CsI:Tl scintillator with a structure resulting from the hot-wall epitaxial growth on a specially designed substrate[15, 16].

X-Ray diffraction pattern of the hot-wall epitaxially grown film of cesium iodide (CsI) doped with thallium (Tl) is performed at MIT, Cambridge, MA, in a Rigaku X-ray diffractometer (model Rotaflex RTP 500) with copper kα radiation. Intensity in arbitrary units is plotted against $2\theta$ value and is shown in Fig. 7.5.

## 7.3 Halides of Heavy Metals

These halides of heavy metals like lead iodide ($PbI_2$) [17], mercuric iodide ($HgI_2$) [18], and thallium bromide (TlBr) [19] have high atomic number (Z) together with their density lead to excellent detection efficiency. As for example, the attenuation coefficient of $PbI_2$ is ten times higher at some energies than that of germanium (Ge). Due to their high stopping power, these compound semiconductor crystals of modest thickness will achieve a full energy peak detection efficiency that is equivalent to that of a substantially thicker high-purity germanium (HPGe). The

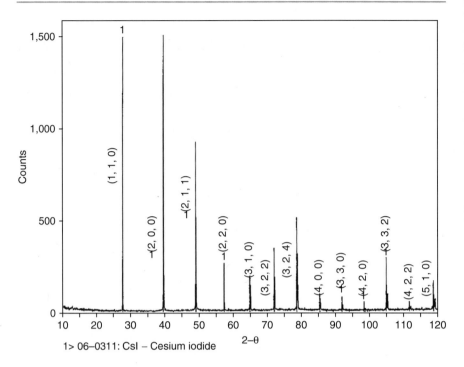

**Fig. 7.5** Shows X-ray diffraction pattern of a hot-wall epitaxially grown CsI:Tl film

advantages of high band gap for mercuric iodide (2.1 eV) and lead iodide (2.3 eV) have made operation of these devices at room temperatures, and their operations as direct solid-state detector avoid the use of cumbersome scintillator–photomultiplier combinations. As a result, several methods to fabricate the devices either from single crystals or from polycrystalline films have been identified. These methods are challenging in the sense of cost effectiveness and their viability as radiation detectors for use in nuclear medicine.

### 7.3.1 Lead Iodide (PbI$_2$)

Increasing interest has been focused on layered semiconductor materials like PbI$_2$ and HgI$_2$. These materials have been used for medical imaging and as high-energy resolution room temperature X-ray and gamma ($\gamma$) detectors. Their high density and high atomic number allow one to manufacture small, compact detectors with very good volume efficiency ratio [20]. PbI$_2$ as a photodetector shows low leakage current and high quantum efficiency (>60 %) in 350–500 nm region [21]. Table 7.4 shows some of the properties of PbI$_2$.

#### 7.3.1.1 Crystal Growth

Single-phase PbI$_2$ polycrystalline material is first synthesized directly from pure lead (Pb) and iodine (I) in a stoichiometric ratio with excess of Pb by two-

## 7.3 Halides of Heavy Metals

**Table 7.4** Properties of lead iodide (PbI$_2$)

| Property | Value |
|---|---|
| Band gap | 2.3 eV |
| Crystal structure | Hexagonal |
| Melting point | 408 °C |
| Vapor pressure at 200 °C | $10^{-5}$ torr |
| Dielectric constant | 21 |
| Density | 6.2 g/mL |
| Resistivity ($\rho$) | $>10^{13}$ Ω-cm |
| Electron/hole product ($\mu\tau$) | $10^{-5}/2 \times 10^{-6}$ cm$^2$/V |

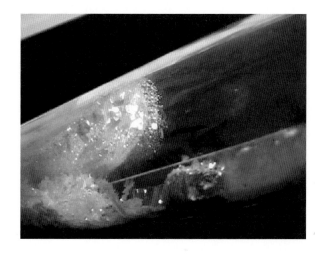

**Fig. 7.6** The picture of a vapor-phase-grown single crystal of lead iodide (PbI$_2$) (Reproduced with permission, SPIE)

temperature vapor transport method (VTM). The polycrystalline material is then purified through zone refining. The purified material is loaded in a vertical Bridgman apparatus and single crystal of PbI$_2$ has been grown [22–24]. During the growth of the crystal, the maximum temperature of the hot zone in the furnace is maintained at ~450 °C, and the temperature gradient (d$T$/d$x$) is maintained at 5°/cm rate. Following these parameters, the growth rate of PbI$_2$ single crystal is typically around 10 mm/h. The grown single crystal shows resistivity typically between $10^{10}$ and $10^{11}$ Ω-cm [25, 26]. Figure 7.6 shows the picture of a vapor-phase-grown single crystal of lead iodide (PbI$_2$).

### 7.3.1.2 Vacuum-Deposited Film

Vacuum evaporation is another method of depositing polycrystalline PbI$_2$ films. During resistive vacuum evaporation, almost 70–80 % of the source material is lost on the walls of the chamber. We designed a hot-wall resistive vacuum evaporation and have been able to recover almost 50–60 % of the lost material inside the vacuum chamber. The three-dimensional sketch below shows the relation of the deposition rate of PbI$_2$ with wall and source temperatures (Fig. 7.7).

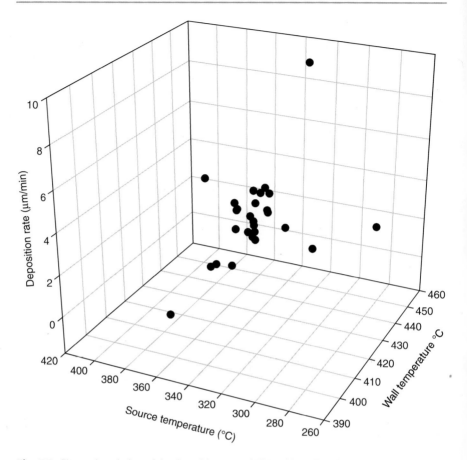

**Fig. 7.7** Shows the relation of the deposition rate of PbI$_2$ with wall and source temperatures

X-ray diffraction of the hot-wall epitaxially grown lead iodide (PbI$_2$) film is performed at MIT, in a Rigaku X-ray diffractometer (model Rotaflex RTP 500) with copper k$\alpha$ radiation. Intensity in arbitrary units is plotted against 2$\theta$ value and is shown in Fig. 7.8.

### 7.3.1.3 Screen-Printed Film

Extensive research in recent years has shown that the flat-panel X-ray image detectors based on a large-area thin-film transistor (TFT) or switching diode scanned active matrix array (AMA) is the most promising digital radiographic technique and suitable to replace the conventional X-ray film/screen cassettes for diagnostic medical digital X-ray imaging applications (mammography, chest radiography, and fluoroscopy) [27, 28]. Recent development in the flat-panel imaging matrix is expected to produce commercial large area of panels that can extend an area of 40 cm × 40 cm. To cover such a large area is not possible by small single crystal or vacuum-evaporated film. In order to cope with the technical difficulty,

## 7.3 Halides of Heavy Metals

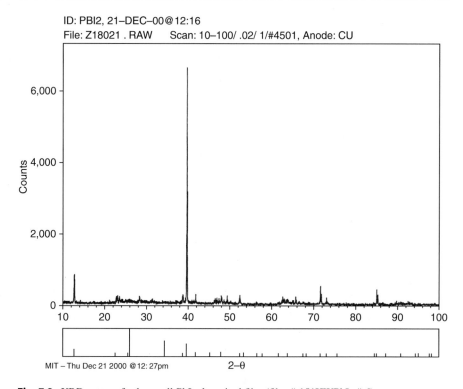

**Fig. 7.8** XRD pattern for hot-wall PbI$_2$-deposited film (film # 151HWPbI$_2$ # 6)

research organizations have come up with screen-printed films. The technology is old and the method of screen printing goes back to early 1960 [29].

The lead iodide (PbI$_2$) slurry (bright yellow in color) is prepared by mixing PbI$_2$ powder with a proper binder. The slurry thus prepared is used for screen printing on a previously cleaned ITO-coated glass substrates. After printing, the films are subjected to air drying for 30 min before sintering them in an oven. Figure 7.9 shows a PbI$_2$ film deposited on an ITO glass.

Following deposition of the PbI$_2$ films on ITO glass, large-area lead iodide films are deposited over the TFT array (Fig. 7.9b), and extensive characterization of the physical and electrical properties has been carried out. The bulk film density had also been estimated by measuring the film dimensions and its weight. In addition, evaluation of the surface morphology and grain size has been conducted using scanning electron microscopy (SEM). Film morphology gives an idea about the grain boundaries and defects that can be scattering centers for charge carriers. Figure 7.10a and b shows the scanning electron micrograph of the vacuum-deposited and the screen-printed films, respectively. The pictures show that PbI$_2$ is a layered structure with alternate Pb and I$_2$. The particle sizes for vacuum-deposited film vary between 0.25 and 1 μm, while the grain sizes of the screen-printed film are larger and vary between 0.75 and 2.5 μm.

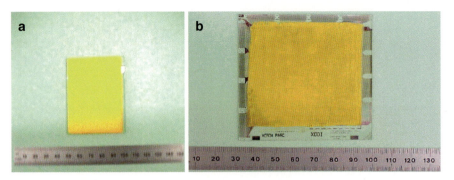

**Fig. 7.9** (a) Screen printed PbI$_2$ film deposited on an ITO glass. (b) TFT array covered with screen-printed PbI$_2$ film

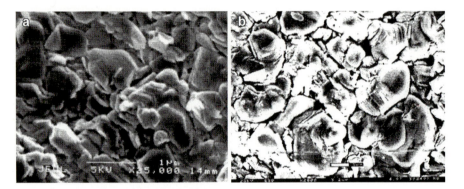

**Fig. 7.10** Scanning electron micrograph of PbI$_2$ films. (a) Vacuum-deposited film, (b) screen-printed film

### 7.3.2 Mercuric Iodide (HgI$_2$)

*Mercuric iodide* (HgI$_2$) due to its high atomic number (Hg-80, I-53), large band gap (2.15 eV), and required low energy ($W_{\text{eff}}$) to create electron hole pair has been a subject of interest for medical imaging. It has been studied for the past 50 years owing to its efficiency as a nuclear spectrometer at room temperature [30, 31]. The potential of HgI$_2$ nuclear detector technologies will certainly be more fully realized as more progress is made toward the understanding of the material properties responsible for wide variability in detector performance [32].

Figure 7.11 shows the picture of a single crystal of HgI$_2$. In HgI$_2$ configuration, each mercury (Hg) atom is tetrahedrally bonded to four-iodine (I$_2$) atoms in a covalent bond. These tetrahedral bondings are stacked in the $z$-direction, resulting in weak I–I bonds. Using calculated values for electronic band structure and optical matrix elements, one can model the absorption coefficient as a function of the incident photon energy [33].

**Fig. 7.11** Shows a single crystal of mercuric iodide (HgI$_2$) (Photo courtesy: Constellation Technology, Florida)

Room temperature operation combined with a high resolution and good energy resolution opens the way to miniaturization and portability of the detectors and spectrometers [34]. The polymeric form of HgI$_2$ can reside in two phases, red α phase, which is stable but transforms into yellow orthorhombic β phase above 127 °C. However, the orange α-phase tetragonal structure of HgI$_2$ has much attention as a potential candidate for medical imaging and synchrotron applications. The layered structure of HgI$_2$ exhibits a pronounced anisotropy of thermal and mechanical properties.

Figure 7.12 shows the alpha-phase crystal structure of mercuric iodide. The polymeric forms of HgI$_2$ are known as red α-HgI$_2$ and yellow β-HgI$_2$ [35]. The tetragonal HgI$_2$ is stable up to ~127 °C, and at higher temperatures, it transforms into yellow orthorhombic β-HgI$_2$, which exhibits a tendency to be present in a metastable state on the surface of the α-HgI$_2$.

### 7.3.2.1 Growth of Single Crystal of Mercuric Iodide (HgI$_2$)

Successful growth of high-quality α-HgI$_2$ single crystals is largely dependent upon the purity of the starting material. The repeated sublimation procedure is followed by sealing the quartz ampoule inside a Bridgman furnace with a temperature gradient along the length of the furnace tube. The technique of the crystal growth can be described as *physical vapor transport* (PVT) technique.

The initial starting material (reagent grade HgI$_2$) is first purified by sublimation process which is repeated for three to four times. The material thus obtained is undergone zone refining for further purification. Single crystal of HgI$_2$ is obtained by applying vapor-phase methodology.

Thermodynamic calculations of the vapor-phase stoichiometric composition of HgI$_2$ in a closed system (when the temperature inside is maintained ~200 °C) show that in equilibrium condition, the vapor phase contains only HgI$_2$ [36, 37].

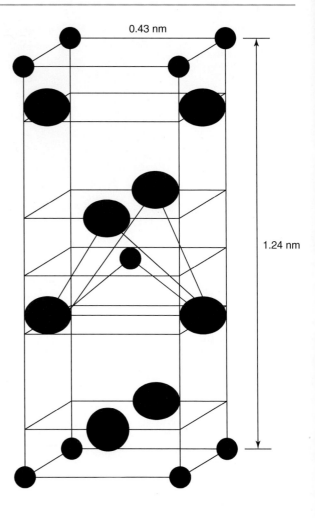

**Fig. 7.12** Shows the elemental cell of tetragonal alpha-phase (α) mercuric iodide (HgI$_2$) mercury ● iodine ●

Figure 7.13 shows the schematic of a crystal growth furnace of mercuric iodide (HgI$_2$). It has been reported that the furnace is capable of growing single crystal of HgI$_2$ up to 1,000 g in weight [38].

An alternative method for direct synthesis of HgI$_2$ has been reported by Pencotka and Kaldis [39] which has been carried at a temperature exceeding 300 °C in a stream of a carrier gas saturated with the vapors of mercury and iodine. However, precipitation from aqueous solutions by reactions of potassium or sodium iodide (KI or NaI) with mercury chloride (HgCl$_2$) or nitrate ((HgNO$_3$)$_2$) seems to be simpler, but the product after reactions shows the presence of nitrate or chloride.

Vapor-phase-grown mercuric iodide (HgI$_2$) single crystal has been reported by Ponpon et al. [40]. The material is synthesized by reaction of the constituent

## 7.3 Halides of Heavy Metals

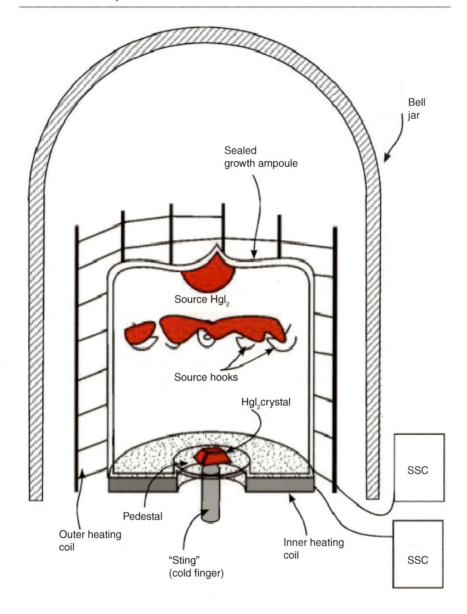

**Fig. 7.13** Shows the schematic of a crystal growth furnace of mercuric iodide (HgI$_2$)

elements with excess of iodine. The synthesis is followed by double stage purification (zone refining under reduced pressure). The growth method is controlled by vapor condensation of the purified charge. After natural seeding during heating and condensation, the crystal grown over the seed is pulled slowly at 1–10 mm/day (Fig. 7.14).

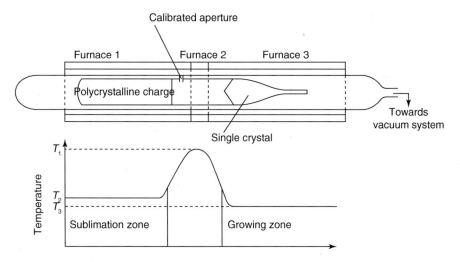

**Fig. 7.14** Shows the schematic of a vapor phase growing HgI$_2$ single crystal

### 7.3.2.2 Screen-Printing System

In mid-2001, Siemens Medical System offered a 41 × 41 cm$^2$ detection area electronic portal devices (EPIDs), following the commercial 30 × 40 cm$^2$ detection area EPIDs of Varian medical system [41]. To deposit ~133 mg cm$^{-2}$ over such a large area with uniform thickness and stoichiometry is difficult, challenging, and expensive, especially some materials which change their phase at a higher temperature. A more cost-effective and easy way of covering such a large area is by adopting screen-printing technology—an old and matured technology [29]. We are going to present a very promising scintillating material (HgI$_2$), which has shown much promise [42]. Indeed, with its high atomic number, mass density, and relatively high detection efficiencies at diagnostic energies, screen-printed thick film (300–500) of HgI$_2$ is comparable to other potential scintillating materials that are being used in commercial radiation detectors [43, 44].

### 7.3.2.3 The Technology for HgI$_2$ Deposition

Figure 7.15 shows the schematic and the photograph of an HgI$_2$ screen-printing system. The system uses HgI$_2$ particle in binder, where the particles of pure HgI$_2$ powder are mixed with a binder prepared from a proper polymer dissolved in a solvent. The viscosity of the binder is controlled so that the slurry/ink can easily pass through the screen mesh. At the same time, the resistivity of the binder becomes compatible with the resistivity of HgI$_2$, to facilitate the flow of X-ray-induced charge across the boundaries between the HgI$_2$ grains and the binder. To prevent the corrosion of the local aluminum alloy connecting lines, a barrier layer is deposited over TFT arrays, before HgI$_2$ deposition. The barrier layer allows anisotropic current flow (only in perpendicular direction of the flat panel) and minimizes the cross talk between the pixels.

7.3 Halides of Heavy Metals 331

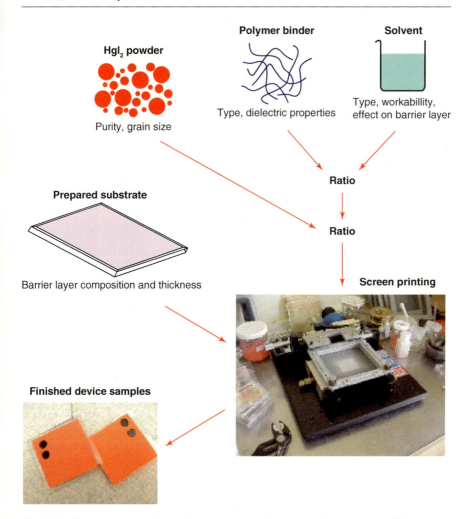

**Fig. 7.15** Shows the schematic and the photograph of the screen-printing system of $HgI_2$

### 7.3.3 Thallium Bromide (TlBr)

The average atomic mass of thallium (Tl) is comparable with that of lead (Pb) and mercury (Hg). Therefore, it will not be unrealistic to place TlBr in the halides of heavy metals. TlBr crystals have the CsCl-type simple cubic structure with density of 7.56 g/cm$^3$. The compound melts congruently at 480 °C and, therefore, is well suited for melt-based growth methods such as Bridgman and Czochralski processes. Table 7.5 shows some of the properties of thallium bromide (TlBr).

#### 7.3.3.1 Growth of Single Crystal of Thallium Bromide (TlBr)

Thallium bromide (TlBr) is dense and has high atomic number with wide band gap and has potential as an efficient and compact room temperature nuclear radiation

**Table 7.5** Properties of thallium bromide

| | |
|---|---|
| Molecular weight | 284.31 |
| Melting temperature | 460 °C |
| Crystal structure | Cubic, CsCl-type structure |
| Density | 7.453 g/mL |
| Transmission range | 0.44–40 μ |
| Hardness | 11.9 mhos |
| Solubility in water | 0.05 g/100 g |
| Cleavage planes | No cleavage |
| Band gap energy | ~2.7 eV |
| Resistivity | $10^{12}$ Ω-cm |

**Fig. 7.16** Shows a 26.5-cm (in length) single crystal of TlBr grown by BS method

detector [45]. Pixelated thallium bromide (TlBr) detectors have been successfully used in gamma-ray (γ-ray) spectroscopy and medical imaging [46].

Single crystal of TlBr is grown by Bridgman–Stockbarger (BS) method in vertical two-zone furnace [47]. Distilled and zone-refined high-purity, ultradry TlBr was placed in an evacuated quartz ampoule and moved through a two-zone furnace very slowly. Typical growth rates for BS process are 1–10 mm/h; however, certain layered structures require slower rates. The temperature of the upper zone is kept at about 600 °C while that of the lower zone is kept at 400 °C, which acts as an annealing temperature for the grown crystal. The dropping rate of the ampoule determines how close the conditions at the solid–liquid interface come to equilibrium. At very slow rates, the system operates close to equilibrium, which is desirable for growth of defect-free crystals. Figure 7.16 shows a 26.5-cm (in length) single crystal of TlBr grown by BS method. Experimental observations show that lead (Pb) and indium (In) impurities degrade the electrical and optical properties of TlBr crystal [48]. There are some reports of growing single crystal of TlBr by electrodynamic gradient process (EDG) [49].

**Fig. 7.17** Picture of a single crystal of LaCl$_3$ grown by Bridgman–Stockbarger method

## 7.4 Lanthanide Halides

In the beginning of the twentieth century, lanthanum halide scintillators are attracting much attention of the scientific community in nuclear spectroscopy, because of their almost ideal scintillation properties, like optimal energy resolution (<3 % at *662 keV*), excellent time resolution (~300 ps), and good efficiency [50–52]. The crystals are available commercially from Saint-Gobain Crystals under the brand name BrilLianCe$^{TM}$. These scintillators have high light yield, fast response, and good energy resolution [53, 54]. The *single crystal* of lanthanum halides can be grown as big as *1,000 mL in volumes*. Recent survey shows that the lanthanide crystals are more attractive than high-purity (HP) germanium (HPGe) for gamma-ray (γ-ray) measurements. These detectors are commonly used in nuclear and high-energy physics research, medical imaging, diffraction, nondestructive testing, nuclear treaty verification and safeguards, and geological exploration [55, 56].

### 7.4.1 Cerium-Activated Lanthanum Chloride (LaCl$_3$:Ce)

Lanthanum chloride crystals have hexagonal structure, and its density is ~3.9 g/cm$^3$. The melting point of the material is ~860 °C and can be grown by two-zone Bridgman–Stockbarger process [57, 58]. Lanthanum chloride (Lacl$_3$, mp ~848 °C) and cerium chloride (CeCl$_3$, mp ~849 °C) (10 mol%) after drying under argon atmosphere are loaded inside a quartz ampoule and is sealed under vacuum. The lower and the upper temperatures are maintained at 450 and 960 °C, respectively, with a temperature gradient ($dT/dx \approx 4°/cm$). The boule is pulled very slowly (1.5 cm/h). Typical growth rate in BS method is approximately 5–10 mm/h. Figure 7.17 shows the picture of a LaCl$_3$ single crystal grown by Bridgman–Stockbarger (BS) technique. The impurities are seen to have segregated at the bottom and top of the crystal. Further purification of the crystal is done by zone refining in an argon atmosphere. Table 7.6 shows some of the properties of lanthanum chloride (LaCl$_3$).

**Table 7.6** Properties of lanthanum bromide (LaCl$_3$)

| | |
|---|---|
| Crystal structure | Hexagonal |
| Density (g/mL) | 3.89 |
| Hygroscopic | Yes |
| Effective Z | 37 |
| Index of refraction | 1.90 |
| Light yield (photons/MeV) | 49,000 |
| Peak wavelength (nm) | 350 |
| Principal decay time (ns) | 25 |
| Attenuation length (cm) | 2.8 |
| Energy resolution | 3.2 |
| Melting temperature (°C) | 862 |

**Fig. 7.18** Picture of a LaBr$_3$ single crystal grown by BS method (Courtesy: RMD)

### 7.4.2 Cerium-Activated Lanthanum Bromide (LaBr$_3$:Ce)

The material LaBr$_3$ is supposed to be superior to CsI and NaI in quality like light output, as well as high proportionality [52]. Single crystal of LaBr$_3$ activated with cerium (Ce$^{3+}$) is grown by Bridgman method. LaBr$_3$ and CeBr$_3$ powders (99.99 %) are loaded in quartz ampoules and are sealed in an inert atmosphere [59]. The ampoule is passed slowly (0.5 mm/h) through a two-zone furnace. The upper zone of the furnace is kept at a temperature of 850 °C, which is higher than the melting point of LaBr$_3$ (783 °C). The lower zone of the furnace is kept ~680 °C. The growth rate of crystal is observed approximately 0.5 mm/h when the temperature gradient inside the furnace is kept around 28 °C/cm. The orientation of the growth of the crystal is seen to depend on the rate of the movement of the ampoule inside the furnace [60]. Figure 7.18 shows a single crystal of LaBr$_3$ grown by

## 7.4 Lanthanide Halides

**Table 7.7** Properties of lanthanum bromide ($LaBr_3$)

| Crystal structure | Hexagonal |
|---|---|
| Density (g/mL) | 5.3 |
| Hygroscopic | Yes |
| Effective $Z$ | 46.9 |
| Index of refraction | 1.88 |
| Light yield (photons/MeV) | 80,000 |
| Peak wavelength (nm) | 380 |
| Principal decay time (ns) | 35 |
| Attenuation length (cm) | 2.2 |
| Energy resolution | 2.9 |
| Melting temperature ($^\circ$C) | 783 |

Bridgman–Stockbarger (BS) method in our laboratory. Table 7.7 shows some of the properties of lanthanum bromide ($LaBr_3$).

### 7.4.3 Rubidium Gadolinium Bromide (RGB) (Cerium Doped) $RbGd_2Br_7$:Ce

RGB crystals have an orthorhombic structure with density of 4.7 $g/cm^3$. The melting point (mp) of the material is ~590 $^\circ$C, so it can be grown by Bridgman method. The material is first synthesized by mixing ultradry material of 99.99 % purity of RbBr (mp ~693 $^\circ$C) and $GdBr_3$ (mp ~770 $^\circ$C) in appropriate ratios (1:2). $CeBr_3$ (mp ~733 $^\circ$C) is used as dopant to introduce Ce into the system. The powders are well mixed and loaded inside a quartz ampoule which is evacuated to ~$10^{-2}$ torr. The ampoule is heated to a temperature of about 850 $^\circ$C in a tube furnace. After cooling, the solid-phase RGB formed is used as a feed material for Bridgman growth method [61].

During the Bridgman growth, the evacuated quartz ampoule with the synthesized RGB material is placed in a two-zone furnace. The upper zone temperature is kept at 650 $^\circ$C, while that of the lower zone is kept at 550 $^\circ$C. This allowed the growth of RGB crystal as the ampoule moved slowly from the higher temperature zone to cooler zone. Typical growth rate is observed as 1–10 mm/h. Crystals are hygroscopic in nature; therefore, the polishing $Al_2O_3$ girt is mixed in mineral oil to polish the crystal. Immediately after polishing and cleaning, the crystal is packaged to avoid moisture.

### 7.4.4 Cerium-Doped Lutetium Iodide ($LuI_3$:Ce)

Another lanthanide halide is lutetium iodide ($LuI_3$). Cerium-activated $LuI_3$ has remarkable light yields, but its resolution is not as good as cerium-activated $LaBr_3$ [62]. Activated lutetium iodide has high atomic number and average atomic

**Fig. 7.19** Shows the single crystal of LuI$_3$:Ce grown by BS method

**Table 7.8** Properties of cerium-activated LuI$_3$

| | |
|---|---|
| Density (g/mL$^3$) | 5.6 |
| Attenuation length (cm) | 1.82 |
| Light yield | 90,000 |
| Decay time (ns) | 6–140 |
| Emission (nm) | 474 |
| Hygroscopic | Yes |
| Melting temperature (°C) | 1,050 °C |
| Effective atomic number (Z) | 62 |
| Crystal structure | Hexagonal |

mass is 151. The rare-earth halide scintillator is used in gamma-ray spectroscopy. High-purity ultradry LuI$_3$ and CeI$_3$ are mixed in appropriate amount. The mixture melts at 1,050 °C when it passes through the hot zone of the BS furnace, which is controlled at 1,075 °C. As the melt passes through the cold zone, the end of the melt solidifies and acts as a seed for the rest of the melt and ultimately forms a big crystal. In our laboratory, we have grown crystals of LuI$_3$:Ce as large a 1 cm$^3$ [63] and is shown in Fig. 7.19. Table 7.8 shows some of the properties of cerium-activated lutetium iodide (LuI$_3$).

### 7.4.5 Cerium-Activated Lutetium Oxyorthosilicate (LSO)

Lutetium oxyorthosilicates (LSO): Cerium-activated (Ce$^{3+}$) Lu$_2$SiO$_5$ (LSO) scintillating material has been developed for several years. Following several years of development for application in positron emission tomography (PET), LSO:Ce scintillators have reached a rather mature stage. Literature survey reveals that single crystal or synthesized LSO:Ce produces high light yield (30,900 photons/MeV) and fast decay time (~43 ns). LSO:Ce went into large-scale commercial production in the late 1990s.

LSO (Lu$_2$SiO$_5$) has high light output and short decay time. It has high density and is nonhygroscopic. It suggests that it will be a very good candidate for positron emission tomographic (PET) medical imaging system. Unfortunately, the melting

7.4 Lanthanide Halides 337

point of the oxides of lutetium ($Lu_2O_3$), which is a starting material for LSO, is very high (~2,500 °C). As a result, the BS method cannot be applied to grow single crystal of LSO.

The alternative method to BS is Czochralski (CZ) technique, which will be appropriate to grow single crystal of LSO. The $Lu_2O_3$, $SiO_2$ (ratio 1:1), and $CeO_2$ (0.25 mol%) with 99.99 % purity were chosen for the single crystal growth. Crystal growth was initiated with seed crystal and was controlled via an automated system. Iridium (Ir) crucible was used to load the charge materials and was inductively heated (at a temperature ~2,150 °C) [64].

### 7.4.5.1 Single Crystal Growth of Lutetium Oxyorthosilicate (LSO)

Single crystal of LSO activated with Ce has been grown by Czochralski method by using iridium crucible with the charge (high purity $Lu_2O_3$, $SiO_2$, and $CeO_2$) inside an inductively heated furnace. The atmosphere inside the furnace is nitrogen with 0.7 % oxygen. The crystal seed is normally <100> with a pull rate of 1.5 mm/h with a rotational speed of the seed ~10–15 rpm. The crystals are normally doped with 0.05–0.5 % Ce while controlling other impurities at low levels.

Experimental evidence shows that $Ce^{3+}$ is seen to be active and no evidence or least activity is observed for $Ce^{4+}$ [65]. During pulling, the rod carrying the seed was rotated, and argon gas was supplied inside the furnace to prevent oxidation of the iridium crucible. The stoichiometry of the mixture is kept tight throughout the experiment (as far as practicable). The CZ technique can grow near transparent crystal as large as 10 cm in length with 6 cm in diameter very easily.

Considerable work has been done by several research workers on the use of co-doping (Ca and/or Zn) LSO to improve the efficiency. 0.1 at % of Ca and Zn doping is seen to improve the light yield by 25 and 7.4 % and 14 and 30 % in decay-time recovery [66]. Figure 7.20 shows the picture of a single crystal of cerium-activated LSO. And Table 7.9 shows some of the properties of cerium-activated lutetium oxyorthosilicate ($Lu_2SiO_5$).

### 7.4.5.2 Solgel Technique to Procure LSO Powder/Film

As the Czochralski method requires high temperature and the technology is expensive, alternate method like precipitation method to grow LSO has been reported. The method is called *solgel technique* where tetraethyl orthosilicate (TEOS) is used as precursor [67]. Solution of the nitrates of lutetium and cerium is prepared from their respective oxides, washed, and dissolved in alcohol. The coprecipitate of the mixture is mixed with the precursor. The gel formed from the mixture is heated at 1,200 °C and fine powder (nanoscale particles) of LSO is collected. The advantage of the solgel method is that the grain sizes are very small (~50– 200 nm) depending upon the conditions of the process. When the grain sizes are smaller, the hot-pressed ceramic will exhibit macro hardness because the grain boundary instability does not deteriorate the hardness in extremely fine (nanoscale) microstructures. At the same time, the high-purity nanoscale powder when hot-pressed becomes transparent, too [68].

**Fig. 7.20** Shows the picture of a single crystal of cerium-activated LSO (Courtesy: SIPAT)

**Table 7.9** Properties of cerium-activated $Lu_2SiO_5$

| Crystal structure | Trigonal |
|---|---|
| Melting point (°C) | 2,047 |
| Density (g/mL$^3$) | 7.4 |
| Emission wavelength (nm) | 418 |
| Decay constant (ns) | 40 |
| Light yield (ph/MeV) | 25,000 |
| Attenuation length (cm) | 1.2 |
| Hygroscopic | No |
| Refractive index | 1.82 |

### 7.4.5.3 Solid-State Reaction to Produce LSO Powder

The traditional method to synthesize LSO powder in solid-state reaction requires ball milling the mixture of $Lu_2O_3$ and $SiO_2$ and high-temperature calcination (normally 1,400 °C) [69]. In our laboratory, we have prepared cerium-activated LSO powder by coprecipitation of the nitrates of cerium and lutetium together with TEOS at 1,200 °C. The cerium LSO powder thus obtained is hot-pressed and ceramic tablets are made out of the powder. The emission and the light output of the LSO ceramics are comparable to single crystal.

Several low-cost alternative methods have been suggested from time to time as an alternative source of fabricating the material. For LSO, we will describe here a method, which has proven to be a very cost-effective method and can be applied to LSO preparation without compromising the quality of a commercially grown single crystal.

The material is synthesized from the coprecipitate of rare-earth oxides. Appropriate amounts of the oxides of cerium ($CeO_2$) and lutetium ($Lu_2O_3$) are mixed in

## 7.4 Lanthanide Halides

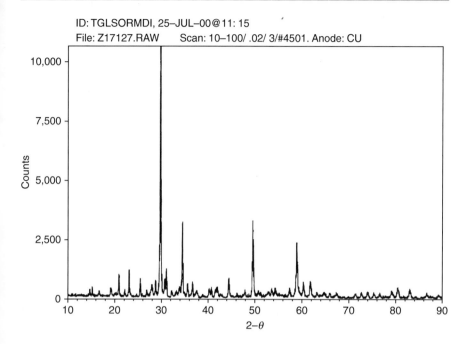

**Fig. 7.21** Shows the X-ray diffraction pattern of synthesized LSO powder activated with cerium

nitric acid to prepare the nitrates of the oxides. After adjusting the pH (slightly acidic), an aqueous solution of oxalic acid is added to the solution and the resultant white precipitate is separated out. The white precipitate is then mixed with silicic acid in the proper ratio in presence of ammonium fluoride as flux. The mixture is transferred to a quartz crucible and heated in a furnace ~1,200 °C for 3 h. The furnace is cooled down at 5 °C/min rate, and the cooled LSO powder is collected from the furnace. The synthesized LSO powder is seen to have almost similar scintillating properties of a single crystal [70].

Figure 7.21 shows almost 74 peaks of lutetium oxyorthosilicate ($Lu_2SiO_5$) ceramic. The strong lines for $2\theta$ values are plotted along $x$-axis against the arbitrary intensity of the lines along $y$-axis. An important property of any scintillator is that it must have minimal absorbance in the wavelength region of the scintillation emission so that the emission intensity is not reduced as the light passes through the crystal. The emission spectra of the synthesized powder and that of a single crystal powder are in Fig. 7.22.

Figure 7.23 shows time vs. emission spectra of a good and a bad LSO crystal along with the emission spectrum of the synthesized ceramic at ~427 nm (X-ray energy: 30 kV/15 mA). The *induction effect* measurements can identify the quality of the crystals by correlating trap density with the steady-state luminescence properties of the samples, because as the X-ray excitation begins, the carrier starts filling the traps and, therefore, is not available for luminescence. The area of

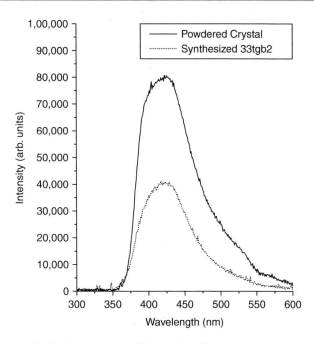

**Fig. 7.22** X-ray-excited emission spectra of the powder of synthesized and single crystal of LSO

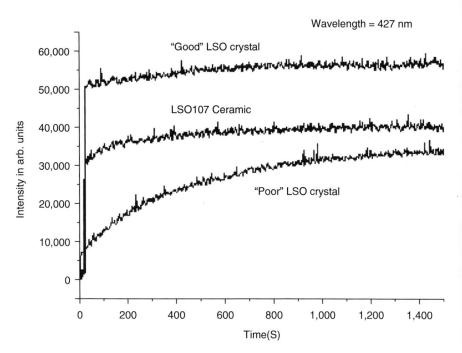

**Fig. 7.23** Shows time vs. intensity curve for LSO samples at room temperature. Samples are irradiated with X-ray at 427 nm (X-ray-irradiated power ~30 kV/15 mA)

## 7.4 Lanthanide Halides

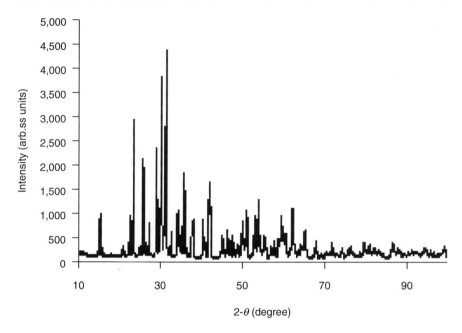

**Fig. 7.24** Shows the X-ray diffraction pattern of LSO synthesized by hydrothermal synthesis

luminosity $\{L_\infty - L(t)\}$ from these luminescence growth curves can be represented as a function of the number of trapping centers [70].

### 7.4.5.4 Hydrothermal Synthesis for LSO

In this method, silicon dioxide ($SiO_2$) is mixed with lutetium nitrate ($LuNO_3$) and sodium hydroxide in the molar ratio 1:2 and 6:1 to prepare $Na_2SiO_3$ solution. The pH of the solution is adjusted between 7 and 8 by adding hydrochloric acid solution. A mixture of Lu $(NO_3)_3$ and Ce $(NO_3)_3$ was added to $Na_2SiO_3$ solution and mixed thoroughly. The mixture is put into an acid digestion bomb to undergo hydrothermal treatment at 200 °C for 10–12 h. The cooled powder is washed in DI water and finally in ethyl alcohol and dried in an oven under 100 °C [71].

In compound semiconductor, one of the most important and complicated parameters to be controlled during the growth of the crystal is the stoichiometry. Most of the time, it is observed that lack of stoichiometric composition enhances the generation of intrinsic point defects, like vacancies, interstitials, and antisites. These defects contribute to compensation level, type of conductivity, carrier concentration, absorption behavior, diffusivity, efficiency of dopant incorporation, etc. Moreover, the interruption of growth during supercooling and interface instability affects the heat transfer between solid and liquid interface [72, 73]. Most of the time, the high melting temperature together with the sophistication of the instrument with special processing conditions ultimately increases the cost of single crystals. Figure 7.24 shows the X-ray diffraction pattern of synthesized LSO powder activated with cerium.

## 7.5 Complex Oxides with High Atomic Number

An improved production technology has been developed to grow single crystals based on complex oxides with large atomic number like gadolinium silicate (GSO), bismuth germanate (BGO), cadmium tungstate (CWO), and lead tungstate (PWO). Scintillators based on these crystals have good energy resolution, high light output, detection efficiency, and radiation stability and good mechanical strength, and above all, these crystals are not hygroscopic. One of the main mechanisms of the crystal growth depends on the melting points of the charge materials, and our present complex oxide materials are having high melting point. As a result, single crystals of these complex oxides are grown following Czochralski (CZ) technique, and crucible used for the charge materials is either iridium or platinum in an inert ambient with controlled movement along the vertical axis and azimuth, ensuring reliable visual control. Computer-controlled growth initiation and inductively heated charge result in a highly perfect crystal with great structural perfection [74].

### 7.5.1 Cerium-Activated Gadolinium Silicate (GSO, $Gd_2SiO_5$)

Cerium-activated gadolinium silicate (GSO) offers high light output, short decay constant, and high absorption coefficient [75, 76]. Due to these excellent properties, GSO has attracted attention of the nuclear medicinal physicists to explore the experimental techniques to grow single crystal of GSO with minimum defects. At the beginning, it was difficult to grow large crystals without cracks because of its poor mechanical properties and the yield was low. As a result, the price of a good crystal of GSO was relatively high compared to bismuth germanate (BGO). It is therefore important to increase the yield and to develop an economical production technique for GSO to be used in positron emission tomography (PET) system.

#### 7.5.1.1 Growth of Single Crystal

The melting points of the starting materials $Gd_2O_3$ (mp ~2,310 °C), $SiO_2$ (mp ~1,500 °C), and $CeO_2$ (mp ~490 °C) (0.5–1.0 mol%) to grow single crystal of GSO are high. As a result, the cerium-activated GSO crystal is grown by Czochralski technique. The material is characterized by its high density ($\rho = 6.71$ g/cm$^3$), high effective atomic number ($Z = 59$), and high radiation detection index ($\rho Z^4_{eff} = 84 \times 10^6$). Additionally, due the presence of cerium ($Ce^{+3}$) ion activator, GSO:Ce exhibits fast response (~60 ns) [77].

The charge is loaded in an iridium (Ir) crucible in an ambient of $N_2$ to prevent thermal etching of the crucible which will introduce some micro-cracking and ultimately triggers large crack in the boule. The seed crystal is pulled at 3–4 mm/h with a rotational rate of the seed ~30–50/min [78, 79]. The grown crystal is cooled very slowly inside the furnace. It has been reported that the flow of $N_2$ with 1 % $O_2$ prevents any crack of the boule [80]. Large crystal of GSO with 65 mm in diameter and 350 mm in length can be grown by CZ technique very easily. Figure 7.25 shows the picture of a single crystal of cerium-activated gadolinium silicate, and Table 7.10 shows some of the properties of gadolinium silicate (GSO).

## 7.5 Complex Oxides with High Atomic Number

**Fig. 7.25** Picture shows the single crystal of cerium-activated gadolinium silicate (Courtesy: Hitachi Chem)

**Table 7.10** Properties of cerium-activated GSO ($Gd_2SiO_5$)

| Parameter | GSO |
|---|---|
| Density | 6.71 |
| Hygroscopic | No |
| Emission wavelength (nm) | 430 |
| Decay constant (ns) | 30–60 |
| Refractive index | 1.85 |
| Primary decay time (ns) | 30–60 |
| Light yield (photons/MeV) | $8 \times 10^3$ |
| PMT photocathode yield | 20 % of NaI |
| Melting point (°C) | 1,950 |

### 7.5.1.2 Solution Combustion Synthesis (SCS) of GSO Powder/Film

Solution combustion technique is a simple exothermic reaction of the mixture of gadolinium nitrate [Gd$(NO)_3]_3$, cerium nitrate [$(CeNO_3)_3$ : $6(H_2O)$], and fumed silica mixed in stoichiometric ratio in hexamine nitrate [$(CH_2)_6N_4$] solution. The powder is dried in an oven at 150 °C and loaded in a non-glazed alumina crucible. The crucible is combusted inside a muffle furnace preheated at 650 °C. The combusted material is grounded and post-annealed at 1,000 °C. Transmission electron microscope revealed nano-size particles typically between 20 and 80 nm [81, 82]. The powder can be used to make ceramic or to screen printing for making thick films.

### 7.5.2 Bismuth Germanate (BGO, $Bi_4 Ge_3 O_{12}$)

BGO is relatively hard with high Z ($Z_{eff} = 83$ and density = 7.13 g/cm$^3$) value and nonhygroscopic. The single crystal of BGO does not cleave and can be machined to

**Fig. 7.26** Shows the picture of a CZ-grown single crystal of BGO

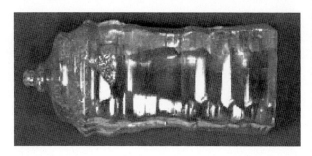

**Table 7.11** Properties of BGO ($Bi_3Ge_4O_{12}$)

| Parameter | BGO |
| --- | --- |
| Density | 7.13 |
| Hygroscopicity | No |
| Emission wavelength (nm) | 480 |
| Decay constant (ns) | 300 |
| Refractive index | 2.15 |
| Crystal structure | Cubic |
| Light yield (photons/MeV) | $8.2 \times 10^3$ |
| PMT photocathode yield | 20 % of NaI |
| Melting point (°C) | 1,050 |
| Attenuation length (cm) | 1.21 |

various shapes and sizes. Due to its high Z value, the photofraction for γ-ray absorption is high. The detection efficiency of BGO is also high. As a result, BGO scintillators are used for PET scanners and Compton suppression spectrometers. Recently, BGO crystal has been used in a fast neutron coincidence counter (FNCC) [83].

Its extremely low afterglow makes it useful for applications such as X-ray computed tomography (CT). The scintillation emission maximum of BGO is around 480 nm. The scintillation intensity of BGO is a strong function of temperature, and the crystals are susceptible to radiation damage starting at radiation doses between 1 and 10 Gray ($10^2$–$10^3$ rad).

### 7.5.2.1 Single Crystal Growth of BGO

Transparent single crystal of BGO is grown by Czochralski technique from a stoichiometric melt of $Bi_2O_3$ (mp ~817 °C) and $GeO_2$ (mp ~1,086 °C). The melt temperature inside the high-purity platinum (Pt) crucible is maintained at 1,150 °C. The pulling and the rotation of the spindle carrying the seed (BGO) are maintained at 0.2 mm/h and 15 rpm, respectively. The single crystal of BGO is cubic and melting temperature of the material is ~950 °C. Due to the presence of $Bi^{3+}$ ion, BGO does not need any activator. The $Bi_3^+$ ion is thought to scintillate by the same mechanism as $Tl^+$ ion [84]. Figure 7.26 shows the picture of a single crystal, and Table 7.11 shows the properties of bismuth germanate (BGO).

## 7.5 Complex Oxides with High Atomic Number

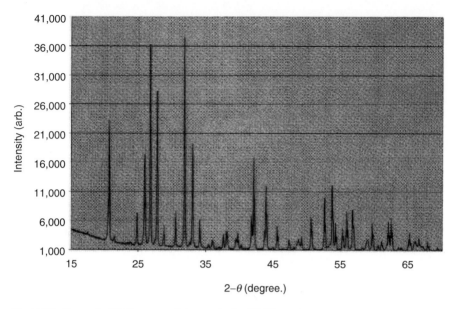

**Fig. 7.27** Shows the XRD pattern of the synthesized BGO

### 7.5.2.2 Chemical Synthesis of BGO

Oxides of metals generally have high melting temperatures. As a result, most of the time, single crystal of these materials is grown by Czochralski method. In order to avoid the cost of growing single crystal, several other techniques are developed to synthesize the material by some other chemical methods.

The growth of BGO, however, is performed using $Bi_2O_3$ and $GeO_2$ charges with stoichiometric composition (approximately 2:3). The powder is well mixed and grounded and loaded in an alumina crucible. The crucible is loaded in a furnace in an ambient of air. The charge is heated to 1,200 °C for 2 h and cool down to room temperature at 5°/min. X-ray diffraction pattern is performed and shown in Fig. 7.27. The synthesized powder is exposed to X-ray, and the emission spectrum of the material is measured and compared with a standard grown crystal. The powder shows almost similar behavior to that of the crystal.

The BGO powder synthesized in the manner described above is analyzed by X-ray diffraction with an X-ray unit (Rigaku model Rotaflex # RTP-500), and the corresponding spectrum is shown in Fig. 7.27. The pattern is representative of the cubic structure of the material

To further quantify the material, emission spectrum of the synthesized material is performed. The powder is excited with radiation from a Philips X-ray tube with a copper target, with power settings of 30 KVp and 15 mA. The scintillation light is passed through a McPherson 0.2-m monochromator and detected by Hamamatsu C31034 photomultiplier tube with a quartz window. Prior to the measurement, the system is calibrated with a standard light source to enable correction for sensitivity

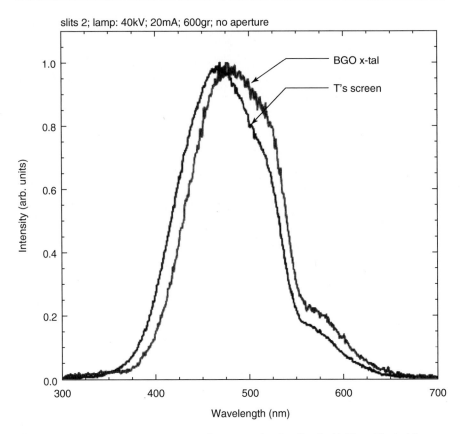

**Fig. 7.28** Shows the wavelength vs. intensity curves when irradiated with X-ray. The inside curve (T's screen) represents the characteristic of the screen printed thick film made out of the ground powder of a single crystal BGO, and the outside curve is the representative of the ground powder

variations as a function of wavelength. Figure 7.28 shows emission spectrum of the synthesized powder screen printed on an ITO glass as well as the spectrum of a reference powder prepared by pulverizing a high-quality single crystal. In both cases, the emission peak is at around 475 nm, which is characteristic of BGO.

### 7.5.2.3 Solution Combustion Technique

Solution combustion technique (SCT) or self-propagating high-temperature synthesis (SHS) is a unique technique to prepare a number of refractory cermets: oxide boride, oxide carbide, and oxide silicide. The process is exothermic, so synthesis does not required high temperature initially. The process is a powerful tool to produce many high-temperature inorganic compounds supplying only a fraction of heat energy necessary to melt the product [85]. Typically, SHS route produces nano-size particles, but unfortunately, the process results in high degree of agglomeration of the particles. However, by adjusting the pH, the degree of agglomeration can be minimized [86].

## 7.5 Complex Oxides with High Atomic Number

**Table 7.12** Properties of cadmium tungstate ($CdWO_4$)

| Parameter | CWO |
| --- | --- |
| Density | 7.9 |
| Hygroscopicity | No |
| Emission wavelength (nm) | 470–540 |
| Decay constant | 20 ns–5 ms |
| Refractive index | 2.2–2.3 |
| Crystal structure | Monoclinic |
| Light yield (photons/MeV) | $1.2 \times 10^4$ |
| Melting point (°C) | 1,325 |
| Attenuation length (cm) | 1.21 |

To synthesize BGO powder by SCT, the cationic precursors used are $GeO_2$, $Bi(NO_3)_3$, $5H_2O$, and $CON_2H_4$ (urea) which is the fuel [87]. The stoichiometric composition of the redox mixture is calculated on the basis of the oxidizing and reducing valences of the oxidizer and the fuel. The precursors are mixed with water to make a paste, and the pH of the paste is adjusted adding ammonium hydroxide ($NH_4OH$). The mixture is heated at temperature 500 °C, when exothermic decomposition of the fuel started. After cooling down the product, it is calcined and sintered. In order to prevent the grain growth and to preserve the nanostructure of the synthesized compound, simultaneous calcination and sintering are performed [88]. Experimental observations show that silicon- and germanium-containing compounds have a tendency of grain growth with calcination and sintering.

### 7.5.3 Cadmium Tungstate ($CdWO_4$ or CWO)

Cadmium tungstate has high density and high $Z$ with relatively higher light output than BGO. Unfortunately, the crystal cleaves very easily and is therefore very susceptible to mechanical and thermal shock. Furthermore, since some self- absorption of the scintillation light occurs in $CdWO_4$ crystals, it is not very encouraging to grow a very large crystal like BGO and other inorganic materials. However, the material shows a very good radiation resistance for gamma rays ($\gamma$ rays). Table 7.12 shows some of the properties of cadmium tungstate ($CdWO_4$).

#### 7.5.3.1 Growth of Single Crystal of Cadmium Tungstate (CWO)

Cadmium tungstate ($CdWO_4$) single crystal is grown by Czochralski (CZ) technique. The charges CdO and $WO_3$ are mixed together. In order to compensate the volatile component loss during crystallization process, the CWO melts for crystal growth contain excess of CdO in concentration of about 1.0–1.2 % mass. The pulling and rotational rates are maintained between 0.2–0.3 mm/h and 12–15 rpm, respectively. The temperature of the melt crystal is optimized in order to keep high

**Fig. 7.29** Shows the picture of a single crystal of CWO grown by Czochralski technique

structural perfection. Post-thermal treatment and the optimization of the growth rate result in a homogeneous and transparent crystal of CWO. The technological yield of the crystals is high. Figure 7.29 shows the picture of a single crystal of cadmium tungstate (CdWO$_4$) grown by CZ technique.

### 7.5.3.2 Solgel Technique

To grow single crystal of CWO by CZ technique is expensive, and it is difficult to eliminate all the defects from single crystal. Moreover, it is rather difficult to use single crystal of CWO for the fabrication of micro- and nanoscale devices that are used in micro-electromechanical (MEMS) and nano-electromechanical (NEMS) systems. On the other hand, nano-structured polycrystalline of complex oxides synthesized from solution growth method has many advantages [89, 90].

The precursors used in solgel process are tungsten chloride (WCl$_6$) and cadmium acetate {Cd(OOCCH$_3$)$_2$} · 6H$_2$O. First, WCl$_6$ is dissolved in ethanol (C$_2$H$_5$OH). The solution is further diluted in ethanol and refluxed in air at 50–60 °C to remove completely hydrochloric acid from the solution. The gel is then dried at 100 °C to form powder and sintered at temperature 500 °C.

### 7.5.4 Lead Tungstate (PbWO$_4$ or PWO)

Lead tungstate (PbWO$_4$) is another popular, fast, self-activated scintillator. Its luminescence is due to the rapid quenching of an excited state by electron transfer in the WO$^{-2}$ anion. PbWO$_4$ is dense, fast, and low cost compared to similar

## 7.5 Complex Oxides with High Atomic Number

**Table 7.13** Properties of lead tungstate ($PbWO_4$)

| Parameter | PWO |
|---|---|
| Density (g/mL$^3$) | 8.28 |
| Hygroscopicity | No |
| Emission wavelength (nm) | 450 |
| Decay constant (ns) | 40 |
| Refractive index | 2.16 |
| Crystal structure | Tetragonal |
| Light yield (photons/MeV) | $25 \times 10^4$ |
| Melting point (°C) | 1,123 |
| Attenuation length (cm) | 1.21 |

scintillator like $BaF_2$ and is used in high-energy physics [91]. PWO is used as radiation detector in high energy, in Compact Muon Solenoid experiment at Large Hadron Collider (HRC), to investigate the mechanism responsible for electroweak symmetry breaking [92].

### 7.5.4.1 Single Crystal Growth of PWO

Single crystal of lead tungstate (PWO) is grown by Czochralski (CZ) technique using platinum crucible. The initial stoichiometric powder or preset composition was prepared by high-temperature solid-state synthesis of the lead oxide powder (PbO), melting point 888 °C, and tungstate oxide powder ($WO_3$), melting point 1,500 °C. The synthesized powder was zone refined for 29 passes. The zone-refined powder is loaded in the platinum crucible as charge and inserted inside the CZ furnace with proper attachments. Table 7.13 shows some of the properties of lead tungstate ($PbWO_4$).

### 7.5.4.2 Chemical Method

The tetragonal scheelite structures $PbWO_4$ powder or solution can be prepared by *polymeric precursor* method [93]. Tungsten citrate is formed by mixing tungstic acid ($H_2WO_4$) with aqueous solution of citric acid ($NH_4O_2CH_2C(OH)(CO_2H)$ $CH_2CO_2NH_4$). Lead acetate ($Pb(OOCCH_3)_2$) is added to the homogenized solution of tungsten citrate. The pH of the solution is adjusted by adding ammonium hydroxide ($NH_4OH$). The viscosity of the solution was then adjusted by slow heating (~70–80 °C), to spin coat the solution or to screen print on a proper substrate. The spin-coated substrate or screen-printed substrate is heated to 150 °C and annealed ~500 °C in air.

### 7.5.5 Cerium-Activated Lutetium Gadolinium Silicon Oxide ($Lu_{0.4}Gd_{1.6}SiO_5$:Ce)

Recently, $Lu_{0.4}Gd_{1.6}SiO_5$:Ce single crystal has been reported with better scintillation properties than GSO alone. The single crystal has been grown from $Gd_2O_3$

| Material | LUAP |
|---|---|
| Density (g/mL$^3$) | 8.4 |
| Hygroscopic | No |
| Emission wavelength (nm) | 365 |
| Energy resolution ($^{137}$Cs) | 9.3 |
| Refractive index | 1.94 |
| Primary decay time (ns) | 16 |
| Light yield (photons/MeV) | $11 \times 10^3$ |
| PMT photocathode yield | 20 % of NaI |

**Table 7.14** Characteristics of LuAP crystal

(purity 4N), $Lu_2O_3$ (4N), $SiO_2$ (6N), and $CeO_2$ (4N) mixed at stoichiometric composition [94].

## 7.5.6 Cerium-Activated Ortho-aluminates (Perovskites)

The pseudo binary phase diagrams $Al_2O_3$–$RE_2O_3$ (RE stands for a rare-earth element from La to Lu or Y) contain up to four intermediate compounds. It is reported that $LuAlO_3$ decomposes upon heating to the garnet $Lu_3Al_5O_{12}$ and the monoclinic $Lu_4Al_2O_9$ or even $Lu_2O_3$ [95, 96]. The versatility of this crystal is enhanced by adding dopant.

### 7.5.6.1 Growth of Single Crystal by Czochralski Technique

As the melting temperatures of the materials like $Lu_2O_3$, $Al_2O_3$, and $CeO_3$ are all very high, Czochralski method has been found to be most suitable for growing single crystal. Iridium crucible is used as charge carrier, and iridium wire is used to nucleate the crystal. Small defect-free single crystal of LUAP is used as a seed. Cerium concentration can be varied between 0.25 and 0.75. The crystal is pulled very slowly from the melt. YAP has been found to grow with less defects than the other lanthanide ortho-aluminates, but the crystal during growth has a tendency to incorporate Ce in different oxidation states [97, 98].

### 7.5.6.2 Single Crystal Grown by Bridgman Method

Single crystal of LuAP has been successfully grown by Bridgman method [99]. The charge is a mixture of $Al_2O_3$ and $Lu_2O_3$, which is loaded inside molybdenum (Mo) crucible together with $Ce^{3+}$ (doping concentration $\leq 1$ %). To limit the amount of oxygen interaction with the charge, hydrogen flow at a rate of $\leq 30$ % is maintained inside the furnace. The pulling rate is maintained in between 0.5 and 5 mm/h. The pulling rate is very important in the sense that higher pulling rate may results inclusion of gas bubble and cracks. LuAP is also very sensitive to the inside temperature of the furnace [100, 101]. Table 7.14 shows the characteristics of LuAP crystal.

# 7.6 Compound Semiconductor

## 7.6.1 Cadmium Telluride (CdTe) and Cadmium Zinc Telluride (CZT)

CdTe and CZT are compound semiconductors that have been used as gamma-ray detectors. The single crystals of CdTe and CZT can be grown either by traveling heater method (THM) [102–104] or by Bridgman method. The Bridgman method can be vertical or horizontal according to the layout of the tube furnace. In the high-pressure Bridgman (HPB) technique, the pressure inside the furnace can be several hundred atmosphere at a temperature about 1,600 °C. The HPB technique can be vertical or horizontal according to the layout of the furnace.

The quality of the radiation detector depends on the quality of the grown crystal, which ultimately depends on the purity of the starting material (charge). As a result, the charge material is undergone zone melting process. The method is based on the fact that the solubility of a given impurity is generally higher in the liquid $C_L$ than in the solid $C_S$ (the segregation coefficient $K = (C_S/C_L)$), and it is more efficient for purification of the material [105].

## 7.6.2 CdTe Crystals

### 7.6.2.1 Traveling Heater Method (THM) for Growing CdTe Crystals

In THM, chlorine-doped crystals are grown from cadmium telluride solution. The ratio of Cd:Te determines the melting temperature of the powder. In general, the temperature of the melt is maintained between 650 and 700 °C. High-quality crystals can be grown by THM method, but the growth rate is very slow (only few millimeters per day). Recently, CdTe single crystals with 75 mm diameter and 200 mm in length have been reported [106]. The Cl-doped CdTe material shows greater than $1 \times 10^9$ $\Omega$-cm resistivity. Though some improvements have been made on crystal quality, some experts think that THM is not a very economic method for the production of single crystals of CdTe.

### 7.6.2.2 Bridgman Method for Growing CdTe Crystal

CdTe crystals grown by low-pressure Bridgman (LPB) method give high-quality crystals. The evacuated ampoule is loaded with cadmium (Cd) with excess tellurium [107]. In both *vertical* and *horizontal* Bridgman method, the ampoule moves along a region with temperature gradient that extends from above to below the melting temperature of powder. The growth process involves continuous material transfer from solid to liquid and from liquid to vapor. The constant vapor pressure keeps constant liquid composition and balanced amounts of cadmium and tellurium within the crystal.

Figure 7.30 shows the schematic of the Bridgman technique for growing single crystals of CdTe and CZT. The figure on the left-hand side shows the vertical movement of the ampoule, whereas on the right-hand side, it shows the horizontal movement of the ampoule. Annealing the pure crystals in presence of cadmium

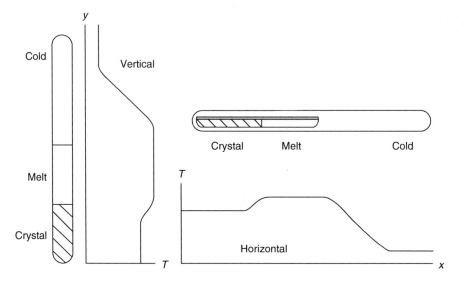

**Fig. 7.30** Shows a schematic of vertical and horizontal Bridgman method (Courtesy: Lachish)

**Fig. 7.31** Shows the picture of a CdTe crystal grown by BS technique

transforms it into an n type with medium resistivity $10^5$–$10^6$ Ω-cm. Tellurium excess will transform the crystal to p-type material with resistivity on the order of $10^3$–$10^5$ Ω-cm. Figure 7.31 shows the picture of a CdTe single crystal grown by BS Method.

### 7.6.2.3 High-Pressure Bridgman (HPB) Method

In high pressure Bridgeman (HPB) method, CdTe crystals are grown with an excess of cadmium which generates high vapor pressure that requires special design of the furnace [108, 109] so that it can allow a fine control of the stoichiometry of the CdTe ingot by using molten Cd source to control the Cd partial pressure in the ampoule during growth. The crystal is grown from the polycrystalline material obtained after zone refining the CdTe materials of high purity (6N). The zone refining can control the undesired impurities like lithium (Li), sodium (Na), and copper (Cu) which are shallow acceptors that cause uncontrolled doping of the lattice.

**Fig. 7.32** Picture shows a single crystal of CZT grown by THM method (Courtesy: RMD)

## 7.7 Cadmium Zinc Telluride (CZT)

### 7.7.1 Growth of Single of Cadmium Zinc Telluride (CZT)

For CZT ($Cd_{1-x}Zn_x Te$) crystals, however, zinc (Zn) in the 8–15 % concentration range is added to the CdTe materials. Depending on the percentage volume of zinc (Zn) in CZT, Zn increases band gap, resistivity (~1.0–4.0 × $10^{10}$ Ω-cm), and cadmium vacancies ($V_{Cd}$) and reduces the density of Te antisites ($Te_{Cd}$) of the crystal. The mobility ($\mu$)–lifetime ($\tau$) for electrons and holes of the crystals grown with 10 % concentration of Zn are expected to be in the range of ($\mu\tau_e = 0.5$–$5.0 \times 10^{-3}$ cm$^2$/V) and ($\mu\tau_h = 0.2$–$5.0 \times 10^{-5}$ cm$^2$/V) [110, 111]. The high resistivity of the crystals, however, can be due to intrinsic defects and residual impurities inside the forbidden band [112, 113].

Figure 7.32 shows the picture of a single crystal of CZT grown by THM method. Most of the time, the single crystals grown by THM method show the mole fraction of zinc (Zn) to vary from crystal to crystal. The variation is due to the fact that the segregation coefficient of Zn has a value of 1.3; consequently, there is a gradual change in the value of $x$ from tip to tail of the solidified boule, which results in adverse conditions for producing large volume CZT crystals. The variation of Zn along the length of the boule causes a variation in the band gap of the material and in the number of photo-generated electron–hole (e–h) pairs per incident gamma ray [113].

Table 7.15 shows some of the properties of cadmium telluride (CdTe) and cadmium zinc telluride (CZT).

**Table 7.15** Properties of CdTe and CZT

| Properties | CdTe | CZT |
|---|---|---|
| Atomic numbers | 48, 52 | 48, 30, 52 |
| Average atomic number | 50 | 49.1 |
| Density (g/mL$^3$) | 5.85 | 5.78 |
| Resistivity (Ω-cm) | $10^9$ | $3 \times 10^9$ |
| Band gap (eV) | 1.5 | 1.572 |
| Dielectric constant | 11 | 10.9 |
| Electron mobility (cm$^2$/V · s) | 1,100 | 1,000 |
| Electron lifetime (s) $\tau_e$ | $3 \times 10^{-6}$ | $3 \times 10^{-6}$ |
| Hole mobility (cm$^2$/V · s) | 100 | 50–80 |
| Hole lifetime (s) $\tau_h$ | $2 \times 10^{-6}$ | $10^{-6}$ |
| $(\mu\tau)_e$ (cm$^2$/V) | $3.3 \times 10^{-3}$ | $(3–5) \times 10^{-3}$ |
| $(\mu\tau)_h$ (cm$^2$/V) | $2 \times 10^{-4}$ | $5 \times 10^{-5}$ |

## 7.8 Elemental Semiconductor

The identification of new semiconducting material for radiation detection has been time-consuming and exhaustive [114]. The synthesis or the growth of single crystal of a semiconducting material is challenging from the point of achieving crystallinity, single-phase, and stress- and defect-free compound. The stress during the growth introduces phase transformation, which is susceptible to the loss of light inside the compound crystal. The loss of light inside the scintillator (single crystal/polycrystalline synthesized compound) could result degradation in resolution and could affect the overall resolution of the image.

Elemental semiconductors mostly used in medical imaging systems are *silicon* (Si), *germanium* (Ge), and *diamond*, and the problems related to the growth of a single crystal or deposition of the elemental semiconductor by CVD method are not severe and complicated. Moreover, tuning the properties of these elemental semiconductors as radiation detectors is largely a matter of doping and processing techniques though most of the time the doping concentration along the length of the crystal is not uniform. It is true for both elemental and compound semiconductor.

Silicon (Si) and germanium (Ge) have long been the most widely used semiconductors for radiation detection and thus serve as bench mark materials. Since 1970, the purity ($\sim 10^{12}$ Ω-cm) of these two elemental semiconductors (Si and Ge) has reached to a status where further purification is almost impossible. Thus, physically, they are unique for their relatively high purity and for the fact that they are mono-elemental semiconductors. The other elemental semiconductor that is being used is diamond. It has high band gap and is a suitable choice as a radiation detector for high-energy physics. Unfortunately, diamond radiation detector presents an efficiency challenge due to its limited size.

# 7.8 Elemental Semiconductor

Commercially available Ge detectors now exceed 150 % relative efficiency compared to NaI(Tl) [115]. The feature implies that when imaging with different isotopes simultaneously, increased organ or lesion specificity can be realized. The high resolution also implies better rejection of tissue scattering, which in turn helps to acquire better image. Unfortunately, the cryogenic atmosphere for the operation of Ge detectors primarily restricted the use of this elemental semiconductor in many national security needs.

## 7.8.1 Diamond

As methods for growing synthetic (nonnatural) diamond, both at high pressure and by chemical vapor deposition (CVD), have improved the quality, diamond has become an important rugged radiation detector in high-energy physics. The scientists and the engineers have taken advantage of the rugged elemental semiconductor and found that diamond's superlative properties are boundless—from super electronics to strong optical windows to unscratchable surfaces to rugged radiation detector. As a matter of fact, synthetic diamond has become an obvious choice for radiation detectors in high-energy physics.

*Diamond* has been used for years as radiation detector for ionizing radiation because of its *high electron and hole mobilities, high dielectric breakdown strength, high resistance to radiation damage, high-temperature operation, fast time response*, and *tissue equivalence* [114–116]. Figure 7.34 shows the picture of a PECVD-grown single crystal of diamond (size $3 \times 3 \times 0.9$ mm$^3$) on a <100> surface of a type 1b diamond substrate synthesized by the high-pressure/high-temperature (*HP/HT*) method.

The highest quality diamond synthesized in the laboratory involves the growth in a flux of molten alloys under high temperature (HT $> 1,400$ °C) and high pressure (*HP*) (55,000 atmosphere). The *HPHT* method recreates the actual geological conditions that occur 150 km deep in Earth's mantle under unique and powerful combinations of high temperature (~2,500 °C) and extreme pressure (~60,000 atmosphere). When the pressure and the temperature reach ideal conditions, carbon from the upper capsule source dissolves in the metal and is driven to the bottom. During cooling, carbon precipitates out of the solution on to the seed crystal in the form of a diamond [117].

Diamond single crystal can be grown by following the most sophisticated and versatile plasma-enhanced chemical vapor deposition (PECVD) method. In CVD, instead of pressurizing carbon into creating diamonds, CVD frees carbon atoms to allow them to join together to create a diamond. The diamond film grown by PECVD method is inferior in quality to HPHT diamond. But PECVD-grown diamond contains less nitrogen and boron impurities. Figure 7.33 shows the schematic of the physical and chemical interactions that are assumed to take place during CVD diamond growth [118], and Fig. 7.34a shows the picture of a PECVD grown diamond film, and Fig. 7.34b shows diamond wafers prepared by microwave

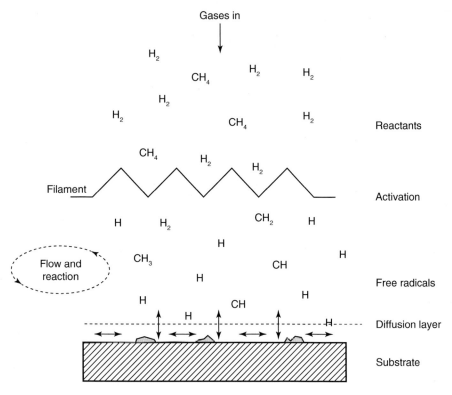

**Fig. 7.33** Schematic of the physical and chemical interactions that are assumed to take place during CVD diamond growth

plasma CVD: Blue disk is boron doped, transparent is optical grade, next one is mechanical grade and the last one is an unpolished disk.

### 7.8.2 Silicon (Si)

As a base material for semiconductor devices, silicon is the single most used semiconduction material in semiconductor industry. As a matter of fact, since the creation of the first integrated circuit (IC) in 1960, there has been an ever-increasing demand of silicon, not to talk about its application in nuclear medicine as a radiation detector. One of the primary factors in the performance of the radiation detector is the quality of the starting material. In contrast to germanium, where the breakthrough in the technology as a detector grade material came with the development of the high-purity germanium specifically for detectors at the General Electric and Lawrence Berkeley Laboratories [119, 120], silicon detectors derive from the material development carried out for other electronic devices as mentioned

## 7.8 Elemental Semiconductor

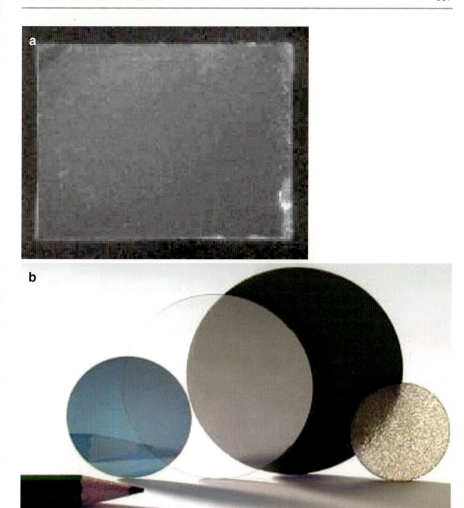

**Fig. 7.34** (a) is a PECVD grown diamond film (size $3.0 \times 3.0 \times 0.9 \times$ mm$^3$), and (b) shows the wafers of diamond grown by microwave plasma CVD (Reproduced with permission, Chris. Nobel, IAF, Fraunhofer, Germany)

earlier. Figure 7.35 shows a flow diagram for the production of semiconductor grade silicon.

The metallic impurities in silicon is separated by using the principle that the metallic elements of smaller segregation factors at the silicon solid/silicon liquid phase boundary are discharged into liquid phase via the one-directional solidification method. For polycrystalline silicon, the presence of grain boundaries creates crystal defects and/or complicates the micros at the grain boundaries due to

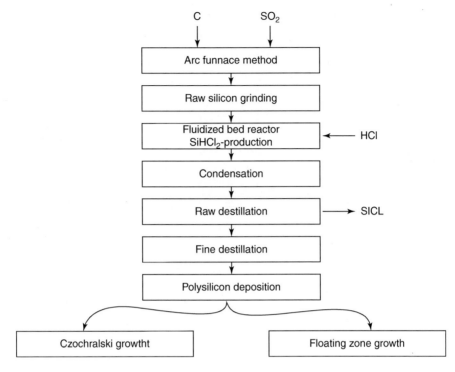

**Fig. 7.35** A flow diagram for the production of semiconductor grade silicon

**Table 7.16** Properties of diamond vs. silicon

| Property | Diamond | Silicon |
|---|---|---|
| Band gap (eV) | 5.5 | 1.12 |
| Energy to create e–h pairs (eV) | 13 | 3.6 |
| Electron mobility (cm$^2$/V · s) | 1,800 | 1,350 |
| Hole mobility (cm$^2$/V · s) | 1,200 | 480 |
| Saturation velocity (μm/ns) | 220 | 82 |
| Dielectric constant | 5.6 | 11.9 |
| Resistivity (Ω-cm) | >10$^{11}$ | <10$^5$ |
| Mass density (g/mL$^3$) | 3.5 | 2.33 |
| Thermal conductivity (W/m · k) | 1,000–2,000 | 150 |

impurities. As a result, the impurity-related defects limit the performance of the device as a radiation detector.

Ordinary silicon PIN (p-type, insulator, and n-type) photodiodes can serve as radiation detectors for X-ray and gamma (γ)-ray photons. For a wafer thickness of 300 μ (which is a typical wafer thickness available in the market), the detection efficiency is close to 100 % at 10 keV, falling to approximately 1 % at 150 keV (ignoring attenuation in the diode window or package) [121]. Table 7.16 represents some of the properties of diamond and silicon.

## 7.8 Elemental Semiconductor

**Fig. 7.36** Picture of a 150-mm single crystal of silicon cut from a CZ-grown ingot

Figure 7.36 shows the picture of a 150-mm single crystal of silicon cut from a single crystal of silicon (Si) ingot grown by Czochralski method.

### 7.8.3 Germanium

High-purity germanium (HPGe) with a reverse biased p–n junction has been used as high-resolution gamma ($\gamma$)-ray detector for a long time. The biggest advantage of germanium detector compared to detectors such as scintillation detectors is its excellent energy resolution, which is about 0.2 % FWHM [122]. Germanium (Ge) detector used for nuclear radiation detector is required to have a purity (the net electrical impurity concentration should be greater than 10 cm$^{-3}$) and crystal perfection (free of charge trapping defects), which is unsurpassed by any other solid material.

The melting point of germanium (Ge) is 937 °C. So Czochralski technique is suitable to grow single crystal of Ge. The charge material is loaded in a synthetic silica crucible in hydrogen (H$_2$) atmosphere [119, 123, 124]. The difficulty about a crystal growing from the melt is that it can never be completely free of thermal stress. In order to minimize the stress, the growing interface is kept flat, and the melt–solid interface appears convex when viewed from the melt. Once these are achieved, the pulling of the seed material out of the melt started. After few seconds, the pulling rate of the seed is increased until the heat of crystallization balances the radial heat loss due to radiation and convection. To initiate the crystal growth, a low dislocation density seed is selected, and it is necked down by fast pulling until only few dislocations remain. However, when crystal diameter is large, growth stability is threatened because the facets increase thermal radiation due to large surface area that may lead to dendritic growth. Thus, to grow a perfect single crystal of Ge for nuclear radiation detector requires a great deal of experience. Figure 7.37 shows an HPGe coaxial detector, which is made from a single crystal ingot of HPGe.

Among the impurities that are associated with the single crystal during growth are aluminum (Al) and boron (B) (as acceptors). Aluminum (Al) is by far the most troublesome impurity to remove because it does not seem to segregate. Copper (Cu) on the other hand has low solubility and small segregation coefficient but has a very high diffusion coefficient at low temperatures and becomes a deep level hole trap center at 77 k. Following table represents some of the physical properties of silicon (Si) and germanium (Ge). Table 7.17 shows some the physical and electrical properties of silicon (Si) and germanium (Ge).

**Fig. 7.37** Picture of a HPGe coaxial detector

**Table 7.17** Physical properties of silicon (Si) and germanium (Ge)

| Parameter | Silicon (Si) | Germanium (Ge) |
|---|---|---|
| Density (g/mL$^3$) | 2.33 | 5.33 |
| Average atomic number(s) | 14 | 32 |
| Band gap (eV) | 1.12 | 0.67 |
| Energy required to produce e–h pairs (eV) | 3.62 | 2.96 |
| Electron mobility (cm$^2$/V · s) | 1,400 | 3,900 |
| Hole mobility (cm$^2$/V · s) | 1,900 | 1,900 |
| Electron lifetime (s) | $>10^{-3} - 5 \times 10^{-7}$ | $>10^{-3}$ |
| Hole lifetime (s) | $10^{-3} - 7 \times 10^{-7}$ | $2 \times 10^{-3} - 10^{-7}$ |
| Electron $\mu\tau$ product (cm$^2$/V) | $>1$ | .1 |
| Hole $\mu\tau$ product (cm$^2$/V) | $\approx 1$ | $>1$ |
| Crystal structure | Cubic | Cubic |
| Lattice constant (Å) | 5.4309 | 5.646 |
| Melting point (°C) | 1,412 | 958 |
| Dielectric constant | 11.7 | 16 |
| Resistivity (Ω-cm) | $<10^4$ | 50 |
| 1/e abs. depth (mm) at 10 keV/100 keV | 0.127/23.30 | 0.05/3.51 |
| Typical FWHM $\Delta E$ (keV) at 60 keV | 0.4 | 0.53 |
| Fano noise at 60 keV | 0.415 | 0.25 |
| Typical thickness (mm) | 0.3 | 20 |

# References

1. Hofstadter R (1975) Twenty years of scintillation counting. IEEE Trans Nucl Sci NS-22:13
2. Derenzo SE, Weber MJ, Bourret-Courchesne E, Klintenberg MK (2003) The quest for ideal scintillator. Nucl Instrum Methods A 505:111
3. Nassalski A et al (2007) Comparative study of scintillators for PET/CT detectors. IEEE Trans Nucl Sci 54(1):3
4. Jaffray DA, Siewerdsen JH, Wong JW, Martinez AA (2002) Flat panel cone beam computed tomography for image guided radiation therapy. Int Radiat Oncol Biol Phys 53(5):337
5. van Eijk CWE (2002) Inorganic scintillators for medical imaging. Phys Med Biol 47:R-85
6. Rando R et al (2004) Radiation testing of GLAST LAT tracker ASIC. IEEE Trans Nucl Sci 51(3):1067
7. Adamson P et al (2002) The MINOS scintillator calorimeter system. IEEE Trans Nucl Sci 49(3):861
8. Gozani T (2004) The role of neutron based inspection techniques in the post 9/11/01 era. Nucl Instrum Methods B-213:460

# References

9. Leo WR (1994) Techniques for nuclear and particle physics experiments, 2nd edn. Springer, Berlin
10. Nagarkar VV, Tipnis SV, Gupta TK, Miller SR (1998) High speed X-ray imaging camera using a structural CsI (Tl) scintillator. IEEE Trans Nucl Sci 45(3):492
11. Parlog M et al (2002) Response of CsI (Tl) scintillators over a large range in energy and atomic number of ions (part I). Nuclr Instrum Meth A 482:674
12. Valais IG et al (2007) Luminescence properties of (Ly, Y) $SiO_5$:Ce, and $Gd_2SiO_5$:Ce. IEEE Trans Nucl Sci 54(1):11
13. Balamurugan N et al (2006) Growth and characterization of undoped and thallium doped cesium iodide single crystals. J Cryst Growth 286:294
14. Brisson O, Ganaoui ML, Simonnet A, Launay JC (1999) Experimental determination of physical parameters and analysis of the B-S solidification for the growth of $AgGaSe_2$ crystal. J Cryst Growth 204:201
15. Gupta T et al (1996) CCD based non-destructive testing system for industrial applications. IEEE Trans Nucl Sci 43(3):1559
16. Gupta T et al "X-ray Security System" , NIST funded Jt. Program, Varian Med. System, Palo Alto, CA., Xerox Research, PlaoAlto, CA, and RMD, Watertown, MA. Partial report to DARPA contract #DAAH01-95-C-R188 and NIH contract #2R44CA65213-02
17. Oliveria JB, Costa FE, Armelin MJ, Cardoso LP, Hamada MM (2002) Purification and growth PbI2 crystals. IEEE Trans Nucl Sci 49(4):1968
18. Ugucioni JC, Ferreira M, Fajardo F, Mulato M (2006) Growth of mercuric iodide crystals. Braz J Phys 36(2A):274
19. Kozlov V, Laskela M, Sipila H (2005) Annealing and characteristic of TlBr crystals for detector applications A. Nucl Instrum Methods A 546(1–2):200
20. Li Z, Li W, Liu J, Zhao B, Zhu S, Xu H, Yuan JH (1995) Improved technique for $HgI_2$ purification. J Cryst Growth 156(5):86
21. Shah KS et al (1997) Lead iodide optical detectors for gamma ray spectroscopy. IEEE Trans Nucl Sci 44(3):448
22. Bennett P, Dmitriev Y, Gupta T, Klugerman M, Partain L, Paryluchova R, Shah KS, Squillante M (2003) Polycrystalline lead iodide films for digital X-ray sensors. Nucl Inst Methods Phys Res Sec A 505:269
23. Zhu XH, Zhu BJ, Zhu SF, Jin YR, He ZY, Zhang JJ, Huang Y (2006) Synthesis and characterization of $PbI_2$ polycrystals. Cryst Res Technol 41:239
24. Matuchova M, Zdansky Z, Zavadil J, Maixiner J, Alexiev D, Prochokova D (2006) Study of lead iodide semiconductor crystals doped with Ag. J Cryst Growth 9(1–3):394
25. Shoji T, Ohba K, Suchiro T, Hiratate Y (1995) Fabrication of radiation detector using $PbI_2$ crystal. IEEE Trans Nucl Sci 42(4):659
26. Lund JC, Shah KC, Olschner F, Jang J, Moy L, Medrick S, Squillante MR (1992) $HgBr_xI_{2-x}$ photodetectors for use in scintillation spectroscopy. Nucl Instrum Methods Phys Res A 322:509–513
27. Yaffe MJ, Rowlands JA (1997) X-ray detectors for digital radiography. Phys Med Biol 42:1
28. Kabir MZ (2005) Ph.D. thesis, University of Saskatchewan, Saskatoon
29. Gupta TK (2003) Handbook of thick and thin film hybrid microelectronics. Wiley, Hoboken
30. Wiling WR (1971) Mercuric iodide as a gamma ray spectrometer. Nucl Instrum Methods Phys Res 96:615
31. Schieber M, Zuck A, Lukach M (2001) Spectra in strontium hexa aluminate. J Opt Electron Adv Mater 3(3):757
32. Bao XJ, Schlesinger TE, James RB (1995) Electrical properties of mercuric iodide. In: Schlesinger TE, James RB (eds) Semiconductors for room temperature nuclear detector applications. Academic, San Diego
33. Chang YC, James RB (1992) Electronic and optical properties of $HgI_2$. Phys Rev B46:15040
34. Piechotka M (1998) Mercuric iodide for room temperature radiation detectors. Mater Sci Eng R-181:1

35. Piechotka M (1997) Mercuric iodide for room temperature radiation detectors, synthesis, purification, crystal growth, and defect formation. Mater Sci Eng R18:1–98
36. Lamonds HA (1983) Review of mercuric iodide development program in Santa Barbara. Nucl Instrum Methods 213:5
37. Schieber M, Schnepple WF, van den Berg L (1976) Vapor growth of $HgI_2$ by periodic source or crystal temperature oscillation. J Cryst Growth 33:125
38. van den Berg L (1993) Growth of single crystals of mercuric iodide on the ground and space. Mater Res Soc Proc 302:73
39. Kadish KM, Smith KM, Guilard R (eds) (1987) The porphyrin handbook, vol 3. Academic press, San Diego
40. Ponpon JP, Stuck R, Siffert P (1975) Properties of vapor phase grown mercuric iodide single crystal detectors. IEEE Trans Nucl Sci NS-22:182
41. Granfors PR et al (2003) Performance of a $41 \times 41$ $cm^2$ amorphous silicon flat panel x-ray detector designed for angiography and R & F imaging applications. Med Phys 30:2715
42. Hartugh NE, Iwanczyk JS, Patt BE, Skinner NL (2004) Imaging performance of mercuric iodide polycrystalline films. IEEE Nucl Sci 51:1812
43. Antonuk LE et al (2005) Systematic investigation of the signal properties of polycrystalline $HgI_2$ detectors under mammographic, radiographic, fluoroscopic and radiotherapy irradiation conditions. Phys Med Biol 50:2907
44. Antonuk LE et al (1990) Development of hydrogenated amorphous silicon sensors for high energy photon radiotherapy imaging. IEEE Trans Nucl Sci 37(2):165
45. Shorohov M et al (2009) Recent results in TlBr detector crystals performance. IEEE Trans Nucl Sci 56(4):1855
46. Hitomi K, Matsumoto M, Muroi O, Shoji T, Hiratate Y (2001) Characterization of thallium bromide crystals for radiation detector applications. J Cryst Growth 225(2–4):129
47. Kim II, Cirignano L, Churilov A, Ciami G, Higgins W, Olschner F, Shah K (2009) Developing larger TlBr detectors – detector performance. IEEE Trans Nucl Sci 56(3):819
48. Dmitriev Y, Bennett P, Cirignano L, Gupta TK, Higgins W, Shah K, Wong P (2007) Doping impact on the electro-optical properties of a TiBr crystal. Nucl Instrum Methods A 578(3):510
49. Zhou D et al (2009) A novel method to grow thallium bromide single crystal and crystal habit discussion. Cryst Growth Des 9(10):4296
50. Crespi FCL et al (2009) Alpha gamma discrimination by pulse shape in $LaBr_3$:Ce and $LaCl_3$:Ce. Nucl Instrum Methods A 602:520
51. Menge A et al (2007) Performance of large lanthanum bromide scintillators. Nucl Instrum Methods 579:6
52. Shah KS, Glodo J, Klugerman M, Higgins WH, Gupta T, Wong P (2004) High resolution scintillation spectrometer. IEEE Trans Nucl Sci 51(5):2395
53. Birowosuto MD, Dorenbos P, van Ejik CWE, Kramer KW, Gudel HU (2005) Scintillation properties of $LuI_3$:$Ce^{3+}$ high light yield scintillators. IEEE Trans Nucl Sci 52(4):1114
54. Lee W, Wehe DK, Kim B (2005) Comparative measurements on $LaBr_3$ (Ce) and $LaCl_3$ (Ce) scintillators coupled to PSPMT. IEEE Trans Nucl Sci 52(4):1119
55. Knoll G (1999) Radiation detection and measurement, 3rd edn. Wiley, New York
56. Kleinknechi K (1998) Detectors for particle radiation, 2nd edn. Cambridge University Press, Cambridge
57. Brice JC (1986) Crystal growth processes. Blackie Halsted Press, Glasgow/New York
58. Allier CP, van Loef EVD, Dorenbos P, van Ejik CWE, Kramer K, Gudel HU (2000) Readout of $LaCl_3$ ($Ce^{3+}$) scintillator crystal with large avalanche photodiode. In: IEEE nuclear science symposium, Lyon, Oct 2000
59. Higgins WM, Glodo J, Van Loef E, Klugerman M, Gupta T, Cirignano L, Wong P, Saha KS (2006) Bridgman growth of $LaBr_3$:Ce and $LaCl_3$:Ce crystals. J Cryst Growth 287 (2):239
60. Shi H et al (2010) The $LaBr_3$:Ce crystal growth by self seeding Bridgman technique and its scintillation properties. Cryst Growth Des 10(10):4433

# References

61. Shah K, Cirignano L, Grazioso R, Klugerman M, Bennet PB, Gupta TK, Moses WW, Weber J, Derenzo SE (2002) RbGd$_2$Br$_7$:Ce scintillators for gamma-ray and thermal neutron detection. IEEE Trans Nucl Sci 49(4):1655
62. Birowsuto MD, Dorenbos P, van Eijk CWE, Kramer KW, Gudel HU (2006) High-light output scintillator for photodiode readout: LuI$_3$:Ce$^{3+}$. J Appl Phys 99:123520
63. Gupta T et al (2004) LuI$_3$:Ce – a new scintillator for gamma ray spectroscopy. IEEE Trans Nucl Sci 51(5):2302
64. Ren G, Qin L, Li H, Lu S (2006) Investigation on defectsin Lu$_2$SiO$_5$:Ce crystals grown by Czochralski method. Cryst Res Technol 41(2):163
65. Melcher CL, Friedrich S, Cramer SP, Spurrier MA, Szupryczynski P, Nutt R (2005) Cerium oxidation state in LSO:Ce scintillators. IEEE Trans Nucl Sci 52(5):1809
66. Spurier MA, Szupryczynski P, Rothfuss H, Yang K, Carey AA, Melcher CL (2007) The effect of co-doping on the growth stability and scintillation properties of LSO:Ce. In: 15th international conference on crystal growth, Salt Lake City, 2007
67. Mansuy C, Mahiou R, Nedelec M (2003) A new sol–gel route to Lu$_2$SiO$_5$ (LSO) scintillator. Chem Mater 15(17):3242
68. Krell A, Klimke J, Hutzler T (2009) Advanced spinel and sub-μm Al$_2$O$_3$ for transparent armour applications. J Euro Ceram Soc 29:275
69. Farhi H et al (2009) Single crystal growth by LHPG technique and optical characterization Ce$^{3+}$-doped Lu$_2$O$_3$. Opt Mater 30(9):146
70. Gupta TK et al (1971) Unpublished documents, optical ceramic scintillator for digital radiography, grant awarded from NIH 2000–2004 and Pankove JI (1971) Optical process in semiconductors, Dover Publication, New York, Chapter 17
71. Ping Y, Ying S, Ding Z, Jianjun X (2009) Hydrothermal synthesis of Ce:LSO scintillator powders. J Rare Earths 27(5):801
72. Lisa L, Yeckel A, Daoutidis P, Derby JJ (2005) Decreasing lateral segregation in cadmium zinc telluride via ampoule tilting during vertical growth. J Cryst Growth 291(2):348
73. Yeckel A, Compere G, Pandy A, Derby JJ (2004) Three dimensional imperfections in a model vertical Bridgman growth system for cadmium zinc telluride. J Cryst Growth 263(629)
74. Burachas SF et al (2000) Advanced scintillation single crystals based on complex oxides with high atomic numbers. Semicond Phys Quatum Electron Optoelectron 3(2):237
75. Takagi K, Fukazawa T (1983) Cerium activated Gd$_2$SiO$_5$ single crystal scintillator. Appl Phys Lett 42:43
76. Choi J, Tseng TK, Davidson M, Holloway PH (2010) Synthesis and characterization of scintillating Gd2SiO5:Ce nanoparticles. In: IEEE NSS conference, Knoxville, 30th Oct–6th Nov 2010
77. Valais I et al (2005) Luminescence efficiency of Gd$_2$SiO$_5$:Ce scintillator under X-ray excitation. IEEE Trans Nucl Sci 52(5):1830
78. Shimura N et al (2006) Zr doped GSO:Ce single crystals and their scintillation performance. IEEE Trans Nucl Sci 53(5):2591
79. Kurashige K et al (2004) Large GSO single crystals with diameter of 100 mm and their scintillation performance. IEEE Trans Nucl Sci 51742
80. Kurata Y, Kurashige K, Isibashi H, Susa K (1995) Scintillation characteristics of GSO single crystal grown under O$_2$ containing atmosphere. IEEE Trans Nucl Sci 42(4):1038–1040
81. Muenchausen RE et al (2008) Science and application of oxyorthosilicate nanophosphors. IEEE Trans Nucl Sci 55(3):1532
82. Moriarty P (2001) Nanostructured materials. Rep Prog Phys 64:297
83. Miller MC et al (1999) Neutron detection and applications using a BC454/BGO array. Nucl Instrum Methods A 422:89
84. Weber MJ, Monchamp RR (1973) Luminescence of Bi$_4$Ge$_3$O$_{12}$ spectral and decay properties. J Appl Phys 44(12):5495
85. Macedo ZS, Ferari CS, Hermandes AC (2004) Self-propagating high temperature synthesis of bismuth titanate. Powder Technol 139:175

86. Luan W, Gao L (2001) Influence of pH value on properties of nanocrystalline $BaTiO_3$ powder. Ceram Int 27:645
87. de Jesus FAA et al (2009) Effect of pH on the production of dispersed $Bi_4Ge_3O_{12}$ nanoparticles by combustion synthesis. J Eur Ceram Soc 29:125
88. Macedo ZS, Ferrari CR, Harnades AC (2004) Impedance spectroscopy of $Bi_4Ti_3O_{12}$ ceramic produced by self-propagating high temperature synthesis technique. J Eur Ceram Soc 24:2567
89. Lio HW, Wang YF, Liu X-M, Le YD, Qian YT (2000) Preparation and characterization of luminescent $CdWO_4$ nanotubes. Chem Mater 12:2819
90. Lennstrom K, Limmer SJ, Cao G (2003) Synthesis of cadmium tungstate films via sol–gel processing. Thin Solid Films 434:55
91. Novotny R (2005) Inorganic scintillators – a basic material for instrumentation in physics. Nucl Instrum Methods A 537:1
92. Zang L, Bailleux D, Bornheim A, Zhu K, Zhu R-Y (2005) Performance of the monitoring light source for the CMS lead tungstate crystal calorimeter. IEEE Trans Nucl Sci 52(4):1123
93. Pontes FM et al (2003) Preparation, structural and optical characterization of $BaWO_4$ and $PbWO_4$ thin films prepared by chemical route. J Eur Ceram Soc 23:3001
94. Usui T et al (2007) 60 mm diameter $Lu_{0.4}Gd_{1.6}SiO_5$:Ce (LGSO) single crystals and their improved scintillation properties. IEEE Trans Nucl Sci 54(1):19
95. Szupryczynski P, Spurrier MA, Rawn CJ, Melcher CL, Carey AA (2005) Scintillation and optical properties of LuAP and LuYAP crystals. In: IEEE nuclear science symposium, San Juan, Puerto Rico, 23–29, Oct p 1305
96. Ding D, Lu S, Qin L, Ren G (2007) Research on the phase decomposition of $LuxY_{1-x}AlO_3$:Ce crystals at high temperatures. Nucl Instrum Methods A 572:1042
97. Limpicki A, Randles MH, Wisniewski D, Balserzyk M, Brecher C, Wojtowicz A (1995) $LuAlO_3$:Ce, and other aluminate scintillators. IEEE Trans Nucl Sci 42(4):280
98. Gumanskaya EG, Egorcheva OA, Korzhik MV, Smirnova SS, Pavlenko VB, Fedorov AA (1992) Opt Spectrosc 72:215
99. Petrosyan AG et al (1998) Bridgman growth and characterization of $LuAlO_3$-$Ce^{3+}$ scintillator crystals. Cryst Res Technol 33:241
100. Petrosyan A et al (2006) J Cryst Growth 293:74
101. Klimm D (2010) The melting behavior of lutetium aluminum perovskite $LuAlO_3$. J Cryst Growth 312:730
102. Triboulet R (1994) Prog Cryst Gr Char Mater 128:85 and Funaki M, Ozaki T, Satoh K, Ohno R (1999) Growth and characterization of CdTe single crystals for radiation detectors. Nucl Instrum Methods A 436:120–126 or by Bridgman (HPB) method Raiskin E, Butler JF (1988) IEEE Trans Nucl Sci NS-35:82
103. Parmham K (1996) Recent progress in $Cd_{1-x}Zn_xTe$ radiation detectors. Nucl Instrum Methods A 377:487
104. Fougeres P et al (1998) Properties of $Cd_{1-x}Zn_xTe$ crystal by high pressure Bridgman method. J Cryst Growth 184:1313
105. Pfann WG (1958) Zone melting. Wiley, New York
106. Shiraki H, Funaki M, Ando Y, Tachibana A, Kominami S, Ohno R (2009) THM growth and characterization of 100 mm diameter CdTe single crystal. IEEE Trans Nucl Sci 56(4):1717
107. Zanio K (1978) Growth of high purity CdTe single crystals. In: Willardson RK, Beer AC (eds) Semiconductors and semimetals, vol 13. Academic, Orlando FL
108. Raiskin E, Butler JF (1988) CdTe low level gamma detectors based on a new crystal growth method. IEEE Trans Nucl Sci NS-35:81
109. Szeles CS, Eissler EE (1998) MRS Symp Proc 487:3
110. Szeles C, Eissler EE, Reese DJ, Cameron SE (1999) Radiation detector performance of CdTe single crystals grown by the conventional vertical Bridgman technique. In: SPIE conference, Denver, 1999

# References 365

111. Chu M, Terterian S, Ting D (2004) Role of zinc in CdZnTe radiation detectors. IEEE Trans Nucl Sci 51(4):2405
112. Zanio K (1978) Cadmium telluride, vol 13. Academic, New York, pp 139, 220
113. Fochuk R et al (2010) Study of point defects in Cd1-xZnxTe crystals. In: IEEE Transactions on nuclear science sympoisum, Knoxville, 30th Oct–6th Nov 2010 and James RB, Schlessinger TE, Lund J, Schieber M (1995) Cd1-xZnx Te Spectrometers for gamma and X-ray applications. In: James RB, Schlessinger TE (eds) Semiconductors for room temperature nuclear detector applications. Academic, San Diego
114. Hollingsworth MS (2010) Resolution studies of single crystal CVD diamond pixel detectors. In: IEEE nuclear science symposium, Knoxville, 30th Oct–6th Nov 2010 and also Lutz G (2007) Semiconductor radiation detectors. Springer, New York
115. Kaneko JH et al (2003) Radiation detector made of a diamond single crystal grown by chemical vapor deposition method. Nucl Instrum Methods A 505:187 and also Milbrath BD et al (2008) J Mater Res 23(10):2666
116. Pan LS et al (1993) Electrical properties of natural IIA diamond using photo-and particle excitation. In: James RB, Schlesinger TE, Siffert P, Franks L (eds) Semiconductors for room temperature radiation detector applications. MRS Publication, Pittsburgh, p 145
117. Abbaschian R, Zhu H, Clarke C (2005) High pressure high temperature growth of diamond crystals using split sphere. Diamond Relat Mater 14(11–12):1916
118. May PW (2000) Diamond thin films: a 21th century material. Philos Trans R Soc Lond A 358:473
119. Hall RN, Soltys TI (1971) High purity Germanium detector fabrication. IEEE Trans Nucl Sci NS-18(1):160
120. Hansen Nuclr WL (1970) High purity germanium crystal growing. Instrum Methods 94:377
121. Ftzgerald JJ, Brownell GL, Mahoney JJ (1967) Mathematical theory of radiation dosimetry. Gordon Beach Science Publication, New York
122. Luke PN, Amman M (2007) Room temperature replacement for Ge detectors – are we ready. IEEE Trans Nucl Sci 54(4):834
123. Hansen WL, Haller EE (1972) High purity germanium observations on the nature of acceptors. IEEE Trans Nuclr Sci NS-19:1
124. Hansen WL, Haller EE (1983) High purity germanium crystal growing. In: Haller EE, Kraner HW, Higinbotham WA (eds) Nuclear radiation detector materials. North Holland Publication, Amsterdam

# Characterization of Radiation Detectors (Scintillators) Used in Nuclear Medicine

**8**

## Contents

8.1   Introduction ........................................................................... 368
8.2   Compound Halides ................................................................... 368
8.3   Alkali Halides ......................................................................... 370
    8.3.1   Activated Sodium Iodide .................................................... 370
    8.3.2   Cesium Iodide (CsI)—Cesium Iodide (CsI) Activated with Either Thallium (Tl) or Sodium (Na) ........................................................... 375
8.4   Halides of Heavy Metals ............................................................ 384
    8.4.1   Lead Iodide (PbI$_2$) .......................................................... 387
    8.4.2   Mercuric Iodide (HgI$_2$) ..................................................... 391
    8.4.3   Thallium Bromide (TlBr) .................................................... 401
8.5   Lanthanide (Ln) Halides ............................................................ 404
    8.5.1   Cerium-Activated Lanthanum Chloride (LaCl$_3$:Ce) ...................... 406
    8.5.2   Cerium-Activated Lanthanum Bromide (LaBr$_3$:Ce) ...................... 409
    8.5.3   Rubidium Gadolinium Bromide (RGB) (Cerium-Doped) RbGd$_2$Br$_7$:Ce ...... 412
    8.5.4   Cerium-Doped Lutetium Iodide (LuI$_3$:Ce) ................................ 414
8.6   Cerium-Activated Lutetium Oxyorthosilicate (LSO, Lu$_2$SiO$_5$) ................ 417
    8.6.1   Synthesized LSO ............................................................ 418
8.7   Complex Oxides with High Atomic Number ...................................... 420
    8.7.1   Cerium-Activated Gadolinium Oxyorthosilicate (GSO,Gd$_2$SiO$_5$:Ce$^{3+}$) ...... 420
    8.7.2   Bismuth Germanate (BGO, Bi$_4$Ge$_3$O$_{12}$) .................................... 421
    8.7.3   Cadmium Tungstate (CWO, CdWO$_4$) ..................................... 421
    8.7.4   Lead Tungstate (PWO, PbWO$_4$) ........................................... 422
8.8   Cadmium Telluride (CdTe) Crystals ............................................... 422
8.9   Cadmium Zinc Telluride (CZT) .................................................... 426
8.10  Elemental Semiconductor ........................................................... 429
    8.10.1  Diamond ..................................................................... 429
    8.10.2  Silicon ....................................................................... 435
    8.10.3  Silicon Strip Detector ...................................................... 436
    8.10.4  Germanium .................................................................. 437
References .................................................................................. 442

---

T.K. Gupta, *Radiation, Ionization, and Detection in Nuclear Medicine*,
DOI 10.1007/978-3-642-34076-5_8, © Springer-Verlag Berlin Heidelberg 2013

368    8   Characterization of Radiation Detectors (Scintillators) Used in Nuclear Medicine

## 8.1    Introduction

Since the invention of planar technology, semiconductor radiation detector technology has been developed on planar technology for a wide range of applications in medical imaging, high-energy physics applications, and photon detection. The vertex detectors used in high-energy physics consists of both single-sided and double-sided silicon strip detectors arranged in a cylindrical shape around the beam collision point [1–3]. The success of silicon detectors in position-sensitive particle detection in accelerator experiments was soon used to the ideas of applications in other field of research such as medicine and industry.

Since digital X-ray imaging system offered several advantages over analog film-based imaging system, immense effort and a lot of money have been invested to develop digital X-ray imager with inorganic scintillating materials (e.g., selenium (Se), mercuric iodide ($HgI_2$), cesium iodide (CsI)) on flat-panel pixelated substrate [4–6].

Small field detectors that may be combined for scanning configurations are mainly based on charge-coupled devices (CCD) [7, 8], but directly converting systems with crystalline semiconductors, CMOS, and bump bonding technology have also been developed [9, 10].

The X-ray imaging system for medical use requires higher X-ray energy, and Z value materials higher than 40 are preferable for high stopping power. The lattice should have a close-packed geometry (such as body-centered cubic) to optimize density. The material should have a low dielectric constant to ensure low capacitance and system noise. From electronic point of view, the band gap energy should be greater than 0.14 to ensure that there is negligible thermal generation of carriers at room temperature, and the resistivity should be greater than $10^8$ $\Omega$-cm to allow larger biases to be applied, resulting in faster drift velocities and deeper depletion depth. Finally, electron and hole mobility-lifetime product ($\mu\tau$) should be better than $10^{-2}$ and $10^{-3}$, respectively, to ensure good carrier transport and therefore spectral performance.

According to some observers, CdTe and CdZnTe (CZT) are more promising in medical gamma ($\gamma$)-ray imaging than GaAs and $HgI_2$ because of improved spatial and energy resolution [11]. However, recent studies with screen-printed $HgI_2$ films are showing not only promising results but these devices are cost-effective [12, 13].

## 8.2    Compound Halides

After Roentgen's discovery of X-rays, the search for detectors to detect the radioactivity in materials became number one priority. As a result, the discovery of the inorganic compound $CaWO_4$ came into the market. In 1940, the discovery of

## 8.2 Compound Halides

scintillation in naphthalene led to the discovery of thallium-activated sodium iodide (NaI). In 1950, there was a burst of alkali halide compounds that seemed very interesting, and their performance as radiation detectors was investigated. Among them, lithium-containing compounds to detect neutrons are worth mentioning. Activated cesium iodide (CsI:Tl) came afterward along with barium fluoride $BaF_2$. The iodide scintillators like NaI, NaI:Tl, and CsI:Tl have wide emission spectra with maximum ranging from 330 for NaI:Tl to 550 nm for CsI:Tl. The most interesting thing about these two compound halides is their electron response functions that are very much similar in shape but the magnitudes are different [14].

The other compound halides that we will discuss are $PbI_2$, $HgI_2$, TlBr, $LaBr_3$ $LaCl_3$, $RbGd_2Br_7$, and $LuI_3$. There are too many inorganic compound halide scintillators that are in the state of developments. It is not possible to include each and every one of the scintillators.

Both mercuric iodide ($HgI_2$) and lead iodide ($PbI_2$) have high atomic numbers which is one of many characteristics of a scintillating material used as a radiation detector. Mercuric iodide ($HgI_2$) has received the attention of solid-state physicists because of its optical properties, especially the excitonic effects in correspondence of the fundamental absorption edge and interesting photoconductive properties. In 1971, Willing and Roth first pointed out that red variety of $HgI_2$ is a good material for radiation detector because of its high atomic number (80/53), large band gap (~2.2 eV), and low electron–hole pair creation energy (~4.2 eV) [15]. Lead iodide ($PbI_2$) on the other hand has comparable atomic number (82/53) and band gap (2.3 eV), but requires more energy (~8.4 eV) to create electron-hole (e-h) pair. However, the e–h pair creation energy depends on the quality of the grown crystals (both $HgI_2$ and $PbI_2$).

Thallium bromide (TlBr) has almost all the properties that should be possessed by a scintillating material to be used as a radiation detector. It has wide band gap (2.68 eV) and high atomic number (81/35), so the Fano noise has low and high resistivity ($10^{12}$ $\Omega$-cm), meaning larger bias can be applied resulting in faster drift velocities and deeper depletion depth. However, from the graph, we can see that the compound halides such as $HgI_2$, $PbI_2$, and TlBr are displaced from the line of other bulk semiconductors giving a 30 % reduction in mean pair creation energy for a given band gap compared to main branch (Fig. 8.1) [16].

Both $LaBr_3$:Ce and $LaCl_3$:Ce doped have high light yield ($\geq$50,000 and 40,000 ph/MeV), fast response, and good energy resolution. On the other hand, $LuI_3$ $Ce^{3+}$ has also shown high light yield, fast response, and good energy resolution [17].

The cerium-activated (with 10 % cerium) rubidium gadolinium bromate ($RbGd_2Br_7$:Ce), the new cerium-doped halide scintillator has a very high light output (56,000 photons/MeV) almost comparable to $LaBr_3$:Ce and $LaCl_3$:Ce, a very good linearity in energy response and over a wide range of energy, and fast principle decay constant (43 ns) [18].

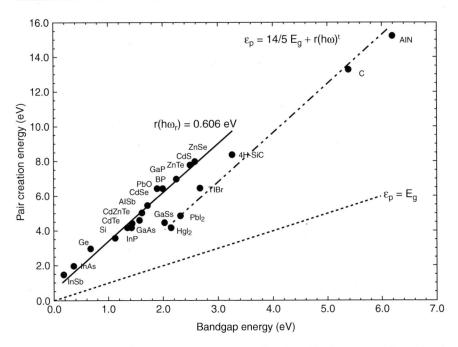

**Fig. 8.1** The energy required for creating e–h pair as a function of band gap energy (Reproduced after Owens and Peacock, with permission, Elsevier Pub.)

## 8.3 Alkali Halides

### 8.3.1 Activated Sodium Iodide

Activated sodium iodide (NaI) with thallium (Tl) has been used as a gamma (γ)-ray detector for a long time. The Tl concentration is maintained about $10^{-3}$ mole fraction. As a matter of fact, thallium-doped sodium iodide (NaI:Tl) material is the most widely used scintillation material in spite of three decades of research and development of several new scintillating materials. NaI:Tl scintillating material is used traditionally in nuclear medicine, environmental measurements, geophysics, and medium-energy physics. NaI:Tl crystals are also used to manufacture X-ray detectors of high spectrometric quality with increased thallium concentration.

The high light output, *convenient emission range* coincident with maximum efficiency region of photomultiplier tube (PMT) with bi-alkali photocathodes, no significant *self-absorption* of the scintillated light, the possibility of *large size production* of *single crystals*, and their low cost and toxicity compared to the other scintillation materials mainly compensate to a great extent for the main

## 8.3 Alkali Halides

thallium-doped sodium iodide (NaI:Tl) crystals disadvantage, namely, the hygroscopicity.

Some of the properties of NaI (Tl) are given below:

Properties of thallium-activated sodium iodide (NaI:Tl)

Afterglow = approx. 4 % after 2 ms and 8 % after 1 ms

Cleavage planes = <100>

Decay constant ($\mu$s) = 0.23

Density (g/cm$^3$) = 3.70

Spectral emission range (nm) = 325–525

Gamma and X-ray absorption coefficients (cm$^{-1}$) = 0.29 at 660 keV and 5.5 at 100 keV

Light escape cone to air (0°) = 44.7

Light escape cone to glass (0°) = 56.5

Light escape from one face to air (%) = 8.4

Light escape from one face to glass (%) = 22.4

Melting point (°C) = 651

Optical transmission range = 250 nm–35 $\mu$m

Peak scintillation wavelength (nm) = 410

Photons/MeV = 40,000

Radiation length (cm) = 2.59

Refractive index at peak emission = 1.80

Structure = bcc

Stability = very hygroscopic

Temperature coefficient of light output (%K$^{-1}$) = 3.47 at 298 K

1. *Emission Spectra of Activated NaI*: Emission spectra of NaI:Tl under study are measured at room temperature, irradiating the sample with an X-ray beam characterized by a mean energy of 30 keV (peak energy at 50 keV) and an intensity of 2 kRad/min. The single crystal of NaI:Tl doped (0.001 mole fraction) and grown by Bridgman–Stockbarger (B–S) method shows a broad peak around 420 nm, corresponding to the thallium (Tl) emission. The hump in the shoulder around 480 nm is presumed to be due to impurities (Fig. 8.2) [19].

   Thallium-activated sodium iodide (NaI) is notable for high light yield (more than 40,000 Photons/MeV) at room temperature, and about 25 eV is required to produce single photon. The light yield is maximum at room temperature (~27 °C). However, light yield significantly reduces below 0 °C and above 60 °C (Fig. 8.3). Lowering of temperature leads to the deterioration of monocrystal self-resolution caused by inhomogeneity of light yield. It has been observed that activator concentration also affects the temperature dependency of the light output to some extent [20].

   The high concentration of thallium (higher than 0.1 mole fraction) in sodium iodide crystal influences the excitation and luminescence spectra of the crystal and ultimately the energy spectra due to the transfer of the absorbed energy in the dimmer bands [21].

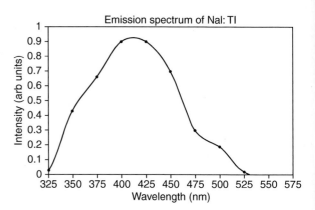

**Fig. 8.2** The emission spectrum of thallium-activated sodium iodide (NaI:Tl)

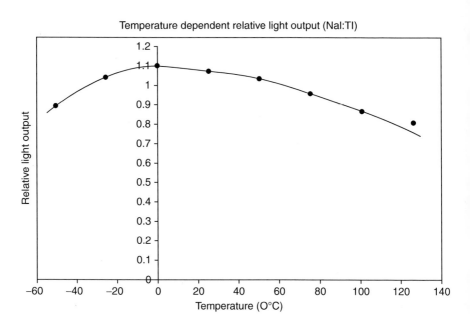

**Fig. 8.3** A typical temperature-dependent light output plot of a single crystal NaI:Tl

2. *Absorption Phenomenon in Activated NaI*: At a temperature less than 80 K, NaI:Tl shows four (*A*, *B*, *C*, and *D*) absorption bands at 285, 253, 245, and 210 nm, and at liquid helium temperature excitation in the *A*, *B*, and *C* bands of NaI:Tl produces two emission bands at 330 and 420 nm [22, 23].

The graph (Fig. 8.4) is based on the calculations of the exponential function for certain values of $x$ (thickness) for NaI. The number of photons absorbed by a certain thickness of the absorber is calculated on the basis of the ratio of $[(I_0/I)/I_0]$, where $I_0$ is the initial intensity of photons and $I$ is the intensity after being

## 8.3 Alkali Halides

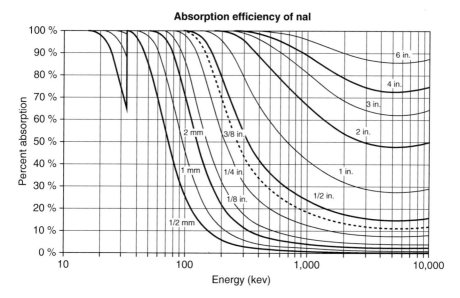

**Fig. 8.4** The absorption efficiency of NaI:Tl at room temperature (Courtesy of Saint-Gobain Crystals)

traveled through the absorber. From the above graph, we can say that NaI crystal of quarter of an inch thick will absorb ~20 % of the 500-keV photons incident on its surface.

3. *Electron Energy Resolution in Activated NaI*: At high-energy excitation, the self-trapped excitons' emissions at 295 nm are observed in addition to the activator bands. In order to account for transport of carriers and excitons, in terms of how they transfer energy to the activators with competition from non-radiative decay pathways, several theories and experiments have been put forward by several investigators [24, 25].

The incident gamma ($\gamma$) ray on the scintillator produce fast electrons, and the photon energy resolution is the convolution of electron spectrum. The final signal produced by the photomultiplier (PM) dynodes as a result of multiplication depends on the total number of photons produced by the scintillator (considering the non-proportionality), the total photoelectron collection by the cathode, and the photoelectrons collected by the first dynode. Though there are several papers on the non-proportional light yield (Fig. 8.2), which ultimately affects the energy resolution of thallium-activated NaI, the science is still not well understood.

Figure 8.5 shows thallium-activated NaI electron energy resolution $\eta_e$ measured following modified Compton coincidence technique (MCCT) [25]. The activated sodium iodide crystal is coupled to two PMTs for dark current reduction. The

**Fig. 8.5** Electron energy resolution in thallium-activated NaI (Reproduced from [25], with permission from IEEE)

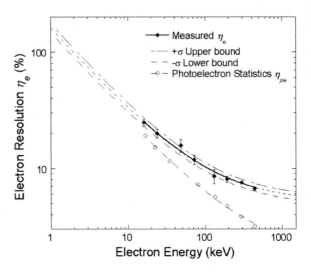

electron energy deposited on NaI:Tl crystal was determined by subtracting scattered energy from the source energy (662 keV). Non-proportionality is noticed to be minimum in strontium iodide (SrI$_2$) doped with europium (Eu). The thallium-doped NaI shows almost double amount of non-proportionality compared to Eu-doped SrI$_2$ [24] while offering very high light output (~100,000 photons/MeV).

4. *Decay Constant of Thallium-Activated NaI*: The most notable property of NaI:Tl is its high excellent light yield property, but the dominant decay time of scintillation pulse of 230 ns is not comfortable for a scintillating material when high speed is the main criterion.

In addition to the excellent light yield, phosphorescence with characteristic of 0.15 s *decay time* has also been measured in NaI:Tl crystal, which contributes 9 % to the overall light yield [26].

The decay time measurements are performed by irradiating the NaI:Tl single crystal with an X-ray beam of mean energy 10 keV. Aluminum foil is used to reduce the presence of soft X-ray in the beam. The calculated rise time and the decay time ($\tau$) calculated from the decay time signal (Fig. 8.6) are 49 and 250 ns, respectively. The temperature-dependent decay time is measured and presented in Fig. 8.7. When increasing the temperature, the decay time decreases very sharply up to 150 ns at 80 °C and then slowly to 100 ns. NaI:Tl exhibits several decay time constant components. As temperature increases, the longer time constant components decrease in intensity.

In addition to the excellent light yield, phosphorescence with characteristic of 0.15 s decay time has also been measured in NaI:Tl crystal, which contributes 9 % to the overall light yield [26].

8.3 Alkali Halides

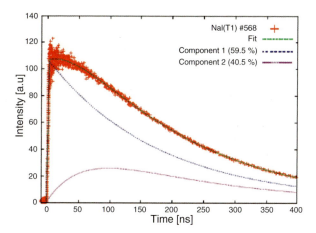

**Fig. 8.6** Decay time signal of NaI:Tl single crystal (Reproduced after Mengesha et al, with permission from IEEE)

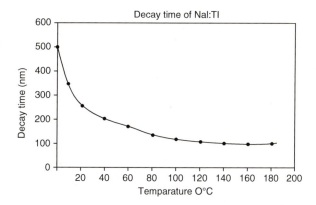

**Fig. 8.7** Temperature-dependent decay time constant ($\tau$) in NaI:Tl crystal

## 8.3.2 Cesium Iodide (CsI)—Cesium Iodide (CsI) Activated with Either Thallium (Tl) or Sodium (Na)

Cesium iodide (CsI) is another alkali halide scintillating material and has been known since 1950 [27]. The crystal exhibits two emission bands of 290 and 338 nm with decay constants of 0.1 and 0.7 μs, respectively. At room temperature, the CsI crystal exhibits a wide (~50 nm) emission band at 305 nm with decay constants of 10 and 36 ns, and the CsI/NaI:Tl light output ratio is measured between 0.05 and 0.08 for various samples [28]. Commercially CsI is available with thallium (Tl) or sodium (Na) as activator, and different scintillator properties are observed in two cases.

The emission spectra of CsI crystal and doped crystals (Tl and Na) are shown in Fig. 8.8. The undoped crystal shows the maximum peak around 315 nm, whereas

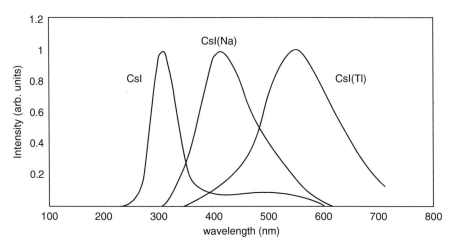

**Fig. 8.8** The emission spectrum of CsI and doped (Tl and Na) CsI crystals (Courtesy: SCIONIX)

the sodium (Na)- and the thallium-doped crystals show maximum peaks approximately at 420 and 550 nm, respectively.

Both the peaks of the doped crystals are broader in comparison with the undoped crystal. The broad peak of the Tl-activated crystal around 550 nm does not match well to the sensitivity of a (bi-alkali) photocathode of a photomultiplier tube (PMT). As a result, the photoelectron yield for gamma ($\gamma$) rays amounts ~45 % of the value of NaI:Tl. However, CsI:Tl scintillation is very well matched with silicon PIN photodiodes, yielding a wavelength-averaged internal quantum efficiency (QE) of approximately 90 % and external QE of about 70 % [29]. On the other hand, the sodium-activated crystal has a peak around 420 nm and is well matched with a (bi-alkali) photocathode of a PMT. Therefore, the photoelectron yield for $\gamma$-rays accounts almost 85 % of that of NaI:Tl. However, the photofraction of Tl-doped CsI is higher than Tl-doped NaI because of its higher Z value.

CsI:Tl-doped material shows highest gamma-ray absorption coefficient per unit size of any scintillator. It is less brittle than NaI:Tl and thus can withstand more mechanical shock and vibration, and the crystals combine the advantages of high stopping power and high reliability for large thickness, good energy resolution, and the possibility of light particle isotopic identification using the pulse shape discrimination technique. CsI:Tl has variable decay time for various exciting particles, and thus pulse shape discrimination techniques can be used to differentiate among various types of radiation. The material is less hygroscopic than NaI:Tl. Table 8.1 shows some of the scintillator crystals' data.

The cesium iodide (CsI) when activated with sodium (Na) shows higher light output and shows two decay components at 0.46 and 4.18 μs [30]. Table 8.2 shows the properties of cesium iodide (CsI) scintillators without and with activators (Na and Tl).

## 8.3 Alkali Halides

**Table 8.1** Some of the data of the scintillator crystals used in medical imaging

| Material | NaI:Tl | CsI:Tl | CaF$_2$:Eu | BaF$_2$ | BGO | YAG:Ce | YAP:Ce | GSO:Ce |
|---|---|---|---|---|---|---|---|---|
| Chemical formula | NaI:Tl | CsI:Tl | CaF$_2$:Eu | BaF$_2$ | Bi$_4$(GeO$_4$) | Y$_3$Al$_5$O$_{12}$ | YAlO$_3$ | Gd$_2$SiO$_5$ |
| Density (g/cm$^3$) | 3.67 | 4.51 | 3.18 | 4.89 | 7.13 | 4.57 | 5.37 | 6.71 |
| Hardness (Moh) | 2 | 2 | 4 | 3 | 5 | 8.5 | 8.6 | 5.7 |
| Hygroscopic | Yes | Slightly | No | No | No | No | No | – |
| Crystal structure | Cubic | Cubic | Cubic | Cubic | Cubic | Cubic | Rhomb | Mono |
| Thermal expansion (ppm) | 47.5 | 50 | 19.5 | 18.4 | 7.0 | 8–9 | 4–11 | 4–12 |
| Melting point (C) | 651 | 621 | 1,360 | 1,280 | 1,050 | 1,970 | 1,875 | |
| Integrated light output (%NaI:Tl) | 100 | 45 | 50 | 20/2 | 15–20 | 15 | 40 | 20–25 |
| Wavelength of maximum emission (nm) | 415 | 550 | 435 | 325/220 | 480 | | 550 | 370 | 440 |
| Decay constant (ns) | 230 | 900 | 940 | 630/0.6 | 300 | 70 | 25 | 30–60 |
| Afterglow (% at 6 ms) | 0.5–5 | <2 | <0.3 | – | <0.005 | <0.005 | <0.005 | <0.005 |
| Radiation length (cm) | 2.9 | 1.86 | 3.05 | 2.03 | 1.1 | 3.5 | 2.7 | 1.38 |
| Photon yield at 300 K-10$^3$pH/MeV | 38 | 52 | 23 | 10 | 2–3 | 8 | 10 | 8–10 |

**Table 8.2** Properties of cesium iodide scintillator (activator)

| Scintillator (activator) | CsI | CsI (Tl) | CsI (Na) |
|---|---|---|---|
| Density (g/ml) | 4.51 | 4.51 | 4.51 |
| Hygroscopic | Slightly | Slightly | Yes |
| Emission wavelength (nm) | 315 | 550 | 420 |
| Lower cutoff | 260 | 320 | 300 |
| Refractive index | 1.95 | 1.79 | 1.84 |
| Primary decay time μ-s | 0.016 | 1.0 | 0.63 |
| Light yield photons/MeV | $2 \times 10^3$ | $54 \times 10^3$ | $42 \times 10^3$ |
| Crystal structure | Cubic | Cubic | Cubic |

Figure 8.9 shows the energy-dependent percentage absorption in CsI:Tl crystal. It has been found that thallium-activated cesium iodide (CsI:Tl) crystal is an efficient absorber of gamma ($\gamma$) rays, with a radiation length of 1.65 cm at 662 keV.

The measured absorption with the energy of the gamma ray further shows that absorption is almost 100 % up to the energy of 200 keV. With further increase of energy, the absorption falls very rapidly until the energy reaches 400 keV. After 400 keV the variation in absorption of gamma ($\gamma$) ray is almost exponential.

Figure 8.10 shows the electron energy response of CsI crystal activated with thallium (Tl) and sodium (Na). The energy response curve (top) of thallium-activated NaI is also given for comparison. The Compton coincidence technique (CCT) has been implemented for the purpose of photon response in

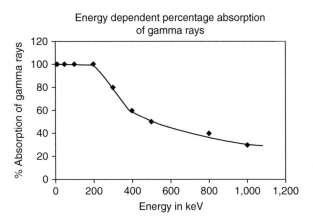

**Fig. 8.9** The energy-dependent percentage absorption in CsI:Tl scintillator

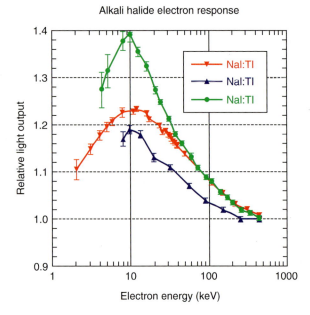

**Fig. 8.10** The electron energy response of activated (Tl 2nd and Na 3rd) CsI crystal. The energy response curve (*top*) of thallium-activated NaI is also given for comparison (Courtesy of Mengesha et al. [31], with permission from IEEE)

activated cesium iodide to the energy of the incident gamma ($\gamma$) ray. For comparison, the response of the photon in activated sodium iodide (NaI) is also given. The photon response, which is defined as the ratio of the light yield to the energy deposited by the photon, has a maximum peak at about 10 keV and then monotonically decreases by about 20 % as the photon energy increases from 10 to 440 keV. Below 10 keV, the photon yield decreases very sharply with decreasing incident energy. Accurate CCT measurements are difficult below 4.0 keV due to interfering thermal noise of the photomultiplier tube

## 8.3 Alkali Halides

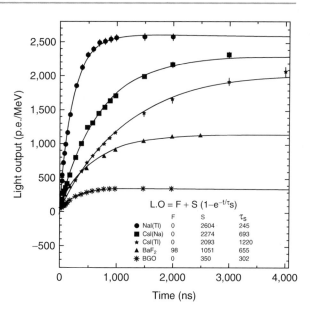

**Fig. 8.11** The comparative light output of doped CsI with other inorganic scintillators as a function of the integration time (Courtesy of Mao et al. [34], reproduced with permission from IEEE)

(PMT). The convolution measurements of the electron response to discrete energies ranging from the lowest energy (0.71 keV for CsI) to the largest photon energy (typically 2 MeV) are made. The results shown above are from the CCT measurements, and it agrees well (almost 85 %) with the calculated data of Gwin and Murray [32, 33].

Figure 8.11 shows the light output of doped cesium iodide with other inorganic scintillators as a function of the integration time which is plotted by fitting the following equation [34]:

Light output as a function of time $(LO/t) = F + S\{1 - \exp(-t/\tau_d)\}$ (8.1)

where $F$ and $S$ are the fast and slow components of the scintillation light and $\tau_d$ is the decay time constant. The number gamma-ray-excited decay modes for thallium- and sodium-activated cesium iodide (CsI) have slow and fast modes at room temperature [35].

At room temperature, the fast decay time $(\tau_{d1})$ constant is approximately 679 ± 10 ns (63.7 %) and slow decay time $(\tau_{d2})$ constant is approximately 3.34 ± 0.14 μs (36.1 %). The observed $(\tau_{d1})$ luminescent state is populated by an exponential process with a rising time constant of 19.6 ± 1.9 ns at room temperature. In thallium-activated CsI, the longer component has a significant effect on pulse processing and energy resolution and light yield [36]. For excitation with highly ionizing particles, such as α-particle or protons, the ratio between the intensities of

**Fig. 8.12** Afterglow in CsI crystals when compared to ZnWO₄ and CdWO₄ crystals. Crystal nos. 1 and 2 are activated with Tl and Na and no.3 has no dopant

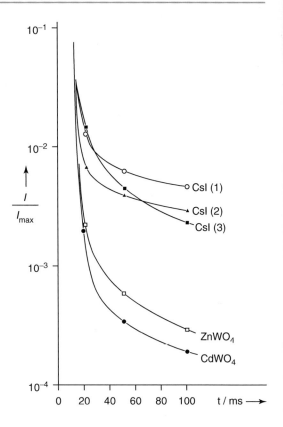

these two decay components varies as a function of the ionizing power of the absorbed particle.

The *scintillation light yield* is dependent on the molar concentration of thallium iodide. However, decay time of CsI(Tl) does not show any such relation with the concentration of thallium in cesium iodide crystals. According to Grassman et al. [37], energy efficiency of CsI(Tl) crystal is higher compared to NaI(Tl) crystal. It has been found that 1-MeV energy was absorbed within NaI(Tl) crystal to generate $4 \times 10^4$ photons, whereas the same energy generates $4.5 \times 10^4$ photons in CsI(Tl) crystal. Unfortunately, the afterglow in activated CsI is very high, almost 5 % after 3–6 ms which significantly limits the counting rate.

Figure 8.12 shows afterglow in CsI crystals. Crystal nos. 1 and 2 are activated with Tl and Na, respectively, and crystal no. 3 is not activated with any dopant. For comparison, afterglow for zinc tungstate (ZnWO₄) and cadmium tungstate is also plotted on the same graph. Afterglow is thought to be due to some deep traps within the crystals (>0.4 eV) which have their glow at temperature >200 K [38]. In order

## 8.3 Alkali Halides

to minimize the afterglow effect, some special co-doping is necessary. Recently co-doped activated cesium iodide with samarium (Sm) has been used to suppress the afterglow effect. The model suggests that Sm electron traps scavenge electrons from thallium traps and that electrons subsequently released by Sm recombine non-radiatively with trapped holes, thus suppressing afterglow [39]. Similar co-doped experiment has been done with NaI:Tl by N.Shiran et al. They found that when the thallium-activated NaI crystal was doped with europium (Eu), it showed increased light yield with shaping time growth [40].

A charged particle passing through a scintillator forms a large number of electron–hole (e–h) pairs. There are several reasons that the total number of $e$–$h$ pair created cannot be captured because of impurity sites or an activator site. These impurity sites or activator sites capture some of these electrons. For the second case, radiationless transitions are possible between some excited states formed by electron capture and the ground state, in which case no visible photon results. Such processes are called *quenching*.

The differential quenching efficiency of CsI (Tl) can be expressed as a function of scintillation efficiency and quenching factors as

$$(\mathrm{d}L/\mathrm{d}E) = S_{\mathrm{e}} \left(1/1 + q_{\mathrm{f}}\right) |\mathrm{d}E/\mathrm{d}x|_{\mathrm{e}}) \tag{8.2}$$

where $S_{\mathrm{e}}$ is scintillation efficiency, $q_{\mathrm{f}}$ is quenching factor, and $(\mathrm{d}E/\mathrm{d}x)_{\mathrm{e}}$ is electronic stopping power, which is also called nuclear recoil charge density $(\mathrm{d}E/\mathrm{d}x)$.

The stopping power $(\mathrm{d}E/\mathrm{d}x) = e_{\mathrm{s}} + n_{\mathrm{s}} = e_{\mathrm{s}} \left(1 + e_{\mathrm{n}}/e_{\mathrm{s}}\right)$, and when it is multiplied by crystal density $\rho$, it gives rise to stopping power per unit length. The scintillation light output is expected to be reduced (quenched), while timing profile of the pulse is different relative to some energy deposition by maximum ionizing particles [41–43]. The cesium iodide (CsI) when activated with sodium (Na) shows higher light output and shows two decay components at 0.46 and 4.18 µs [30]. Figure 8.13 shows the quantum efficiency of a thallium-activated CsI crystal.

Cesium iodide also has a useful property when grown by special technique in thin layers on specially prepared substrates. The deposited material can be formed columnar to make channels for the light to pass through them very easily. The gap between the columns is filled with reflective coating to minimize loss or cross talk between the columns.

*Vacuum-Deposited Columnar Film*: Hot-wall epitaxial (HWE) technology has been very popular in semiconductor industries and has been used to grow columnar-structured thallium (Tl)-activated cesium iodide film to maximize light output from the film. These films have excellent spatial resolution and high detection efficiency and have been used in gamma ($\gamma$)-ray detector, medical imaging system, nondestructive testing (NDT), astronomy, and macromolecular crystallography [44, 45]. Figure 8.14 shows the picture of a columnar-structured film grown by hot-wall epitaxial technology. The 2,000-µm-thick columnar-structured film preserves film

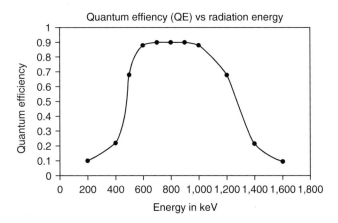

**Fig. 8.13** The quantum efficiency of a thallium-activated CsI crystal

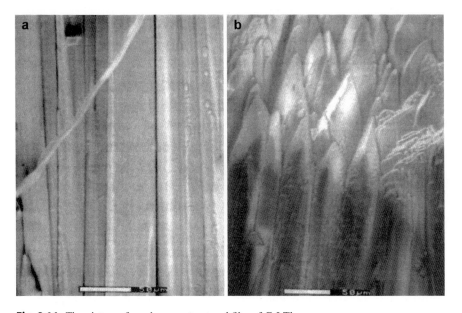

**Fig. 8.14** The picture of a columnar-structured film of CsI:Tl

light output maximum ensuring excellent optical transmission and good spatial resolution. Filling the gap between the columns with an appropriate reflective material increases the light output by 80 %.

Figure 8.15 shows the light output conversion efficiency of a CsI:Tl vacuum-deposited plate after reflective and absorptive coatings. The luminescence properties of these columns differ from those of CsI (Tl) scintillation crystals

**Fig. 8.15** Light output conversion efficiency of a CsI:Tl vacuum-deposited plate after reflective and absorptive coatings

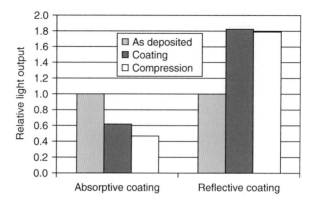

**Fig. 8.16** The spatial resolution versus % modulation transfer function (MTF) of CsI(Tl) films

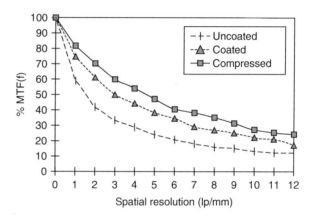

without coatings [46]. In addition to the standard emission bands at 400 nm (Tl emission) and 550 nm (strongly perturbed thallium-bound excitation emission), an additional band is observed at 460 nm at liquid nitrogen temperature, which is attributed to a weekly perturbed thallium-bound excitation center [47].

The modulation transfer function (MTF) of a 150-μm film is measured for spatial frequencies in the range of 0–12 lines per millimeter, and calculations are done by following fast Fourier transform (FFT). The detector is exposed to a flood field of 30-kV W X-rays. The isostatically compressed (30 Kpsi) film shows the highest percentage MTF compared to untreated and treated film (Fig. 8.16).

The detective quantum efficiency (DQE) for these films is measured using a formula given by Hillen et al. [48] and Roehrig et al. [49] and presented in Fig. 8.17.

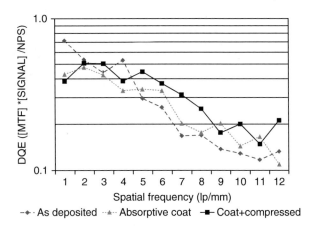

**Fig. 8.17** Detective quantum efficiency versus spatial frequency of a PVD CsI(Tl) film

## 8.4 Halides of Heavy Metals

In this section, we will discuss mostly the heavy metals of halides and metals that have tellurium in the metal matrix. These compounds are semiconductor materials and can be called photoconductors. These materials convert X-ray energy directly to electrical signal. In section 5B.1 we have discussed two alkali halides, namely, NaI and CsI. These two materials are scintillation material and convert energy of X-ray to photons to visible light. According to the principles of operations of the radiation detectors, they are classified as indirect and direct radiation detectors.

Figure 8.18 (a) and (b) show the schematic of the working principle of the indirect and direct scintillators; (c) and (d) are the schematic of the actual arrangements of the scintillators and the electronic circuitry. The fabrication of room temperature nuclear radiation detectors requires a semiconducting material that has *high atomic number (Z)* for *efficient radiation* or *atomic interaction, large enough band gap* for *high resistivity*, and *low leakage currents*. In addition, the material should have small band gap so that the electron–hole (e–h) ionization energy is small (<5 eV). Finally, the device (scintillator/radiation detector) fabricated out of the material should have high intrinsic μτ product, high purity, homogeneous, and defect-free which can be matched with a proper ohmic contact electrodes [50].

Table 8.3 shows the *band gap*, the *μτ product, the resistivity*, and the absorbed *ionizing radiation energy* required to create electron hole pairs (e–h) in different compound semiconductors we are interested in, for example, lead iodide ($PbI_2$), mercuric iodide ($HgI_2$), cadmium telluride (CdTe), and cadmium zinc telluride (CZT) and thallium bromide (TlBr). Later on we will also discuss some elementary semiconductors like silicon (Si) and germanium (Ge) that are used as radiation detectors and their pros and cons.

## 8.4 Halides of Heavy Metals

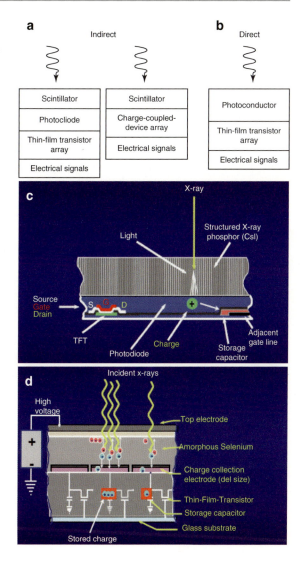

**Fig. 8.18** (a, b) Schematic of indirect and direct working principle of the scintillators. (c, d) Schematic of the actual arrangements of the scintillators and the circuitry (Courtesy: J.A. Scipert, UC, Davis)

**Table 8.3** Some of the properties of compound semiconductors

| Material | Atomic number | Density (g/cm$^3$) | Band gap (eV) | $E_{pair}$ (eV) | Resistivity ($\Omega$-cm) | $\mu\tau$ (e) product (cm$^2$/V) | $\mu\tau$ (h) product (cm$^2$/V) |
|---|---|---|---|---|---|---|---|
| PbI$_2$ | 82, 53 | 6.2 | 2.32 | 4.9 | $10^{12}$ | $8 \times 10^{-6}$ | $\sim 10^{-5}$ |
| HgI$_2$ | 80, 53 | 6.4 | 2.13 | 4.2 | $10^{13}$ | $10^{-4}$ | $4 \times 10^{-5}$ |
| TlBr | 81, 35 | 7.56 | 2.68 | 6.5 | $>10^{10}$ | $10^{-3}$ | $4 \times 10^{-5}$ |
| CdTe | 48, 52 | 6.2 | 1.44 | 4.43 | $10^9$ | $3 \times 10^{-3}$ | $2 \times 10^{-4}$ |
| CdZnTe | 48, 30, 52 | 6.0 | 1.5–2.2 | $\sim 5.0$ | $>10^{10}$ | $3 \times 10^{-3}$ | $10^{-5}$ |

386    8  Characterization of Radiation Detectors (Scintillators) Used in Nuclear Medicine

The charge collection efficiency ($\eta$) of a photoconducting insulator is given by Hecht's equation [51]:

$$\eta = (\mu\tau E)/L \ [1 - \exp(-L/\mu\tau E)] \tag{8.3}$$

where $\mu$ is mobility of the carriers, $\tau$ is bulk trapping or recombination time, $E$ is the electric field applied, and $L$ is the length of the photoconducting detector. Experimentally, the $\mu\tau$ product is extracted from the measured field of half-maximum charge collection $E_{\mathrm{mcc}}$ in the collected charge versus electric field plot. However, when these charges are collected near the surface, surface recombination of charge carriers sets in, in addition to the bulk trapping [52]. Therefore, Hecht's equation must be corrected for a near-electrode (near surface) carrier generation to take account of the surface recombination. As a result, the Eq. (8.3) will be modified as

$$\eta = [1/(1 + s/\mu E)] \ (\mu\tau E)/L \ [1 - \exp(-L/\mu\tau E)] \tag{8.4}$$

where $s$ is the surface recombination velocity of the corresponding charge carrier. The transport kinetics of holes and electrons will be different when negative or positive polarity is applied to the irradiated electrode. The other important parameter is the de-trapping time of the carrier, which will affect the transient time across the sample and thereby the voltage signal. Assuming that the carrier de-trapping times are long compared to their transit time $t_{\mathrm{t}} = (L/\mu E)$ across the detector, the voltage signal $V_{\mathrm{s}}$ can be expressed mathematically as

$$V_{\mathrm{s}} \begin{cases} = 0 & t < 0 \\ = \{(Q_0 L/A\varepsilon_0\varepsilon_{\mathrm{r}})\} \cdot \{1/(1 + s/\mu E)\} \cdot \{\tau/t_{\mathrm{t}}(1 - \exp(-t_{\mathrm{t}}/\tau)\} & 0 < t < t_{\mathrm{t}} \\ = \{(Q_0 L/A\varepsilon_0\varepsilon_{\mathrm{r}})\} \cdot \{1/ \ (1 + s/\mu E)\} \cdot \{\tau/t_{\mathrm{t}}(1 - \exp(-t_{\mathrm{t}}/\tau)\} & t > t_{\mathrm{t}} \end{cases} \tag{8.5}$$

where $Q_0$ is the total charge collected, $\varepsilon_0$ is the permittivity of free space, $\varepsilon_{\mathrm{r}}$ is the relative dielectric constant of the detector, and $A$ is the area of the detector. The ratio of the induced charge ($Q$) to the total charge collected ($Q_0$) at the electrode is dependent on a number of parameters as follows [53]:

$$(Q/Q_0) = \eta = 1/Q_0 \int_0^T \mathrm{d}t \cdot \int_\Omega n \cdot \mu\nabla\varphi \cdot \nabla\psi \cdot \mathrm{d}\Omega \tag{8.6}$$

where $n$ is the free electron density, $\varphi$ is the applied bias, $\psi$ is the weighting function, and $T$ is the time interval.

The other important parameter that should be considered as a *quality factor* of a radiation detector is the *energy resolution*. Sometimes, we observe broadening of the energy spectrum or shift in the peak of the energy spectrum. The cause of the broadening of the spectrum or shifting of the peak is attributed not only to the poor

## 8.4 Halides of Heavy Metals

charge collection but also to electronic noise (*Fano noise*, noise due to leakage current, and noise due to carrier *trapping or polarization*). In addition, compound semiconductors have wide range of band gaps, and the resistivities are also higher than the elementary semiconductors. As a result, these compound semiconductors show minimum leakage current [16, 54]. These compound semiconductors thus minimize most of the noise that are not possible to eliminate in elementary semiconductors. As for example, in cadmium zinc telluride (CZT), it has been possible to optimize the noise performance by alternating the zinc (Zn) fraction in the compound semiconductor where the range of possible alloys moves along the line between CdTe and ZnTe due to alteration of Zn.

### 8.4.1 Lead Iodide ($PbI_2$)

The development of digital (pixelated) radiation detectors that are used in X-ray and gamma ($\gamma$)-ray spectrometers and medical imaging system has prompted interest in semiconductors that may function as direct converters, by passing the radiation source to optical photon conversion process. Lead iodide ($PbI_2$) is one of the many materials that has shown promising results and has attracted the attention of the medical professionals and the scientists practicing in medical imaging and radiation therapy. Its properties include high atomic number (82, 53), wide band gap (2.3 eV), low melting point, and adequate charge carrier capabilities ($\mu\tau$ product $\sim 10^{-5}$ cm$^2$/V) as we can see from the Table 8.6.

Recent result shows that the radiographic and memographic sensitivities at 72-kVp and 26-kVp measurements, respectively, for a physical vapor-deposited (PVD) polycrystalline $PbI_2$ film are $20 \times 10^6$ and $2 \times 10^6$ e/mm$^2$/mR, respectively [55]. However, the line-spread function (LSF) with $PbI_2$ crystal degrades with increasing thickness, and the image lag is higher compared to $HgI_2$ film [56]. The long image lag has been attributed to deep impurity states in the forbidden gap of the photoconductor.

Figure 8.19 shows room temperature $^{137}$Cs spectrum with a $PbI_2$ photodetector when coupled to a LSO scintillator. The detector is irradiated with isotopic source such as $^{137}$Cs, and measurements are taken with standard nuclear instrumentation. The resolution of 662-keV peak is measured and found to be approximately 12 % (FWHM). The signal magnitude corresponding to 662-keV gamma ($\gamma$)-ray interaction is estimated around 3,500 electrons when the energy scale is calibrated with direct X-ray detection in $PbI_2$. The electronic noise is measured with a test pulser, and recorded noise is 9 % (FWHM) which corresponds to 300 electrons (FWHM) or 130 electrons (rms).

In order to testify the temperature effects on the dark current, experiment is performed and the result is shown in Fig. 8.20. From the figure, one can notice that the variation of dark current is almost linear. For 75 °C the leakage current is only about 10 pA. For flat panel imager, the magnitude of the dark current generated in

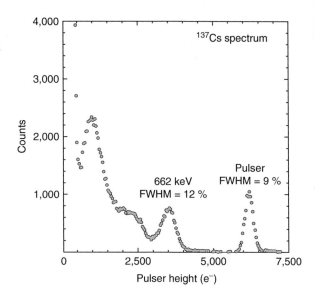

**Fig. 8.19** $^{137}$Cs spectrum with a PbI$_2$ photodetector coupled to LSO scintillator at room temperature

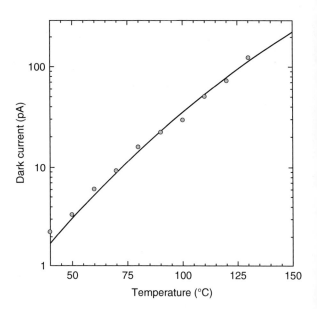

**Fig. 8.20** Temperature-dependent leakage current in PbI$_2$ photodetector

the photoconductor is an important determinant of medical imaging performance [57]. However, at room temperature, the leakage current is only about 1.8 pA, which is a good number for a radiation detector for medical imaging.

## 8.4 Halides of Heavy Metals

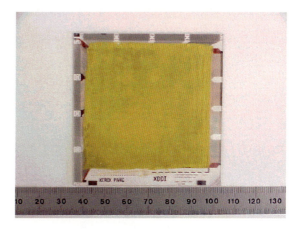

**Fig. 8.21** Picture of a TFT array with lead iodide PIB film (screen printed)

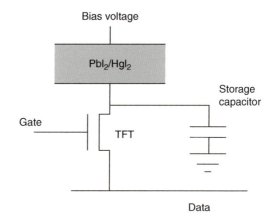

**Fig. 8.22** The schematic diagram of the pixel circuit with TFT and storage capacitor

The TFT array (dpiX®, 5 × 5 cm² area) used in the measurements have 127-μm pixel size in a 768 × 768 format. PbI$_2$ powder (5N) is mixed with a proper binder (PIB). The viscosity of the slurry (ink) is adjusted to control the resistivity and the flow through the screen mesh. The screen-printed PbI$_2$ film over the TFT array is shown in Fig. 8.21. Vacuum-deposited palladium (Pd) film is deposited over the PbI$_2$ film leaving 1 cm area from the edges of the PIB film for local ohmic contact. Finally, Pd wires 2 mm diameter is attached to the vacuum-deposited metal (Pd) film for global connections.

Figure 8.22 shows the schematic diagram of the pixel circuit of the TFT array and the storage capacitor with PbI$_2$ film deposited on the active surface of the array. Bias voltage is applied on the film, and the electrical characterization of the film is performed. Table 8.4 presents comparative data of PbI$_2$ films with that of a single crystal.

**Table 8.4** Comparison of lead iodide properties

| Material | Grain size (microns) | Density (g/cm$^3$) | Resistivity ($\Omega$-cm) | $\mu\tau$ product (cm$^2$/V) |
|---|---|---|---|---|
| Vacuum-deposited PbI$_2$ | 10–30 | 4.5–5.8 | 2–8 × 10$^{12}$ | 1–2 × 10$^{-6}$ |
| Screen-printed PbI$_2$ | 25 | 5.2 | 5 × 10$^{12}$ | 5–8 × 10$^{-7}$ |
| Single crystal PbI$_2$ | >10,000 | 6.3 | 5 × 10$^{13}$ | 1 × 10$^{-5}$ |

**Fig. 8.23** Image of a resolution target taken with a PbI$_2$ array

**Table 8.5** Properties of TlBr detector in comparison to other semiconductor detectors

| Material | Atomic number | Density (g/cm$^3$) | Band gap (eV) | $E_{pair}$ (eV) | Resistivity (25 °C) $\Omega$-cm | $\mu\tau$ (e) product cm$^2$/V | $\mu\tau$ (h) product cm$^2$/V |
|---|---|---|---|---|---|---|---|
| Si | 14 | 2.33 | 1.12 | 3.62 | Up to 10$^4$ | >1 | ~1 |
| Ge | 32 | 5.33 | 0.67 | 2.96 | 50 | >1 | >1 |
| GaAs | 31, 33 | 5.32 | 1.43 | 4.2 | 10$^7$ | 8 × 10$^{-5}$ | 4 × 10$^{-6}$ |
| CZT | 48, 30, 52 | 6 | 1.5–2.2 | ~5.0 | >10$^{10}$ | 3 × 10$^{-3}$ | 10$^{-5}$ |
| CdTe | 48, 52 | 6.2 | 1.44 | 4.43 | 10$^9$ | 3 × 10$^{-3}$ | 2 × 10$^{-4}$ |
| PbI$_2$ | 82, 53 | 6.2 | 2.32 | 4.9 | 10$^{12}$ | 8 × 10$^{-6}$ | |
| HgI$_2$ | 80, 53 | 6.4 | 2.13 | 4.2 | 10$^{13}$ | 10$^{-4}$ | 4 × 10$^{-5}$ |
| TlBr | 81, 35 | 7.56 | 2.68 | 6.5 | >10$^{10}$ | 10$^{-3}$ | 4 × 10$^{-5}$ |

Figure 8.23 shows the image of a lead phantom resolution pattern imaged by ~300-μm-thick PbI$_2$ photoconductor on a 127-m pixel. The line pair resolution pattern is obtained with a XDDI TFT array, (768 x 768 pixel arrays, with 127 μm pitch, procured from dpiX, corporation, USA) coated with lead iodide (PbI2) film is shown in (Fig. 8.21) coated with lead iodide (PbI$_2$) film. Fluoroscopic exposure conditions are 60 kVp with 0.1 mm added copper filtration, and exposure rate is 15 frames per second and bias on the photoconductor is 15 V (0.14 V/μm). Figure 8.23 shows that the PbI$_2$ detector can resolve 3.7 line pair per mm almost nearly equal to Nyquist frequency limit of the 127-μm pixel size [58].

X-ray detection characteristics of the screen-printed films are also evaluated by irradiating them with exposure from an X-ray tube. A comparison of the X-ray response of the screen-printed and vacuum-sublimed films (of same thickness, 0.13 mm; area, 28 mm$^2$) under identical exposure conditions is shown on Table 8.5.

## 8.4 Halides of Heavy Metals

As seen from the table, the response of the vacuum-deposited film is higher by a factor of two, which is mostly due to its higher charge transport parameters. However, the preliminary results for screen-printed films are impressive considering the fact that no special attention was paid to material purity.

In summary, our preliminary investigation indicates that the screen-printing method is a very promising technique to fabricate thick X-ray-sensitive layers of lead iodide, which can be applied to medical imaging. The potential of easy scale-up to larger sizes is also evident.

### 8.4.2 Mercuric Iodide ($HgI_2$)

Mercuric iodide ($HgI_2$) is a layer structure halide of heavy metal. The red variety $\alpha$-phase is of interest for electronic and nuclear medicine applications. Because of its high density (6.28 g/cc) and high effective $Z$ values (~66), it is a useful material for gamma-ray detection, and when it is doped with boron-rich material/s it can be a potential candidate for thermal neutron detector [59]. The dark conductivity in $HgI_2$ is affected by the ionic conductivity, which is due to the movement of iodine vacancies. However, as a room temperature radiation detector, the ionic conductivity is not dominant. Tunneling is not a factor for $HgI_2$ but the bulk-controlled current is determined by the Poole–Frenkel effect (PFE). However, the leakage current is difficult to attribute to the PFE alone. The bulk resistivity is around $3 \times 10^{13}$ $\Omega$-cm considering electron concentration of $2 \times 10^3$ $cm^{-3}$. The dielectric relaxation time for the material at room temperature has been reported about 25 s [60]. Polycrystalline mercuric iodide ($HgI_2$) is a direct detection material superior to selenium (Se). The effective ionization energy $W_{ef}$, which is a measure of sensitivity, is much lower than selenium (between 4.9 and 5.1 eV compared to 25 eV of Se at 80 kVp).

The electrical properties, the energy resolution, the peak to background (peak to valley) ratio, and the efficiency of $HgI_2$ are dependent on the number of defect centers and trapping and de-trapping effects. These effects are mostly inherent properties of the material, and the present challenges in $HgI_2$ sensor technology are to increase control of the detector fabrication processes and to modify the detector processing strategy. The presence of the extended defects also affects the uniform charge collection efficiency in $HgI_2$ photo-sensors [61].

Figure 8.24 shows temperature dependence of the electron and hole mobilities in mercuric iodide ($HgI_2$) photodetector. Figure 8.24 (a) represents the electron mobility and (b) represents hole mobility, parallel to the $c$-axis [62]. Both electron and hole mobilities are found to decrease with increasing temperature. Mercuric iodide ($HgI_2$) crystals have layered structure, and the carrier mobilities are dependent on the crystallographic orientation. Experimental observations show that the electron mobility parallel to the $c$-axis is always greater than the mobility perpendicular to the $c$-axis [63].

The electro-absorption spectrum of $HgI_2$ has been studied by Chester and Coleman [64] as a function of electric field and temperature. Figure 8.25 shows

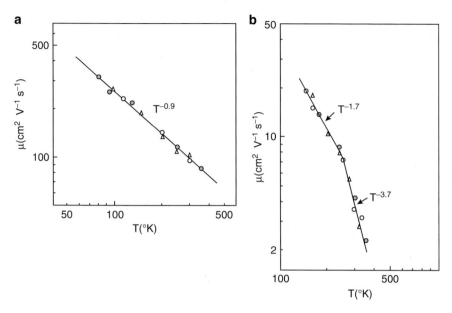

**Fig. 8.24** Temperature dependence mobility of (a) electrons and (b) holes (Reproduced with permission from AIP)

**Fig. 8.25** The electro-absorption signal in HgI$_2$ at 300 K (Reproduced with permission from Elsevier Pub.)

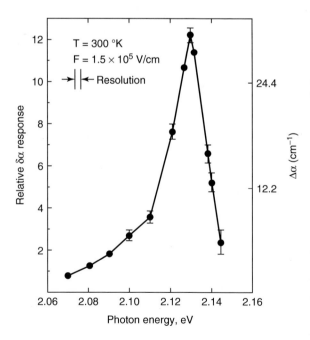

## 8.4 Halides of Heavy Metals

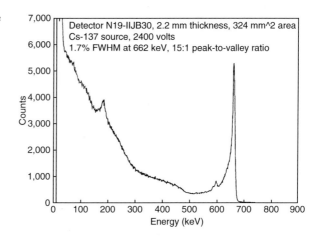

**Fig. 8.26** A spectrum of the 662-keV gamma ray from a $^{137}$Cs source taken with a HgI$_2$ detector at room temperature (Reproduced with permission from Elsevier Pub.)

the electro-absorption signal in HgI$_2$ crystal at 300 k for a field of 1.5 × 10$^5$ V/cm. From the electro-absorption curve, we can see that the peak occurs around the energy of 2.127 eV, which is less than the band gap energy of the material (2.14 ± 0.005 eV at room temperature). It has been found that when the photon energy exceeds the band gap energy of HgI$_2$, the absorption coefficient is quite large (>10$^5$ cm$^{-1}$) (Fig. 8.25).

Figure 8.26 shows the gamma-ray spectrum of the 662 keV from a $^{137}$Cs source taken with a HgI$_2$ detector at room temperature [65]. From the graph, we can find that the resolution is 1.7 % and the peak to valley ratio is 15:1. Experimental observation shows that the crystal has better luminescence property in the band 1 region, which corresponds to near band edge wavelength. Based on these findings, it seems that larger emissions from deeply bound states are indicators of poor detector grade crystals [66].

In recent years, flat-panel X-ray imagers have been developed with primary application to medical imaging. The technology also is becoming increasingly popular for security screening for its higher specificity. The development of high-sensitivity large-area CT scanners at an affordable cost will be a top priority in medical fields. The most widespread sensors in present flat-panel medical imagers use a phosphor screen coupled with an amorphous silicon photodiode array [67].

Today, the most burning question in medical imaging is to deposit the detector film over a large area of the thin film transistor (TFT) array (active area of the future array is expected to reach 25 × 25 cm$^2$ within the year 2015). It is considered to be the most challenging job since large crystal of HgI$_2$ is difficult to grow. As a result, scientists have paid attention to deposit large-area polycrystalline films either by (1) *physical vapor transport* (PVT) or (2) screen printing which is also referred to as *particle in binder method* (PIB) [68]. The challenge with the PVD system is the loss

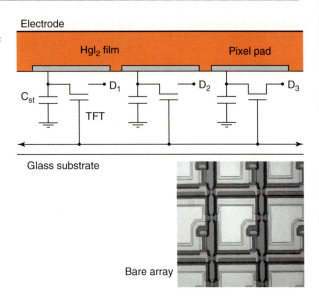

**Fig. 8.27** The circuitry of the TFT array with a top HgI$_2$ screen-printed layer over barrier layer. *Bottom picture is a bare TFT array*

of material, difficulty to make very thick film, and the temperature issue (mercuric iodide changes from alpha phase to beta phase around 127 °C). On the other hand, the second method is capable of making thick film (up to 1 mm) at room temperature and the technology and processing are cost-effective.

As mercuric iodide (HgI$_2$) is an aggressive chemical compound and attacks almost all metals, so a *barrier layer* has to be applied to cover the metallic local interconnections (an alloy of aluminum with copper and silicon) of the TFT array. The resistivity of the barrier layer should be compatible with the resistivity of the host material, and it should be chemically inert to HgI$_2$. A suitable barrier layer material has been developed, which is inert to the HgI$_2$ and the *current conduction* through the barrier layer is *anisotropic* meaning that current conduction is seen only through the perpendicular direction of the flat panel surface. The *anisotropic current conduction* also minimizes the *cross talk* between the pixels. Figure 8.27 shows the electronic circuitry with TFT array carrying the HgI$_2$ layer on the top of the *barrier layer*. The picture underneath is a bare TFT array.

Figure 8.28 shows the picture of mercuric iodide film (red color) over *a barrier layer* (gray color on the background) deposited on a TFT array (ND-10, 6087-10E®) by screen-printing technology. The active area is ~10 × 10 cm with a 768 × 768 pixel format and 127 pitch. The edges show the aluminum–Si–Cu alloy interconnects for global connections. The bottom picture is a 5 × 5 cm TFT array (dpiX®) covered screen-printed film of HgI$_2$ over a *barrier* layer. The thickness of the HgI$_2$ layer at different places over TFT array is shown in the Fig. 8.29. The aluminum alloy interconnections are covered with photoresist (edges of the TFT array) to protect the local alloy contacts during the deposition of PIB HgI$_2$ film.

## 8.4 Halides of Heavy Metals

**Fig. 8.28** Picture of a screen-printed HgI$_2$ layer (*red*) over a barrier layer (*gray*) and the picture on the right hand side shows the scanning electron micrograph (SEM) of the same film (mercuric iodide)

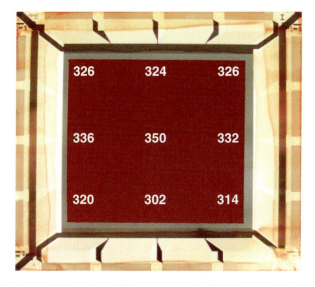

**Fig. 8.29** The picture of a 5 × 5 cm TFT array covered with HgI$_2$ over the barrier layer

The dark current of a detector is one of the most significant parameters of an X-ray imager. Dark current causes excess charge accumulation in a pixel during the exposure time. If the dark current is constant, it can be subtracted from the signal, but if it is too high then it diminishes the available dynamic range and becomes a source of noise. The noise is proportional to the square root of the dark current.

Figure 8.30 shows the typical dark current under conditions of temporal stability at fields 0.2–1.0 V/μm. The dark current is found to increase with increasing field for both the devices (B777-08L and 08I). The positive biased device shows a larger dark current at room temperature compared to negative biased devices. For

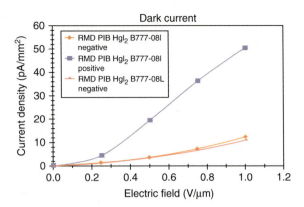

**Fig. 8.30** Dark current of screen-printed HgI$_2$ film with respect to electric field

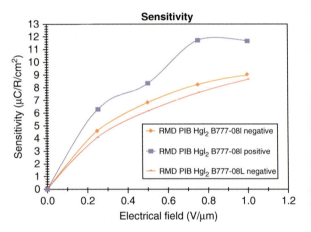

**Fig. 8.31** Average signal plotted as a function of dose for HgI$_2$ PIB film on TFT array

negative bias the dark current for 1.0 V/μm applied field is almost the same for B777-08I and 08L devices, and it is approximately 10 pA/mm$^2$, which is a good number of the device for digital mammography application.

The second most important parameter in X-ray imager is sensitivity because in an imager, we need a highly sensitive detector to generate more electron–hole (e–h) pairs per incident X-ray photon than the noise level of the readout electronics a number of electronic systems. From this point of view, both PbI$_2$ and HgI$_2$ have theoretically much higher sensitivity than a-Se or CsI [69].

To evaluate X-ray sensitivity, it is measured in each case at 60 kVp with 0.1-mm copper filtering. Both the PIB (particles in binder) HgI$_2$ imagers B777-08I and 08L show almost similar behavior (an asymptotic), and the sensitivity values for both of them are ~9 μC/R/cm$^2$ at negative bias. However, at positive bias B777-08I sample behaves very differently. The device with higher sensitivity shows higher dark current too. Figure 8.31 shows average signal plotted as a function of X-ray dose for HgI$_2$ PIB films on TFT array.

## 8.4 Halides of Heavy Metals

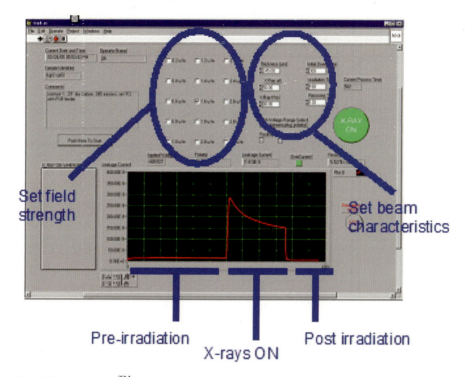

**Fig. 8.32** A Labview™ to measure the photocurrent traces from a sample over a duration of several minutes

Figure 8.32 shows a Labview™ to measure the photocurrent over a duration of several minutes. The software controls a low-power Kevex DC X-ray generator and a high-voltage power supply while acquiring current and voltage data (via GPIB, IEEE-488.2GP18 interface for USB) from multimeters. The trace on the computer screen shown in Fig. 8.34 shows current trace with time (for ~10 min) for a screen-printed $HgI_2$ device.

Figure 8.33 shows the typical photocurrent waveforms of a screen-printed (PIB) $HgI_2$ film when different dose of X-ray radiation is applied to the film. The X-ray machine is set at 80 kV and 50 mA setting during the irradiation. The high voltage from the power supply is set to different values (both polarities) depending on the film thickness. The device is kept under bias for 300 s prior to 180 s irradiation. After 300 s of operation, the device is kept under the same bias for another 200 s before the end of irradiation (when X-ray generator is turned off).

Along the y-axis the current in amperes (A) is plotted against the time in seconds (s). The film is 431 μm thick over a 2–3-μm barrier layer with the palladium (Pd) as ohmic contact having an area of 0.31 $cm^2$. Current waveforms for different PIB $HgI_2$ films have been taken under radiographic conditions at 2.1, 6.3, 10.4, and

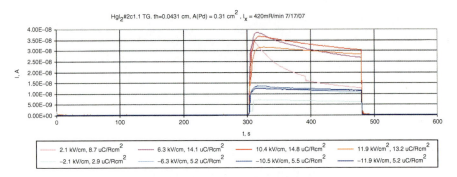

**Fig. 8.33** The typical photocurrent waveforms in a PIB (HgI$_2$)-coated film on a TFT array

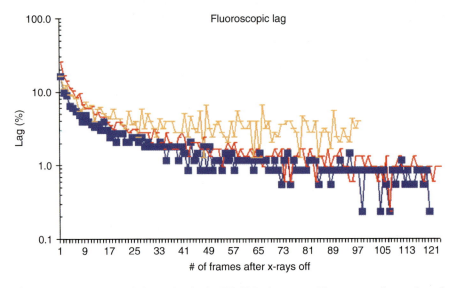

**Fig. 8.34** The fluoroscopic image lag in the PIB HgI$_2$ detectors with respect to the number of frames (*Top* one is for positive bias) (Courtesy: RMD and Varian Med.)

11.9 kV/cm electric fields. In Fig. 8.35, we have presented the current waveforms vs. time of one of the films (ID # Hg$_2$#2CT1.1TG). For all applied fields, the data exhibit an initial surge with a stabilized level photocurrent depending on the applied field. However, for lower fields, the current decay curves show a sharp fall. The effect is presumed to be due to the *polarization effects* in the detector where uncollected charges effectively reduce the applied electric field. This results in the reduction of photocurrent until an equilibrium between charge and trapping and de-trapping is reached leading to a stable current level. From Fig. 8.34, we can see that the polarization effect is minimum when the electric field is 10.5 kV/cm and

## 8.4 Halides of Heavy Metals

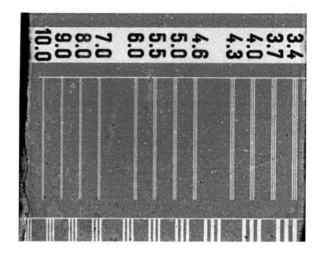

**Fig. 8.35** The image of a resolution target taken with an HgI$_2$ array (Courtesy: RMD and Varian Med.)

above. Similar results have been reported by other investigators [70]. Polarization is the result of the buildup of trap or fixed charges in semiconductors [50].

The charge trapping in photoconductors, incomplete charge collection, and large fluctuations due to charge trapping and de-trapping can cause temporal artifacts such as image lag and ghosting [71, 72]. The *image lag* is a phenomenon of charge carried over from the image charge generated by previous X-ray exposures into subsequent image frames, whereas *ghosting* is the change in sensitivity (long-term image persistence) as a result of previous exposures. Figure 8.34 shows the fluoroscopic image lag in the PIB HgI$_2$ detectors with respect to number of frames (top one is for positive bias).

From the plot (Fig. 8.34), we can see that the percentage lag is almost 3–4 % for the negative biases (two bottom curves) compared to 5–7 % for the positive bias (top curve). After about 3 s (45th frame), the percentage lag is 3 %. The fluoroscopic lag for both the films is almost similar when negative bias is applied. From the above results, we can assume that the mean drift length ($\mu\tau F$, where $\mu$ and $\tau$ are the mobility and the lifetime of the charge carriers, respectively, and $F$ is the electric field intensity) is comparable to the thickness of the photoconductor, meaning that charge trapping and de-trapping events are minimum in the device.

Figure 8.35 shows the line pair resolution pattern of a lead phantom acquired with a TFT array (XDDI device from dpiX Corporation) with 768 × 768 pixel (50 µm pitch) coated with 300-µm-thick PIB mercuric iodide (HgI$_2$) film. Fluoroscopic exposure conditions are 60 kVp with 0.1 mm added copper filtration with an exposure rate of 15 frames per second and a bias (on the photoconductor) of 15 V (0.14 V/µm). From Fig. 8.36a we can see that the screen-printed (PIB) HgI$_2$

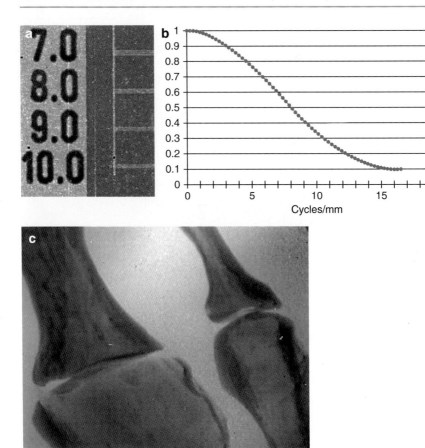

**Fig. 8.36** (a) The resolved line pair per mm and (b) the measured MTF versus cycles/mm in screen-printed HgI$_2$ film (c) The picture of a foot phantom taken by PIB HgI$_2$ detector (Courtesy: RMD and Varian Med.)

detector can resolve 10 line pair per mm which is almost nearly equal to Nyquist frequency limit of the 50-μm pixel size. Similar type of film resolution has been found with the PIB PbI$_2$ films [58].

Another important imaging parameter is the *spatial resolution*, which is measured as *modulation transfer function* (MTF) (Fig. 8.36b). The MTF is one of the most useful means of characterizing the optical performance of an imaging system. It can be defined as *the modulus of the Fourier transform of the line-spread function* (LSF) and mathematically it can be expressed as:

$$\text{MTF}(v) = |F\{\text{LSF}(x)\}| = (\text{MTF})_{\text{pix}} = \text{Sinc}(dv) = \{\text{Sin}(dv\pi)\}/\{(\pi dv)\} \quad (8.7)$$

## 8.4 Halides of Heavy Metals

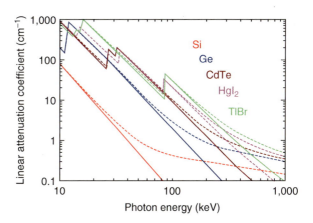

**Fig. 8.37** Linear attenuation coefficients for various semiconductor nuclear detector materials

where $v$ is the spatial frequency, $x$ is the spatial distance, pix meaning when pixelated TFT array is being used, and $d$ is the pixel size (in our present case the pixel size is 50 µM).

Figure 8.36 (c) shows the image of a foot phantom that supports the quality of the X-ray image procured by the PIB HgI$_2$ film. The procured picture of the foot phantom is a measure of the parameter called *resolution* or *MTF* of the film. The foot phantom image shows sharper bone boundaries and fine structures.

### 8.4.3 Thallium Bromide (TlBr)

At room temperature, thallium bromide (TlBr) is an ideal semiconductor nuclear detector. The compact, lightweight detector has easy portability, low-power requirements, good energy resolution, high detection efficiency, and low manufacturing cost. Purification, zone refining, and careful monitoring of crystal growth can lead the device as a useful tool for high-performance gamma-ray spectrometer. The instrument with TlBr detector can be a potential instrument for homeland security systems such as handheld radioisotope identifiers and personal radiation detection devices. Eventually, TlBr gamma-ray spectrometers will also be useful in many other applications such as environmental monitoring, medical imaging, nuclear and particle physics research, gamma-ray astronomy, and geophysical exploration. Some of the properties of TlBr as a detector in comparison to the other existing photoconductors are tabulated in Table 8.5.

The atomic number (Z) value of thallium bromide (TlBr) is comparable to that of lead iodide and mercuric iodide (PbI$_2$ and HgI$_2$). As a result, the material is becoming attractive for high gamma ($\gamma$)-ray *stopping power*. Due to its *high stopping power*, TlBr crystals of modest thickness will achieve a full-energy peak detection efficiency equivalent to that substantially thicker high-purity germanium (HPGe) detector. Figure 8.37 shows the attenuation coefficient ($\alpha$) against the

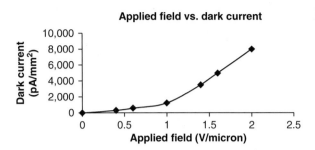

**Fig. 8.38** Dark current plotted as a function of electric field for a typical TlBr crystal

photon energy. The total attenuation coefficients over the energy range (20 keV to >1 MeV) of TlBr supersede silicon (Si), germanium (Ge), cadmium telluride (CdTe), and mercuric iodide (HgI$_2$). The very high photoelectric coefficient is particularly important since it will yield detectors with high photo-peak to Compton ratios. This is important for isotope identification application, which is a number one priority for homeland security department.

From Fig. 8.37, one can extract the number of absorbed photons in a given thickness $x$ of a material used as radiation detector. Conversely, if one knows the number of absorbed photons, one can extract the sensitive depleted layer thickness ($x_d$) of the detector. The ratio of the absorbed number of photons $I_p$ to the initial number of emitted photons $I_0$ can be written mathematically as

$$x_d = (I_p/I_0) = 1 - \exp(-\alpha x) \tag{8.8}$$

or

$$x = -\{\ln(1 - x_d)/\alpha\} \tag{8.9}$$

This can be obtained by using a calibrated gamma-ray source or any radioactive source that can be calibrated by a known fully depleted detector controlled by classical methods.

The measurements of dark current (pA/mm$^2$) under conditions of temporal stability are performed for a polished TlBr crystal at electric fields ranging from 0.2 to 2.0 V/μm, and the results are shown in Fig. 8.38. The dark current is seen to increase with the applied electric field in a non-ohmic, superlinear manner. In the TlBr crystal, dark current can be influenced by either Poole–Frenkel effect or Richardson–Schottky effect, which lower the Coulombic potential barrier in the bulk photoconductor or at the interface of the electrode and photoconductor, respectively. The ionic conductivity in TlBr is also responsible for polarization effect and which deteriorates with elapsed time after applying the electric field [73].

The dark electrical resistivity of TlBr is large enough (>$10^{10}$ Ω-cm, which is estimated by measuring the dark current and applied voltage and noting the sample geometry) to limit dark current (and thereby, shot noise) in TlBr detectors [74].

## 8.4 Halides of Heavy Metals

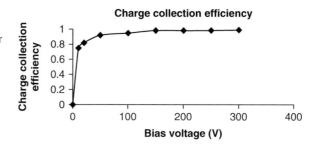

**Fig. 8.39** Charge collection efficiency as a function of bias voltage for a planar TlBr detector. *Solid line* represent the Hetcht equation and the *solid diamonds* represent the measured data

Figure 8.39 shows charge collection efficiency as a function of bias voltage. A commonly used method to determine the mobility-lifetime product is to measure the charge collection efficiency of injected charge carriers as a function of the applied bias voltage [75]. The charge collection efficiency for a photon interaction depth (d in cm) is given by

$$\eta(d) = (Q_{max})/(Q_0) \qquad (8.10)$$

where $(Q_{max})$ is the total charge collected and $(Q_0)$ is the total electron charge (in coulombs) generated by photon interaction. The value of $(Q_{max})$ for both electron and hole can be written mathematically as

$$(Q_{max}) = Q_0(\mu_e \tau_e V/L^2)\{1 - \exp(-(L-d)L/(\mu_e \tau_e V))\}$$
$$+ Q_0(\mu_h \tau_h V/L^2)\{1 - \exp(-(L-d)L/(\mu_h \tau_h V))\} \qquad (8.11)$$

where $\mu_e \tau_e$ and $\mu_h \tau_h$ are the mobility-lifetime product of electrons and holes respectively, $V$ is the applied bias in volts, and $L$ is the thickness in cm of the planar detector (TlBr).

The planar device fabricated from an ultrapure material (purified by zone refining) gives a better mobility-lifetime ($\mu_e \tau_e$) product (~5–7 × $10^{-3}$ cm²/V) compared to THM grown crystal (~$10^{-5}$ cm²/V). Different values of ($\mu\tau$) have been reported by many investigators. Charge trapping causes variations in the charge collection efficiency for different photon interaction depth, which results in spectral peak broadening, asymmetrical peak shapes, and, possibly, reduced detection efficiency. Typically, hole trapping limits the performance of a spectrometer during the detection time of a highly penetrating gamma radiation [76].

### 8.4.3.1 Pulse Height Spectra

Pulse height spectra of TlBr crystal are recorded by irradiating the crystal from an [241]Am source (59.5 keV primary emission) for a total time of 5 h with an amplifier shaping time of 4 µs. The spectra are recorded at an interval of 1 h and are shown in Fig. 8.40. The 59.5-keV peak and a peak due to photons in the 20-keV region are clearly visible in each spectrum.

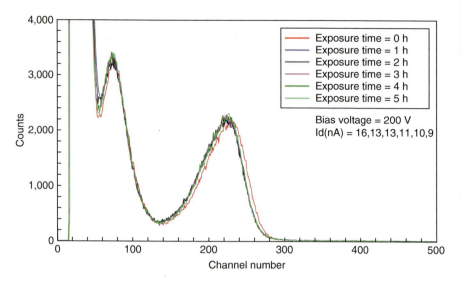

**Fig. 8.40** Pulse height spectra of a TlBr device irradiated continuously for more than 5 h

Since TlBr is an ionic conductor [77, 78], Tl$^+$ ions and bromine ions are accumulated under cathode and anode of TlBr detector, respectively, when bias is applied for a long time (more than 24 h). The accumulated ions induced charges on the surface of the detector and decrease the electric field inside the detector. The phenomenon is known as *polarization*, defined as the time dependence performance of the detector. The *polarization effect* actually degrades the X-ray spectra [79, 80]. Recently, Hitomi and his coworkers have reported that using a better electrode (Au/Tl/TiBr/Tl/Au) can suppress the effect [81].

Figure 8.41 represents the energy spectra of planar and pixelated device when exposed to irradiation from a $^{57}$Co source. Figure 8.41 shows the primary emission around 122 keV in both the devices. However, for both spectra, the 136-keV peak is discernible. The peaks between channels 200 and 400 are due to photoelectric events for which fluorescent photons escape from the detector.

## 8.5 Lanthanide (Ln) Halides

Large numbers of inorganic scintillators are used in gamma-ray detection. The important properties for these scintillation crystals are *very high light output*, *high stopping power*, *fast decay time*, *good linearity*, and *low cost*. The most commonly used scintillator in gamma rays for medical imaging is *thallium* (Tl)-activated sodium iodide (NaI:Tl). Recently, for nuclear security threat, there is a growing interest in employing a large volume of scintillation material for identification of isotopes. To distinguish between different isotopes will require the scintillation

## 8.5 Lanthanide (Ln) Halides

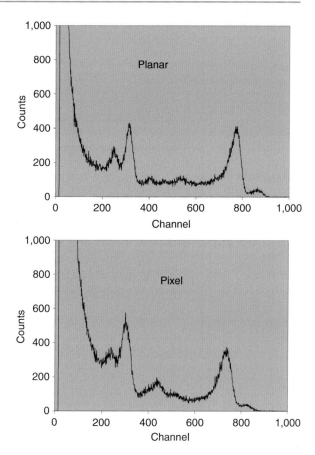

**Fig. 8.41** Energy spectra of a $^{57}$Co source measured for planar (*top*) and pixelated (*bottom*) TlBr devices

material to have excellent energy resolution. *Lanthanide halides* offer sharper resolution and fast response than conventional sodium iodide (NaI:Tl) scintillators. The energy resolution of LaCl$_3$:Ce and LaBr$_3$:Ce has about 3 % FWHM energy resolution for 662-keV gamma rays compared to 5–6 % FWHM for 662-keV gamma rays in NaI:Tl [82–84]. As a matter of fact, Ce$^{3+}$-activated Ln halides' luminosities can reach their theoretical values [85]. However, Ln halides are brittle, hygroscopic, and difficult to produce in large volume. Recently, Ln halide crystals of larger volume with high-energy resolution have been reported [86].

The search for dense ionic crystals with high light output and small decay time with high resolutions is necessary for X-ray detectors in computed tomography (CT), positron emission tomography (PET), and single photon emission tomography (SPECT). The most important part of these applications is the linearity of the energy and flux response of the scintillators. The non-proportional response (nPR) of the lanthanide scintillators is a limiting factor for their applications in medical imaging system. We can present the energy resolution of a scintillator $R$ as a

406     8 Characterization of Radiation Detectors (Scintillators) Used in Nuclear Medicine

function of full width ($\Delta E$) of the full absorption peak height spectrum at FWHM divided by its energy mathematically as [87, 88]

$$\{(\Delta E)/(E)\} = R^2 = R_{lid}^2 + R_{sci}^2 + R_{noise}^2$$
$$\approx (1/LE\eta)\,(1 + \varepsilon) + R_{sci}^2 + R_{noise}^2 \qquad (8.12)$$

where $R_{lid}$ is effect on $R$ due to the photoelectron statistics and the value for PMT readout is 2.4 % FWHM at room temperature, whereas a cooled (250 K) Si-APD is 1.6 % FWHM. $R_{sci}$ is the broadening effect due to inhomogeneities, and its value for the PMT at 300 K is 0.9 %, and for Si-APD, it is 0.9 % FWHM at 250 K. The $R_{noise}$ represents the effect of noise on R, and its value for a PMT at 300K is 0 and for a Si-APD at 250K is 1.6 [86].

By using plane wave pseudopotential density functional theory with Hartree–Fock exact exchange, the trapping of polarons and excitons in undoped $LaCl_3$ and $LaBr_3$ has been reported by Ginhoven et al. [89]. As a matter of fact, the self-trapped excitons (STE) are implicated in energy transport and non-proportionality in scintillation materials. STE luminescence is observed in both $LaCl_3$ and $LaBr_3$ when materials are excited with X-rays. Experimental observations show that both the light output from the scintillation centers and the light output attributed to the STE indicate that these two luminescence mechanisms are in competition; increase from one source is related to a decrease from the other. From our studies, we have found that the phenomena related to STE are most probably related to the excitation energy [86]. Systematic measurements on several crystals of La halide show that these detectors can be used in future planetary missions because of their radiation hardness and gamma-ray resistance [90].

### 8.5.1 Cerium-Activated Lanthanum Chloride ($LaCl_3$:Ce)

Cerium-activated lanthanum chloride ($LaCl_3$:Ce) has been an attractive scintillator because of its high light output and fast principle decay time constant. Figure 8.42 shows that the light output of a single crystal $LaCl_3$ with 10 % cerium (Ce) concentration has the highest light output ~49,000 photons/MeV [91]. The energy spectrum further shows that the higher concentration of the activator shifts the energy peak at a lower energy.

Cerium ($Ce^{3+}$)-activated $LaCl_3$ crystals offer an alternative to thallium-activated sodium iodide (NaI(Tl):Tl) for gamma ($\gamma$)-ray spectrometer applications. As a matter of fact, gamma ray above 200 keV shows energy resolution of $LaCl_3$:Ce superior to NaI:Tl. Figure 8.43 shows the typical percentage energy resolution of $LaCl_3$:Ce and NaI:Tl.

The scintillation process in $LaCl_3$ is controlled by two mechanisms. The fast mechanism is controlled by the trapping of holes and electrons by cerium leading to an independent scintillation process. A thermally activated process involved hole and electron trapping by *self-trapped excitation (STE)* followed by subsequent

## 8.5 Lanthanide (Ln) Halides

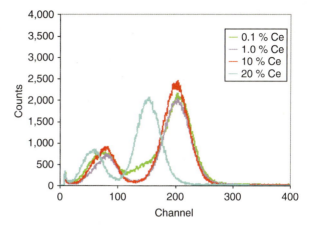

**Fig. 8.42** Energy resolution of LaCl$_3$ crystal with different concentration of the cerium dopant

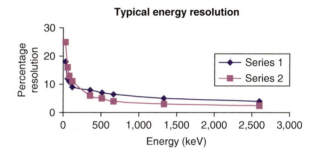

**Fig. 8.43** The energy resolution of LaCl$_3$:Ce (series 2) and NaI:Tl (series 1)

energy transfer to cerium. The photon absorption inside the crystal results in the formation of electron–hole (e–h) pairs. The electron is promoted to the conduction band from valence band (Cl$^-$ → Cl$^0$ + e$^-$), and the final result is the formation of a V$_k$ center (Cl$^-$ → Cl$^0$ → Cl$_2^{-1}$). No crystals grown in the laboratory are perfect, and we can always assume that these crystals have defects like dislocations and twins. Due to these defects, electrons are trapped by an La ion or a Ce ion in LaCl$_3$: Ce crystals. The situation is completely different in alkali halide crystals where electrons are trapped by defects. Due to electron trapping, we can expect a picture like (La$^{+3}$ + e$^-$ → La$^{+2}$) which ultimately gives rise to La$^{+2}$ + Cl$^-$ → LaCl$^+$. Now thermally activated STE process involved transformation of (LaCl$^+$ + Cl$_2^{-1}$ to LaCl$_3$) or (LaCl$^+$ + Cl$_2^{-1}$ to LaCl + Cl$_2$), because LaCl$^+$ complex is not stable. In the STE process which reaction will predominate is dependent on the available energy [92].

The second process is highly dependent on the concentration of the activator and the temperature. The temperature-dependent light yield of LaCl$_3$ crystal irradiated with a gamma-ray ($^{137}$Cs) source with two shaping times of the amplifier is shown in Fig. 8.44.

**Fig. 8.44** Percentage light yield as a function of temperature. Series 1 and series 2 are with 4-μs and 12-μs shaping times, respectively

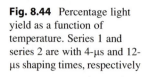

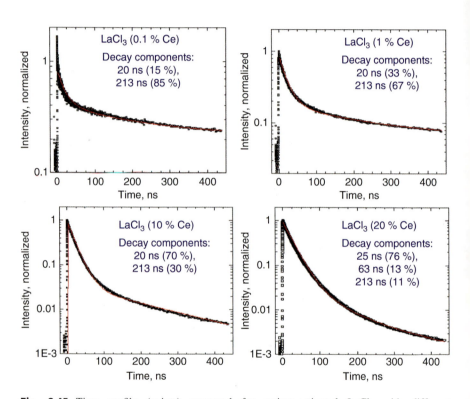

**Fig. 8.45** Time profile (points) measured for cerium-activated LaCl$_3$ with different concentrations of cerium (Ce$^{3+}$) along with multi-exponential fits (*lines*)

Figure 8.44 shows percentage light yield as a function of temperature. Series 1 and series 2 are with 4-μs and 12-μs shaping times, respectively. For 4-μs shaping time, the variation is almost linear in the temperature range between −25 and 75 °C. However, the plotted data for 12-μs shaping time shows an increase in light yield in a hyperbolic fashion with a decreasing trend after 0 °C. The study reveals that at high temperatures (0–100 °C), cerium-activated LaCl$_3$ will be a good detector with increased light yield.

## 8.5 Lanthanide (Ln) Halides

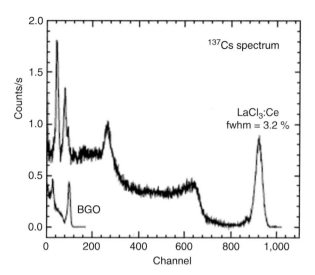

**Fig. 8.46** Energy spectra of bismuth germanate (BGO) and cerium (10 %)-activated LaCl$_3$

The Fig. 8.45 shows time profile (points) measurements for cerium-doped LaCl$_3$ crystals doped with different percentage of cerium (Ce$^{3+}$) concentrations. The principal decay constant is higher when concentration of the activator is higher in the host lattice. The fast decay constants (20 and 25 ns) is probably due to direct electron hole capture on Ce$^{3+}$ site, since the observed decay time constant is the characteristics for Ce$^{3+}$ luminescence [91].

Figure 8.46 shows the energy spectrum of cerium (10 %)-activated LaCl$_3$ crystal. For comparison, the energy spectrum of a BGO crystal is presented in the same plot. The 662-keV gamma ($\gamma$)-ray spectrum of LaCl$_3$:Ce has been recorded with a shaping time of 4 μs. The FWHM calculated for 662-keV peak is calculated and found to be approximately 3.2 % at room temperature. The lanthanum (La) and chlorine (Cl) escape peaks are also visible in the spectrum [91].

### 8.5.2 Cerium-Activated Lanthanum Bromide (LaBr$_3$:Ce)

The other member of lanthanide halide is cerium-activated lanthanum bromide (LaBr:Ce). The burgeoning interest in the subject of Ce$^{3+}$-activated luminescence in lanthanide halides (chloride and bromide) is due to their recent breakthrough in new scintillators with high light output, fast radiative decay time, and high-energy resolution. Low melting temperature of the material is an added advantage for growing single crystal in a cost-effective way. Recently, LaBr$_3$ grown by Bridgman method in our laboratory has shown almost 80,000 photons per MeV [82]. The Pr$^{3+}$-activated LaBr$_3$ crystals have shown broadband spectrum centered at 250 nm and

**Fig. 8.47** The radioluminescence spectrum of a LaBr$_3$:0.5 mol% Ce crystal

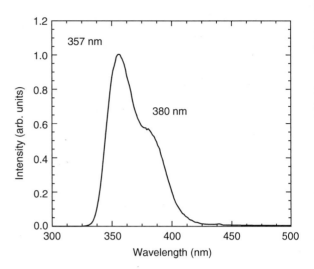

has been ascertained as the peak due to the charge transfer transition associated with Br$^-$Pr$^{3+}$ [93].

The cerium-activated (Ce$^{3+}$) lanthanum bromide (LaBr$_3$) crystal 2 cm$^3$ with 0.5 % (by mole) cerium concentration is grown by Bridgman method [86]. The polished crystal is encapsulated, and emission spectrum is taken under X-ray excitation using Phillips X-ray tube with a copper target and operating at 40 kVp and 20 mA. The emitted light is passed through a McPherson monochromator and detected by a Hamamatsu R2059 photomultiplier tube with a quartz window. The recorded data are plotted and presented in Fig. 8.47.

Figure 8.47 shows the radioluminescence spectrum of a LaBr$_3$:0.5 mol% Ce crystal. There are two peaks observed in the spectrum: the well-defined peak around 360 nm and a small peak-like hump on the shoulder at 380 nm. Both are the characteristics of Ce$^{3+}$ luminescence.

We have characterized the time profiles of the crystal (activated with 0.5 mol% Ce$^{3+}$). During measurements, the sample is coupled to a PMT and irradiated with 511-keV gamma rays ($^{22}$Na source). Nanosecond time profiles are measured by using a standard single photon counting method [94]. The results are plotted and presented in Fig. 8.48 along with an exponential fit with decay time constant of 26 ns and rise time constant of 7 ns.

Figure 8.49 shows the $^{137}$Cs gamma-ray spectrum together with a BGO spectrum for comparison. The light output of the BGO crystal is estimated by the single-electron method and compared with that of the LaBr$_3$:Ce. The cerium-activated LaBr$_3$ crystal gives almost 16 times higher signal than BGO crystal.

The energy resolution for gamma ray with 662-keV energy is measured with an avalanche photo diode (APD) cooled at 250 K. The measured data are shown in

## 8.5 Lanthanide (Ln) Halides

**Fig. 8.48** The decay time constant of LaBr$_3$:0.5 % scintillator

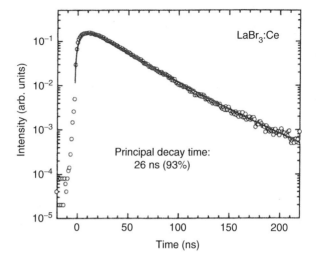

**Fig. 8.49** The $^{137}$Cs gamma-ray spectrum together with a BGO spectrum for comparison

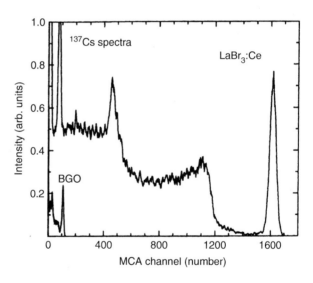

Fig. 8.49. The energy resolution (FWHM) for $^{57}$Co source is calculated approximately 5.0 % and for $^{137}$Cs source is measured approximately 2.4 % (Fig. 8.50a, b). The energy resolution is very high with APD, and it appears that with proper purification of the material and device fabrication, cerium-activated LaBr$_3$ can be a promising scintillator for nuclear nonproliferation, medical imaging, nuclear and space physics research, and nondestructive studies.

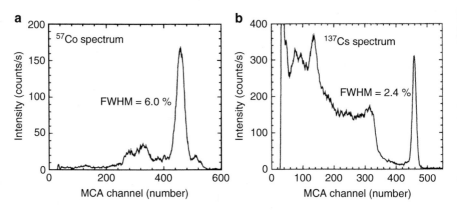

**Fig. 8.50** (a) The $^{57}$Co spectrum and (b) $^{137}$Cs spectrum of LaBr$_3$ coupled to a Si-avalanche photodiode (APD)

### 8.5.3 Rubidium Gadolinium Bromide (RGB) (Cerium-Doped) RbGd$_2$Br$_7$:Ce

The cerium-activated rubidium gadolinium bromide (RGB:Ce) crystals are finding a place in the scintillator book because of its high light output ($\approx$56,000 photons/MeV), fast principal decay constant (43 ns), very good linearity in energy response over a wide range, and decent gamma-ray detection efficiency. Moreover, presence of gadolinium (Gd) has made the device for thermal neutron detector. One of the isotope of Gd, $^{157}$Gd, has a very high cross section of absorption for thermal neutrons (25500b), and the Q-value for thermal neutron with $^{157}$Gd is also very high [95].

The RGB:Ce$^{3+}$ crystals are grown in our laboratory by Bridgman method by melting RbBr, GdBr$_3$, and CeBr$_3$ in proper stoichiometry. The concentration of cerium is varied between 0.1 and 10 %. The emission spectrum of RGB crystal with 10 % Ce concentration is measured by exciting the sample from a Philips X-ray tube having a copper target, with power settings of 30 kVp and 15 mA. The scintillation light is passed through a monochromator and detected by a PMT with a quartz window. The collected data are plotted and presented in Fig. 8.51. The emission peak is observed around 430 nm.

The gamma ($\gamma$)-ray energy resolution of our lab-grown cerium-activated RGB crystal (10 %) is measured by irradiating the crystal in 662-($\gamma$) ray. The crystal is coupled to a PMT with a multialkali S-20 photocathode. The signal is recorded with a preamplifier (Canberra#2005) with a shaping time of 0.2 µs. The pulse height data from $^{137}$Cs source is recorded and plotted in Fig. 8.52. The energy resolution for 662-keV peak is found to be about 5.2 % (FWHM) at room temperature.

The coincidence timing resolution of the cerium-activated RGB crystal is measured at LBNL. The measurement involves irradiation of a BaF$_2$ and an RGB crystal in a coincidence setup with a 511-keV positron annihilation ($\gamma$) ray pair (emitted by a $^{22}$Na source). The time difference between the two channels with

## 8.5 Lanthanide (Ln) Halides

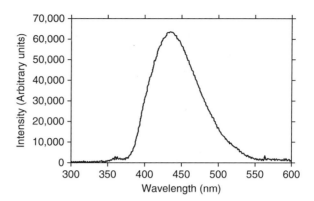

**Fig. 8.51** The emission spectrum of RbGd$_2$Br$_7$:Ce (10 %)

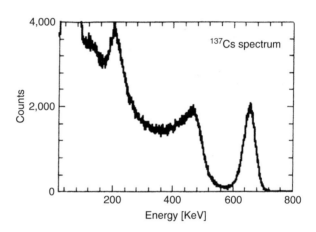

**Fig. 8.52** A $^{137}$Cs spectrum recorded with a cerium-activated RGB crystal

fixed delay between them is observed in a timing spectrum and recorded in an MCA (Fig. 8.53).

The measured coincidence timing resolution is approximately 330 ps (FWHM). The timing resolution with the BaF$_2$ crystal is recorded to be 273 ps (FWHM) confirming the coincidence timing resolution for RGB to be 267 ps [95].

Since the RGB crystal has gadolinium (Gd) as one of the host material, it can be used as a thermal neutron detector. Two isotopes of Gd, $^{157}$Gd and $^{155}$Gd, have very high thermal neutron absorption cross sections of about 2.55 × 105b and 6.1 × 104b, respectively. About 80 % of the neutron capture events in Gd occur in 157Gd and about 18 % occur in 155Gd (2555000b). Therefore, it is expected that an RGB crystal of 200 μm thick will be able to stop 90 % of the thermal neutrons, while a 500-μm-thick crystal will have the stopping efficiency of 100 %.

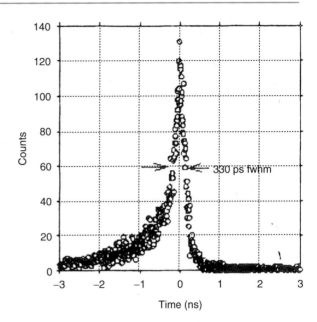

**Fig. 8.53** The coincidence timing resolution with RGB and BaF$_2$ crystals

### 8.5.4 Cerium-Doped Lutetium Iodide (LuI$_3$:Ce)

The general requirements for the scintillation crystals that are used as radiation detectors in medical imaging, high-energy physics, nondestructive testing, nuclear treaty verification, and geological exploration [96, 97] are having high Z (greater than 40 to have higher absorption value), high resistivity (to achieve low dark current), and high mobility-lifetime product (should be better than 10–3 to ensure good carrier transport and spectral performance). The cerium-activated lutetium iodide (LuI$_3$:Ce) met almost all the requirements and as a detector is comparable to the other lanthanide halides we have discussed earlier (50,000 ph/Mev compared to 60,000 ph/MeV in LaBr$_3$:Ce).

Cerium-activated (varied concentration) hexagonal crystals of LuI$_3$ 1 cm$^3$ are grown in our laboratory by two-zone Bridgman method. The crystal growth in detail has been discussed elsewhere in the book. The emission spectrum is performed by exciting the crystal by X-ray source (Philips) with power settings of 40 kV and 20 mA. McPherson monochromator and a Hamamatsu R2059 photomultiplier have been used during measurements [98]. The data for the emission measurement are collected and plotted in Fig. 8.54. The peak emission wavelength for the particular LuI$_3$:Ce is around 474 nm. Smaller peaks around 577 and 600 nm are assumed to be due to the impurities (e.g., Tb$^{3+}$).

The time profiles of a 5 % cerium-activated crystal of LuI$_3$ is shown in Fig. 8.55. The principal decay constant which covers about 70 % of the integrated light is derived from the fit and found to be ~23 ns. The second component of the decay is

**Fig. 8.54** The emission spectrum of LuI$_3$:Ce with 0.5 % Ce$^{3+}$

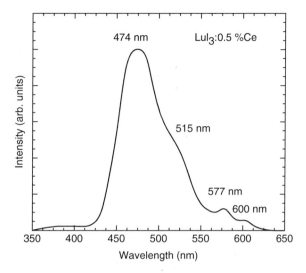

**Fig. 8.55** Time profiles of LuI$_3$:Ce crystal (5 % Ce) scintillation (points). The *solid line* is the exponential fit of the experimental data

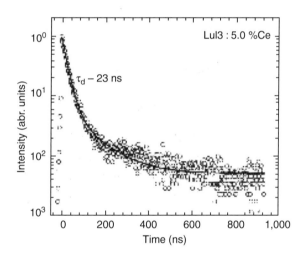

found to be 120 ns. When the concentration of cerium is reduced to 0.5 %, the principal decay constant becomes 31 ns.

Gamma-ray spectroscopy measurement is conducted using LuI$_3$:Ce crystal and irradiating the crystal with 662-keV gamma source ($^{137}$Cs). The procedure is the same as described elsewhere. The energy resolution of the 662-keV peak is estimated from a double Gaussian fit to be about 11 % (FWHM) at room temperature. Figure 8.56 shows the spectrum of LuI$_3$:Ce collected by using a $^{137}$Cs source.

The coincidence timing experiment is performed together with a BaF$_2$ crystal. The signal from each detector is processed using two channels of a Tennelec

**Fig. 8.56** The spectrum of LuI$_3$:Ce collected by using a $^{137}$Cs source

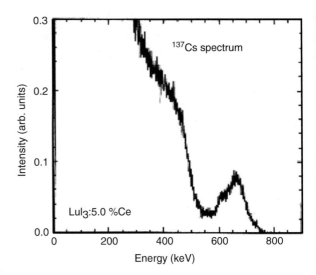

**Fig. 8.57** The coincidence timing resolution plot for LuI$_3$:Ce

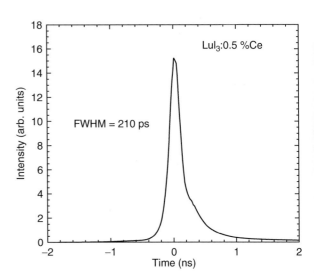

TC-454 CFD that has been modified for use with fast rise time PMTs. Data are accumulated until coincidence timing distribution has more than 10,000 counts in the maximum bean. Figure 8.57 shows the collected data and the plot for the crystal with 0.5 % Ce$^{3+}$ concentration. The coincidence timing resolution measured from the plot is 210 ps (FWHM).

Figure 8.58 shows the proportionality of response as a function of gamma-ray energy. The non-proportionality (nPR) of a scintillator at $E_x$ is defined as the

## 8.6 Cerium-Activated Lutetium Oxyorthosilicate (LSO, Lu₂SiO₅)

**Fig. 8.58** Proportionality of response as a function of gamma-ray energy for LuI₃:0.5 % Ce crystal

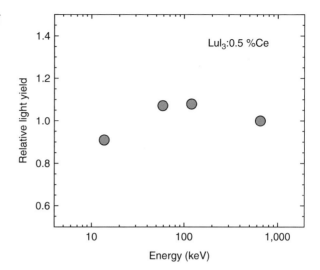

number of $N^{PMT}_{phe}$/MeV observed at energy $E_x$ divided by the number $N^{PMT}_{phe}$/MeV observed at 662-keV energy [87]. Thus, non-proportionality (as a function of energy) in light yield is one of the important reasons for degradation of energy resolution in some established scintillators such as NaI:Tl and LSO:Ce [99]. The non-proportionality calculated from the plot for our LuI₃ crystal with 0.5 % Ce³⁺ shows only 10 % compared to 35 % for LSO:Ce and 20 % for NaI:Tl [100].

## 8.6 Cerium-Activated Lutetium Oxyorthosilicate (LSO, Lu₂SiO₅)

The discovery of calcium fluoride (CaF) and BaF₂ scintillators and the shortcomings of these two detectors lead to the discovery of a high-density (7.4 g/cm³) inorganic cerium (Ce₃₊)-activated lutetium oxyorthosilicate (LSO:Ce). Compared to the other cerium-activated scintillators like gadolinium oxyorthosilicate (GSO:Ce) and yttrium oxyorthosilicate (YSO:Ce), LSO:Ce could be considered as a good compromise between fast response (40 ns) and high light yield (≥26,000 photons/MeV). A significantly lower stopping power of cerium-activated lanthanide halides (LnX₃:Ce) compared to cerium-activated Lu₂SiO₅ (LSO:Ce) suggested returning again to study the fast timing with LSO to learn the present state of the art and possible further improvements of the crystal and the photomultipliers [101]. The high-resolution research tomography (HRRT) is a three-dimensional (3-D) brain tomography, which employed LSO crystal as a detector [102]. The unfortunate side of the lutetium-based scintillators is the cost of the material. Moreover, the melting point of

**Table 8.6** Characteristics of cerium-activated LSO powder

| Parameters | LSO:Ce |
| --- | --- |
| $Ce^{3+}$ content | 0.22 % |
| Light output (ph/MeV) | 27,300 $\pm$1,400 |
| Lifetime $\tau$ (ns) | 47.2 |
| Density (g/cm$^3$) | 7.4 |
| $\lambda$p (nm) peak emission | 420 |
| Refractive index | 1.82 |
| Energy resolution for $^{22}$Na (511 keV) | 11 % |
| Time resolution $^{60}$Co (ps) | 160 |
| Background | $^{176}$Lu |

$Lu_2O_3$, one of the starting ingredients for making single crystal of LSO, is very high (2,487 °C). Thus, to grow single crystal, one has to adopt Czochralski technology.

## 8.6.1 Synthesized LSO

In order to make the processing of LSO more simple and cost-effective, we have successfully synthesized oxyorthosilicate ($Lu_{2\,(1-x)}\,Ce_{2x}\,(SiO_4)$ O or LSO) powder. The advantage of the method over single crystal growth by Czochralski method is that it is cost-effective, and the powder can be used either to fabricate a large screen by screen printing, adopting thick film technology [65]. The fabrication method has been discussed in a separate chapter of the book. Table 8.6 shows the characteristics of cerium-activated LSO powder.

The emission spectra of the synthesized powder and that of a single crystal powder are excited with X-rays (30 kVp, / 15 mA), and the scintillation light is allowed to pass through Jarrell-Ash 82–410 monochromator with 600 grooves/mm grating blazed at 300 nm. The monochromator is equipped with Hamamatsu C31034/76 dry-ice-cooled photomultiplier tube (PMT) with quartz window. Single photon counting is recorded from the signal coming out of the dry-ice-cooled PMT. The wavelength dependence light yield is recorded and presented in Fig. 8.59.

Figure 8.59 is a plot of the emission intensity versus wavelength of the ceramic fabricated from the synthesized powder. To testify the emission of the ceramic with that of a single crystal, the emission spectrum of the ground powder of the crystal is also plotted in the same plot. The evidence of trapping centers for minority carriers is determined usually by mobility or lifetime measurement in order to determine the quality of the LSO powder (synthesized 33tgb$_2$ and ground single crystal). We will identify the experiment as *induction effect measurements*.

The time versus emission intensity of the X-ray (~427-nm wavelength and X-ray energy 30 kV/15 mA)-irradiated single crystal and the ceramic made out of the synthesized powder is shown in Fig. 8.60. As the X-ray excitation begins, the carrier starts filling the traps and therefore, is not available for luminescence. After the traps are filled, the luminescence intensity reaches a steady state. Hence, at room

## 8.6 Cerium-Activated Lutetium Oxyorthosilicate (LSO, Lu$_2$SiO$_5$)

**Fig. 8.59** X-ray excited emission spectra of lutetium oxyorthosilicate (LSO:Ce)

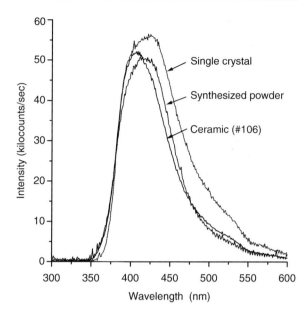

temperature where the carriers are not thermally excited out of traps, the area between the growth curve and the steady-state luminescence can be correlated with trap density.

Consider a step excitation which pumps carriers at the rate $I$ into a volume $V$ of active region in the LSO, the traps of which are initially empty. If the luminescence growth curve is presented by a function of $L(t)$, the trap density is [103]

$$N(t) = [(1/(V\eta)] \int_0^\infty [L_\infty - L(t)] \, dt \qquad (8.13)$$

where $\eta$ is the quantum efficiency for radiative recombination; $L(t)$ is expressed in photons/s.

Figure 8.60 shows time versus intensity curve for LSO samples at room temperature. As the area between $[L_\infty - L(t)]$ is proportional to trap density, from the above curves, one can notice that the poor crystal has more traps compared to good crystal and the ceramic fabricated from the synthesized LSO powder.

In our study [86], we have detected that the non-proportionality measured over the range 60–127 keV for LaBr$_3$:Ce is 6 %, whereas with LSO:Ce crystal over the same range of energy the non-proportionality is 35 % [104]. The study of *induction effect measurements* reveals that the trapping–de-trapping might be one of the causes of non-proportional response of the crystal besides the contribution from the irradiation source and PMT [105].

**Fig. 8.60** Time versus intensity curve for LSO samples at room temperature

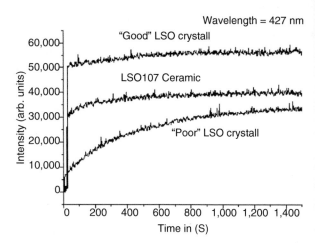

## 8.7 Complex Oxides with High Atomic Number

The technology, the science, and the advanced research have developed the complex oxide scintillating materials with high atomic number like cerium-activated gadolinium oxyorthosilicate (GSO,Gd$_2$ SiO$_5$:Ce$^{3+}$), bismuth germanate (BGO, Bi$_4$Ge$_3$O$_{12}$), cadmium tungstate (CWO, CdWO$_4$), and lead tungstate (PWO, PbWO$_4$). Scintillators based on these crystals have good energy resolution, reasonable light output, and high detection efficiency. Moreover, these crystals are not hygroscopic, but having high radiation stability, and high mechanical strength. The X-ray stopping power (attenuation coefficient) of the scintillating materials is expected to be high because of their high densities ($\rho$) and the effective atomic numbers. Considering only the interaction through photoelectric effect, the X-ray stopping power is considered to be proportional to $(\rho Z_{eff})^{3-4}$ [106].

### 8.7.1 Cerium-Activated Gadolinium Oxyorthosilicate (GSO, Gd$_2$SiO$_5$:Ce$^{3+}$)

GSO and LSO crystals have different physical and scintillation properties. The differences in properties come not only from different rare earth atoms but also from the difference in crystal structure [107]. The cerium-activated GSO is so far the best scintillation material for detection of gamma-ray radiation in precision electromagnetic calorimeters, in positron emission tomography (PET), in gamma camera system, and in well logging [108]. Compared to BGO, it has lower effective atomic number and density, but it has shorter decay time, higher light output, and better energy resolution. The peak in fluorescence spectrum is 420 nm, and the light yield is maximum when the cerium concentration is ~0.5 mol% and it is weakly dependent on temperature.

## 8.7 Complex Oxides with High Atomic Number

**Table 8.7** The comparative analysis of the properties of the complex oxides with high atomic numbers

| Properties | GSO (Ce)$Gd_2SiO_5$ | BGO($Bi_4Ge_3O_{12}$) | $CdWO_4$ | $PbWO_4$ |
|---|---|---|---|---|
| Effective atomic number | 59 | 74 | 66 | 73 |
| Density (g/cm$^3$) | 6.71 | 7.13 | 7.9 | 8.2 |
| Decay constant (ns) | 60/600 | 60/300 | 1.7, 10, 38 | <10 |
| Light output | 10,000 ph/MeV | 8,600 ph/MeV | 15,000 ph/MeV | 490 ph/MeV |
| Energy resolution % | 7.8 ($^{137}$Cs) | 9.5 ($^{137}$Cs) | 9.0 | <3 |
| Radiation hardness | 10$^9$ | 10$^{5-7}$ | | |
| Hygroscopic | No | No | No | No |
| Emission max. (nm) | 420 | 480 | 480 and 570 | 410–420 |
| Attenuation length (cm) | 1.50 | 1.11 | 1.21 | 0.96 |

## 8.7.2 Bismuth Germanate (BGO, $Bi_4Ge_3O_{12}$)

BGO has higher density and effective atomic number than GSO. The peak luminescence spectrum is ~480 nm. Thus, BGO can be used with PMT and also with photodiodes. Moszynski et al. measured few BGO crystals 9 mm in diameter and 4 mm thick cooled to liquid nitrogen temperature (LN$_2$). The crystals have been coupled to large-area avalanche photodiodes. The measured average light output and the energy resolution of the BGO crystals at 662 keV with $^{137}$Cs source are found to be ~14,000 ± 300 electron–hole pairs and 6.5 % ± 0.2 %, respectively. The escape peak (KX) observed in the energy spectrum at liquid nitrogen (LN$_2$) temperature is mainly created by electrons. They further observed that the non-proportionality of light output and energy resolution is not dependent on temperature [109–111]. However, the scintillation decay time is seen to be strongly dependent on temperature.

Because of high effective atomic number, pulse amplitude spectra of BGO are characterized not only by a high photo-peak but also by a very large peak to valley ratio. The short radiation length, density, high effective $Z$ value, and good energy resolution have made BGO an efficient scintillator in particle physics, geophysical research, PET, anti-Compton spectrometers, and electromagnetic measurements and as optical fiber sensor [112]. A comparative analysis of the properties of these crystals is tabulated in Table 8.7.

## 8.7.3 Cadmium Tungstate (CWO, $CdWO_4$)

According to some experts, new high $Z$ scintillators even these are not having high light output are needed to improve current border monitoring instrumentation utilizing plastic detectors [113]. As a matter of fact, one of the most sensitive 2β experiments has been performed with the help of enriched $CdWO_4$ crystal scintillators [114].

The emission spectrum of $CdWO_4$ (CWO) under optical excitation shows two bands, the blue band at 480 nm and yellow band at 570 nm. The decay time has two components: 1.1 and 14.5 µs. For this reason, it can be used in slow spectrometers. The light is very weakly dependent on temperatures. A good energy resolution of 6.8 % has been reported with [137]Cs source for a 2.5 cm in diameter and 1.2-cm-thick CWO crystal. Cadmium tungstate can be a good candidate for X-ray detection in computer tomography (CT) because it has a very low afterglow [115]. The non-proportionality characteristic in light output and energy resolution has been attributed due to the presence of the heavy metal oxides. The $CdWO_4$ (CWO) crystal as scintillator has many applications like computed tomography (CT), dosimetry, oil logging, and nuclear spectroscopy [116].

### 8.7.4 Lead Tungstate (PWO, $PbWO_4$)

Lead tungstate (PWO) crystal is used as radiation detectors for high-energy physics. It shows the emission peak around 420 nm. The luminescence decay constant (85 % intensity) is less than 10 ns and light yield is 490 ph/MeV. Temperature dependence of light yield of PWO scintillator has been found to be around 1.9. The energy resolution of EM-shower deposited by electrons in a $3 \times 3$ crystal matrix amounts to $\sigma/E = 2.39\,\%/\sqrt{E} + 0.20\,\%$ ($E$ is in GeV). The time resolution of $\sigma \leq 130$ ps allows photon/particle discrimination via time of flight measurement [117]. The PWO crystals have proven to be appropriate scintillators for electromagnetic calorimetry at medium energies below 1-GeV energy down to a few tenths of MeV [118]. The studies on PWO crystals show that scintillation mechanism of these crystals is not affected by radiation and the loss of light output is only because of radiation induced absorption [119]. The high melting point of these materials (GSO, BGO, $CdWO_4$, and $PWO_4$) and the processing to grow single crystals cause complications (due to formation of different types of defects that are being incorporated during crystal growth). As a result, the overall yield is low and the cost of growing quality crystals for radiation detection and measurements becomes high.

## 8.8 Cadmium Telluride (CdTe) Crystals

Cadmium telluride (CdTe) is another member of the scintillator family and has been the object of recent studies because of its potential use as a material at room temperature gamma ($\gamma$)-ray detectors, Gunn-effect devices, electro-optical modulators, and infrared windows [120–123]. The experimental results reveal that the trapping–de-trapping times of electrons are dependent on the electric field, and it ultimately affects the charge collection efficiency of the detector. Therefore, an understanding of trapping–de-trapping phenomena is important for the users to obtain maximum energy resolution from the device [124].

## 8.8 Cadmium Telluride (CdTe) Crystals

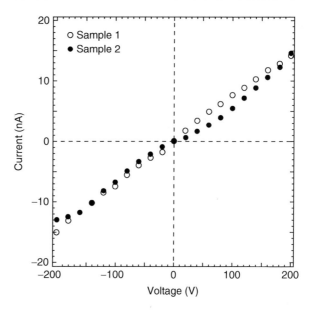

**Fig. 8.61** The current–voltage relation of two CdTe crystals

The charge transport properties of CdTe is very much dependent on the purification and processing of the material. As a matter of fact, the electrical properties of CdTe depend on the two native defects such as cadmium interstitial as divalent donor and cadmium vacancy as divalent acceptor [125].

In purified CdTe crystals, we can relate the resistivity with electron–hole (n, p) concentrations, the electric charge ($q$), and the mobility of electrons ($\mu_e$) and holes ($\mu_h$) as

$$1/\rho = nq\mu_e + pq\mu_h \tag{8.14}$$

The typical resistivity of a CdTe crystal grown in chlorine (Cl)-compensated tellurium (Tl)-rich solution is measured from current–voltage characteristics (Fig. 8.61) and is found to be ~$2 \times 10^9$ $\Omega$-cm. However, several investigators have reported different values of the resistivity of the CdTe crystals from time to time because of different processing [126, 127].

The dark current is found to increase with the bias voltage due surface leakage. Different investigators have performed a large variety of surface treatments in order to improve dark current. Schottky CdTe diode detectors suffer from a polarization effect, which is characterized by degradation of spectral properties over time following exposure to high bias voltage. This is attributed to charge accumulation at deep acceptor level [128]. As a matter of fact, trapping center responsible for polarization effect is located not in the middle of the band gap but closer to conduction band [129].

**Fig. 8.62** $^{57}$Co energy spectrum measured with CdTe single crystal (Courtesy of IEEE)

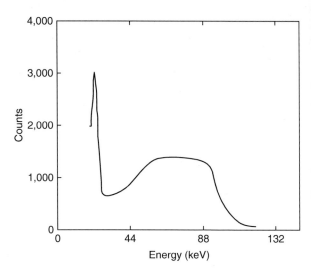

Figure 8.62 shows the energy-dependent intensity spectrum of a CdTe crystal with $^{57}$Co source. The absence of radiation peak in the highly resistive (~2 × 10$^9$ Ω-cm) CdTe crystals with the $^{57}$Co source is the evidence of charge carrier trapping–de-trapping phenomenon [130]. Since the ionization rates associated with the deep level are much slower than the deionization rates, a very slow increase in the number of holes trapped near the positive electrode results, but the associated de-trapping is quite fast [131]. Multiple zone refining coupled with distillation of the initial compound can produce a very high quality detector with negligible polarization [132].

The effect of trapping and de-trapping of carriers is identified as the possible cause of polarization in semiconductor detectors when these detectors are irradiated with X-ray and gamma ray. Because of the polarization effect, the charge collection efficiency is altered correspondingly through the changes in the average drift length of carriers or the effectiveness of moving carriers away from the surface recombination centers.

Figure 8.63 shows the energy spectrum of a lapped and chlorine-compensated CdTe crystal using $^{57}$Co source. Here, we can see the 122-keV peak which is absent in Fig. 8.62 due to more trapping–de-trapping phenomenon. As a result, polarization effect is minimum here with the lapped and chlorine-compensated CdTe crystal. Polarization effect is very severe in many semiconductor detectors, and the detectors exceeding some critical resistivity exhibited a progressive reduction in both signal amplitude and counting rate as a function of time after the detector is biased [133, 134].

The ionization by X-ray induces defects that trap electrons or holes, and the result is the reduction in the charge collection efficiency and energy resolution. The

## 8.8 Cadmium Telluride (CdTe) Crystals

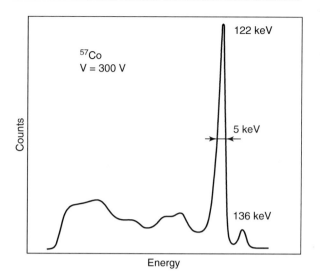

**Fig. 8.63** The energy spectrum of chlorine-compensated CdTe crystal (Reproduced with permission from IEEE)

capturing of the electrons or holes or both can induce distortion in the electric field distribution [135, 136]. Localized charge buildup can produce internally depleted zones in which no collection occurs. Because trap filling after ionization is not that of the equilibrium state, the return to equilibrium results in temporal dependence of the detector response and of dark current [137–139]. Therefore, the decay time of the signal of a CdTe crystal will depend on the trapping–de-trapping consequences.

Flat-panel image sensors with CdTe crystals have been studied by many investigators. Although selenium (Se) is one of the most attractive materials, yet it can not compete with CdTe when $W_{en}$ values (a-Se = 50 ev, CdTe = 5) are considered. At the same time, the $\mu\tau$ product (a-Se = $1 \times 10^{-6}$, CdTe = $4 \times 10^{-3}$ cm$^2$/V) which is the measure of the efficiency of a detector is higher than selenium. The difficulty with CdTe crystal is that it has high temperature of crystallization (above 500 °C) and its deposition over flat panel is challenging [140].

Figure 8.64 shows the linear attenuation coefficient of several nuclear scintillating materials. As can be seen from the figure, proton stopping power of the materials depends on the effective Z value of the material. CdTe has a wide band gap, which suppress the thermal excitation of carriers and hence permits CdTe detectors to have low noise performance at room temperature. However, the band gap of CdTe is less than TlBr and PbI$_2$. Accordingly, the linear attenuation coefficient of CdTe secured third place in the list. Table 8.8 shows some of the properties of the compound semiconductors that are promising radiation detectors.

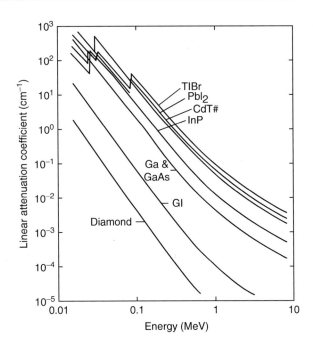

**Fig. 8.64** The linear attenuation coefficient of several nuclear scintillating materials

**Table 8.8** Some of the properties of compound semiconductors

| Material | Atomic number | Density (g/cm$^3$) | Band gap (eV) | $E_{pair}$ (eV) | Resistivity ($\Omega$-cm) | $\mu\tau$ (e) product (cm$^2$/V) | $\mu\tau$ (h) product (cm$^2$/V) |
|---|---|---|---|---|---|---|---|
| PbI$_2$ | 82, 53 | 6.2 | 2.32 | 4.9 | 10$^{12}$ | 8 × 10$^{-6}$ | ~10$^{-5}$ |
| HgI$_2$ | 80, 53 | 6.4 | 2.13 | 4.2 | 10$^{13}$ | 10$^{-4}$ | 4 × 10$^{-5}$ |
| TlBr | 81, 35 | 7.56 | 2.68 | 6.5 | >10$^{10}$ | 10$^{-3}$ | 4 × 10$^{-5}$ |
| CdTe | 48, 52 | 6.2 | 1.44 | 4.43 | 10$^9$ | 3 × 10$^{-3}$ | 2 × 10$^{-4}$ |
| CdZnTe | 48, 30, 52 | 6.0 | 1.5–2.2 | ~5.0 | >10$^{10}$ | 3 × 10$^{-3}$ | 10$^{-5}$ |

## 8.9 Cadmium Zinc Telluride (CZT)

The small band gap, requirement of cryogenically cooled temperature, bulky detector size has made germanium detector applications in the areas of medical imaging and therapy, homeland security, high-energy physics, and industrial applications in spite of its high-energy resolution property. Thus, detectors based on wide band gap compound semiconducting materials with high densities and large effective atomic numbers are logically choices to fulfill the requirements of the technological needs [141]. Cadmium zinc telluride (CdZnTe) as a compound semiconducting material is well studied and has been a choice in the detector community [142].

## 8.9 Cadmium Zinc Telluride (CZT)

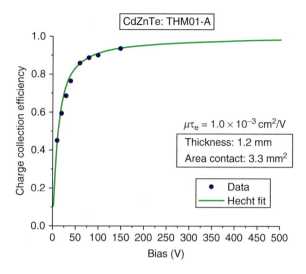

**Fig. 8.65** Mobility-lifetime ($\mu\tau$) product of electrons on a $Cd_{0.85}Zn_{0.15}Te$ detector (Courtesy of RMD)

For producing high-resistivity material of CZT, it has been experimentally found out that 10 % of zinc (Zn) in CZT matrix gives the best result. The A center ionization energy increases slightly with increasing Zn concentration in CZT and reaches a typical value of ZnTe which is between (Ev + 0.14) eV and (Ev + 0.19) eV. However, at this concentration, CZT is unable to identify the peaks below 122 keV with $^{57}$Cs source. However, CZT crystals doped with 15–20 % of Zn can show these peaks (Fig. 8.66). The role of zinc-induced compound is to reduce the density of tellurium (Te) antisites ($Te_{Cd}$), to increase the density of ($V_{Cd}$), and to enhance the diffusion rate of $V_{Cd}$. It has been found that the increased amount of Zn doping shifted the transmission spectra toward longer wavelengths [143].

Mobility-lifetime product ($\mu \times \tau$) is considered as one of the fundamental figure of merit for a semiconductor X-ray or gamma-ray spectrometer. The electron mobility ($\mu_e = 1{,}000–800 \text{ cm}^2/\text{V s}$) and lifetime ($\tau_e = 1–5 \times 10^{-6}$ s) are relatively high but the hole mobility ($\mu_h = 80–30 \text{ cm}^2/\text{V s}$) and lifetime ($\tau_h = 10^{-6}$ to $10^{-7}$ s) are typically very low in CZT [144].

The product can be obtained by measuring the photo-peak shift as a function of applied bias [145]. The detector is excited by using radiation (e.g., alpha particle) near the surface region, and the induced pulse can be fit to the Hecht equation to determine the mobility-lifetime product of each carrier [146]. We perform the mobility-lifetime measurement on the crystals that are grown in our laboratory by traveling heater method (THM), and the mobility-lifetime product ($\mu\tau_e$) is found to be equal to $1 \times 10^{-3}$ cm$^2$/V (Fig. 8.65). Recent result reported by Amman et al. [141, 147] shows the value of mobility-lifetime product for electrons is $1.6–1.0 \times 10^{-3}$ cm$^2$/V.

It has been found that better THM processing and consecutive zone refining have resulted in better crystals with high values of $\mu\tau$ product. Larger values of mobility-lifetime product together with the continued increases in the size of the grown

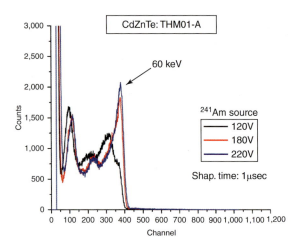

**Fig. 8.66** The gamma-ray spectra with $^{241}$Am source with different bias voltages

crystal and delectability of CZT spectrometers will make CZT detector as one of the invaluable member in radiation measurements. Figure 8.70 shows the gamma-ray spectra with $^{241}$Am source with different bias voltages.

The effects of increased fluences of 700-keV protons on the batch of CZT detectors have been investigated, and the results relative to $^{241}$Am (main photo-peak at 60.0 keV) are reported in Fig. 8.66. The spectrum also shows peaks at 14.0, 18.0, and 26.50 keV and escape peaks relating to the 60.00 main peak. Different spectra are recorded with bias voltages 120, 180, and 220 V. It is evident from the 241-Am spectra that the charge-collecting properties of the detectors deteriorate after the last delivered fluence of $5 \times 10^{12}$ p/cm$^2$. The gamma-ray spectra evolution with fluence is similar for LE and HE neutrons, but the trend differs from one to another [147]. Increased bias voltages increase the photo-peak but the change is not appreciable. Similar pulse height spectra reported by Hess et al. at 59.5 main photo-peak show the FWHM about 4.2 % [148].

Typically in compound semiconductor, one charge carrier is trapped more than the other. The ideal situation for higher energy resolution will be to fabricate a planar device. Detectors designed to sense the motion of only one carrier type over the other have shown dramatic improvement over the simple planar design [149]. One way to improve the energy resolution and photo-peak detection efficiency at room temperature is by designing a three-electrode CZT device. The cathode of the device is grounded and the two anodes are connected to high bias voltages. Planar devices with the three electrodes have resulted to a photon energy resolution of 8.3 and 3.5 % (FWHM) for gamma-ray energies of 122 and 662 keV, respectively [150]. The geometrically weighted Frisch grid approach has shown excellent single polarity sensing [151]. However, for Frisch collar devices, there exists an optimal dielectric layer thickness for best performance of a CZT device [152–154].

## 8.10 Elemental Semiconductor

Elemental semiconductors, as a class of radiation detection materials, tend to be used in applications with the most demanding energy resolution requirements because they often have more compact configurations and better energy resolution which involves the statistical character of the signal itself. The stochastic nature of the energy cascade determines the intrinsic variance of the elemental semiconductor signal, whereas it is non-proportional scintillation and nonuniform light collection processes that contribute most of the variance of the scintillator signal. Semiconductor variance is generally far lower.

The energy (band) gap of the semiconductor is an important parameter, and larger band gap materials increase the average energy required to create information carriers and thus increase variance. The other important parameter is the mobility-lifetime ($\mu\tau$) product. The carrier mobility is dependent on the materials and the defect centers that play a role in charge trapping–de-trapping or recombination in the crystal and affect their electron–hole (e–h) mobilities. The disadvantages about these materials are the growth process, operational complexity, need for high purity, and defect-free crystals to promote complete charge transfer over large distance. For security applications and for compact handheld instruments, they have limited applications because of the instrumental designs and environment to operate these detectors.

### 8.10.1 Diamond

*Diamond*: Diamond has long held a special place in the hearts and minds of the public and scientists at large. To the public, the word diamond conjures up images of brilliant gemstone for jewelry and special occasions. It has a very fast response and very high radiation hardness when exposed to β-radiation. As a result, it can be used in high-energy physics experiment besides its use as radiation detectors in nuclear medicine. Although in principle low-energy $\beta$ and $\gamma$ pulse counting are possible using state-of-the art, DC-coupled, low-noise, charge-sensitive preamplifiers, diamond application for electron detection is restricted to radiation dosimetry only [153].

Diamond can detect any radiation (UV, X-rays, and gamma ($\gamma$) rays) and charged particles, including very high energy particles, neutrons, and pions that generate free carriers (electron–hole pairs) in the material. It is an extremely hard and rugged material with high resistivity and has a wide band gap (5.5 eV). The fundamental mechanism of radiation detection in diamond is independent of the exciting radiation as long as it is more energetic than the band gap.

Obviously, given its many unique properties, it is possible to envisage many other potential applications for diamond as an engineering material. Unfortunately, because of its high melting point and scarcity of natural diamond, many ideas of the scientist and engineers have been hampered. The industrial diamond has been synthesized using high pressure (50–100 kbar) and high temperature (1,800–2,300 K). However,

**Table 8.9** Comparison of diamond with that of 4H-SiC and Si

| Property | Diamond | 4H-SiC | Si |
|---|---|---|---|
| Band gap (eV) | 5.5 | 3.3 | 1.12 |
| Breakdown field (V/cm) | $10^7$ | $4 \times 10^6$ | $3 \times 10^5$ |
| Resistivity (Ω-cm) | $>10^{11}$ | $10^{11}$ | $2.3 \times 10^5$ |
| Intrinsic carrier density (cm$^{-3}$) | $<10^3$ | | $1.5 \times 10^{10}$ |
| Electron mobility (cm$^2$/V/s) | 1,800 | 800 | 1,350 |
| Hole mobility (km/s) | 220 | 200 | 82 |
| Mass density (g/cm) | 3.52 | 3.21 | 2.33 |
| Atomic charge | 6 | 14/6 | 14 |
| Dielectric constant | 5.7 | 9.7 | 11.9 |
| Displacement energy (eV/atom) | 43 | 25 | 13–20 |
| Energy to create electron–hole pair (eV) | 13 | 8.4 | 3.6 |
| Radiation length (cm) | 12.2 | 8.7 | 9.4 |
| Spec. ionization loss (MeV/cm) | 4.69 | 4.28 | 3.21 |
| Average signal created/100 µm (e) | 3,600 | 5,100 | 8,900 |
| Average signal created/0.1 % $X_0$ (e) | 4,400 | 4,400 | 8,400 |

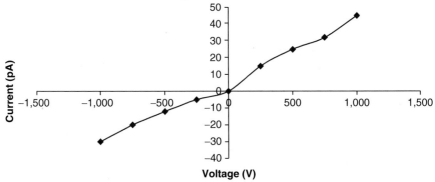

**Fig. 8.67** The typical current–voltage relation of a CVD-grown diamond device

recent progress to produce polycrystalline diamond and its film over single crystal silicon by *chemical vapor deposition* (CVD) techniques has increased further interest of the scientific communities [154]. Table 8.9 shows a comparative analysis of diamond with that of a hydrogenated silicon carbide (4H-SiC) and silicon (Si).

Diamond is very attractive not from its ornamental value but also for its hardness and ruggedness. It has a wide band gap (6.6 eV), has high resistivity, and has been used as picosecond photoconductors and for analysis of fast gamma-ray pulses over a wide dynamic range [155–158]. The best quality diamond gave a highest *photocurrent* of 500 nA at 50 V/mm and 2.75 Gy/min [159].

The *I–V* characteristics of a chemical vapor deposition (CVD) diamond have been plotted in Fig. 8.67. The detectors are fabricated from 6-mm square and 750-µ

## 8.10 Elemental Semiconductor

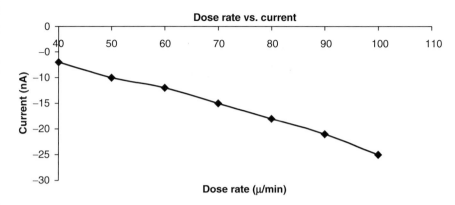

**Fig. 8.68** The photocurrent as a function of dose in a *diamond* detector

m-thick pieces of CVD-grown diamonds. The ohmic contacts have been made by vacuum-evaporated chromium (100 μm thick) and gold (1,000 μm thick) metals. The global connections are made with platinum wire 2 mils in diameter. The samples are annealed at 500 °C for 15 min before taking current–voltage (*I–V*) data. The device is kept in a dark enclosure and typical current–voltage (*I–V*) data are taken using a Keithley 237 instrument. The leakage current is very low (~30–50 pA) depending on the voltages. Initial investigations of photoconductive gain as a function of applied field, waveform, and photon energy have provided insight into performance of state-of-the-art single crystal diamond.

The single crystal of diamond with reasonable purity has high resistivity, meaning *low dark current,* and high atomic number, meaning *high attenuation*; as a result, it is a potential candidate for medical purposes, particularly for X-ray and *gamma dosimetry*. In addition to those properties, diamond has a unique advantage of being a *tissue equivalent material* (Z is close to human body) [160]. As these detectors are used in dosimeters for use in radiotherapy (beam calibration and profiling, *in situ* dose measurements, etc.), diamond detectors are tested and appropriately packaged in a clinical environment, using clinical apparatus and following clinical procedures [161]. For radiotherapy, an ideal dosimeter should have high precision, reproducibility, and ability to detect over a wide range of dose. In order to testify the diamond's ability to fulfill these requirements, we performed several experiments like dark current including the measurements of steady-state photocurrent as a function of dose rate. Figure 8.68 shows the experimental results.

Photocurrent is expected to depend on the dose rate (*D/t*, where *D* is dose and *t* is time in second). From the photocurrent verses dose rate, we can see that the relation follows a power law relationship $\{I_{ph} \propto (D/t)^x\}$, where the exponent *x* usually lies between 0.5 and 1.0. When the exponent value reaches ~1.0, one can expect that the detector has too many traps [162].

In reality, the charge measured for a fixed dose is not independent of the dose rate. As a matter of fact, there is a time lag in between the time the dose is applied

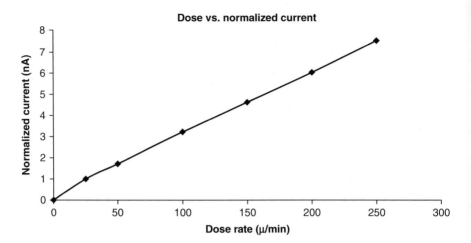

**Fig. 8.69** The normalized dose versus current in the *diamond* detector

and the time to record the full dose. The higher the dose, the longer is the time lag. In order to normalize the dose with time, we have plotted charge measure divided by delivery time in Fig. 8.68. The curve is fitted to a power law, the exponent ($x$) is seen to vary between 0.94 and 0.98.

Figure 8.69 shows the normalized dose versus current in a typical diamond detector. Raman spectroscopy is performed in CVD-grown diamond to detect the non-diamond carbon and to make a rough estimate of defect density. Photoluminescence (PL) spectroscopy has also been used to identify and monitor intrinsic and extrinsic defects in natural and synthetic diamonds [163].

It has been found the morphology of the polycrystalline CVD-grown film is very sensitive to the precise growth conditions. As for example, CVD-grown film under slow growth conditions, low $CH_4$ partial pressure, and low substrate temperature shows microcrystalline film with triangular {111} orientation along with many twin boundaries. However, when the pressure of $CH_4$ is high, microcrystallinity almost disappears.

A linear relation is observed between the average grain size and the depth $x$ measured from the substrate size. The charged particle-induced conductivity (CPIC) measurements with Ti/Pt/Au contacts show a collection distance behavior as $d_{CPIC} \approx 0.51 \pm 0.07\, d_{PC}$, when electric field is applied to the contacts [164]. It has been observed that placing contacts on two opposite cube {100} faces gives a higher *photocurrent* than on a pair of octahedral {111} faces [165]. Experimental studies also reveal that deep charge trapping states in synthetic diamond have favorable photoluminescent dosimetry properties for high-dose $\beta$ and $\gamma$ fields using radiophotoluminescence (RPL) techniques [166].

*Dark current* measurement from a IIa natural diamond reveals 30 fA for an electric field of 1 V/µm and a very high resistivity, approximately $10^{14}$ Ω-cm. In sub-picosecond mode, the measured decay time is found to be ~100 ps. The X-ray sensitivity was found to be linear with the absorbed X-ray energy, and the measured sensitivity was detected to be around 37.5 nC/J.V [167].

## 8.10 Elemental Semiconductor

**Fig. 8.70** Changes of mobility with carrier density in natural IIa crystal and synthetic sample (Reproduced with permission, MRS, Pittsburgh, Vol. 302, 1993, p 302)

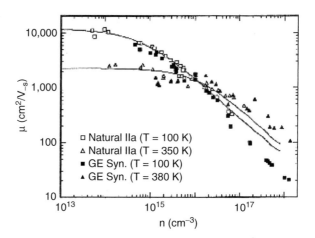

**Fig. 8.71** Collection distance of the IIa natural sample and the synthetic sample (Reproduced with permission, MRS, Pittsburgh, Vol. 302, 1993, p 302)

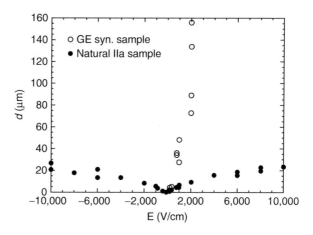

The electrical transport properties of a natural IIa-type diamond and a synthetic sample of CVD-grown sample are measured using transient photoconductivity. The excitation was achieved by a dye laser pumped by a mode-locked Nd:YAG laser with photon energy 6.1 eV. The incident photon pulses were less than 5 ps in duration, containing up to 50 μJ of energy pulse. The decay time in the synthetic sample was nearly 50 times longer than the natural IIa diamond crystal. The electrical mobility ($\mu$) is less than 100 cm$^2$/V-s when the field ($E$) was 200 V/cm (Fig. 8.70) and the lifetime ($\tau$) is between 100 and 500 ps. The average collection distance ($d = \mu\tau E$) is 150 μm. Figure 8.70 shows changes of mobility with carrier density in natural IIa crystal and synthetic diamond.

Figure 8.71 shows collection distance of the IIa sample and the synthetic sample. From the figure we can see that the collection distance in the synthetic

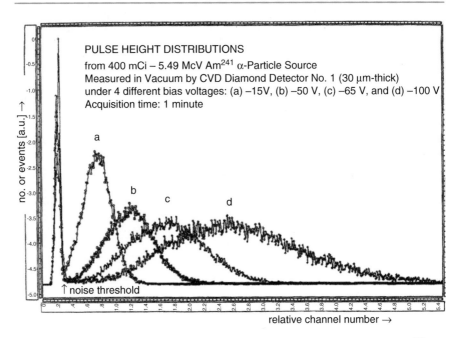

**Fig. 8.72** Pulse height spectra of a CVD-grown diamond detector with α-radiation ($^{241}$Am) (Courtesy of Elsevier Publications, 1997)

diamond exceeds that of a natural diamond due to its significantly longer lifetime ($\tau$). The mobility-lifetime product defines the figure of merit of a semiconductor material. For elemental semiconductors, this is of the order of unity for both electrons and holes. It is also an indicator for a detector about the number of defect centers the device possesses. Poorer $\mu\tau$ products result in short drift lengths, which in turn limit the maximum size and therefore the energy range of the detector. Limitations to charge collection efficiency such as recombination and charge trapping have been investigated quantitatively using quasi-continuous tunable synchrotron radiation under flexible biasing schemes as well using detailed Monte Carlo simulations [168].

A sealed 400 μC $^{241}$Am isotope is used as a source of alpha (α)-radiation. The source is placed at a distance of 2 in. from the entrance hole to the detector mount. The sample, 30 μm thick, is placed in a vacuum chamber at a pressure of 100 mtorr. Figure 8.72 shows the typical pulse height distribution generated by 5.49-MeV α-particles with four different bias voltages ($a = -15$ V, $b = -50$ V, $c = -65$ V, and $d = -100$ V). Along the x-axis, relative channel numbers and, along y-axis, number of events are plotted in arbitrary units. The position of the pulse height distribution shows more or less linear dependence on the bias voltage. The broadening in the spectrum is thought as a polarization phenomenon in the diamond detector at a relatively higher relative channel number with higher bias voltage [169].

**Fig. 8.73** A picture of a 20 × 20 orthogonal strip lithium-drifted silicon detector (Photo courtesy of Lawrence Berkeley Laboratory, CA)

## 8.10.2 Silicon

The name silicon comes from Latin silex or *silicis* meaning *flint*. Since the creation of the first integrated circuit in 1960, silicon (Si) has become an indispensable semiconductor not only for electronics industry but also in modern medical nuclear medicine (as radiation detector and in medical imaging, e.g., Si-TFT array) and high-energy physics.

The lithium-drifted silicon (SiLi) detectors are used in gamma-ray Compton telescope. It is also very popular for low-energy X-ray spectroscopy studies because of the low anode capacitance (~50fF) and low leakage current (~10 pA). Commercially available p-type silicon (Si) is compensated with lithium (Li) which acts as an interstitial donor, in order to produce material with lower carrier concentration [170–172].

Figure 8.73 shows a picture of a 20 × 20 orthogonal strip lithium-drifted silicon detector. Lithium-drifted silicon (SiLi) orthogonal strip detectors are being developed at Lawrence Berkeley National Laboratory (LBNL) for the use in Compton telescopes. The detection technology of the 3D array of Si (Li) orthogonal strip is based on Compton scattering. Individual detectors are having good position resolution (≤2 mm) and low noise (≤2 keV FWHM). For ohmic contacts, n-type contact is made of a diffused lithium layer and p-type is made with gold surface barrier.

For laboratory measurements, guard ring detectors with an outer annular ring of ~1.5 mm wide and with an inner region for center contact are used. A gap of ~1.5 mm is formed between the guard ring and the center contact [173].

Figure 8.74 shows the temperature dependence of leakage current of the Si (Li) detector. For comparison, the leakage current vs. temperature of the Si (Li) orthogonal strip detector (OSD) has been plotted along with the guard ring detector. At room temperature, the standard guard ring detector shows lower leakage current in comparison to the OSD.

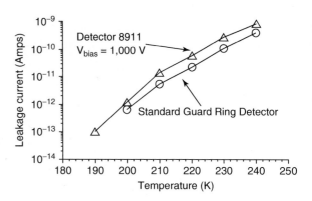

**Fig. 8.74** The temperature dependence of leakage current of the Si (Li) detector

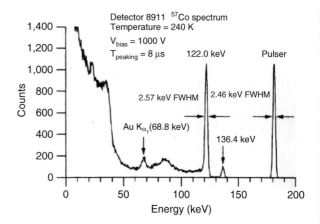

**Fig. 8.75** The energy spectrum of the OSSD with $^{57}$Co source at 240 K (Reproduced with permission from Elsevier Science, 2003)

The energy spectrum of the orthogonal silicon strip detector (OSSD) is plotted in Fig. 8.75. $^{57}$Co has been used as a radiation source. A FWHM of 2.57 keV is obtained for 122-keV peak [174, 175].

## 8.10.3 Silicon Strip Detector

The first silicon strip detectors manufactured with planar technology were single-sided simple p–i–n-type diode detectors processed on an n-type silicon wafers. These detectors were arranged in a cylindrical shape around the beam that provided only $r\varphi$ position information of traversing particles. To overcome manufacturing problems without losing the $r\varphi z$ measurement ability, a single-sided stereo detector (SSSD) has been successfully fabricated (Fig. 8.76). Every second p + strip of

## 8.10 Elemental Semiconductor

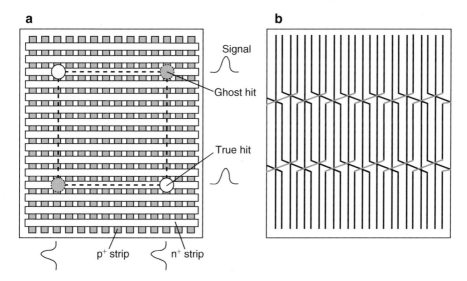

**Fig. 8.76** (a) Schematic of a double-sided strip detector. (b) Schematic of a single-sided stereotype strip detector

SSSD is interrupted at an interval of 1–3 mm. The resolution of Z coordinate of the SSSD is not as good as that of the double-sided strip detector (DSSD) but the ambiguous ghost hit region is smaller. Later capacitively coupled strip detectors are developed to eliminate the offset variation between strips due to nonuniform detector leakage current.

In double-sided strip detectors, $n^+$ phosphor is implanted perpendicular to the $p^+$ boron-doped strips on the opposite side of the wafers. A double-sided strip detector (DSSDs) provides space point information of passing particles. However, to achieve high yield especially with additional metal layer on the ohmic side is difficult and expensive. To achieve shorter charge collection time, solid-state detectors with 3D electrode design have been proposed [176]. Table 8.10 shows comparative characteristics of silicon and germanium.

### 8.10.4 Germanium

The biggest advantage of germanium (Ge) detectors compared to other gamma ($\gamma$)-ray detector is their excellent energy resolution of about 0.2 % FWHM at 662 keV. Despite some recent advances with inorganic scintillators, such as Ln halides that can achieve ~2.5 % FWHM, fundamental photon statistics will limit energy resolution of scintillators to not much better than 2.0 % FWHM when read out using PMT [177, 178]. Figure 8.77 shows the superiority of germanium detector when compared with NaI:Tl, CZT, and plastic scintillator.

The line shape of 1.332-MeV gamma ray from $^{60}$Co source with Ge detector shows 22 % efficiency with respect to sodium iodide (NaI) scintillation detector

**Table 8.10** Comparative characteristics of silicon and germanium

| Property | Germanium (Ge) | Silicon (Si) |
|---|---|---|
| Band gap (eV) | 0.665 | 1.12 |
| Breakdown field (V/cm) | $10^7$ | $3 \times 10^5$ |
| Intrinsic resistivity ($\Omega$-cm) | $2.3 \times 10^5$ | 47 |
| Atomic number | 32 | 14 |
| Electron mobility (cm$^2$/V/s) | 3,900 | 1,350 |
| Hole mobility (cm$^2$/V/s) | 1,900 | 480 |
| Mass density (g/cm$^3$) | 5.33 | 2.33 |
| Atomic weight | 72.60 | 28.09 |
| Intrinsic carrier density (cm$^{-3}$) | $2.4 \times 10^{13}$ | $1.5 \times 10^{10}$ |
| Displacement energy (eV/atom) | 43 | 13–20 |
| Energy to create electron–hole pair (eV) | 2.96 | 3.6 |
| Radiation length (cm) | 12.2 | 9.4 |
| Dielectric constant | 16 | 12 |
| Average signal created/100 μm (e) | 3,600 | 8,900 |
| Average signal created/0.1 % $X_0$ (e) | 4,400 | 8,400 |

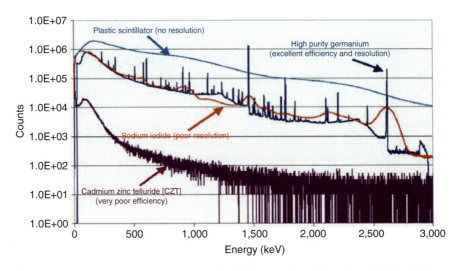

**Fig. 8.77** A comparison of natural background radiation as collected from four different radiation detectors (Courtesy: ORTEC)

[179]. However, there are some drawbacks with Ge detectors such as (1) it requires low temperature for better resolution and (2) susceptible to radiation damage and (3) energy shift and effective gain shift with positions along the longitudinal HPGe crystal axis. Figure 8.78 shows the energy resolution (FWHM) and effective gain shift of an HPGe crystal with respect to position when exposed to gamma ray from a collimated $^{241}$Am source [180]. Similar effect has also been reported by Wolf et al. [180].

## 8.10 Elemental Semiconductor

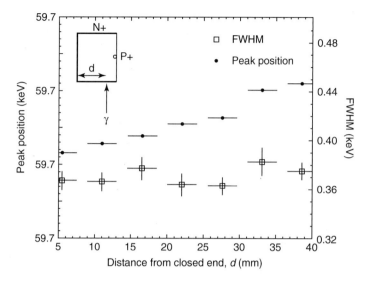

**Fig. 8.78** Energy resolution and effective gain shift with distance in a HPGe detector (Courtesy of IOP)

Collimator-based gamma-ray measurements are realized as hevimet or lead plates with parallel hole with a compromise between efficiency and resolution and image degradation due to scattering in and penetration through the collimator. In addition, Compton scattering which leads to multiple interactions degrades the collimator-based measurements. In most of the beam experiments with the tracking arrays, in addition to the gamma ($\gamma$) rays, fast neutrons are also emitted in the reactions. Experimental and simulated HPGe energy spectra of the planar and coaxial detectors are shown here in Fig. 8.79 [181].

In HPGe detectors, the main energy deposition mechanisms of neutrons with energies between 0.5 and 10 MeV include both elastic and inelastic scattering. However, elastic scattering is more probable within the energy range of 3 MeV. Above 11 MeV, the elastic cross section decreases rapidly. For elastic scattering, the largest number of Ge recoils is produced with energies below 30 keV ($E_R$). Only 25–30 % of the recoil energy in the range 10–100 keV is converted to ionization energy which can create electron–hole (e–h) pair in the crystal, while the rest of the energy is converted to photons. Experimental and simulated results of the HPGe energy spectra of the planar and coaxial detectors are shown in Fig. 8.79.

Pulse height spectra from gamma-ray detectors often show a peak in the vicinity of 0.2–0.25 MeV called backscatter peak. The peak is caused by the gamma rays from the source that has first interacted by Compton scattering in one of the materials surrounding the detector. Similarly when plastic and HPGe detectors are shielded with lead to avoid the direct gamma rays, it gives rise to the structure observed between the prompt gamma-ray peak and the neutron distribution in the time of flight spectra. Figure 8.80 illustrates the neutrons that first backscatter in the planar HPGe detector [181].

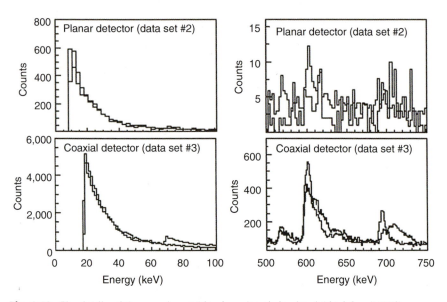

**Fig. 8.79** The details of the experiment (*thin layers*) and simulated (*thick lines*) HPGe energy spectra in the elastic (*left*) and nonelastic regions (Courtesy of Elsevier Publications)

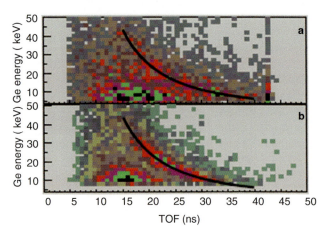

**Fig. 8.80** The neutrons that first backscatter in the planar HPGe detector (Courtesy of Elsevier Publications)

Two dimensionally segmented HPGe detectors are built in either coaxial or planar geometry. Figure 8.81 shows a $25 \times 25$ cm$^2$ orthogonal strip germanium detector built in Lawrence Berkeley Laboratory, CA, for gamma-ray imaging. Gamma-ray imaging has emerged as an essential tool in the areas such as biomedical research and nuclear medicine [182], exploration of chemical composition of planets and extraterrestrial materials [183], and nondestructive assay method to

## 8.10 Elemental Semiconductor

**Fig. 8.81** A 25 × 25 cm² orthogonal strip germanium detector for gamma-ray imaging (Photo courtesy of Lawrence Berkeley Laboratory, CA)

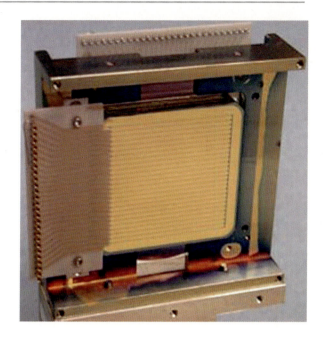

identify chemical warfare agents [184]. Recently, double-sided HPGe strip detector with interstrip interpolation measurements has been shown to exploit charge splitting and charge loss, yielding interaction position with lateral resolution of 200-μm FWHM at 356 keV and ~300-μm FWHM at 662 keV [185].

High-purity germanium (Ge) with impurity concentration (~$10^{10}$ atom/cm³) is considered to achieve full detection and can be used as X-ray detector at liquid nitrogen (LN$_2$) temperature. But to grow single crystal of such purity is very difficult even when the grown crystal is passed through several zone refining processes [186]. N-type coaxial HPGe detector is found to increase the fluence threshold for energy degradation by more than an order of magnitude than p-type coaxial detector [187].

Lithium-drifted smaller Ge detectors are generally used to minimize the leakage current and to achieve smaller contact resistance. When the detector is fabricated from p-type germanium material, an n$^+$-type diffused lithium (Li) back contact and a p$^+$ front contact are made by implanting boron. The bare surfaces are treated with a protective layer to have any surface oxidation or to minimize polarization. The device is mounted on a vertical dipstick to be immersed inside a container with LN$_2$. All parts that are cooled also need vacuum for thermal insulation. Thus, the detector is covered with end cap which is sealed to a vacuum flange containing the necessary feed through and a pump-out port. The best performance of a Li-drifted Ge detector with boron K-α-X-ray peak at 185 eV produced by electron beam excitation is 57-eV FWHM, which is comparable about 15 eV lower than the best achievable from a practical Si(Li) device [188]. Experimental observation shows that the key to successful performance of the detector for X-ray applications is the entrance window, which determines the dead layer thickness and therefore the magnitude of any peak distortions. The dead layer is dependent on the surface preparation and the type of contact used [189–191].

## References

1. Bingefors N et al (1993) The Delphi microvertex detector. Nucl Instrum Methods Phys Res A 328:447–471
2. Betancourt C, Wright J, Patak N, Fadeyev V, Sadrozinski HFW (2010) Punch through effect and collapse of the electric field in silicon strip detectors. NSS and Medical imaging conference, Knoxville, 30th Oct–6th Nov 2010
3. Tindall C, Amman M, Luke PN (2004) Large area Si (Li) orthogonal-strip detector. IEEE Trans Nucl Sci 51(3):1140
4. Gupta TK, Shah K, Benett P, Partain L, Green M, Street R (2004) Novel X-ray security systems, fast accurate and affordable, NIST Project
5. Antonuk L (2002) Electronic portal imaging devices. Phys Med Biol 47:R-31
6. Street R et al (2002) Comparison of $PbI_2$ and $HgI_2$ for direct detection active matrix X-ray image sensors. J Appl Phys 91:3345
7. Yaffe M, Rowlands JA (1997) X-ray detectors for digital radiography. Phys Med Biol 42:1–39
8. Henry JM et al (1995) Solid state detectors for digital mammography. SPIE 2432:392–401
9. Mandez JA, Balzer SJ, Watson SA, Reich RK, O'Mara DM (2009) A multiframe, megahertz CCD imager. IEEE Trans Nucl Sci 36(3):118
10. Gupta TK (2003) Hand book of thick and thin film hybrid microelectronics. Wiley, Hoboken
11. Scheiber C (2000) CdTe and CdZnTe detectors in nuclear medicine. Nucl Instrum Methods Phys Res A 448:513–524
12. Schieber M et al (2001) Thick films of X-ray polycrystalline mercuric iodide detectors. J Cryst Growth 225:118
13. Gupta TK, Bennett PB, Shah KS, Research Partain L, Green M, Street R (2004) Unpublished document of the awarded research from National Institute of Standards and Test, (NIST)
14. Mengesh W, Taulbee TD, Rooney BD, Valentine JD (1998) Light yield non-proportionality CsI(Tl) CsI(Na) and YAP. IEEE Trans Nucl Sci 45:512
15. Ponpon JP et al (1975) Properties of vapor phase grown mercuric iodide single crystal detectors. IEEE Trans Nucl Sci NS-22:182
16. Owens A, Peacock A (2004) Compound semiconductor radiation detectors. Nucl Instrum Methods Phys Res A 531:18
17. Shah KS, Glodo J, Klugerman M, Wong P, Higgins W, Gupta T, Wong P, Moses WE, Drenzo SE, Weber MJ, Dorenbos P (2004) $LuI_3:Ce^{3+}$ – a new scintillator for gamma ray spectroscopy. IEEE Trans Nucl Sci 51(5):2302
18. Shah KS, Crignano L, Grazioso R, Klugerman M, Bennett PB, Gupta TK, Moses WW, Weber MJ, Derenzo SE (2002) $RbGd_2Br_7:Ce$ scintillators for gamma ray and thermal neutron detection. IEEE Trans Nucl Sci 49(4):1655
19. Hull G et al (2007) Measurements of NaI:Tl electron response using SLYNCI. IEEE nuclear science symposium and medical imaging conference Honolulu, 27th Oct–3rd Nov 2007
20. Cirlin YA, Globus ME, Sysoeva EP. Optimization of gamma ray detection by scintillation crystal
21. Fontana MP, van Sciver WJ (1968) Energy transfer and optical properties of T+ centers in NaI(Tl) crystals. Phys Rev 168(3):960
22. Aluker ED, Yu Lusis D, Chernov SA (1979) Electronic excitations and radioluminescence of alkali halide crystals (in Russian). Zinatne, Riga, Chap. 6

# References

23. Rodnyi PA (1997) Physical process in inorganic scintillators. CRC Press, New York, Chap. 4
24. Pyne SA et al. Nonproportionality of scintillator detectors: Theory and experiment, Work performed at LLNL and LBNL National Labs. Under contract DE-AC52-07NA27344 and DE-AC02-05CHI1231
25. Mengesha W, Valentine JD (2002) Benchmarking NaI(Tl) electron energy resolution measurements. IEEE Trans Nucl Sci 49(5):2420
26. Koicki S, Koicki A, Ajdacic V (1973) The investigation of 0.15s phosphorescent of NaI(Tl) and its application in scintillation counting. Nucl Instrum Methods Phys Res A 108:297
27. Birks JB (1964) The theory and practice of scintillation counting. Pergamon Press, London, Chaps. 4 and 11
28. Kubota S, Sakuragi S, Hashimotoand S, Ruan(Gen) J (1988) A new scintillation material: pure CsI with 10 ns decay time. Nucl Instrum Methods Phys Res A 268:275
29. Geist J, Gladden WK, Zalewski EF (1959) Physics of photon flux measurements with silicon photodiodes. J Opt Soc Am 72(8):1068
30. Keszthelyi-Landori S, Hrehuss G (1969) Scintillation response function and decay time of CsI(Tl) to charge particles. Nucl Instrum Methods Phys Res A 68:9
31. Mengesha W et al (1998) Light yield non-proportionality of CsI (Tl), CsI (Na) and YAP. IEEE Trans Nucl Sci NS-45:456
32. Gwin R, Murray RB (1963) Scintillation process in CsI(Tl)I. Comparison with activation saturation model. Phys Rev 3(2):501
33. Rooney BD, Valentine JD (1997) Calculating non-proportionality of scintillator photon response using measured electron response data. IEEE Trans Nuclr Sci 44(3):509
34. Mao R, Zhang L, Zhu R-Y (2008) Optical and scintillation properties of inorganic scintillators in high energy physics. IEEE Trans Nucl Sci 55(4):2425
35. Valentine JD, Jordannov VT, Whe DK, Knoll GF (1992) Charge collection of CsI(Tl)/photodiode spectroscopy systems. Nucl Instrum Methods Phys Res A A314:119
36. Valentine J, Moses WW, Derenzo SE, Wehe DK, Knoll G (1993) Temperature dependence of CsI(Tl) gamma-ray excited scintillation properties. Nucl Instrum Methods Phys Res A 325:147
37. Grassman H, Lorenz E, Moser HG (1985) Properties of CsI(Tl). Nucl Instrum Methods Phys Res A 228A:323
38. Grabmaier BC (1984) Crystal scintillator. Renaissance of an old scintillation material. IEEE Trans Nucl Sci NS-31:372
39. Kappers LA et al (2010) A tunneling model for afterglow suppression in CsI:Tl, Sm scintillation materials. Radiat Meas 45:426
40. Shiran N et al (2009) Seventh international conference on information technology, SCINT 2009, Institute for scintillation materials, Kharkov
41. Birks JB (1964) The theory and practice of scintillation counting. Pergamon Press, London
42. Kudryavtsev VA et al (2001) CsI(Tl) for WIMP dark matter. Nucl Instrum Methods Phys Res A A-456:272
43. Kim HJ et al (2008) Development of low background CsI (Tl) crystals and search for WIMP. IEEE Trans Nucl Sci 55(3):1420
44. Nagarkar VV, Gupta TK, Miller SR, Klugerman Y, Squillante MR, Entine G (1995) Structured CsI (Tl) Scintillators for X-ray imaging applications, ARPA and NIH contracts #DAAH01-95-C-R-188 and #2R44CA65213-02
45. Nagarkar VV, Gordon J, Vasile S, Gothoskar P, Gupta TK, Squillante M, Entine G (1996) CCD based non-destructive testing system for industrial applications. Trans Nucl Sci 43(3):1559
46. Garnier N et al (2000) Spectroscopy of CsI(Tl). Proceeding of 5th international conference on inorganic scintillators and their applications Scint 99, M V Lomonosov Moscow State University, p 394
47. van Eijk CWE (2002) Inorganic scintillators in medical imaging. Phys Med Biol 47:R85
48. Hillen W, Eckenbanch W, Ouadfliey P, Zaengel T (1987) Imaging performance of a digital storage phosphor system. Med Phys 14(5):744–751

49. Roehrig H, Fajardo L, Fu T, Schem WS (1987) Signal noise and detective quantum efficiency. Proc IEEE 70(7):715
50. Schlesinger TE, James RB (1995) Semiconductors for room temperature nuclear detector applications, vol 43. Academic Press, San Diego
51. Hecht HK (1932) Zum mechanisums des lichtelektrischen primastromers in isolierenden kristallen. Z Phys 77:235
52. Levi A, Schieber M, Burshtein Z (1983) Carrier surface recombination in $HgI_2$ photon detectors. J Appl Phys 54:2472
53. Shockley WS (1938) Currents to conductors induced by a moving point charge. J Appl Phys 9:635 and also Ramo S (1939) Current induced by electron motion. Proc IRE 27:584
54. Toney JE, Schlesinger TE, James RB (1999) Optimal band gap variants of $Cd_{1-x}Zn_xTe$ for high resolution X-ray and -ray spectroscopy. Nucl Instrum Methods Phys Res A 428:14
55. Kang Y et al (2005) Examination of $PbI_2$ and $HgI_2$ photoconductive materials for direct detection, active matrix, flat panel imagers for diagnostic X-ray imaging. IEEE Trans Nucl Sci 52(1):38
56. Zentai G et al (2006) Comparison of mercuric iodide and lead iodide X-ray detectors for X-ray imaging applications. IEEE Trans Nucl Sci 53(5):2506
57. Anotonuk LE (2003) a-Si:H TFT based active matrix flat panel imagers for medical X-ray applications. In: Kuo Y (ed) Thin film transistors, vol 1. Kluwer Academic Publishers, Boston, pp 395–484, Chap.10
58. Gupta T et al (2003) Polycrystalline lead iodide films for digital X-ray sensors. Nucl Instrum Methods Phys Res A A-505(269)
59. Bell ZW, Pohl KR, van den Berg L (2004) Neutron detection with mercuric iodide. IEEE Trans Nucl Sci 51(3):1163
60. Levi A, Schieber MM, Burshtein Z (1985) Dark current transients in $HgI_2$ single crystals used as and X-ray spectrometers. J Appl Phys 57:1944
61. Camarda GS et al (2010) Correlation between extended defects and uniformity charge collection in $HgI_2$ material. IEEE nuclear science symposium and medical imaging conference, Knoxville, 30th Oct–6th Nov 2010
62. Minder R et al (1974) Measurements of drift velocity of charge carriers in mercuric iodide. J Appl Phys 45:5074
63. Ottaviani G, Canali C, Quaranta AA (1975) Charge carrier transport properties of semiconductor materials suitable for nuclear radiation detectors. IEEE Trans Nucl Sci NS-22:192
64. Chester M, Coleman CC (1971) Electroabsorption in $HgI_2$. J Phys Chem Solids 32:223
65. Mertz JL, Wu ZL, van den Berg L, Schnepple WF (1983) Low temperature photoluminescence of detector grade $HgI_2$. Nucl Intrum Meth 213:51
66. Bao XJ, James RB, Schlesinger TE (1995) Optical properties of red mercuric iodide. In: Schlesinger TE, James RB (eds) Semiconductors for room temperature nuclear detector applications. Vol. 43, Academic Press, San Diego, CA
67. Street RA (2000) Large area image sensors arrays. In: Street RA (ed) Technology and applications of amorphous silicon. Springer-verlag, Heildelberg, Germany, Chap. 4
68. Gupta TK (2003) Hand book of thick and thin film hybrid micro electronics. Wiley, Hoboken
69. Zentai G, Partain L, Pavlyuchkova R, Proano C, Breen BN, Dagan O, Schieber M, Gilboa H (2004) Mercuric iodide medical imagers for low exposure radiography and fluoroscopy. Proceeding of SPIE conference on medical imaging, San Diego, 14–19 Feb 2004, p 200
70. Su Z et al (2005) Systematic investigation of the signal properties of polycrystalline $HgI_2$ detectors under mammographic, radiographic, fluoroscopic and radiotherapy irradiation conditions. Phys Med Biol 50:2907
71. Bloomquist AK, Yaffe JM, Mawdsley GE, Hunter DM, Beideck DJ (2006) Lag and Ghosting in clinical flat panel selenium digital mammography system. Med Phys 33(8):2998
72. Kim HK (2006) Analytical model for incomplete signal generation in semiconductor detectors. Appl Phys Lett 88(13):132112

# References

73. Onodera T, Hitomi K, Shoji T (2006) Spectroscopic performance and long term stability of thallium bromide radiation detectors. Nucl Instrum Methods Phys Res A 568:433
74. Gazizov LM, Zaletin VM (2010) The sensitivity of pure and doped TlBr crystals. IEEE nuclear science symposium and medical imaging conference, Knoxville 30th Oct–6th Nov 2010
75. Mayer JW (1967) Evaluation of CdTe by nuclear particle measurements. J Appl Phys 38 (1):296
76. Gerrish VM (1995) Characterization and quantification of detector performance. In: Schlesinger TE, James RB (eds) Semiconductors for room temperature nuclear detector application. Academic Press, San Diego
77. Samara GA (1981) Pressure and temperature dependencies of the ionic conductivities of thallus halides TlCl, TlBr, and TlI. Phys Rev B23:575
78. von Herrmann P (1964) Fehlor dnungserscheinungen infestem thullium bromid. Z Physik Chem (Leipzig) 227:338
79. Bao XJ, Schlesinger TE, James RB (1995) Electrical properties of mercuric iodide. In: Schlesinger TE, James RB (eds) Semiconductors for room temperature nuclear detector applications. Academic Press, San Diego, p 160
80. Kozorezov A, Gostillo V, Owens A, Quarati F, Shorohov M, Webb MA, Wigmore JK (2010) Polarization effects in thallium bromide X-ray detectors. J Appl Phys 108(6):064507
81. Hitomi K, Shoji T, Niizeki Y (2008) A method for suppressing polarization phenomena in TlBr detectors. Nucl Instrum Methods Phys Res A A-585:102
82. Higgins WM, Glodo J, van Loef E, Klugerman M, Gupta T, Cirignano L, Wong P, Shah KS (2006) Bridgman growth of $LaBr_3$:Ce and $LaCl_3$:Ce crystals for high-resolution gamma-ray spectrometers. J Cryst Growth 287(2):239
83. van Loef EVD, Dorenbos P, van Ejik CWE, Kramer K, Gudel H (2000) Appl Phys Lett 77: 1467
84. van Loef EVD, Dorenbos P, van Ejik CWE, Kramer K, Gudel H (2001) High energy resolution scintillator: $Ce^{3+}$. Appl Phys Lett 79: 1573
85. Dorenzo SE, Weber MJ, Bourret-Courchesne E, Klintenberg MK (2003) The quest for the ideal inorganic scintillator. Nucl Instrum Methods Phys Res A A-505:111
86. Shah KS, Glodo J, Higgins WM, Gupta T, Wong P (2004) High energy resolution scintillation spectrometers. IEEE Trans Nucl Sci 51(5):2395
87. Khodyuk IV, Dorenbos P (2010) Non-proportional response of $LaBr_3$:Ce and $LaCl_3$:Ce scintillators to synchrotron X-ray irradiation. J Phys Condens Matter 22:485402
88. van Ejik CWE et al (2010) Energy resolution of some new inorganic-scintillator gamma ray detector. Radiat Meas 33:521
89. vanGinhoven RM, Jaffe JE, Kerisit S, Rosso KM (2010) Trapping of holes and excitons in scintillators: CsI and $LaX_3$ (X=Cl, Br). IEEE Trans Nucl Sci 57(4):2303
90. Kraft S et al (2007) Development and characterization of large La-halide gamma ray scintillators for future planetary mission. IEEE Trans Nucl Sci 54(4):873
91. Shah KS et al (2003) $LaCl_3$:Ce scintillator for gamma ray detection. Nucl Instrum Methods Phys Res A A-505:76
92. Mcllwain ME, Gao D, Thompson N (2007) First principle quantum description of the energetics associated with LaBr3, LaCl3 and Ce doped scintillators, NSS-MIC-07. US Department of Energy, INL
93. Dorenbos P et al (2006) Level location and spectroscopy of $Ce^{3+}$, $Pr^{3+}$, $Er^{3+}$, and $Eu^{3+}$, in $LaBr_3$. J Lumin 117:147
94. Bollinger LM, Thomas GE (1961) Measurement of the time dependence of scintillation intensity by a delayed coincidence method. Rev Sci Instrum 32:1044
95. Shah KS, Cirignano L, Grazioso R, Klugerman M, Bennett PR, Gupta T, Moses WW, Weber MJ, Derenzo SE (2002) $RbGd_2Br_7$:Ce Scintillators for gamma ray and thermal neutron detection. IEEE Trans Nucl Sci 49(4):1655
96. Knoll G (1999) Radiation detection and measurements, 3rd edn. Wiley, New York

97. Kleinknecht K (1998) Detectors for particle radiation, 2nd edn. Cambridge University Press, Cambridge
98. Shah K, Glodo J, Klugerman M, Higgins W, Gupta T, Wong P, Moses WW, Derenzo SE, Weber MJ, Dorenbos P (2004) LuI3:Ce – a new scintillator for gamma ray spectroscopy. IEEE Trans Nucl Sci 51(5):2302
99. Moses WW (2002) Current trends in scintillator detectors and materials. Nucl Instrum Methods Phys Res A A-487(1223)
100. Noel OG et al (1999) Scintillation properties of $RbGd_2Br_7$:Ce advantages and limitations. IEEE Trans Nucl Sci 46:1274
101. Moszynski M et al (2006) New prospects for time of flight PET with LSO scintillators. IEEE Trans Nucl Sci 53(5):2484
102. Erikson L et al (2005) The ECAT HRRT: an example of NEMA scatter estimation issues for LSO-based PET systems. IEEE Trans Nucl Sci 52(1):90
103. Pankov JI (1971) Optical properties of semiconductors. Dover Pub, New York, p 371
104. Noiel OG et al (1999) Scintillation properties of $RbGd_2Br_7$:Ce advantages and limitations. IEEE Trans Nucl Sci 46:1274
105. Bao XJ, Schlesinge T, James RB (1995) Electrical properties of mercuric iodide. In: Schlesinger T, James RB (eds) Semiconductors for room temperature nuclear detector applications. Academic Press, San Diego
106. van Ejik CWE (2002) Inorganic scintillators in medical imaging. Phys Med Biol 47(R85)
107. Usui T et al (2007) 60 mm diameter $Lu0.4Gd1.6SiO_5$:Ce (LGSO) single crystals and their improved scintillation properties. IEEE Trans Nucl Sci 54(1):19
108. Yamamoto S et al (2010) Development of a pixelated GSO gamma camera system with parallel hole collimators for single photon energy. Nuclear science symposium and medical imaging conference, Knoxville, 30th Oct–6th Nov 2010
109. Moszynski M et al (2004) Intrinsic energy resolution and light yield non-proportionality of BGO. IEEE Trans Nucl Sci 51(3):1074
110. de Voigt MJA et al (1995) A novel Ge-BGO Compton suppression spectrometer. Nucl Instrum Methods Phys Res A A-356(2–3):362
111. Lam S et al (2010) Cryogenic pulse height spectrometer for non-proportionality studies in BGO and Ce:YAG. Nuclear science symposium and medical imaging conference, Knoxville, 30th Oct–6th Nov 2010
112. Williams PA et al (1996) Optical, thermo-optic, electro-optic, and photoelectric properties of bismuth germanate ($Bi_4Ge_3O_{12}$). Appl Opt 35(19):3562
113. Moszynski M et al (2004) $CdWO_4$ crystal gamma ray spectrometry. IEEE nuclear science symposium and medical imaging conference, Rome, 2004
114. Danevich FA et al (2003) $\alpha$-activity of natural tungsten isotopes. Phys Rev C 67(1):014310
115. Kobayashi M, Ishii M, Usuki Y, Yahagi H (1994) Cadmium tungstate scintillators with excellent radiation hardness and low background. Nucl Instrum Methods Phys Res A A-349 (2–3):407
116. Burachas SF et al (2000) Advanced scintillation single crystals based on complex oxides with large atomic number. Semiconductor Phys Quantum Electron Optoelectron 3(2):237
117. Novotny R et al (1997) Response of a $PbWO_4$ scintillator array to electrons in the energy regime below 1GeV. IEEE Trans Nucl Sci 44(3):477
118. Hoek M, Doring W, Hejny V, Lohner H, Metag V, Novotony R, Wortche H (2002) Charged particle detection with $PbWO_4$. IEEE Trans Nucl Sci 49(3):946
119. Zhang L, Bailleux D, Bornheim A, Zhu K, Zhu RY (2005) Performance of monitoring light source for the CMS lead tungstate crystal calorimeter. IEEE Trans Nucl Sci 2(4):1123
120. Canali C, Martini M, Ottaviani C, Zanio KR (1971) Transport properties of CdTe. Phys Rev B 4(2):422
121. Sato G et al (2010) Recent developments of Schottky CdTe diodes and applications to medical images. IEEE nuclear science symposium and medical imaging conference, Knoxville, 30th Oct–6th Nov 2010

# References

122. Senio T, Takahashi I (2007) CdTe detector characteristics at 30C and 35C when using the periodic bias reset technique. IEEE Trans Nucl Sci 54(4):777
123. Wantable S et al (2002) Stacked CdTe gamma ray detector and its application to range finder. Nucl Instrum Methods Phys Res A A-505:118
124. Sorodo SD et al (2009) Progress in development of CdTe and CdZnTe semiconductor radiation detector for astrophysical and medical applications. Sensors 9:3491
125. Grill R et al (2002) High temperature defect structure of Cd and Te rich CdTe. IEEE Trans Nucl Sci 49(3):1270
126. Niraula M et al (2007) Characterization of CdTe/n+−Si heterojunction diodes for nuclear radiation detectors. IEEE Trans Nucl Sci 54(4):817
127. Ali MH, Siffert P (1995) Characterization of CdTe nuclear detector materials. In: Schlesinger TE, James RB (eds) Semiconductors for room temperature nuclear detector applications. Academic Press, San Diego
128. Sato G et al (2011) Study of polarization phenomena in Schottky CdTe diodes using infrared light illumination. Nucl Instrum Methods Phys Res A A-10:1016
129. Stoneham AM (2001) Theory of defects in solids: electronic structure of defects in insulators and semiconductors. Oxford University Press, Oxford
130. Chu M, Terterian S, Ting D (2004) Role of zinc in CdZnTe radiation detectors. IEEE Trans Nucl Sci 51(5):2405
131. Serreze HB, Entine G, Bell RO, Wald FV (1974) Advances in CdTe gamma ray detectors. IEEE Trans Nucl Sci 21:404
132. Triboulet R, Marfaing Y, Cornet A, Siffert P (1974) Undoped high resistivity cadmium telluride for nuclear radiation detectors. Nat Phys Sci 245:12
133. Hage Ali M, Siffert P (1995) CdTe nuclear detectors and applications. In: Schlesinger TE, James RB (eds) Semiconductors for Room Temperature Nuclear Detector Applications, Vol. 43, Academic Press, San Diego, CA
134. Koike A, Okunoyama T, Ito T, Morii H, Neo Y, Mimura H, Aoki T (2010) Carrier transportation and polarization properties in CdTe diode detectors. Nuclear science symposium and medical imaging conference, Knoxville, 30th Oct–6th Nov 2010
135. Castaldini A, Cavallini A, Fraboni B, Fernandez P, Piqueras J (1998) Deep energy levels in CdTe and CdZnTe. J Appl Phys 83:2121
136. Burger A et al (2000) Defects in CZT crystals and their relationship to gamma ray detector performance. Nucl Instrum Methods Phys Res A A-448:586
137. Greaves CM, Brunett BA, Van Scyoc JM, Schlesinger TE, James RB (1999) Materials uniformity of CdZnTe grown by low pressure bridgman. Nucl Instrum Methods Phys Res A A-458:96
138. Vartsky D et al (1988) Radiation induced polarization in CdTe detectors. Nucl Instrum Methods Phys Res A A-263:457
139. Jahnke A, Matz R (1999) Signal formation and decay in CdTe X-ray detectors under intense irradiation. Med Phys 26:38
140. Adachi S et al (2000) Experimental evaluation of a-Se and CdTe flat panel X-ray detectors for digital radiography and fluoroscopy. Proc SPIE 3977:38
141. Amman M, Lee JS, Luke PN, Chen H, Awadalla SA, Redden R, Bindley G (2009) Evaluation of THM-grown CdZnTe material for large volume gamma ray detector applications. IEEE Trans Nucl Sci 56(3):795
142. Szeles C, Driver MC (1998) Growth and properties of semi-insulating CdZnTe for radiation detector applications. Proc SPIE 3446:2
143. Chu M, Terterian S, Ting D (2004) Role of zinc in CdZnTe radiation detectors. IEEE Trans Nuclr Sci 51(5):2405
144. Szeles C (2004) Advances in the crystal growth and device fabrication technology of CdZnTe room temperature radiation detectors. IEEE Trans Nucl Sci 51(3):1242
145. Eisen Y, Horovitz Y (1994) Correction of incomplete charge collection in CdTe detectors. Nucl Instrum Methods Phys Res A 353:60
146. Fraboni B, Cavallini A, Auricchio N, Bianconi M (2007) Deep trap induced by 700 keV protons in CdTe and CdZnTe detectors. IEEE Trans Nucl Sci 54(4):828

448     8 Characterization of Radiation Detectors (Scintillators) Used in Nuclear Medicine

147. Fraboni B, Cavallini A, Dusi W (2004) Damage induced by ionizing radiation on CZT and CdTe detectors. IEEE Trans Nucl Sci 51(3):1209
148. Hess R, DeAntonis P, Morton EJ, Gilboy WB (1994) Analysis of the pulse shapes obtained from single crystal $Cd_{0.9}$ Zn $_{0.1}$ Te radiation. Nucl Instrum Methods Phys Res A A-353:76
149. Mcgregor DS, Rojeski RA, He Z, Wehe DK, Driver M, Blakely M (1999) Geometrically weighted semiconductor frisch grid radiation spectrometers. Nucl Instrum Methods Phys Res A A-422:164
150. Kim HD, Cirignano L, Shah K, Squillante M, Wong P (2004) Investigation of energy resolution and charge collection efficiency of Cd (Zn) Te detectors with three electrodes. IEEE Trans Nucl Sci 51(3):1229
151. Barrett H, Eskin JD, Barber H (1995) Charge transport in arrays of semiconductor gamma ray detectors. Phys Rev 75(1):156
152. Karger A et al (2009) The effect of the dielectric layer thickness on spectral performance of CdZnTe frisch collar gamma ray spectrometers. IEEE Trans Nucl Sci 56(3):824
153. Grobbelaar JH, Burns RC, Nam TL, Keddy RJ (1991) Miniaturized radiation detector with custom synthesized diamond crystal sensor. Nucl Instrum Methods Phys Res A B-61:553
154. May PW (1995) CVD diamond—a new technology for the future. Endeavour Mag 19(3):101
155. Collins AT (1974) Visible luminescence from diamond Ind Diamond Rev 34:131
156. Malm HI et al (1975) Gamma ray spectroscopy with single carrier collection in high resistivity scintillators. Appl Phys Lett 26:344
157. Malm HL, Litchinsk D, Canali C (1977) Single carrier charge collection in semiconductor nuclear detector. Revue Physique Appliquee Tome 12:303, Fevrier
158. Levita M, Schlesinger T, Friedland SS (1976) LiF disimetry based on radioluminescence. IEEE Trans Nucl Sci NS-23:1
159. Yacoot A, Moore M, Makepeace A (1993) X-ray studies of synthetic radiation counting diamonds. In: James RB, Schlesinger TE, Siffert P, Franks L (eds) Semiconductors for room temperature radiation detector applications. Material Soc, Pittsburgh
160. Vatnitsky S, Jarvinen H (1993) Application of a natural diamond detector for the measuremenyt of relative dose distribution in radiotherapy. Phys Med Biol 38:173
161. Keister JW, Smedley J (2009) Single crystal diamond photodiode for soft X-ray radiometry. Nucl Instrum Methods Phys Res A A-606:774
162. Fowler J (1966) Radiation dosimetry. Academic Press, New York
163. Clark CD, Mitchel EWJ, Parsons BJ (1979) In: Field JE (ed) The properties of diamond. Academic Press, New York
164. Zhao S et al (1993) Electrical properties in CVD diamond films. In: James RB, Schlesinger T, Siffert PE, Franks L (eds) Semiconductors for room temperature radiation detector applications, vol 502. MRS Proceedings
165. Yacoot A, Moore M (1993) Make place, X-ray studies of synthetic radiation counting diamonds. In: James RB, Schlesinger T, Siffert PE, Franks L (eds) Semiconductors for room temperature radiation detector applications, vol 502. MRS Proceeding
166. Keddy RJ, Nam TL, Araikum S (1993) Radiation dosimetry via radio-photoluminescence of synthetic diamond. In: James RB, Schlesinger T, Siffert PE, Franks L (eds) Semiconductors for room temperature radiation detector applications, vol 502. MRS Proceedings
167. Nail M, Gibert PH, Miquel JL, Cuzin M (1993) Experimental results in picosecond and subpicosecond range of Iia type diamond detector in X-UV, visible and IR fields, in CVD diamond films. In: James RB, Schlesinger T, Siffert PE, Franks L (eds) Semiconductors for room temperature radiation detector applications, vol 502. MRS Proceedings
168. Keister JW, Smedley J, Dimitrov D, Busby R (2010) Charge collection and propagation in diamond X-ray detectors. IEEE Trans Nucl Sci 57(4):2400
169. Souw EK, Meilunas RJ (1970) Response of CVD diamond detectors to alpha radiation. Nucl Instrum Methods Phys Res A A-400:69
170. Tindall C, Hau ID, Luke PN (2003) Evaluation of Si(Li) detectors for use in Compton telescope. Nucl Instrum Methods Phys Res A A-505:130

References 449

171. Zampa G, Rashevsky A, Vacchi A (2009) The X-ray spectroscopic performance of a very large area silicon drift detector. IEEE Trans Nucl Sci 56(3):832
172. Goulding FS (1966) Semiconductor detectors for nuclear spectroscopy. Nucl Instrum Methods Phys Res A 43:54
173. Hau ID, Tindall C, Luke PN (2003) New contact development for Si (Li) orthogonal strip detectors. Nucl Instrum Methods Phys Res A A-505:148
174. Hau ID, Tindal C, Luke PN (2003) New contact development for Si(Li) orthogonal strip detectors. Nucl Instrum Methods Phys Res A A-505:148
175. Pahn G (2009) First beam test characterization of a 3-D-stc silicon short strip detector. IEEE Trans Nucl Sci 56(6):3834
176. Parker SI et al (1997) 3-D a proposed new architecture for solid state radiation detectors. Nucl Instrum Methods Phys Res A A395:328–343
177. Luke PN, Amman M (2007) Room temperature replacement of Ge detectors – are we there yet? IEEE Trans Nucl Sci 54(4):834
178. Moses WW (2002) Current trends in scintillator detector and materials. Nucl Instrum Methods Phys Res A A-4487:123
179. Darken LS, Cox CE (1993) High purity germanium technology for gamma and X-ray spectroscopy. Mat Res Soc Symp Proc 302:31
180. Barbeu PS, Collar JI, Tench O (2007) Large mass ultra low noise germanium detectors. JCAP 09:1 and also Wolf EA, Ampe J, Johnson WN, Kroeger RA, Kurfess JD, Philips BF (2002) Depth measurements in a germanium strip detector. IEEE Trans Nucl Sci 49(4):1876
181. Ljungvall J, Nyberg J (2005) A study of fast neutron interactions in high purity germanium detectors. Nucl Instrum Methods Phys Res A A-546:553
182. Pivovaroff MJ et al (2002) High resolution radionuclide imaging focussing gamma ray optics J Nucl Med 43(5 Suppl):231
183. Roemer K et al (2009) A technique for measuring the energy resolution of low scintillators. IEEE nuclear science symposium, Orlando, FL 25–31 Oct 2009
184. Meyer R (1987) Explosives, 3rd edn. Wiley-VCH Pub., Hoboken, NJ
185. Hayward JP, Wehe DK (2009) Interstrip interpolation measurements in a high purity germanium double-sided strip detector. IEEE Trans Nucl Sci 56(3):800
186. Teal GK, Little JB (1950) Growth of Germanium single crystal. Phys Rev 78:647
187. Pehl RH, Madden NW, Elliot JH, Raudrof RW, Trammel RC, Darken LS Jr (1979) Radiation damage resistanceof reverse electrode GE coaxial detectors. IEEE Trans Nucl Sci NS-26:321
188. Darken LS, Cox CE (1993) High purity germanium technology for gamma-ray and X-ray spectroscopy. Mat Res Soc Proc 302:31
189. Llacer J, Haller EE, Cordi RC (1977) Entrance windows in Ge-low energy X-ray detectors. IEEE Trans Nucl Sci NS-24(1):53
190. Cox CE, Lowe BE, Sareen RA (1988) Small area high purity germanium detectors for use in energy range 100 eV to 100 keV. IEEE Trans Nucl Sci 35(1):28
191. Rossington CS, Giauque RD, Jacklevic JM (1992) A direct comparison of Ge and Si(Li) detectors in the 2–20 keV range. IEEE Trans Nucl Sci 39(570)

# Instrumentation and Its Applications in Nuclear Medicine

# 9

## Contents

9.1  Introduction ........................................................................ 451
9.2  Administration of the Radionuclides ............................................ 454
9.3  Preparation of Radionuclides .................................................... 455
    9.3.1  Generator ................................................................ 455
    9.3.2  Reactor .................................................................. 455
    9.3.3  Cyclotron ................................................................ 458
9.4  Radiation Dose .................................................................... 459
9.5  Radiation Therapy ................................................................ 460
    9.5.1  External Beam Therapy (EBT) ............................................ 460
    9.5.2  Proton Beam Therapy .................................................... 461
    9.5.3  Neutron Beam Therapy .................................................. 461
9.6  Intensity-Modulated Radiation Therapy (IMRT) ................................ 461
    9.6.1  Linear Accelerator (LINC) .............................................. 461
    9.6.2  Gamma Knife ............................................................ 462
    9.6.3  Proton Beam ............................................................ 463
    9.6.4  Stereotactic Radiosurgery (SRS) ........................................ 464
9.7  Scintigraphy ...................................................................... 464
    9.7.1  Angiography (CTA, MRA) ................................................ 465
    9.7.2  CT ........................................................................ 467
    9.7.3  PET ...................................................................... 469
    9.7.4  Single-Photon Emission Computed Tomography (SPECT) .................. 476
    9.7.5  Magnetic Resonance Imaging (MRI) ...................................... 478
    9.7.6  Ultrasound (US) (Sonography) .......................................... 481
9.8  Fusion or Hybrid Technology .................................................... 485
9.9  Final Remarks .................................................................... 487
References ............................................................................ 490

## 9.1 Introduction

*Nuclear medicine* is a subspecial branch of radiology, which utilizes radioactive materials to create images of a body anatomy and its function. The *Wikipedia encyclopedia* defines *nuclear medicine* as the medical specialty involving

T.K. Gupta, *Radiation, Ionization, and Detection in Nuclear Medicine*,
DOI 10.1007/978-3-642-34076-5_9, © Springer-Verlag Berlin Heidelberg 2013

**Table 9.1** Radionuclides administered for most frequently performed procedures

| Activities | Radionuclide | Chemical form | Dose (MBq) |
|---|---|---|---|
| Bone scan (planar 98 %) | $^{99m}$Tc | Phosphates | 400–775 |
| SPECT (2 %) | $^{99m}$Tc | | 500–800 |
| Lung perfusion (planar (99.99 %) | $^{99m}$Tc | MAA (macroaggregated albumin) | 50–200 |
| SPECT (0.01 %) | | | 100–100 |
| Myocardium (SPECT 98 %) | $^{99m}$Tc | Tetrofosmin | 250–600 |
| Planar (2 %) | $^{99m}$Tc | Tetrofosmin | 370–400 |
| Lung ventilation | $^{81m}$Kr | Gas | – |
| Kidney | $^{99m}$Tc | MAG3 | 20–200 |
| Kidney | $^{99m}$Tc | DMSA | 23–200 |
| GFR measurement | $^{99m}$Tc | EDTA | 0.2–4 |
| Myocardium (SPECT 87 %) | $^{99m}$Tc | Sestamibi | 388–450 |
| Planar (13 %) | $^{99m}$Tc | Sestamibi | 400–500 |
| Lung ventilation | $^{99m}$Tc | DTPA (diethylenetriamine pentaacetic acid) | 10–2,500 |
| Myocardium (SPECT 98 %) | $^{201}$Tl | Thallous chloride | 55–80 |
| Planar (2 %) | $^{201}$Tl | Thallous chloride | 78–80 |
| Lung ventilation | $^{99m}$Tc | Technigas | 15–300 |
| Thyroid | $^{99m}$Tc | Pertechnetate | 35–180 |
| Thyrotoxicosis therapy | $^{131}$I | Iodide | 185–800 |
| Cardiac blood pool | $^{99m}$Tc | Normal erythrocytes | 370–800 |
| Tumors (PET) | $^{18}$F | FDG (fluorodeoxyglucose) | 222–400 |
| Infection, inflammation, tumors | $^{99m}$Tc | Exametazime | 40–600 |
| Helicobacter pylori test | $^{14}$C | Urea | 0.01–0.2 |
| Kidney | $^{99m}$Tc | DTPA | 12–800 |
| Lung ventilation | $^{133}$Xe | Gas | 200–600 |
| Cerebral blood flow (SPECT 94 %) | $^{99m}$Tc | Exametazime | 72–800 |
| Planar 6 % | $^{99m}$Tc | Exametazime | 200–500 |

Reproduced with permission: Hart and Wall [96]

*diagnostic tests* and *therapeutic examinations*. *Therapeutic examinations* are performed with unsealed sources consisting of *radionuclides* (shown in Table 9.1) or *radiopharmaceuticals* labeled as radionuclides (radiopharmaceuticals) [1].

Radiopharmaceuticals introduced inside a patient's body mostly have two components: a *radionuclide*, an excited state of atom which emits energy so that the atom can convert to a stable form and a *carrier molecule* that travels through the body until it interacts with its target cell tissue or organ system. For example, highly lipophilic molecules (such as exametazime) that penetrate the protective blood

## 9.1 Introduction

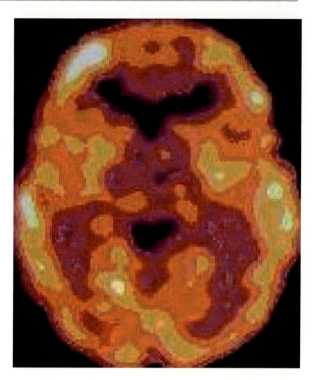

**Fig. 9.1** $^{99m}$Tc exametazime brain imaging (Courtesy, Purdue University, IN)

brain carrier are taken up in brain tissues. Since uptake will be proportional to regional blood flow, the amount of tracer deposited will be indicative of how much blood flow reaches a particular area. Figure 9.1 shows the picture of a brain taken after $^{99m}$Tc (Tc is technetium, its half-life is 6 h, and the end point energy is 0.292 Me, used to *image* skeleton, heart muscle, brain, thyroid, lungs, liver, spleen, kidney, gall bladder, bone marrow, salivary and lacrimal glands) is administered into the body [2, 3].

The characteristics of an ideal radioisotope for therapy are nontoxic, and they are administered in very low concentrations. Not only that, they need careful scrutiny, quite different from those required for *imaging*. For *imaging*, the energy of the radioisotopes is deposited in the camera crystal, without significant absorption in the tissue, whereas the energy of the *therapeutic radioisotope* must be deposited in the tissue to damage the DNA (deoxyribonucleic acid) chains to keep the diseased cells from replicating [4]. DNA plays an important role in cellular multiplication and function, and ionizing radiation can cause deletions or substitutions of bases and/or actual breaks in DNA chains [5].

Unlike magnetic resonance imaging (MRI) [6, 7] and computer tomography (CT), nuclear medicine uniquely provides information about both the structure and the function of body organs within the body. A large portion of the diagnostic work

**Fig. 9.2** Steps showing the final imaging of the embedded radioisotope and the administration of radionuclides

involves imaging, and a small portion of the diagnostic work involves non-imaging parts (the assay of blood samples to measure the glomerular filtration of the kidneys).

No single isotope dominates in radioisotope therapy as $^{99m}$Tc does in nuclear imaging. The design of a successful therapeutic radiopharmaceutical requires (a) selection of the targeting molecule to deliver the radioisotope to the diseased site properly, (b) accurate calculations of the amount of the dose to destroy the affected cells [8], and (c) development of a method for destroying the bad cells without adversely affecting the good cells. The widespread availability of the radioisotope's therapy depends upon the availability of the therapeutic doses, appropriate legends chemistry techniques, emitters at reasonable costs, and long-term therapy without complications such as bone marrow toxicity and renal damage [9].

## 9.2 Administration of the Radionuclides

These radionuclides used in nuclear medicine are administrated into the body either by injection or aggregate form according to the parts of the body or organ that will be studied through the medical imaging [10, 11]. For example, intravenous, subcutaneous is used to investigate the lymphatic system, intrasynovial to take the image of the knee joint, inhalation in gaseous form, like krypton-81 ($^{81}$Kr), for investigating the function of lungs, ingestion ($^{99}$Tc added to scrambled eggs) to investigate gastric emptying, or topical application of the radiopharmaceutical directly to the area to be investigated (e.g., $^{99Tc}$ eye drops to investigate tear duct flow). Depending upon on which type of scan is being performed, the imaging will be done immediately, a few hours later, or even several days after the radiopharmaceutical agent is administered. Figure 9.2 shows the process sequence of the medical imaging and the administration of the radioisotope inside the region of interest (ROI).

Specifically, nuclear medicine can be used to:
(a) Analyze kidney function
(b) Image blood flow and function of the heart
(c) Scan the lungs for respiratory and blood-flow problems
(d) Identify blockage of the gall bladder
(e) Locate the presence of infection

# 9.3 Preparation of Radionuclides

(f) Identify bleeding of the bowel

(g) Measure thyroid function to detect an overactive or underactive thyroid

## 9.3 Preparation of Radionuclides

Most of the radionuclides are made either by bombarding stable atoms or by splitting massive atoms. There are three types of equipment commonly used to introduce radionuclides, namely, generators, cyclotron, and nuclear reactors.

### 9.3.1 Generator

The most commonly used equipment, which is used to make nuclear medicine, is the generator. The most commonly used nuclides used in nuclear medicine is 99 m-Tc ($^{99m}$Tc) which is produced in a highly shielded column of fission product Mo-99 ($^{99}$Mo-parent) bound to alumina ($Al_2O_3$). The $^{99}$Tc (daughter) is milked (eluted) by drawing sterile saline through the column into the vacuum vile. The parent $^{99}$Mo is firmly bound to alumina, and as a result, the eluted $^{99m}$Tc contains negligible amounts of $^{99}$Mo, and the daughter $^{99}$Tc, which is washed away with saline, is collected in a vile (vacuum). From the sterile solution containing $^{99m}$Tc, sodium pertechnetate injection is prepared for applications into the body organs and remains on the column. The generator is called a cow; the daughter nuclide is referred to as milking and surrounding lead shield is called a pig (Fig. 9.3). The $^{99m}$Tc decays to $^{99}$Tc during radioactive decays with principal radiation $\gamma$-ray, having specific $\gamma$-ray constant of 0.19 mGy per MBq-h-1at-1 cm. Figure 9.3 shows a cross-sectional view of the $^{90}$Mo–$^{99m}$Tc generator.

### 9.3.2 Reactor

Radionuclides for the use in nuclear medicine are also produced in nuclear reactor by the process of fission, neutron capture, or transmutation, for example, $^{131}$I, $^{133}$Xe, and $^{99}$Mo. Table 9.2 shows the parent and daughter nuclei and their applications, and the *isotopes* used in nuclear medicine are given in Table 9.3. Atomic nuclei with the same number of protons, but with different numbers of neutrons, are called *isotopes*. *Isotopes* are identified as *stable*, when the *protons* and the *neutrons* in the atom can coexist in a state of peaceful tranquility. On the other hand, the isotopes that spontaneously emit radiation are called radioisotopes. The relative abundance of three isotopes of potassium (K) ($^{39}$K, $^{40}$K, and $^{41}$K, have 19 protons, and (39–19), (40–19), and (41–19) neutrons) are constant, regardless of the source. Even the human body contains almost 150–200 g of potassium (K) of which 20–25 mg exists as a radioisotope ($^{40}$K). The other natural radioisotopes are carbon ($^{14}$C) and hydrogen ($^3$H, tritium) that result from the nuclear reactions of the atmospheric nitrogen (N). During our lifetime, we absorb and excrete $^{14}$C, and as a result, $^{14}$C levels in our tissues gradually increases to an equilibrium level and in course of time it decays due

**Fig. 9.3** Cross-sectional view of $^{99m}$Tc generator (Courtesy: WIKI)

**Table 9.2** Splitting of the nucleus (Parent–daughter relation inside generator)

| Parent | Daughter | Applications |
|---|---|---|
| $^{99}$Mo | $^{99m}$Tc | Radiopharmaceuticals |
| $^{82}$Sr | $^{82}$Rb | Cardiac perfusion imaging |
| $^{81}$Rb | $^{81m}$Kr | Lung ventilation scans |

to its radioactive nature. But the half-life of $^{14}$C is 5,730 years, so the effect of decay is not much noticeable during our lifetime.

A revolution in radiotherapy has been possible for the reactor-based radionuclides, which are β-emitters such as $^{153}$Sm, $^{186}$Re, $^{166}$Ho, $^{177}$Lu, and $^{105}$Rh. These radionuclides are used in more sophisticated radioactive targets including radioactive intra-arterial microspheres, chemically guided bone agents, labeled monoclonal antibodies, and isotopically tagged polypeptide receptor-binding agents [12]. Figure 9.4 shows the picture of the inside of a nuclear reactor.

## 9.3 Preparation of Radionuclides

**Table 9.3** Radioisotopes used in nuclear medicine

| Radioisotopes | Half-life | Application |
|---|---|---|
| Bismuth (Bi)-213 ($^{213}$Bi) | 46 min | Used for targeted alpha therapy (TAT) |
| Chromium-51 ($^{51}$Cr) | 28 days | Use to level red blood cells and quantify gastrointestinal protein loss |
| Cobalt-60 ($^{60}$Co) | 10.5 months | External beam therapy (EBT) |
| Copper-64 ($^{64}$Cu) | 13 h | Study genetic diseases affecting copper metabolism |
| Dysprosium-165 ($^{165}$Dy) | 2 h | Used as an aggregated hydroxide for arthritis |
| Erbium-169 ($^{169}$Er) | 9.4 days | Arthritis joints |
| Holmium-166 ($^{166}$Ho) | 26 h | For diagnosis and treatment of liver tumors |
| Iodine-125 ($^{125}$I) | 60 days | For cancer brachytherapy, diagnoses of the filtrate rate of kidneys, deep vein thrombosis in leg, and radioimmunoassays |
| Iodine-131 ($^{131}$I) | 8 days | Thyroid cancer, liver, renal blood flow, and urinary obstruction |
| Iridium-192 ($^{192}$Ir) | 74 days | Used as wire in internal radiotherapy and cancer treatment |
| Iron-59 ($^{59}$Fe) | 46 days | To study iron metabolism in spleen |
| Lutetium-177 ($^{177}$Lu) | 6.7 days | For small tumors |
| Molybdenum-99 (99Mo) | 66 h | Used as parent in a generator to produce $^{99m}$Tc |
| Palladium-103 ($^{103}$Pd) | 17 days | Prostate cancer |
| Phosphorus-32 ($^{32}$P) | 14 days | Treatment of excess red blood cells (polycythemia vera) |
| Potassium-42 ($^{42}$K) | 12 h | To determine potassium in coronary blood flow |
| Rhenium-186 ($^{186}$Re) | 17 h | Pain relief in bone cancer |
| Samarium-153 ($^{143}$Sm) | 47 h | Treatment in bone, prostate, and breast cancer (Quadramet) |
| Selenium-75 ($^{75}$Se) | 120 days | To study production of digestive enzymes (selenomethionine) |
| Sodium-24 ($^{24}$Na) | 15 h | To study electrolysis within the body |
| Strontium-89 (89Sr) | 50 days | Pain reliever for prostate and bone cancer |
| Technetium-99 ($^{99m}$Tc) | 6 h | Used to image the skeleton, heart muscle, brain, thyroid, lungs, liver, spleen, kidney, gall bladder, bone marrow, salivary, and lacrimal glands |
| Xenon-133 (133Xe) | 5 days | To study pulmonary (lung) ventilation |
| Ytterbium-169 ($^{169}$Yb) | 32 days | To study cerebrospinal fluid in the brain |
| Ytterbium-177 ($^{177}$Yb) | 1.9 h | Progenitor of Lu-177 |
| Yttrium-90 ($^{90}$Y) | 64 days | Used for cancer brachytherapy and pain reliever in arthritis |

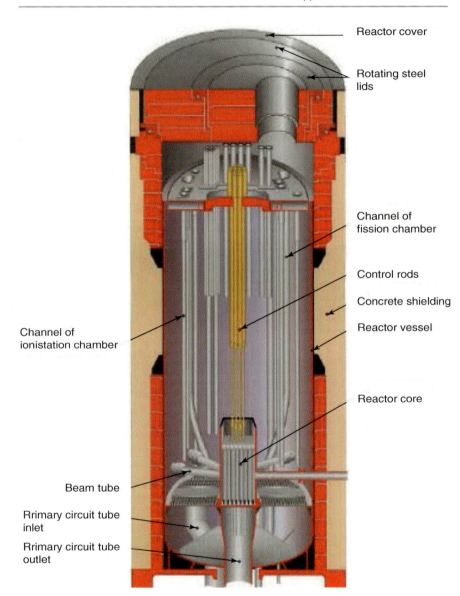

**Fig. 9.4** An LVR-15 light-water moderated and cooled tank nuclear reactor with forced cooling (Photo courtesy UJV)

### 9.3.3 Cyclotron

Ernest Lawrence in Berkeley, CA, reported the invention of cyclotron in 1934 and recognized the possibilities of the machine to produce artificial radioisotopes for the application in medicine. In 1937, Joseph Hamilton was the first to use the radioisotopes to study circulatory physiology [13]. The cyclotron accelerates

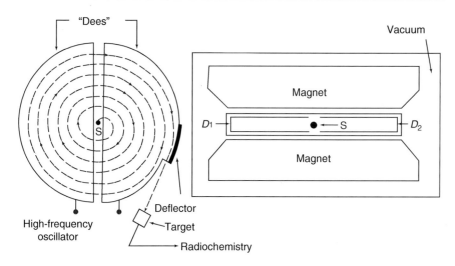

**Fig. 9.5** Schematic representation of a cyclotron

charged particles such as hydrogen nuclei (protons) and heavy hydrogen nuclei (deuterons) to high energies. The molecules of deuterium at the center of the cyclotron are bombarded with electrons whose energy is high enough to produce positive ions during collisions.

Figure 9.5 shows the schematic of a cyclotron. $D_1$ and $D_2$ are the two *dees* made of copper sheet to which a high-frequency oscillator applies the accelerating voltage. The direction of the potential difference across the gap between the *dees* is made to change signs some millions of times per second. The dees are immersed in a magnetic field produced by a large electromagnet, and the space where the ions will move is evacuated to avoid collisions between ions. Table 9.4 shows some of the radioisotopes produced in cyclotron.

## 9.4 Radiation Dose

The direct measurement of photons (*X-ray, gamma rays*, or *bremsstrahlung, beta particles* or *alpha particles*) is often the most accurate method to determine the quantity and location of radioactive material in the human body. The activity of a radioisotope is defined as its rate of decay, $\{(dN/dt)=(\lambda N)\}$, where $N$ is the number of radioactive nuclei, $t$ is the time to decay, and $\lambda$ is the decay constant. The most commonly used unit for radioactivity is expressed in curie (Ci), which means 3.7 $10^{10}$ disintegrations per second. However, the SI unit of radioactive dose is becquerel (Bq).

A patient undergoing a therapy is administered a dose of the nuclear medicine which can range from 0.006 mSv (millisieverts) for a 3 MBq (megabecquerels) of chromium (Cr)-51 (EDTA measurement of glomerular filtration rate) to 37 mSv for

460 9 Instrumentation and Its Applications in Nuclear Medicine

**Table 9.4** Isotopes produced in cyclotron

| Radioisotopes | Half-life | Application |
| --- | --- | --- |
| Carbon-11 ($^{11}$C) | 20.4 min | Used in PET as positron emitters to study neuropharmacology and dementia |
| Cobalt-57 (57Co) | 272 h | Used as a marker to estimate organ size |
| Fluorine-18 ($^{18}$F) | 109.8 min | Used in PET as positron emitters to detect cancers cells |
| Gallium-67 (67Ga) | 78 h | For tumor imaging and location of inflammatory lesions |
| Indium-111 ($^{111}$In) | 2.8 h | Used for diagnostic studies of the brain, infection, and colon transit |
| Iodine-123 ($^{123}$I) | 13 h | Used in the diagnosis of thyroid function |
| Krypton-81m ($^{81}$Kr$^{m}$) | 13 s | To image pulmonary ventilation for asthmatic patients and lung disease |
| Nitrogen-13 ($^{13}$N) | 10 min | Used in PET as positron emitters to study neuropharmacology and dementia |
| Oxygen-15 ($^{15}$O) | 2 min | Used in PET as positron emitters to study neuropharmacology and dementia |
| Rubidium-82 ($^{82}$Rb) | 65 h | To image myocardial perfusion |
| Strontium-92 ($^{92}$Sr) | 65 h | Used as parent in a generator to produce Rb-82 |
| Thallium-201 ($^{201}$Th) | 73 h | To diagnostic studies of coronary artery disease and location of lymphomas |

a 150 MBq of thallium (Tl)-201 for nonspecific tumor imaging. However, for bone scan with 600 MBq of Tc-99 m-MDP, the effective dose is 3 mSv [14, 15].

## 9.5 Radiation Therapy

*Radiation therapies* that are most frequently performed in nuclear medicine can be categorized as (Radiological Society of North America Inc., RSNA):
1. *Brain tumors*
2. *Breast cancer*
3. *Head and neck cancer*
4. *Lung cancer*
5. *Prostate cancer*
The procedure to cure these cancers is done by exposing the cancerous cells by external radiation by adopting any one of the therapies according to the suitability such as (a) *external beam therapy* (EBT), (b) *intensity-modulated radiation therapy* (IMRT), or (c) *stereotactic radiosurgery* (SRS).

### 9.5.1 External Beam Therapy (EBT)

In *external beam therapy* (*EBT*), high energetic X-ray is administered to the location of the affected cells. The beam is generated outside the patient's body

9.6 Intensity-Modulated Radiation Therapy (IMRT) 461

usually by either a *linear accelerator (LINC)*, *orthovoltage X-ray machines*, *Cobalt-60 machines*, *proton beam machines*, or *neutron beam machines*. In EBT no radioactive sources are placed inside the patient's body. EBT can be successfully applied to treat any of the following diseases such as (1) breast cancer, (2) colorectal cancer, (3) head and neck cancer, (4) lung cancer, and (5) prostate cancer as well as many other cancerous cells inside a specific area of the body.

### 9.5.2 Proton Beam Therapy

It is a form of external beam radiation treatment that uses protons rather than X-rays to treat certain types of cancer and other diseases. The physical characteristics of the proton therapy beam allow doctors to more effectively reduce the radiation dose close to nearby healthy tissues.

### 9.5.3 Neutron Beam Therapy

It is a specialized form of external radiation, which consists of neutrons instead of X-rays. Neutron beam therapy is a powerful tool, which is used to treat certain types of tumors that are radiation resistant, meaning that they are difficult to kill by using ordinary X-ray or proton beam therapy. As a matter of fact, neutrons have a greater biological impact on cells than other types of radiation.

## 9.6 Intensity-Modulated Radiation Therapy (IMRT)

In intensity-modulated radiation therapy (*IMRT*), a computer-controlled X-ray accelerator is used to deliver a precise radiation dose to a malignant tumor or to the specific areas of a cancerous cell [16, 17]. The radiation dose is designed to conform to the three-dimensional (3-D) shape of the tumor by modulating or controlling the intensity of the radiation beam [18]. The equipments that are very frequently used to administer the radiation from outside are (1) *linear accelerator (LINAC)*, (2) *Gamma Knife*, and (3) *proton beam*. In IMRT, the ratio of normal tissue dose to tumor dose is reduced to a minimum. As a result, higher and more effective radiation doses can safely be delivered to tumors. Moreover, it reduces the treatment toxicity even when the doses are not increased. The most important of all is that the patterning of the radiation allows doctors to deliver radiation to different places in the body (e.g., paranasal sinuses) that have been difficult with other external radiation therapy.

### 9.6.1 Linear Accelerator (LINC)

LINC is a device mostly used for external beam radiation treatments for cancer patients [19]. In *linear accelerator (LINC)*, electrons are accelerated by microwave technology similar to that used in radar. The waveguide then allows these electrons

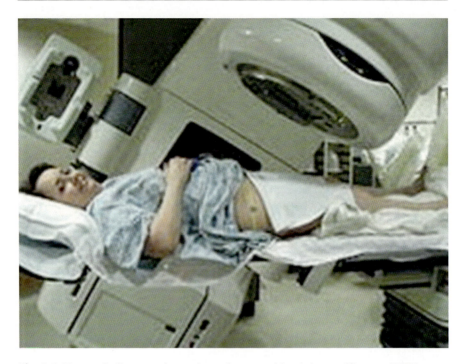

**Fig. 9.6** Picture of a linear accelerator in use for external beam therapy (Courtesy: WIKI)

to collide with a heavy metal target. A portion of these X-rays coming out of the accelerator (called gantry) are collected and shaped into a narrow beam. The beam is usually rotated around the affected area of the cancerous cells of the body (Fig. 9.6).

The safety of the patient is very important during treatment, and the radiotherapist should continuously monitor the patient through a closed-circuit television monitor [20]. The development of a compact and fully digital LINC provides beam accuracy, treatment automation, and patient comfort—translating maximum radiation dose to the tumor, minimum dose to the surrounding tissue, and less time on table for the patients. Modern radiation machines have internal checking systems to provide further safety so that the machine will not turn on until all the treatment requirements are met.

### 9.6.2 Gamma Knife

The Leksell Gamma Knife is a neurosurgical device used to treat mostly brain tumors with radiation therapy. The device was invented by Lars Leksell, a Swedish neurosurgeon, in 1967 at the Karolinska Institute in Sweden. The procedure involves a highly targeted radiation source, which is focused precisely on the target from many different directions, and it has brought a revolutionary type of surgery in

## 9.6 Intensity-Modulated Radiation Therapy (IMRT)

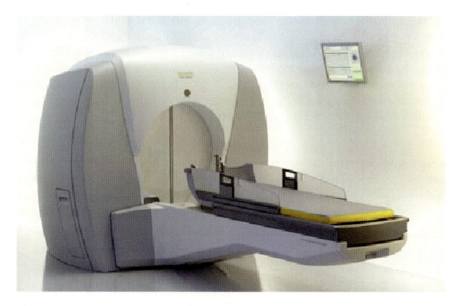

**Fig. 9.7** The detailed drawing of the Gamma Knife (Courtesy: Virginia Med., VA)

the field of nuclear medicine. The individual radiation beam is weak enough to harm the brain tissue it passes through. The beams of the gamma-ray radiation called *blades* are programed to target the lesion at the point where they intersect. The exposure session is generally brief and precise. Thus, the Gamma Knife is able to cut deep into the brain without using a scalpel at all.

*Gamma Knife* surgery is preferred to noninvasive (non-incision) surgery for brain tumors, arteriovenous malformations, and brain dysfunction like neuralgia [21]. It is an alternative nonsurgical operation for many patients for whom the traditional surgery is not an option. The greatest advantage of the Gamma Knife technology is that it allows treatment of inoperable lesions. The risks of Gamma Knife radiosurgery treatment include but are not limited to radiation necrosis, secondary malignancy head frame, paralysis, and death.

Figure 9.7 shows the picture of a Gamma Knife setup showing in details of the machine with the radiation source which can be focused from all directions. The setup also includes computerized data from imaging tests, to pinpoint areas within the brain and destroy them using multiple beams of gamma radiation [22]. Cumulative effect of the gamma rays from all directions produces a high dose of radiation at the exact site of the lesion, without damaging the healthy cells.

### 9.6.3 Proton Beam

The *proton beam therapy* or radiosurgery has properties different from the radiation used by *Gamma Knife* and *LINAC* machines. Proton beams deliver almost all their radiation at a set distance from the radiation source, rather than gradually releasing the energy as the beam travels. This difference eliminates the need for multiple beams from different directions as we have seen in *Gamma Knife* operation. To improve precision, a head frame is used and metallic beads are also inserted under the skin to help the computer triangulate tumor volume. Sometimes, the instrument might have the facility to rotate the body so that the proton beam can be focused on the tumor from different angles. The main drawback with the process is that it required a highly specialized particle accelerator called synchrocyclotron, which made the processing costly [23].

Proton radiation therapy is considered as one of the most precise forms of noninvasive image-guided cancer therapy [24]. It is based on the well-defined range of protons in material, with low entrance dose and a maximum (Bragg peak) and a rapid distal dose falloff, providing better sporting of healthy tissue and allowing higher tumor doses than conventional radiation therapy with photons [25, 26].

### 9.6.4 Stereotactic Radiosurgery (SRS)

Stereotactic radiosurgery (SRS) is a nonsurgical procedure that applies radiation therapy in a highly precise form to treat brain tumors and other abnormalities of the brain. The radiosurgery in SRS has such a dramatic effect in the target zone that the changes are considered as *surgical* [27]. The treatment involves the delivery of a single high dose or smaller multiple doses of radiation from an external source such as (a) *linear accelerator* or *cyclotron*, (b) *Gamma Knife*, and (c) *proton beam*. The SRS does not remove the tumor; instead, it destroys the DNA of tumor cells. It provides an *external frame of reference* for the subsequent radiation treatment planning. The linear accelerator, the cobalt source machines, and proton radiosurgeries are three different types of machines generally used in SRS.

## 9.7 Scintigraphy

The second phase of the test is the *medical imaging* with the help of a detector. It is referred to as *radionuclide imaging* or nuclear *scintigraphy* [28, 29]. In diagnosis, radioactive substance is administered to a patient, and the radiation emitted from the radionuclide is measured. The majority of these diagnostic tests are performed with *gamma camera*. In *diagnostic radiology*, imaging of the affected tissues has brought revolution in the detection of the affected diseased cells. In diagnostic radiology the imaging technologies are different according to the way the procedures are followed. These are:

1. *Angiography*
2. *Computed tomography (CT)*
3. *Positron emission tomography (PET)*
4. *CT/PET*
5. *Interventional radiology (IR)*

9.7 Scintigraphy

6. *Magnetic resonance imaging (MRI)*
7. *Mammography*
8. *Nuclear medicine (NM)* (*general, cardiac, pediatric,* and *PET*)
9. *Ultrasound* (*sonography*)
10. *X-ray radiography*

## 9.7.1 Angiography (CTA, MRA)

### 9.7.1.1 CTA

CTA, the computed tomography angiography as it stands for is the name of a procedure that uses X-ray to produce picture (angiogram) of the area of interest (AOI). Angiography examinations remain the method of choice for vascular diagnostics, although many are replaced by noninvasive methods to visualize blood flow in arterial and venous vessels throughout the whole body. Computerized analysis of the captured images is routinely performed with great precision. Compared to catheter angiography, CTA is much less invasive and patient-friendly procedure [30].

The procedure requires the injection of a radiopaque called *contrast agent* or *dye* that absorbs X-ray. The procedure is noninvasive, and it requires a very tiny tube with a special shape to radiopaque a particular artery or vein that will block the X-rays and will cast a shadow of the injected artery or vein on to the X-ray film or fluoroscope. The image of the artery or the vein is analyzed to detect if there is any alteration, blockage, obstruction, or narrowing present. Figure 9.8 shows (a) arch aortogram oblique and (b) lateral ICA arteriogram showing vascular malformation.

Compared to catheter angiography where a sizable catheter is placed and contrast material is injected into the artery or vein, CTA is much less invasive and most friendly procedure to the patient [31]. It is generally used to (a) examine the pulmonary arteries in the lungs, (b) visualize blood flow in the renal arteries, (c) identify aneurysms and dissection in the aorta, (d) identify arteriovenous malformation, (e) detect atherosclerotic disease, (f) detect thrombosis (blood clotting in the vein), and (g) detect narrowing or obstruction in the pelvis and in the carotid arteries. However, the disadvantage of CTA is its inability to reliably image small twisted arteries or vessels in organs that move rapidly.

Figure 9.9 shows a specialized CT scanner with a circular opening. The scanner rotates inside the frame of device, and it spins around the patient and carries an X-ray tube mounted on one side with a banana-shaped detector opposite to it. The patient is injected with a contrast material, and he or she does not feel any pain on that area. The whole procedure generally takes an hour and half.

### 9.7.1.2 MRA

*MRA* or *magnetic resonance angiography* is a magnetic resonance imaging system that uses radio waves and a magnetic field to visualize the structure of the blood and its flow within blood vessels. The test can reveal narrowing of vessel (stenosis) and obstructions in the flow of blood through the vessels and help physicians to

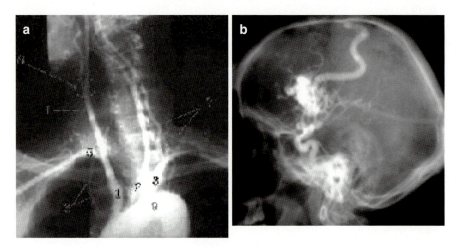

**Fig. 9.8** (**a**) Arch aortogram oblique and (**b**) lateral ICA arteriogram showing vascular malformation (Photo courtesy, The University of the Health Sciences, Bethesda, MD)

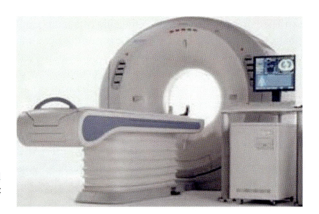

**Fig. 9.9** CT equipment used for CTA operation (Courtesy: WIKI)

diagnose and treat vascular disease. MRA has the similar image resolution of CTA, but it does not require the use of contrast agents [32].

The techniques most commonly used for MRA can be classified into two major categories, time of flight and phase contrast. Both techniques rely on separate physical effects and will result in images with different information about the vasculature. More recently, the use of contrast agents in combination with ultrafast $T_1$-weighted imaging sequences has shown significant improvements in the delineation of the vessel lumen. It goes without saying that proper use of MRA techniques and correct interpretation of the angiographic images need a knowledge of the underlying physical mechanisms of the flow sensitivity in magnetic resonance imaging (MRI) [33, 34].

Physicians working in cardiovascular disease consider magnetic resonance angiography (MRA) an important procedure to noninvasively visualize vascular disease.

It has the potential to replace conventional X-ray angiography (CA) that uses iodinated contrast. For several years, the interest has been simulated by the current emphasis on cost containment, outpatient evaluation, and minimally invasive diagnosis and therapy. Moreover, recent advances in magnetic resonance (MR) technology resulting from fast gradients and use of contrast agents have allowed MRA to make substantial advances in many arterial beds of clinical interest [35]. MRA continues to show rapid evolution because of its ability to portray blood vessels in a projective format similar to CA. The hallmark of the last few years has been the widespread interest in contrast-enhanced MRA (CEMRA), which has essentially displaced many of the flow-dependent MR strategies used previously [36].

## 9.7.2 CT

*CT* or *computerized tomography* uses X-rays and computer to make an image of the sections of the body. It is sometimes called as a *CAT scan* or computed axial tomography *(CT or CAT scan)*. It is a specialized branch of X-ray imaging technique. Since its invention in the year 1970 by Hounsfield's and the introduction of industrial CT in 1980s, the CT imaging system has reached to its maturity [37]. The historical development of the first-generation CT with rotate/translate and pencil beam finally moved to the fifth generation with stationary/stationary with beam scanner with 50 ms scan times with tungsten target. The superimposed of the overlying tissues on the image is overcome by scanning thin slices of body with a narrow X-ray beam, which rotates around the body. A CT scan shows the body organs in greater details than regular X-rays [38]. Moreover, computed tomography can differentiate between tissues with similar density, such as soft tissue and fluid, because of the narrow X-ray beam. For CT scan, the patient receives a contrast-enhancing agent by intravenous (IV) line, which helps to capture clearer images. The benefits of CT do not come without price. Firstly, while CT does excellent job of predicting structures and anatomy, it sometimes fails to detect small or early-stage tumors. Secondly, patient preparation for CT requires certain amount of X-ray dose when magnetic resonance imaging (MRI) can do the job without any X-ray exposure to the patient. Thirdly, CT is more expensive than conventional radiography. Figure 9.10 shows the schematic of the first-generation CT scanners that used a rotational motion.

Figure 9.11 shows the schematic of the setup of a fifth-generation CT scanner. The fifth-generation *cine-CT* scanner does not use a conventional X-ray tube but rather a large ring of tungsten that circles the patient, which lies directly, opposite to the detector ring [39]. The scan time of cine-CT system is 50 ms and it can produce 17-CT slices each second. The recent improvements in X-ray tube design, computer speed with helical scanning and attenuation correction, have brought many exciting results [40]. The motion of the helical scanner is in fact helical not spiral and the speed of the table motion relative to the rotation of the CT gantry is very important.

The slice thickness in CT is controlled by distance between the two lead jaws associated with the X-ray beam. However, the width of the detectors in the detector

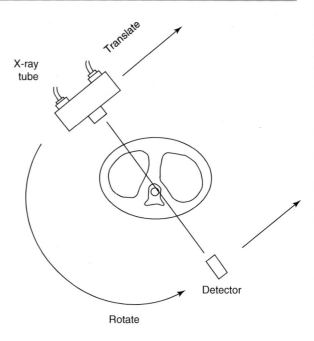

**Fig. 9.10** Schematic of first-generation CT scanners that used a rotational motion (Reproduced after Bushberg et al., with permission from LWW Pub)

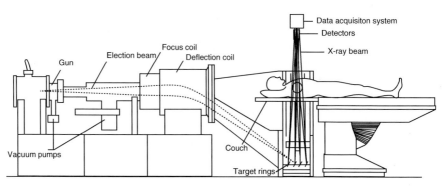

**Fig. 9.11** Schematic of the setup of a fifth-generation CT scanner (Reproduced after Bushberg et al., with permission from LWW Pub.)

array places a maximal limit on the slice thickness [41]. It has been found that narrower slice thickness improves spatial resolution but requires increased tube current per slice to maintain the same statistical integrity (signal to noise ratio, SNR).

There are many parameters that influence image quality in CT. But the spatial resolution in CT is much lower than that of screen film radiography. For example, the Nyquist limit for 0.5-mm pixel in CT is 1.0 line per mm vs. the 10 line pairs per mm that is possible with mammographic screen film system.

9.7 Scintigraphy

In radiography, the resolution characteristics are approximately the same over the entire extent of the image, whereas in CT, the resolution characteristic changes depending on the location in the image. In digital imaging, the size and spacing of the detector will influence resolution. It is also true that the voxel of a given CT image depends upon the thickness of the detector coupled with the slice thickness collimators [42].

The radiation dose in CT is higher than the conventional screen–film radiography and the typical doses in CT are the following: head ~65 mGy (6.5 rad), spine ~23 mGy (2.3 rad), and body ~15–25 mGy (1.5–2.5 rad). The dose near the entrance of the tissue receives the most dose. As it penetrates deeper, the dose is distributed exponentially as $D = (D_0 A/\mu) (1 - e^{-d\mu})$, where $D_0$ is the entrance dose, $d$ is the thickness of the patient, and $\mu$ is the linear attenuation coefficient of the X-ray beam [43, 44].

The X-ray used in CT is polyenergetic even after filtration, which causes problem. As a result, the lower-energy photons are attenuated ($\mu(E)$) more rapidly than the higher-energy ones. Moreover, the effective energy, which is a function of the transmission path, causes *hardening artifacts*. It occurs in soft tissue regions and results in decreased image intensity. The other two artifacts that are observed in CT are *volume artifacts* and *motion artifacts*. The first one is the result of a variety of different tissue types being contained in a single voxel. It is believed that a solid knowledge of cross-sectional anatomy is the best defense against *partial volume artifacts* leading to incorrect diagnosis. On the other hand, the *motion artifacts* arise mainly due to the motion that occurs halfway through the CT scan acquisition. Rapid scan times are the best defense against this type of defect.

CT scan is performed after the injection of a contrast agent. The computer then processes the information to create an image of the affected area of interest (AOI). The images are called sections or cuts because they appear to resemble the cross sections of the body. Figure 9.12 shows a non-contrast axial CT image of tuberous sclerosis.

Recent development in 3.5 CT system will enable users to employ advanced CT viewing and data manipulation tools to any PC in a health-care enterprise. It will also promote the newest version of the CT to reduce dose (in the range of 3 mSv or less) (Fig. 9.13).

### 9.7.3 PET

*PET* or *positron emission tomography* creates an image of a patient's body showing the biological activity like metabolism or the rate at which the body cells breakdown and use sugar. PET is highly sensitive in detecting even the slightest changes in the body cells, but unfortunately, it cannot pinpoint the exact size or location. PET utilizes radionuclide tracer techniques that produce images of the vivo radionuclide distribution, and like CT, the resulting images represent cross-sectional slices through the patient. However, with PET, the image intensity reflects organ function rather than anatomy [45, 46].

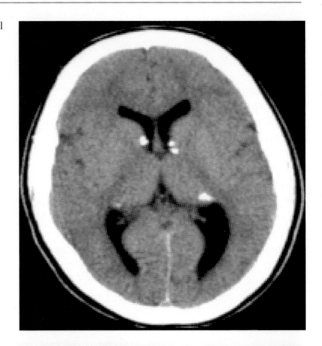

**Fig. 9.12** Non-contrast axial CT (tuberous sclerosis) (Photo courtesy, The University of the Health Sciences, Bethesda, MD)

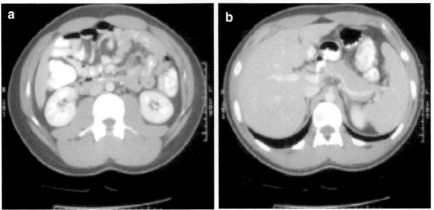

**Fig. 9.13** Normal abdominal CT, oral and IV contrast. (**a**) Level of liver and spleen. (**b**) Level of kidneys (Photo courtesy, The University of the Health Sciences, Bethesda, MD)

The functional information depicted in PET images depends upon the radiopharmaceutical, and the radioisotopes used in PET can reach a more stable configuration by the emission of positron. Compared to single-photon emission computed tomography (SPECT), PET images are not only of higher quality than SPECT but can be accurately quantified in terms of local activity concentration (MBq/ml) [47]. In addition, most clinical PET scanners are full-ring devices that can measure projections at all angles simultaneously, whereas SPECT cameras require time to

## 9.7 Scintigraphy

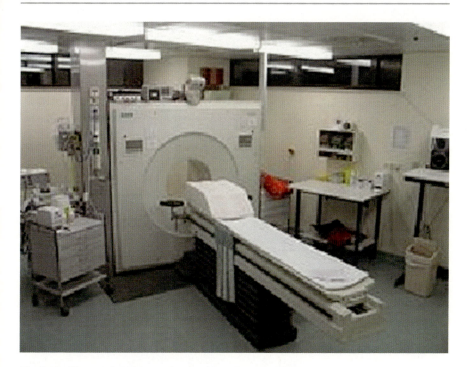

**Fig. 9.14** Picture of a PET scanning machine (Courtesy:WIKI)

rotate the detector heads about the patient. Although some dedicated PET scanners employ sodium iodide (NaI), most tomographs use detector materials with a higher stopping power for energy 511-keV photons. Figure 9.14 shows the picture of a PET scanning machine.

The greatest advantage with PET is that the system can be used in the study like neurological and non-neurological disorders and in biomedical research that quantitatively demonstrates scintigraphic representation of tumor physiology as well as anatomy. Thus, it is a unique advantage over other cross-sectional imaging methods such as CT. Unlike other imaging modalities, PET is a technique used to detect cancerous cells, especially, of the breast, brain, lung, colon, or prostate or lymphoma. It is also used to study brain's blood flow and its metabolic activity. In addition, PET scan helps a doctor to find nervous system problems, such as Alzheimer's disease, Parkinson's disease, multiple sclerosis, transient ischemic attack, amyotrophic lateral sclerosis (ALS), Huntington disease, stroke, and schizophrenia. Figure 9.15 shows some of the pictures acquired from PET scan to study some neurological disorder due to Parkinson's (9.15a) and different pathological alterations of tissues due to Huntington diseases (9.15b).

The energy of the gamma photons in PET is 511 keV, which is too low for pair production. However, Compton scattering process cannot be ignored and its effect can deteriorate image resolution. When detected, the 180° emission of two γ-rays following disintegration (combination of positron and electron to form

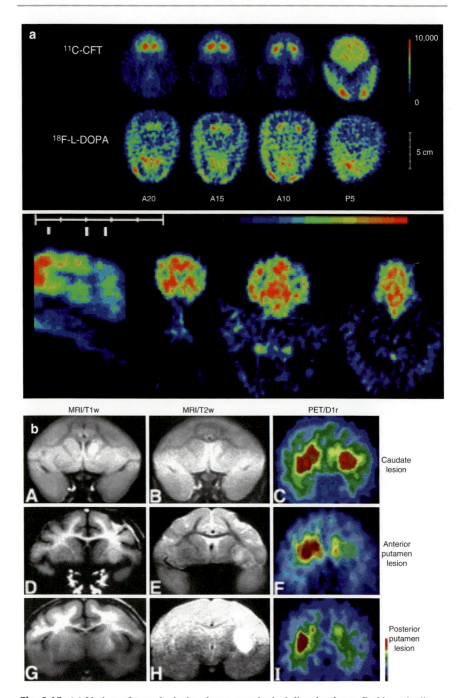

**Fig. 9.15** (**a**) Variety of neurological and non-neurological disorder due to Parkinson's disease (Photo courtesy Dr. A. L. Brownell, MGH, Boston, MA). (**b**) Comparative analysis of the pictures taken by MRI and PET showing different pathological alterations of tissues (lesions) due to Huntington's disease (Photo courtesy Dr. A. L. Brownell, MGH, Boston, MA)

positronium) of positronium is called coincidence line that provides a unique detection scheme for forming tomographic images with PET. In PET study, a patient is administered a positron-emitting drug by injection or inhalation. The isotope circulates through the blood stream to reach the area of interest (AOI). As positron annihilation (Fig. 9.16) occurs, the tomograph detects the isotope's location and concentration.

The ring of squares schematically represents one ring of detectors in a PET scanner (Fig. 9.16), which may, for example, have 16 such rings for simultaneous tomography of many transaxial slices.

These are detected when they reach a scintillator in the scanning device, creating a burst of light, which is ultimately detected by PMT or silicon avalanche photodiode (Si-APD). The ring detector shown in Fig. 9.17 has 16 detector blocks. If detectors on one side of the gantry are only allowed to be in coincidence with the detectors on the opposite side of the same ring, the detected coincidence events are referred to as direct coincidence. To improve the axial sampling and slice sensitivity, detectors can be allowed to be coincident with detectors in the neighboring rings, referred to as cross coincidence. The direct and cross coincidences are combined together to detect the pair of photons moving in approximately opposite directions [48].

Each detector is operated in multiple coincidences with many detectors across from it, thereby defining coincidence-sampling paths over many angles (fan-beam response) (Fig. 9.18) [49]. Many such parallel coincidence result in high linear sampling affecting the final image quality (Fig. 9.19).

Each pair of parallel and opposite detectors produces a coincidence line, which is unique in terms of location and direction. A large number of such coincidence lines form the data. The tomograph's reconstruction software then takes these data measured at all angles and linear positions to reconstruct the image (Fig. 9.20).

The use of the detector material mostly determines the cost of the instrument. Therefore, the material used should be friendly, easy to handle, and cost-effective (e.g., LSO is much more expansive because of the $Lu_2O_3$ material) [50].

For most clinical PET applications, the intrinsic resolution is found to be approximately 6 mm which can reconstruct a high-quality image with final resolution of 8–10 mm. The system also provides a sampling distance of approximately 3 mm in all spatial directions, which is suitable for true 3-D volumetric imaging. In clinical setting, it is unlikely that the raw data will be reprocessed after final images have been produced [51]. It is therefore necessary to store only image data in a form that can be retrieved within a few minutes. Figure 9.21 shows the schematic of dataflow and archiving.

The important aspects of clinical PET study are the necessity of minimum loss in image resolution and fast and efficient processing of software and hardware. The hardware-induced artifacts in PET are most of the time caused by failure in the detection system or the subsequent sorting system is a fanlike ripple (Fig. 9.22).

Another possible artifact introduced in PET is ellipse artifact, which is due to the over or under estimation of the elliptical diameter. Figure 9.23 is a simulated

**Fig. 9.16** PET scan instrument and visualization of positron annihilation events (Photo courtesy Tutorial UCLA, CA)

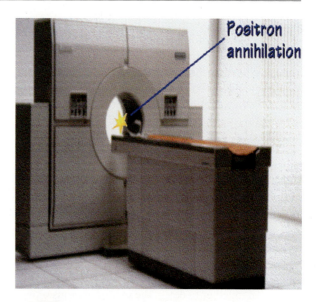

**Fig. 9.17** A PET scanner with one of the ring detectors for tomography (Photo courtesy Tutorial UCLA, CA)

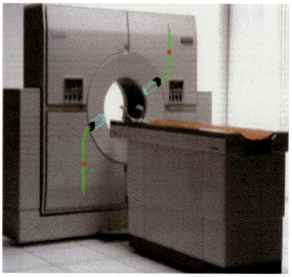

cylinder used to demonstrate the effect of under- and overestimating the ellipse dimension.

Positron emission tomography (PET) has already established its place in noninvasive diagnostic imaging technique, multimodality molecular imaging, to provide metabolic information related to tumor cells, detection of numerous cancerous cells but not limited to breast, head, neck, ovary, and gastrointestinal tract.

The emergence of PET has opened a new era to demonstrate tumor physiology and anatomy. Recent development of PET–CT has not only channeled fusion images but also enable attenuation correction in an approximate manner [52].

## 9.7 Scintigraphy

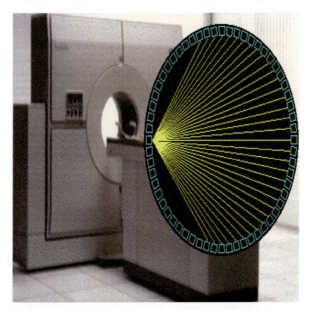

**Fig. 9.18** Fan beam due to multiple coincidence with many detectors (Photo courtesy Tutorial UCLA, CA)

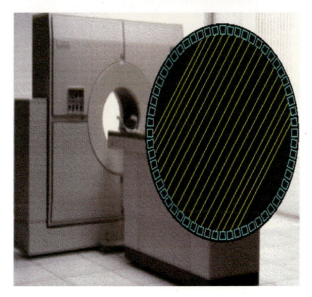

**Fig. 9.19** Parallel coincidence paths at a given angle of incidence of the ray (Photo courtesy Tutorial UCLA, CA)

The scintillation detector to detect nuclear radiation has been fabricated using scintillating materials that can be organic, inorganic, plastics, gases, or glasses. The advantage of inorganic scintillators lies in their greater stopping power due to their higher density and higher atomic number $Z$. They also have some of the highest light outputs (number of photons emitted in the visible range of wavelength). For example, bismuth germanate or BGO ($Bi_4Ge_3O_{12}$) has higher stopping power than barium fluoride ($BaF_2$), but time response is not as high as $BaF_2$. As a result,

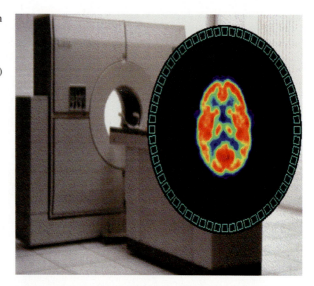

**Fig. 9.20** The reconstruction of the image from the data taken from all angular and linear positions (Photo courtesy Tutorial UCLA, CA)

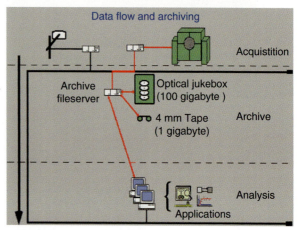

**Fig. 9.21** A schematic of dataflow and archiving (Photo courtesy Tutorial UCLA, CA)

BGO is used for ordinary PET, but in time-of-flight positron emission tomography (TOFPET) applications, $BaF_2$ is a better choice, because of the higher resolution [53–56].

### 9.7.4 Single-Photon Emission Computed Tomography (SPECT)

Single-emission computed tomography (SPECT) is a technique which measures the emission of single photons of a given energy from radioactive tracers to construct images of the distribution of the tracers in the human body [57]. A SPECT scan which integrates two technologies to construct the images primarily (CT and

**Fig. 9.22** Picture of a fanlike ripple due to the failure of the single detector system (Photo courtesy Tutorial UCLA, CA)

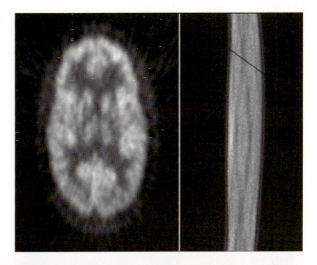

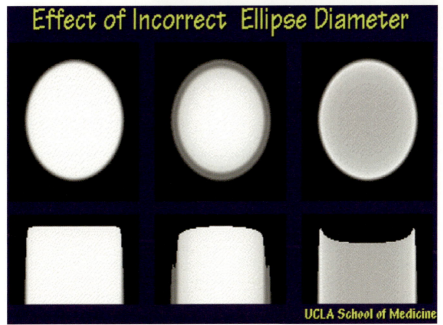

**Fig. 9.23** Demonstrates the artifacts introduced in the ellipse diameter due to under- and overestimating ellipse dimension (Courtesy, UCLA, CA)

radioactive tracer) enables to view how blood flows through arteries and veins of the brain (Fig. 9.24).

Figure 9.24 shows the SPECT image of the flow of blood in the brain [58]. The first one shows the blood flow in normal condition, the second one shows the blood flow when condition attention-deficit/hyperactivity disorder prevails, the third one

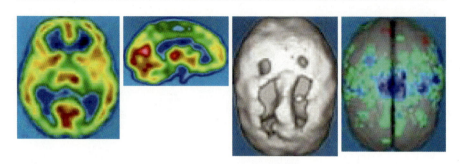

**Fig. 9.24** SPECT image of the flow of blood in the brain (Courtesy:WIKI)

shows the blood flow in an injured brain, and the last one shows the blood flow when a person is suffering from Alzheimer's disease [59]. For example, to diagnose *ischemic heart disease* (IHD), myocardial perfusion imaging (MPI) is performed to study the flow of blood under stress condition. SPECT imaging is performed after the administration of a specific cardiac pharmaceutical, for example, $^{99m}$Tc, followed by induced myocardial stress [60, 61]. Similarly, the blood flow in the brain is tightly coupled to brain metabolism and energy. SPECT imaging can be used to study to assess brain metabolism regionally, in an attempt to diagnose and differentiate casual pathologies of *dementia* and Alzheimer's disease, where multiple strokes cause patchy loss of cortical metabolism [62]. SPECT scan is also useful for presurgical evaluation of origin of medically uncontrolled seizures. As a matter of fact, SPECT can be used to compliment any gamma imaging study, where a true 3-D representation can be helpful.

Figure 9.25 shows a schematic outline of a SPECT camera used to document images. The camera is connected to a large scintillator coupled to multiple photomultiplier tubes (PMTs) or avalanche photodiode (APD) which detects the radiation enhancing the body. The camera can be positioned at different angles around the body accumulating as many as $180°$ views at specific angular intervals. Software then allows integration of all individual projection views into a composite data set, which can be redisplayed as tomographic slices.

One of the major problems in SPECT imaging is the patient-motion artifacts, which cause a major diagnostic image-quality degradation [63, 64]. It has been shown that the polaris stereo-infrared real-time motion tracking system can provide a vendor specified 0.35-mm accuracy and 0.2-mm repeatability. However, the drawback of this method is that the motion during respiratory motion (RM) and rigid body motion (RBM) information are mixed together. Recently, an algorithm based on neural network could be useful in separating the two motions so that each motion can be corrected individually [65].

### 9.7.5 Magnetic Resonance Imaging (MRI)

MRI or magnetic resonance imaging is the nondestructive medical imaging technique which uses radio-frequency waves and a strong magnetic field rather than X-rays to

## 9.7 Scintigraphy

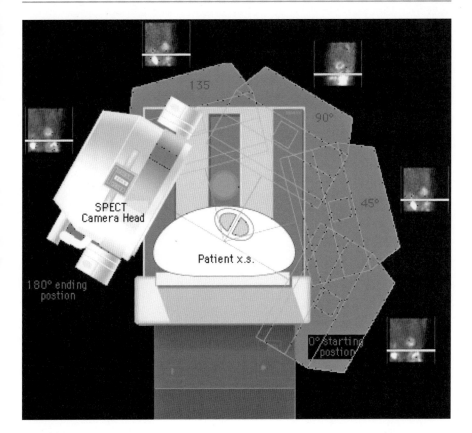

**Fig. 9.25** A SPECT nuclear imaging camera (Photo courtesy Yale University, school of Medicine, CT)

create images. MRI detects the signal that is emitted by the reradiation of the irradiated protons during their return to the ground position or to their initial state. Protons absorb energy when exposed to electromagnetic waves and the amount of energy absorbed by the protons or the strength of the MRI signal depends upon the density of the protons [66].

MRI allows not only diagnostic imaging but it can track some metabolic processes by *tagging* biological molecules with probes and using time lapse imaging. Because MRI can give clear pictures of soft tissue structures near and around bones, it is considered to be one of the most sensitive diagnostic tools for spinal and joint problems. As a matter of fact, it is an alternative to traditional *mammography* in the early diagnosis of breast cancer. Figure 9.26 shows the picture of magnetic resonance imaging (MRI) equipment.

Recently, a technique known as *functional magnetic resonance imaging* (FMRI) or *blood-oxygenation-level-dependent imaging* (BOLD) has been used to differentiate

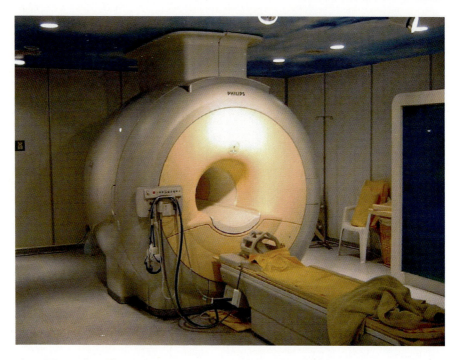

**Fig. 9.26** Picture of a 3 T MRI system (Courtesy Philips Med. USA)

the magnetic susceptibility of oxygenated and deoxygenated hemoglobin, and is now being used routinely to study the basic function of the brain [67, 68].

Most MRI system uses a huge superconducting magnet (SM) with many windings of wire (coils) bathed in liquid helium [69]. In addition to that, it has two small magnets structurally similar to the big magnet, three gradient magnets, and a set of coils that transmit radio-frequency waves into patient's body part/parts being imaged.

The SM coils carry sufficient amount of electricity to create a magnetic field of 0.5 T (~5,000 G) to 2 T (20,000 G) and are insulated from the surrounding by creating a vacuum around the magnet. On the other hand, the two small magnets maintain a magnetic field of 0.3 T and the three gradient magnets provide a magnetic field lower than the two small magnets (from 0.18 T (180 G) to 0.27 T (270 G)).

The three gradient magnets residing inside the main magnet effectively affect the main magnetic field on a local level when they are turned on and off. This helps to pick up an area of interest (AOI) for imaging, which is referred to as the *slice*. Multiple slice acquisition techniques are followed to reduce the average time per slice in spin echo sequence.

Under the influence of a strong external magnetic field, the protons become magnetized and align along the magnetic field, resulting in a net detectable magnetic moment ($B$). The interaction of the proton's magnetic field causes a torque on the proton causing a precession about its axis. The precessional motion occurs at an angular frequency $\omega$, which is the Larmor frequency. It is proportional to the magnetic

# 9.7 Scintigraphy

field $B_0$ and nuclear gyromagnetic ratio $\gamma$, whose value depends on the nuclear species ($\omega = \gamma B_0$) [70, 71]. For hydrogen protons, the gyromagnetic ratio is equal to 42.6/MHz/T (T=tesla).

The magnetization vector $M$ has two components, the longitudinal $M_L$ and the transverse $M_T$. The loss of $M_T$ occurs very quickly (from 0.37 to 0), where as the excited sample takes longer time to come to equilibrium condition [72]. Two signals are obtained from proton's realignment with static magnetic field. These are measured as time constants, $T_1$ (longitudinal time constant) and $T_2$ (transverse time constant). Experimental observations show that the molecular structure and chemistry of the tissues affect $T_1$ and $T_2$ signals.

The spin echo pulse sequence provides proton density $T_2$ or $T_2$ contrast weighting. It consists of a series of $90°$ pulses separated by a period of time known as the time of repetition ($T_R$). The total number of slices that can be obtained is a function of $T_R$. Long $T_R$ acquisition such as proton-/$T_2$-weighted sequences usually have more slices than the short $T_R$ acquisitions with a $T_1$-weighted sequence [73].

The relaxation process of $T_1$ is longer than $T_2$. As field strength increases, the Larmor frequency ($\omega$) increases, which ultimately affects the values of $T_1$ and $T_2$. The local hydrogen concentration gradient which causes differences in spin density and the differences in $T_1$ and $T_2$ relaxation times in different tissues are the key to the exquisite contrast sensitivity of magnetic resonance (MR) images (MRI). Tailoring the pulse sequences, that is, timing, order, polarity, and repetition frequency of the RF pulses and applied gradients, can make the emitted signals dependent upon these relaxation characteristics. As a matter of fact, MR imaging technology mainly relies on three major pulse sequence techniques, namely, *spin echo*, *inversion recovery*, *and gradient-recalled echo sequence* [74, 75].

$T_1$ tends to be long in solids because of limited energy coupling to lattice and the value of $T_1$ depends on the size of the molecules of the species to be studied. For example, the *lipids and fats* have moderately sized molecules and the values of $T_1$ are found moderate. For biological tissues, $T_1$ values range from 0.1 to 1 s and 1 to 4 s in aqueous tissues (cerebrospinal fluids). The strength of the applied magnetic field and the presence of paramagnetic or ferromagnetic materials in the tissue or blood also have greater influence on the value of $T_1$ [76].

The rotational displacement of $M_L$ vector induces a magnetic moment in the transverse plane. After $B_1$ field is removed, the magnetic moment $M_T$ rotates at the Larmor frequency, which a receiver antenna coil can detect. This signal is known as free induction decay (FID) [77]. The decay of the FID envelope is due to loss of phase coherence of the individual spins over time and the loss of $M_T$ occurs very quickly (from 0.37 to 0), whereas the excited sample takes longer time to come to equilibrium condition [78].

## 9.7.6 Ultrasound (US) (Sonography)

In ultrasound (Sonography), the diagnostic medical sonographer uses high-frequency sound waves to perform a wide variety of diagnostic examinations, like harmonic

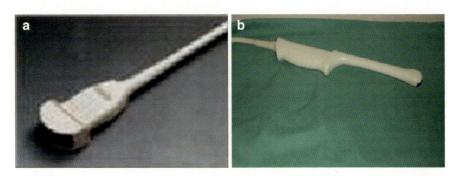

**Fig. 9.27** Picture of (**a**) transabdominal transducer (Courtesy: Parental Diag. Center, VA) and (**b**) transvaginal transducer (Courtesy Virginia Med. VA)

imaging, intracoronary ultrasonic imaging, and low-velocity acoustic streaming for differentiation between solid and liquid cyst in breast cancer [64, 79].

The basic ultrasound equipment has a transducer probe that receives and sends high-frequency sound waves, which penetrates the body and bounces back, and the bounced back wave is then recorded. Different tissues reflect these sound waves differently, causing a signature, which can be measured and transformed into a picture.

The research in the field of transducer materials has improved probe sensitivity and the overall image quality. The latest piezoelectric ceramics has increased the bandwidth with high spatial resolution, while advanced impedance matching and improved acoustic lens material has minimized reverberation and losses. Recently, a capacitive micromachined ultrasound transducer known as CMUT has been able to make ultrasound equipment more versatile and important in image modality. The pictures (Fig. 9.27a, b) show some of the common types of transducers that are used in *US* medical imaging [80].

The two basic arrays that are used in electronically scanned transducers are (1) *sequenced array*, which is subgrouped into *linear* and *curvilinear* arrays and (2) *phased* array, which is subgrouped into *linear* and *annular* arrays [30]. In ultrasound medical imaging where the frequency range is high ~1–20 MHz is produced by applying the output of an electronic oscillator to a thin wafer of piezoelectric material such as lead zirconate titanate $\{PZT, [(PbZr_xTi_{1-x})]O_3\}$. However, some industries also use lithium niobate ($LiNbO_3$) crystal in single element transducer, because of its high longitudinal velocity (7, 340 m/s in the 36E rotated Y-cut) which aids in the fabrication process of the transducer [81]. The crystal also receives the return echo of the sound from the human organ under study. These echoes are converted into electrical signals, which are sent to the console and form a picture.

Figure 9.28 shows the picture of an ultrasound equipment used for medical imaging. There are many advantages to imaging a body with ultrasound. The most important of all is that there is no ionizing radiation as with X-rays. Moreover, liver, spleen, kidneys, and pancreas mostly possess soft tissues, and ultrasound can image the area of interest without injecting any radioisotope. In addition, the entire abdomen and

## 9.7 Scintigraphy

**Fig. 9.28** Picture of an ultrasound equipment for medical imaging (Courtesy: WIKI)

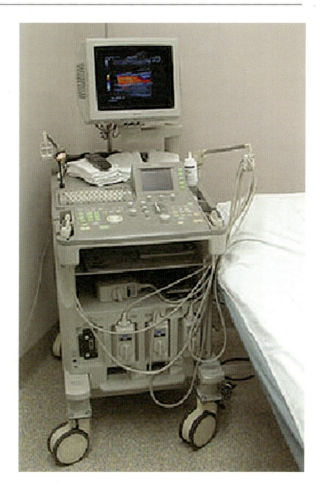

pelvis can be rapidly scanned while the patient is lying on the table, and image can be taken from the area of interest.

However, as the detection principle of ultrasound lies on the echo system produced by the organ under study, organs filled with air such as the lungs, stomach, and intestines are opaque to the sound and are difficult to study. Formerly, it was almost impossible to view the cervix and lower uterus because of their location under the air-filled intestines, which uses to reflect most of the sound waves. This can be done by creating an ideal acoustic window by administering sufficient amount of water into the stomach which pushes the intestine out of the picture of the cervix and uterus. Another barrier is the air gap that exists between the transducer and the skin. In order to overcome the reflections from the air gap, a lubricant is first applied to the patient body to eliminate the air gap.

Figure 9.29 shows different parts and the accessories used to record a picture of a baby inside womb. When the sonography is performed on a flat surface, for example,

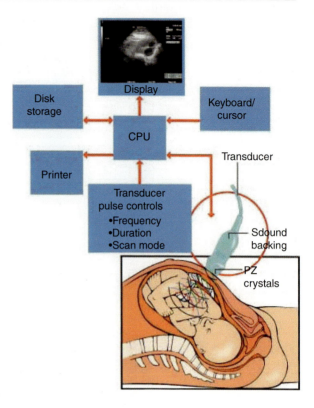

**Fig. 9.29** Different parts of the ultrasound equipment with a picture of a baby inside the womb (Photo courtesy: a discovery company, © 2001 How Stuff Works)

for abdominal examination, the transducer is placed on the flat surface after applying some lubricants to fill up the air gap between the skin and the transducer. Likewise, transvaginal transducer (Fig. 9.27) is used for gynecological examination where the transducer is inserted into the vagina after patient empties her bladder.

Ultrasound beam is a nonionizing longitudinal wave, which recorded signal through reflection (echoes) of the wave from the tissue rather than transmission of the wave. The magnitude of the echo influences the brightness of a display and is coded as gray scale. An echo image is then built up to give an image of a body slice.

The block diagram for the ultrasound image processing is shown in Fig. 9.30. Pulsar consists of a clock and a pulse transmitter. The clock controls all the timing of the transmission pulse, pulse frequency, and image data (*echo*). Transmitter voltage can vary from 5 V to hundreds of volts, but the applied field should not exceed 1 kV/mm. Small echo signals from the transducer are amplified from microvolts to millivolts. Echoes returning from the tissues have a final dynamic range of 50 dB. However, when the echoes undergo a higher depth and suffer loss in intensity due to deep attenuation, the losses are compensated by a time gain compensation (TGC) circuit (Fig. 9.30) (which amplifies the signal depending on the time of arrival). After time gain compensation, large dynamic range of signal cannot be displayed effectively. Thus, it is compressed using logarithmic amplification. Later on, the compressed signal is demodulated. A threshold is placed on the demodulated signals, and the signals above threshold are digitized.

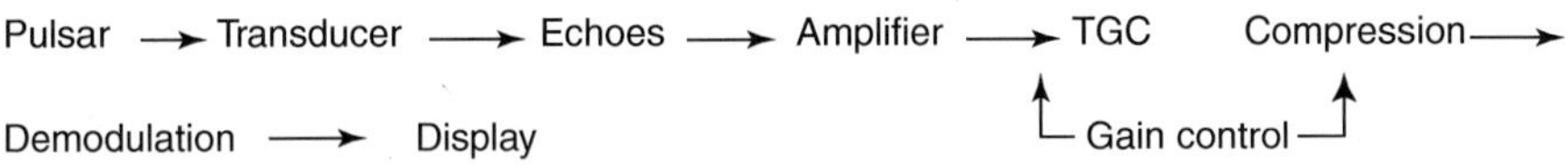

9.8 Fusion or Hybrid Technology 485

Pulsar → Transducer → Echoes → Amplifier → TGC Compression →

Demodulation → Display └ Gain control ┘

**Fig. 9.30** Block diagram of the transducer processing circuit for image processing

There is no doubt that intracoronary ultrasound (ICUS), because of its penetrating nature, represents the most valuable method to asses plaque morphology, where angiography fails to provide adequate distinction between plaque and lumen irregularities and assessing the extent atherosclerotic disease [64, 79].

Undoubtedly, the differentiation of solid and liquid cysts in breast cancer screening is the most important job for cancer screening. It is expected that future generation of US scanners will be able to induce low-velocity acoustic streaming pattern within the cyst and will exhibit Doppler sensitivity which will be able to detect extremely low velocity. The measurements of streaming pattern will provide information about viscosity of the liquid within the cyst. It is hoped that the new generation of high-frequency Doppler machines will be able to offer submillimeter resolution imaging.

Harmonic waves are generated from nonlinear distortion of an acoustic signal as an ultrasound wave insonates in the body [80]. Tissue signature imaging, which is a highly sensitive method for detecting lesions, is based on advanced dynamic flow. Since it utilizes both harmonic and nonlinear fundamental signals, one should not experience the same cluster found with other contrast imaging techniques. Harmonic images have already proved to be capable of providing a degree of detail that clearly surpasses the detail that is available with conventional, fundamental frequency gray scale imaging.

Harmonic imaging is helpful in obese patients because of reduction in deleterious effects of the body wall. However, harmonic imaging is better for demonstrating most abdominal lesions, but at times, it did not have additional benefit over conventional US (Fig. 9.31). It has been reported that in patients with different fatty infiltration in the liver, penetration of the ultrasound beam is better on conventional US images [81].

## 9.8 Fusion or Hybrid Technology

The field of medical imaging has witnessed remarkable advancement and technological achievement. As the modern technology is exploring new avenues for better quality images in the field of nuclear medicine, new technologies are appearing on the horizon. As a result, fusion brain imaging which is the combination of two different modalities has been very successful in improving the amount or quality of information that would gathered from either of the two individual.

All SPECT images are subject to certain degree of spatial distortion, caused by differing scatter and absorption of emitted photons before they reach the detector [81]. On the other hand, the SPECT and PET, a combination or fusion of SPECT and PET technologies, are helping in redefining many aspects of modern diagnostic of medical

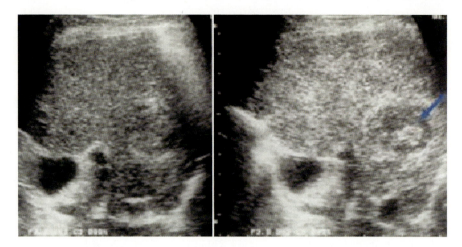

**Fig. 9.31** Hermonic imaging of a well differentiated hepatocellular carcinoma (HCC) of a module in module type. A vascular spot (*arrow*) is seen in a hypovascular HCC (Adopted from: M. Kudo, Contrast Harmonic Imaging:Springer, 2003)

imaging. The primary applications for PET are oncology—soft tissue tumors—which account for 90 % of all clinical utilization. The SPECT, on the other hand, is preferred over PET for most cardiac cases. Clinical utilization of PET has started in early 1990s, while SPECT has been introduced in the beginning 1994.

In the case of heart disease for large patients, PET allows much better clinical image quality and improved diagnostic confidence. But instead of relying on PET or SPECT, the fusion of SPECT–PET provides a tool for obtaining complimentary anatomical and functional information in a single-image session. SPECT technology employs a variety of more traditional agents used in nuclear medicine, depending on the desired application. As opposed to PET's collision-coursed positrons, SPECT tracers rely on a single bundle of ($\gamma$)-radiation to produce images; thus, the combination of two modalities require entirely different hardware. Thus, the price tag of the hybrid machines are more compared to a machine that offers single modality. As a matter of fact, the use of SPECT/CT scanner has been proven feasible and offers a clinically suitable compromise between the improved anatomic details and minimally increased radiation [82–86].

The main thrust of combining two modalities is the combination of structural and functional techniques. The current ongoing work is to extend the work in the area of hybrid technology, the best of both worlds. Thus, the fusion imaging work has been extended to the areas of EEG/MEG, MRI/fMRI, MRI/PET, CT/PET, and CT/SPECT [87]. The most obvious choice is MRI/fMRI, mainly because of the hardware used for each is the same. However, because of the benefits of PET and SPECT over fMRI, other combinations are potentially more useful. For example, SPECT has enormous potential when used alone, but when it is used with PET or CT, the potential benefit is more. One of the potentially beneficial combinations of SPECT with CT in 2003 is that it is beginning to replace SPECT/PET fusion for small animal imaging [88]. According to some research workers, the combination of SPECT with MRI may be more useful than the fusion of MRI with fMRI. The main goal of fusion method should be that the two modalities will work in harmony, and the data acquisition should not be complex and time-consuming [89, 90].

The attenuation correction in cardiac SPECT, by SPECT/CT hybrid scanners, has been possible by using anatomic maps derived from CT. The gold standard for attenuation correction in perfusion imaging at present is rubidium-82 PET. As a matter of fact, attenuation correction during myocardial perfusion imaging has proven the credibility of SPECT/CT scanners [91]. Many neurological applications are also likely to reap benefits from SPECT/CT through superior attenuation correction.

A pet image is color-coded, which shows the activity of different cells, while CT scans can pinpoint the exact locations. Together, a PET/CT scan allows physicians to view metabolic activity and pinpoint where the abnormal lesions are located so that they can target the cells. Thus, the physicians can not only tell which cell is cancerous; they also can precisely target the cells to be repaired (Fig. 9.32). Figure 9.32 shows the images of different organs taken by CT, PET, and CT/PET scans.

Figure 9.33 shows the picture of a PET/CT scanning machine. The additional advantage of CT/PET combination is to take the advantage of the accuracy with which the heart can be positioned within the PET field of view, and the appropriate corrections for the energy difference between CT X-rays and annihilation of photons have allowed CT images to be used for PET attenuation correction [92, 93]. As a result, Ge-68 or Cs-137 scans have been replaced with faster scans which has reduced the overall duration of the scanning process. But one potential problem with faster CT scan for attenuation correction is the motion of the organs during respiration.

## 9.9 Final Remarks

Throughout the medical industry, there are a number of breakthroughs that have made it easier to diagnose and treat many of the serious ailments that affect humanity. One of the most impressive technologies that has been developed and which continues to show improvements is in the area of *nuclear medicine*. *Nuclear medicine* is a medical specialty involving the application of radioactive substances in the diagnosis and treatment of disease. In *nuclear medicine*, diagnosis and therapy are conducted with the aid of radionuclides that are combined with other elements to form chemical compounds, or else combined with existing pharmaceutical compounds, to form radiopharmaceutical. There was a flurry of research and development of new radionuclides and radiopharmaceuticals for use with the imaging devices and for in vitro studies.

Indeed, the whole field of nuclear medicine has progressed rapidly since its beginning half a century ago. According to some historical scientists, the birth date of nuclear medicine can probably be best placed between the discovery of artificial radioactivity in 1934 and the production of radionuclides by Oak Ridge National Laboratory for medicine-related use, in 1946. However, according to some critics, the history of nuclear medicine can go beyond 1900, when Henri Becquerel discovered

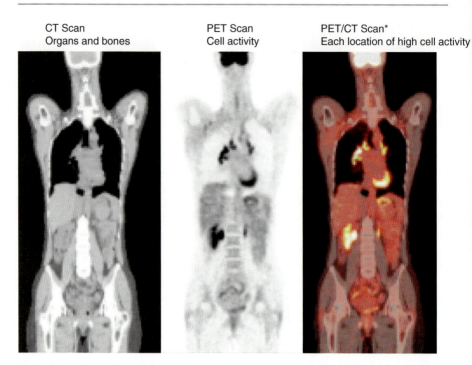

**Fig. 9.32** A comparative analysis of three different pictures taken with different scanners (Photo courtesy, Univ. of Pittsburgh Medical Center, PA)

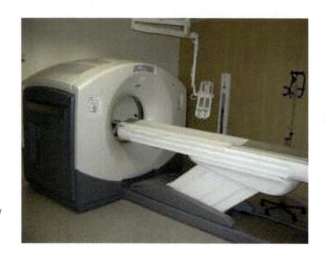

**Fig. 9.33** Picture of a PET/CT scanning machine (Courtesy: WIKI)

mysterious *rays* from uranium in 1896 and Marie Curie named the mysterious rays *radioactivity* in the year 1897. The discovery and research in nuclear medicine went on, and the discovery of iodine-131 by John Livinghood and Glen Seaborg in 1938 and technetium-99 m by Emilio Segre and Glem Seaborg in 1938 brought revolution. The production of technetium-99 m in 1960 by Bookhaven National Laboratory was a

## 9.9 Final Remarks

breakthrough in the history of nuclear medicine when Paul Harper and Katherine Lethrup studied their use for the recovery of the abnormal cells and in the brain, and malfunctioning of the tissues in thyroid and liver in the year 1964. Their study ultimately resulted in the approval by the FDA to useTc-99 m as radiopharmaceutical for brain perfusion in the year 1988, and in 1991, FDA first approved Tc-99 m as the first monoclonal antibody radiopharmaceutical for tumor imaging. Today, there are approximately 100 different *nuclear medicine imaging* procedures which provide information about nearly every organ system. Nuclear medicine is now an integral part of patient care and is extremely valuable in the early diagnosis, treatment, and prevention of numerous medical conditions.

It would not be wrong to call nuclear medicine as "radiology done inside out" or "endo-radiology" because it records radiation emitting from within the body rather than radiation that is generated by external sources like X-rays. The amount of radiation from diagnostic nuclear medicine procedures is kept within a safe limit relative to the established "ALARA" (As Low As Reasonably Achievable) principle. The radiation dose from nuclear medicine imaging varies greatly depending on the type of study. Most diagnostic radionuclides emit gamma rays, while the cell-damaging properties of beta particles are used in therapeutic applications. Refined radionuclides for use in nuclear medicine are derived from fission or fusion processes in nuclear reactor, which produce radionuclides with longer half-lives, or cyclotron, which produce radionuclides with shorter half-lives, or take advantage of natural decay processes in dedicated generators, that is, molybdenum/technetium or strontium/rubidium. A patient undergoing a nuclear medicine procedure will receive a radiation dose. Under present international guidelines, it is assumed that any radiation dose, however small, presents a risk. The radiation doses delivered to a patient in a nuclear medicine investigation, though unproven, is generally accepted to present a very small risk of inducing cancer. Because the *ionizing radiation*, which has sufficient energy to interact with matter, especially in the human body, is capable of producing ions, it can eject an electron from an atom of the DNA (*deoxyribonucleic acid*). As a result, the bonding properties of the DNA atom change, causing a physical change of the DNA. Thus, the activity or the *radioactivity* of an ionizing radiation is very damaging to life. In this respect, it is similar to the risk from X-ray investigations except that the dose is delivered internally rather than from an external source such as an X-ray machine, and dosage amounts are typically significantly higher than those of X-rays.

In the future, nuclear medicine may provide added impetus to the field known as molecular medicine. As our understanding of biological processes in the cells of living organism expands, specific probes can be developed to allow visualization, characterization, and quantification of biological processes at the cellular and subcellular levels. Nuclear medicine is an ideal specialty to adapt to the new discipline of molecular medicine, because of its emphasis on function and its utilization of *imaging agents* that are specific for a particular disease process. In nuclear medicine imaging, radiopharmaceuticals are taken internally, for example, intravenously or orally. Then, external detectors (gamma cameras) capture and form images from the radiation emitted by the radiopharmaceuticals. This process is unlike a diagnostic X-ray where external radiation is passed through the body to form an image.

The remarkable power of medical imaging in providing physicians with sight—and insight—about human disease and physiology is a history in medical science. It enables a range of less invasive, highly targeted cancer therapies that translate into better and more comfortable care of patients. The New England Journal of Medicine calls medical imaging one of the most important medical developments of the past 100 years—ranking with such milestones as the discovery of anesthesia and discovery of antibiotics. The invention of cell phones, personal computers, and the Internet ushered in the digital revolution and altered our existence in ways that were unimaginable just a few decades ago. Now, we are in the verge of a digital medical revolution. Indeed, the digital medical fields are already supplying a wealth of innovation, and one of the recent significant advances in mammography is the development of digital mammography.

Early detection through mammography has been a major factor in the dramatic decrease in breast cancer mortality. In fact, mammography is one of the key factors in early-stage detection—some 63 % of breast cancers are detected at the localized stage for which 5-year survival rate is 97.5 % [94].

In some centers, the nuclear medicine scans can be superimposed, using software or hybrid cameras, on images from modalities such as CT or MRI to highlight the part of the body in which the radiopharmaceutical is concentrated. This practice is often referred to as image fusion or co-registration, for example, SPECT/CT and PET/CT. The fusion imaging technique in nuclear medicine provides information about the anatomy and function, which would otherwise be unavailable, or would require a more invasive procedure or surgery. The modern Biograph mCT (PET–CT) enable the quantification of molecular processes in the body. Current applications include cancer imaging, quantification of absolute myocardial blood flow, and quantification of amyloid deposits in the brain in patient with dementia. According to some experts, the promise of molecular imaging is boundless because of its ability to detect pre-diseases and has been the key component of comprehensive oncology care—an essential tool in cancer therapy and management [95].

## References

1. Hall EJ, Garcia AJ (2006) Radiology for radiologists, 6th edn. Lippincott Williams & Wilkins, Philadelphia, p 216
2. Burgeron P, Carrier R, Roy D, Balis N, Raymond J (1994) Radiation doses to patients in neurointervental procedure. AJNR Am J Neuroradiol 15:1809
3. White RA, Hollier LH (2005) Vascular surgery: basic science and clinical correlations, 2nd edn. Blackwell Futura, Malden, p 325, Chapter 31
4. Arguelles MG, Rutty GA, Sola, Bottlezini DL (collaboration) (1999) Optimization of production and quality control of therapeutic radionuclides. Final report international atomic energy agency, IAEA, Vienna, Sept 1999
5. Darnel J, Lodish H, Baltimore D (1986) Molecular cell biology. Scientific American Books, New York, p 517, Chapter 13
6. Lehman CD (2006) The role of MRI in screening women at high risk for breast cancer. J Magn Reson Imaging 24(5):964, Wiley

# References

7. Toffs P (2003) MRI of the brain. Wiley, Chichester
8. Hamar KA, Sgouros G (1999) A schema for estimating absorbed dose of organs following the administration of radionuclides with multiple unstable daughters. Med Phys 26:2526
9. Canon CL, Canon C (2009) Radiology. McGraw Hill, New York
10. Lashford LS, Clarke J, Kemshead JT (1990) Systematic administration of radionuclides in neuroblastomas as planned radiotherapeutic under vention. J Med Pediatr Oncol 18(1):30
11. Schauer DA, Linton W (1980) Report National council of radiation protection and measurements (NCRP) 65, 15 Apr 1980
12. Ehrhardt GJ, Ketring AR, Ayers LM (1998) Reactor produced radionuclides at the university of Missouri research reactor. Appl Radiat Isot 49(4):295
13. Kahn J (1996) From radioisotopes to medical imaging, history of nuclear medicine written at Berkeley, 9 Sept 1996
14. Roedler HD (1981) Radiation dose to the patient in radionuclide studies. In: Medical radionuclide imaging. vol I, IAEA, Vienna, p 27
15. Lundberg TM, Gray PJ, Bartlet ML (2002) Measuring and minimizing the radiation dose to nuclear medicine technologists. J Nucl Med Technol 30(1):25
16. Chaiken LM, Steinberg ML (2004) Targeting for cure: intensity modulation radiation therapy. PCRI 7(3):30
17. Huang E et al (2002) Late rectal toxicity: dose volume effects of conformal radiotherapy for prostate cancer. Int J Radiat Oncol Biol Phys 54(5):1314
18. Hall E et al (2003) Radiation induced second cancers: the impact of 3-D-CRT and IMRT. Int J Radiat Oncol Biol Phys 83(1)
19. Beddar AS et al (2006) Intraoperative radiation therapy using mobile electron linear accelerators, report of AAPM, radiation therapy task group no. 72. Med Phys 33:1476
20. Daves JL, Mills MD (2001) Shielding assessment of a mobile electron accelerator for intraoperative radio therapy. J Appl Clin Med Phys 2:165
21. Kendzioka D, Flickinger JC, Hudak R (2011) Results following Gamma knife radiosurgical anterior capsulotomies for observe compulsive disorder. Neurosurgery 68(1):28
22. Chung HT et al (2008) Development of a stereotactic device for gamma-knife irradiation of small animals. J Korean Neurosurg Soc 43(1):26
23. Hall EJ (2006) Intensity modulated radiation therapy, proton and the risk of second cancer. Int J Radiat Oncol Biol Phys 65(1):1–7
24. Smith AR (2009) Proton therapy. Phys Med Biol 51:R491
25. Bruzi M et al (2007) Prototype tracking studies for proton CT. IEEE Trans Nucl Sci 54(1):140
26. Sadrozinski HFW et al (2004) Toward proton computed tomography. IEEE Trans Nucl Sci 51(1):3
27. Lo SS, Chang EL, Sloan AE (2006) Role of stereotactic radio surgery and fractionated stereotactic radiotherapy in the management of intracranial ependymoma. Medicine 6(4):501
28. Alexandrides TK et al (2006) The value of scintigraphy and ultrasonography. Hormones 5(1):42
29. Giordano A, Rubello D, Casara D (2001) New trends in parathyroid scintigraphy. Eur J Nucl Med 28:1409
30. Kanistar A et al (2001) Automated vessel detection at lower extremity multislice CTA. Eur J Radiol 11(S1):2001
31. Wink O, Niessen W, Yiergever M (2006) Fast delineation and visualization of vessels in 3-D angiographic images. IEEE Trans Med Imaging 19(4):337
32. Westwood ME et al (2002) Use of magnetic resonance angiography to select candidates with recently symptomatic carotid stenosis for surgery. BMJ 324(7331):198
33. Laub G, Gaa J, Drobnitzky M (1998) Magnetic resonance angiography techniques. Electromedia 66(2):68
34. Parker GL, Yuan C, Blatter DD (1991) MRA by multiple thin slab 3-D acquisition. Magn Reson Med 17:434
35. Hausmann R, Lewin JS, Laub G (1991) Phase contrast MRA with reduced acquisition time: new concepts in sequence design. J Magn Reson Imaging 1:415

36. Collins R et al (2007) Duplex ultrasonography, magnetic resonance angiography. . ..systematic review. BMJ 334:1257
37. Pan X, Siewerdsen J, Riviera PJL, Kalender WE (2008) Development of X-ray computed tomography. Med Phys 35(8):3728
38. Sorensen JA (1979) Techniques for evaluating radiation beam and image slice parameters of CT scanners. Med Phys 6:68
39. Moon JB, Smith WL (1986) Application of cine computed tomography to the assessment of velopharyngeal form and function. In: American cleft palate association meeting, New York, May 1986
40. Alessio AM et al (2007) Cine CT for attenuation correction in cardiac PET/CT. J Nucl Med 48(5):794
41. Hu H (1999) Multi-slice helical CT: scans and reconstruction. Med Phys 26:5
42. Coo HW (2009) State of the art CT imaging techniques for congenital heart disease. Korean J Radiol 11(1):4
43. Lee C et al (2007) Organ and effective doses in pediatric patients undergoing helical multislice computed tomography examination. Med Phys 34:1858
44. Huda W, Scalzati M, Roskopf M (2000) Effective doses to patients undergoing thoracic computed tomography examination. Med Phys 28:838
45. Xie Q et al (2007) Performance evaluation of multipixel photon counter for PET imaging. In: IEEE nuclear symposium, Honolulu, 26th Nov–3rd Oct 2007
46. Bailey DL, Townsend DW, Valk PE, Maisey MN (2005) Positron emission tomography: basic science. Springer, London
47. Wong DF, Gronder G, Brasic JR (2007) Brain imaging research. Int Rev Psychiatry 19(5):541
48. Fahey FH (2002) Data acquisition in PET imaging. J Nucl Med Technol 30(2):39
49. O'Connor YZ, Fessler JA (2007) Fast predictions variance images for fan-beam transmission tomography with quadratic regularization. IEEE Trans Med Imaging 26(3):335
50. Gupta TK et al (2004) Ce-doped lutetium oxyorthosilicate: a fast efficient new scintillator. IEEE Trans Nucl Sci 51:2302
51. Chen S, Tsui BMW (2010) Effect of motion-estimation error on three 4D PET image reconstruction methods with respiratory motion compensation. In: IEEE NSS/MIC/RTSD symposium, Knoxville, 30th Oct–6th Nov 2010
52. Kramer EL, Ko JP, Ponzo F, Mourtzikos K (2008) Positron emission tomography-CT. Taylor & Francis, New York
53. Daube-Awitherspoon ME, Marej S, Werner ME, Surti S, Karp JS (2010) Comparison of list mode and direct approach for time of flight PET reconstruction. In: IEEE NSS/MIC/RTSD symposium, Knoxville, 30th Oct–6th Nov 2010
54. Auffray E et al (2010) Towards a time of flight PET system based on multipixel photon counter readout. In: IEEE symposium on nuclear science. Knoxville, Oct 30th–6th Nov 2010, p 1055
55. Yamaya T, Obi T, Yamaguchi M, Ohyama LN (2011) High resolution image reconstruction method time of flight PET. Phys Med Biol 45(11):3125
56. Saha GB (2006) Physics and radiobiology of nuclear medicine, 3rd edn. Springer, New York
57. Cambhir SS et al (2009) A novel high-sensitivity rapid acquisition single photon cardiac image camera. J Nucl Med 50:635
58. Piccini JP et al (2010) SPECT myocardial perfusion imaging ejection function. J Am Coll Cardiol 56:206
59. Rusina R, Kukal J, Belcek T, Buncova M, Matej R (2010) Use of fuzzy edge SPECT analysis in definite Alzheimer disease – a retrospective study. BMC Med Imaging 10:20
60. Morris TA et al (2004) SPECT of pulmonary embolism and venous thombai using anti-D-dimmer. Am J Respir Crit Care Med 169:987
61. Yucel EK, Anderson CM, Edleman RR, Baum RA, Culebras A, Pearce W (1999) Magnetic resonance angiography. Circulation 100:2284, American Heart Association
62. Elendy A et al (2002) Dobutamine stress myocardial perfusion imaging in coronary artery disease. J Nucl Med 43:1634
63. Frankle WG, Slifstein M, Talbot PS, Larulle M (2005) Neuroreceptor imaging in psychiatry: theory and applications. Int Rev Neurobiol 67:385

# References

64. Beach RD et al (2007) An adaptive approach to decomposing patient-motion tracking data acquired during cardiac SPECT imaging. IEEE Trans Nucl Sci 54(1):130
65. Chen QS, Defrise M, Decononck F, Franken PR, Jonckheer MH (1993) Detection and correction of patient movements in SPECT imaging. J Nucl Med Technol 21(4):198
66. Franklin K, Muir P, Scot T, Wilcocks L, Yates P (2010) Introduction to biological physics for health and life science. Wiley, Sussex, p 419
67. Stephan K, Friston KJ (2010) Analysing effective connectivity with functional MRI. Wires Cogitive Sci 1:446
68. Harrison L, Penny WD, Daunizeau J, Friston KJ (2008) Diffusion- based spatial priors for functional MRI images. Neuroimage 41(2):408
69. Williams JEC (1984) Superconducting magnets for MRI. IEEE Trans Nucl Sci 31(4):994
70. Pooley RA (2005) Fundamental physics of MR imaging. Radiographics 25:1087
71. Maclaren JR (2008) Motion detection and correction in magnetic resonance imaging. Ph.D. thesis, University of Canterbury, Christchurch
72. Wright SM et al (2002) A desktop magnetic resonance system. MAGMA 13:177
73. Reeder SR et al (2006) T1- and T2-weighted fast spin-echo imaging of the brachial plexus and cervical spine with IDEAL water-fat separation. J Magn Reson Imaging 24(4):825
74. Boulant N (2009) $T_1$ and $T_2$ effects during radio frequency pulses in spoiled gradient echo sequences. J Magn Reson 197(2):213
75. Sinha S, Sinha V, Kangar H, Huang H (1992) Magnetic resonance image synthesis from analytical solutions of spin echo and radio frequency spoiled gradient echo imaging. Invest Radiol 27(10):856
76. Cattin HB, Collewet G, Belaroussi B, James HS, Odet C (2005) The SIMRI project: a versatile and interactive MRI simulator. J Magn Reson 173:97
77. Zhou X et al (2004) Experiment on dynamic simulations of radiation damping of linear polarized liquid $^{124}$Xe at low magnetic field in a flow system. Appl Magn Reson 26(3):327
78. Pamilo S (2011) Spatio-temporal segregation of brain circuitries activated during movie viewing. MS thesis, Aalto University School of Science, Helsinki, 11 Mar 2011
79. Mignotte M, Meunier J (2000) Three dimensional blind deconvolution of SPECT images. IEEE Trans Biomed Eng 47(2):274
80. Choudhuri S et al (2000) Comparison of tissue harmonic imaging with conventional US in abdominal disease. Radiographics 20:1127
81. Labbe J (2003) SPECT/CT emerges for shadow of PET/CT. Biophtonics Int 50–57
82. Dickinson RL et al (2011) Hybrid modality fusion planar scintigram...phantom studies. Int J Mol Imaging 2011:1155
83. van der Ploeg IMC, Valdes Olmos RA, Kroon BBR, Nieweg OE (2008) The hybrid SPECT/CT as an additional lymphatic mapping tool in patients with breast cancer. World J Surg 32(9):1930
84. Lodge MA et al (2003) Developments in nuclear cardiology: transition from SPECT to PET/CT. J Invasive Cardiol 17(9):491
85. (Cadmium chloride rb-82) microcardinal perfusion agent named Cardiogen-82,first introduced in the market in 2012, by Branco Diagnostics Inc. Princeton, NJ
86. Lopez A, Molina R, Katsaggelos AK, Mateos J (2001) SPECT image reconstruction using compound prior models. In: IEEE international conference on acoustics speech and signal processing, vol 3, 1909–1912, Salt Lake City, Uttah
87. Horwitz B, Poeppel D (2002) How can EEG/MEG and fMRI/MRI data can be combined? Hum Brain Mapp 17(1):1
88. Brinkmann BH et al (2000) Subtraction ictal SPECT co-registered to MRI for seizure focus localization in partial epilepsy. Mayo Clin Proc 75(6):615
89. Romer W et al (2005) The value of SPECT/spiral CT hybrid imaging in nuclear foci of increased bone metabolism. In: RSNA meeting, Chicago, Nov 2005
90. Koepfli P, Hany TF, Wyss CA et al (2004) CT attenuation correction for myocardial perfusion quantification using a PET/CT hybrid scanner. J Nucl Med 45:537

91. Hunold P, Vogt FM, Schmermund A et al (2003) Radiation exposure during cardiac CT. Radiology 226:145
92. Oskoui-Fard P, Stark H (1988) Tomographic image reconstruction using the theory of convex projections. IEEE Trans Med Imaging 7:45
93. Leahy R (2000) Recent developments in iterative image reconstruction for PET and SPECT. IEEE Trans Med Imaging 19:257
94. American Cancer Society (2005) Cancer prevention and early detection, facts and figures, 2005 and effects of chemotherapy and hormonal therapy for early breast cancer on recurrence and 15 years survival. Lancet 365:1687–1717
95. Weissled R, Mahmood U (2001) Molecular imaging. Radiology 219:316
96. Hart D, Wall BF. A survey of nuclear medicine in the UK during the year 2003–2004. IPA-RPD-003

# Glossary

**Absorption** It is opposite the scattering phenomenon of a medium.

**Absorption coefficient** It measures how quickly a beam of light would lose intensity due to absorption.

**Afterglow** Persistence of light emission long after the radiation source is cut off.

**Air gap** The optimum distance left between the patient and the detector to minimize scattering.

**Air kerma** Kinetic energy released in air from an X-ray beam. The exposure in SI unit is 8.73 mGy/R.

**Alpha rays** High-speed energetic particles sometimes described as extremely high-energy helium nuclei.

**AMLCD** Active matrix liquid crystal display is a flat panel liquid crystal display currently used in notebook computer.

**Angiography** X-ray study of the blood vessels.

**Annihilation** Latin nihil means nothing, and annihilation is defined as total destruction or complete obliteration of an object.

**Attenuation** The loss of photons due to an absorption media placed before the radiation source to improve scattering effect (via elastic or inelastic collisions: Rayleigh/Compton scattering) or conversion of the incident photon into an electron or positron (pair production).

**Attenuation coefficient** It is the quality of the media that characterizes how easily the energy of light, sound, or particle will be weakened.

**Auger electron** When a core electron is removed, leaving a vacancy, an electron from a higher energy level may fall into the vacancy, resulting in a release of energy which can transfer to another electron, and can be ejected from the atom. This second ejected electron is called an Auger electron.

**Avalanche photodiode (APD)** A semiconductor high-speed high-sensitive photodiode utilizes an internal gain mechanism by applying a reverse voltage. The ionization rates in the device are an important factor in determining the avalanche ($\sim$ electric field $2 \times 106$ V/cm) multiplication mechanism.

**Beam hardening** An increase in the effective energy of a polyenergetic radiation due to the attenuation of the lower-energy photons in the spectrum.

**Beta rays** A form of ionizing radiation emitted by radioactive materials, which is more powerful than alpha rays.

**Blurred (image)** Lack of definition or out of focused image.

T.K. Gupta, *Radiation, Ionization, and Detection in Nuclear Medicine,* 495
DOI 10.1007/978-3-642-34076-5, © Springer-Verlag Berlin Heidelberg 2013

**Bremsstrahlung** It is the phenomenon, where a high-energy particle loses energy preferentially by radiation of a photon due to de-acceleration in coulomb field of the nucleus.

**Brewster's angle** Total internal reflection of light occurs in a material at a particular critical angle, which is known as Brewster's angle.

**Bucky factor** The necessary increase in dose to compensate for removing scatter and primary radiation that would otherwise produce a signal on the detector with a grid. Typical bucking factors are from $2\times$ to $4\times$, depending on the characteristics of the grid.

**Cancerous cell** Abnormal cells that grow and divide at an unregulated and quickened space.

**Cardiology** The Greek word *kardia* means *heart*, and *logia* means collections of *saying*. Thus, cardiology can be defined as the medical specialty dealing with disorders of heart.

**CAT** Computed axial tomography in which computer analysis is done from a series of cross-sectional scans made along single slice.

**CCD** Charge-coupled device is a light-sensitive device (integrated circuit) that stores and displays the data of an image.

**Charge-coupled devices (CCD)** Charge-coupled devices (CCDs) are closely spaced arrays of metal oxide semiconductor (MOS) used as imaging devices. CCD converts the photon energy into electrical signal represented by the channel charge.

**Chromosomes** Threadlike structure made out of proteins and a single molecule of deoxyribonucleic acid.

**Cine camera** A camera that takes a sequence of photographs that can give the illusion of motion when viewed in rapid succession.

**Cloud chamber** A chamber containing supersaturated vapor of water or alcohol, which forms a cloud by ionizing radiation to detect ionizing particles.

**Compton effect** It is the result of high-energy photons colliding with a target, which releases loosely bound electrons.

**Compton scattering** Interaction of X-ray (an assembly of photons) with the free electrons of a scattering block transfers partial energy to the electrons and the scattered photons. The scattered photons have lower energy and lower frequency, and the interaction is proportional to the density of the material in the diagnostic energy range.

**Conduction** Transfer of energy through matter from particle to particle without displacement.

**Contrast** It is the difference in the visual properties that makes an object or its representation as image distinguishable from other objects and the background.

**Cosmic rays** Energetic charged particles originating from space.

**Counting efficiency** It is related to the amount of radiation emitted by a radioactive source to the amount measured in a detector.

**Critical energy** When the critical energy loss rate equals to bremsstrahlung rate.

**Cross section (neutrons)** It is the probability of neutrons being absorbed divided by the real time.

# Glossary

**CT** Computed tomography is a medical imaging method employing tomography by computer process.

**Cupping artifact** It is the classic example of beam hardening which occurs in the soft tissue regions.

**Curie** It is the measure of the number of atomic disintegration in a unit time.

**Dead layer** A layer deficit in charge collection very close to the surface of a semiconductor detector.

**Dead time** The time interval between two events (e.g., between two pulses).

**DMSA (dimercaptosuccinic acid)** The scan that is used to find problems in kidney.

**DNA** Deoxyribonucleic acid is composed of two long polymers, the backbones of that are made of sugar and phosphate groups.

**Dose** The total amount of ionizing radiation absorbed by a body or body tissue.

**Dosimeter** The device used to measure the amount of radiation energy absorbed over a given period of time by an object.

**DQE** Detective quantum efficiency is the measure of the performance of a radiation detector (scintillator).

**Electrons** Subatomic negatively charged particle, one of the constituents of atom.

**Endoscopy** Procedure to look into human body with the help of an instrument called endoscope, which has a long thin tube with a tiny camera.

**Energy resolution** The parameter used to measure the energy distribution of the incident radiation.

**Fano factor** The empirical constant, which is multiplied with the predicted variance to convert into the experimentally observed variance.

**Fast Fourier transform (FFT)** Preferred method for reconstruction of algorithms for filtered back projection.

**Fluence** It is the flux (radiative) integrated overtime.

**Fluoroscopy** It is a dynamic X-ray imaging system.

**FONAR (field-focusing NMR)** The system that cannot only resonance signal but also produce an image in a point-to-point scan when the specimen is moved.

**FWHM (full width at half maximum)** It is a parameter commonly used to describe the width of a bump halfway between the peak and the base.

**Galactic cosmic rays** The cosmic rays that have their origin in our galaxy.

**Gamma rays** Gamma rays have shorter wavelengths than X-rays and are more powerful. The frequencies of the rays are in the electromagnetic spectrum.

**Geiger–Muller tube (GM tube)** It detects the radiation by ionization produced in a low-pressure gas confined in the GM tube.

**Gnomic instability** It is a delayed long-lasting effect of ionizing radiation that is caused by high level of reactive oxygen species.

**Grain boundary** It is the interface between two grains or crystallites in a polycrystalline material.

**Gray** It is the SI unit of rad equivalent to 100 rad.

**Gray scale** It is the range of shades of gray without color. It consists of 16 shades of gray in the webmart color palette.

**Half-life** The period of time it takes for a substance (radioactive) undergoing decay to decrease by half.

**Half-value layer** It is the thickness of an absorber of specified composition.

**IC (integrated circuit)** Devices like transistors, diodes, capacitors, and resistors (very rarely inductors) are fabricated on a monolithic base (silicon substrate) by oxidation and preferential diffusion of different types of impurities inside the host base (generally silicon). Local metallization to form ohmic contacts is made by aluminum alloy or copper metal.

**Ionization** It is the physical process of converting an atom or molecule into an ion by adding or removing charged particles such as electrons or ions.

**Ionization chamber** A gas-filled radiation detection chamber used to detect or measure ionizing radiation.

**Isotopes** Nuclides that have the same number of protons but different numbers of neutrons.

**ITO** A layer of indium (In) tin oxide (SnO) is deposited generally over glass to make a conductive layer.

**Kerma** In medical science, it is the kinetic energy released by the medium.

**Magnification** The process of enlarging the image to be able to see in more detail without increasing the physical size.

**Mean life** The average lifetime during which a system, such as an atom, nucleus, or elementary particle, exists in a specified form.

**Moire pattern** The effect of aliasing between the digital sampling and the image of the grid bars.

**Molecular biology** It is the branch of biology dealing with the molecular basis of biological activity.

**MOS (MOSFET)** Metal oxide semiconductor forms the heart of an important family of devices called MOS field effect transistors. MOS technology is also being used to form transistors in ICs.

**MRI (magnetic resonance imaging)** It uses magnetic resonance method to image nuclei of atoms inside a body.

**MTF (modulation transfer function)** It is the discrete Fourier transform of the line-spread function.

**Muons** It is an elementary particle similar to electron with unitary negative electric charge and a magnetic spin equal to half.

**Mutagenic disturbance** The transformation of the normal cells to abnormal cells causing cancer due to ionizing radiation can be described as mutagenic disturbance.

**NEQ (noise equivalent quanta)** The NEQ is a measure of image quality and describes the minimum number of X-ray quanta required to produce a specified signal to noise ratio (SNR).

**Neutrino** Electrically neutral, subatomic weakly interacting particle.

**Neutron scattering** Neutrons primarily interact with nucleus of an atom involving neutron spin and atomic magnetic moment. It is an important part of neutron detection.

# Glossary

**Neutron** It is a neutral particle ($Z = 0$) that makes up the nucleus along with protons with a spin equals to and negative magnetic moment.

**Noise (image)** It is the random variation of the actual information of the image of an object.

**Optical density** For a given wavelength an expression of the transmittance of an optical element. It is also defined as the ratio of the incident light to transmittance light.

**OSLD** Optically stimulated luminescent dosimeter has a thin film of $Al_2O_3$ which undergoes structural changes as it receives radiation. This structural change is previously calibrated with a known radiation source (green laser light).

**Pair production** It refers to the creation of particles and antiparticles usually from a proton or another neutron (boson).

**Paralyzable** When the interaction rate is proportional to m times the dead time, where m is the recorded time rate.

**PET (positron emission tomography)** A medical imaging method that produces a 3-D image of functional process in the body.

**Phosphorescence and fluorescence** Phosphorescence is a specific type of photoluminescence related to fluorescence. Unlike fluorescence, a phosphorescent material does not immediately reemit the radiation it absorbs.

**Photoelectric effect** It is the result of absorption of electromagnetic radiation and the consequences of the emission of photons.

**Positron** Positively charged subatomic particles.

**Protons** It is a subatomic hadron particle. Along with neutrons it makes up the nucleus.

**Quality factor** The relative damage per each radiation.

**Quantum efficiency** For radiation detectors, it is an accurate measurement of the device's electrical sensitivity to radiation.

**Quenching** Refers to any process, which decreases the fluorescence intensity.

**Radiation** It is a process in which energetic particles travel through a medium. The radiation can be ionizing or nonionizing.

**Radiation therapy** The process where radiation energy is utilized to control or to kill an abnormal cancerous cell.

**Radioactive decay** Certain combinations of neutrons and protons are unstable and break up spontaneously, a process referred to as radioactive decay.

**RBE (radiation biological effectiveness)** It is defined as the absorbed dose in rad times the RBE of radiation.

**REM** It is defined as the amount of biological damage caused by ionization.

**Roentgen** It is the unit of measurement for exposure of ionizing radiation.

**Scintillators** Organic or inorganic compounds that scintillate and are used to detect radioactive sources and processes.

**Selectivity** The ratio of the primary transmission of a grid is called selectivity.

**Shielding** Materials used to protect radiation from the radioactive source.

**Signal to noise ratio (SNR)** It compares the level of a desired signal to the level of a background noise.

**SiLi** Lithium (Li)-doped silicon (Si) radiation detectors.

**Solar particles** Energetic particles coming from sun (detected first in the year 1940).

**Spatial resolution** Resolving power of an image-forming device to discriminate between two adjacent high-contrast objects.

**Specific activity** The activity of a radioactive source per unit mass.

**SPECT (single-photon emission computed tomography)** Tomography that uses gamma rays to produce 3-D images of the internal parts of the body.

**Superconductivity** It is the phenomenon found in certain special types of materials that show almost zero resistance at a particular temperature.

**TFT** Thin-film transistor (TFT) is fabricated by depositing thin films of semiconductor material as well as dielectric layer over a supporting substrate (mostly glass). Large-area amorphous silicon (a-Si) and photodiode technologies are the recent development of digital X-ray image sensor. The TFTs are used as pixel switches, which addressed each row of the array, and photodiodes at each pixel location convert incident light to charge, which is read out by charge amplifiers.

**Thermal neutrons** It is a special class of neutrons when the kinetic energy of the neutrons on an average is similar to the kinetic energy of molecules at room temperature.

**Thermoluminescence** It is a form of luminescence observed in a particular class of material exposed to heat (therm).

**TLD** The thermoluminescent dosimeter (TLD) is a device which linearly responses to a dose which is relatively energy independent.

**Townsend avalanche coefficient** It is defined as the fractional increase in the number of electrons per unit length.

**Ultrasound (US)** US uses high-frequency sound waves to produce pictures.

**Weiner spectrum** It measures the level of noise as a function of spatial frequency and equals to Fourier transform of the autocorrelation function in a uniformly exposed radiographic image.

**X-rays** It is a form of electromagnetic radiation capable of penetrating solids and ionizing gases.

# Index

## A

Absolute efficiency, 105
Absorber/scintillator, 266
Absorption, 23, 266
  cross section, 413
  and emission transitions, 274
  coefficient, 326, 393
  correction, 277
  efficiency, 373
  event, 266
  spectrometer, 95
  target, 277
Accelerator, 368
Acid digestion bomb, 341
Acoustic signal, 485
Acoustic window, 483
Acquisition time (AT), 207
Activator, 263, 371, 375, 376, 407
Active matrix array (AMA), 324
Active-matrix flat-panel imager (AMFPI),
    230, 231
Active matrix liquid crystal display
    (AMLCD), 229
Active matrix organic light emitting diode
    (AMOLED), 232
Adaptive software filtration, 197
ADC. *See* Analog to digital converter (ADC)
AEC. *See* Automatic exposure control (AEC)
AER. *See* Alpha-emitting radionuclides (AER)
Afterglow, 377, 380, 381
Air-filled intestine, 483
Algae, 69
Algebraic reconstruction technique (ART), 246
Aliase(s), 235
Aliased signals, 223
Aliasing, 223
Alkali halide(s), 316, 369
Alkali-halide crystals, 140

Alpha ($\alpha$)
  and beta counting, 81
  decay, 15
  particles, 10, 15, 62–64, 101
  plateau, 74
  radiation, 60, 434
  ray, 74
Alpha-emitting radionuclides (AER), 256, 257
Alzheimer's disease, 471, 478
Americium (Am), 16, 18
Amorphous hydrogenated silicon (a-Si:H), 229
Amorphous silicon, 146
Amorphous silicon photodiode, 230
Amyloid, 490
Amyotrophic lateral sclerosis (ALS), 471
Analog imaging system, 227
Analog to digital converter (ADC), 154, 234
Anesthesia, 195
Angiography, 34, 220, 465, 485
Anionic complexes, 254
Anisotropic, 207, 394
Annihilation, 13
Annihilation radiation, 98
Anthracene, 76
Antibiotics, 195
Antimony (Sb), 7
Aromatic hydrocarbon compound, 255
As low as reasonably achievable (ALARA), 489
Astatine (At), 17
Asymmetrical peak, 403
Asymptotic, 396
Atherosclerotic, 465
Attention-deficit/hyperactivity, 477
Attenuation, 431
  coefficient, 21, 265, 321, 401, 420, 425
  constant, 347
  correction, 467
  length, 264, 268, 334, 338, 344, 349

T.K. Gupta, *Radiation, Ionization, and Detection in Nuclear Medicine*,
DOI 10.1007/978-3-642-34076-5, © Springer-Verlag Berlin Heidelberg 2013

501

502 Index

Auger process, 272
 electron, 30
 electron spectroscopy, 30
 luminescence, 276
Automated dose modulated system, 111
Automatic exposure control (AEC), 33, 35
Avalanche photodiode (APD), 251, 410, 411, 473, 478

**B**

Background radiation, 51
Back projection, 210, 216, 218
Back scatter peak, 97
Bacteria, 71
Bandgap, 253, 322, 354, 358, 384, 393, 426, 429
Bandgap energy, 263
Barium fluoride ($BaF_2$), 316
Barrier layer, 330, 394, 395
BARS diamond, 288
Barytes concrete, 17
Beam
 collimation, 197
 collision point, 368
 filtration, 197
 hardening, 217, 218
Becquerel (Bq), 9, 10
Beta ($\beta$)
 decay, 18
 emission, 17
 emitters, 256
 particle, 9, 15, , 62
 plateau, 74
 radioactive element, 13
 ray, 74
 sources, 166
Bethe–Bloch formula, 270
BGO. *See* Bismuth germanate (BGO)
Bi-alkali photocathode, 370
Binary and ternary alloys, 305
Binding energy, 37
Binominal process, 118
Biological effects, 61, 65
 half time, 109
 life time, 109
Biological molecules, 479
Biological tissue, 481
Bioluminescence, 188
Biomedical engineering, 188
Bipolar junction transistor (BJT), 71
Birl's parameter, 258

Bismuth (Bi), 17
Bismuth germanate (BGO), 263, 274, 341, 343, 346, 409, 410, 420, 421, 475
Blood-oxygenation-level-dependent imaging (BOLD), 479
Blurring, 176, 228, 237
Blurring effect, 216
Body centered cubic (BCC), 140, 368
Body counting, 108
Boiling water reactor (BWR), 46
Boltzmann constant, 172
Bone marrow ablation, 48
Bone scan, 452
Bouncy convection, 304
Branching fraction, 265
Bremsstrahlung, 31, 108, 271, 272, 459
Bremsstrahlung escape, 95
Bridgman
 furnace, 303
 growth, 335
 method, 351, 414
Bridgman–Stockbarger (BS)
 furnace, 336
 method, 296, 301, 318, 319, 326, 332, 333, 350, 371
 process, 293, 295
Broad energy spectrum, 217
Broadening of peak, 386
Bubble chamber, 4
Bucky factor, 222, 223
Bulk screening length, 144
Burst noise, 173

**C**

Cadmium telluride (CdTe), 351, 384, 422
Cadmium tungstate ($CdWO_4$), 263, 316, 342, 347, 420, 421
Cadmium vacancy, 423
Cadmium zinc telluride (CZT), 141, 351, 384, 387, 427, 437
Cancer, 195
Cancerous cells, 125
Cancer therapy, 490
Capacitive Frisch grid design, 150
Capacitive grid detector, 150
Capacitive micromachined ultrasound transducer (CMUT), 482
Carbon (C), 12
Cardiac, 478
Cardiac disease, 195
Cardiology, 199

# Index

C-arm design, 202
Carrier
   diffusion, 275
   drift length, 281
   extraction factor, 157
CAT scan, 467
Cavity chamber, 80
$CaWO_4$, 368
CCD. *See* Charge-coupled devices (CCD)
CdTe single crystal, 351
CellScope, 194
Cellular molecules, 188
Cerebral angiographies, 109
Cerium-activated
   Lan. Bromide ($LaBr_3$:Ce), 334, 369
   lanthanum chlo ($LaCl_3$:Ce), 333, 369, 406
   lutetium gadolinium silicon oxide
      (LuGdSiO5:Ce), 349
   lutetium oxyorthosilicate (LSO:Ce), 336
Cesium (Cs), 7, 44
   iodide(CsI), 318, 368, 375, 376
Characteristic X-ray escape, 95
Charge, 308, 329
   accumulation, 395
   collection, 386
   collection efficiency, 136, 141, 386, 403,
      424, 434
   collection time, 437
   compensating ion, 273
   integration mode, 278
   material, 298, 320, 359
   transfer luminescence, 254
   transport, 146
   trapping, 389, 403, 434
   trapping and detrapping, 137
Charge-coupled devices (CCD), 34, 72, 203, 368
Charged particle induced conductivity
   (CPIC), 432
Charge sensitive amplifier, 429
Chemical agents, 305
Chemical synthesis of BGO, 345
Chemical vapor deposition (CVD), 76, 309,
   310, 354, 430, 432
Chest radiography, 34
Chi-square distribution (CSD), 123, 124
Chromosomes, 62
Cine camera, 198, 199
Cine-cardiography, 34
Close packed geometry, 368
Cloud chamber, 4
Cluster, 290
CMOS. *See* Complementary silicon metal
   oxide semiconductor (CMOS)
Coaxial detector, 145
Coaxial Ge (Li) layer, 144

Cobalt (Co), 7, 18, 44
Coincidence, 95, 412, 473
Collecting electrodes, 151
Collection efficiency, 391
Collimators (low medium and high-energy), 213
Color centers, 278
Columnar film, 381
Columnar structure, 198
Combined modulation (CARE), 111
Combustion flame, 310
Compact high energy neutron spectrometer
   (COLONS), 99
Complementary silicon metal oxide
   semiconductor (CMOS), 71, 89, 368
   imager, 231
Complex activator centers, 273
Compositional homogeneity, 309
Compound semiconductors (C-S), 351, 384,
   385, 425
Compton
   coincidence, 373
   coincidence technique, 377, 378
   collimated imaging, 224
   edge, 97
   effect, 20, 267, 269
   ratio, 402
   scattering, 38, 41, 52, 96, 211, 268, 435,
      439, 471
   spectrometer, 421
   suppression spectrometer, 344
Computed tomography (CT), 34, 41, 43, 71,
   188, 189, 191, 195, 197, 209, 215,
   229, 238, 405, 422, 453, 467, 469,
   471, 490
   acquisition, 218
   images, 210
   numbers, 210, 218
   or PET, 487
   radiography imaging plate, 222
   scanners, 209, 215, 217
   scans, 195, 469
   or SPECT, 486
   system, 219
Computed tomography angiography (CTA),
   465, 466
Computer axial tomography (CAT), 215, 218
Computer controlled X-ray accelerator, 461
Conduction, 292
Coniferous trees, 71
Conservation of energy, 279
Contrast, 237–239, 244
   agent, 465
   reduction factor, 43
   sensitivity, 208, 481
Contrast enhanced MRA (CEMRA), 467

## Index

Control of stoichiometry, 303
Convection, 292
Convection process, 293
Conventional
    angiography (CA), 467
    fluoroscopy imaging, 237
    radiography, 34
Conversion efficiency, 382
Convolution, 216, 379
Convolution kernel, 217
Coplanar geometry, 153
Coplanar grid
    detector, 151
    geometry, 149
Cosmic
    radiation, 63
    rays, 4, 60
Cosmogenetic radionuclides, 63
Coulomb
    blockade, 174
    field, 271
Coulombic potential, 402
Counting
    efficiency, 105
    plateau, 94
    system, 177
Covalent bond, 275
Critical energy, 272
    barrier, 290
Cross section, 412
    concept, 27
    value, 78
Cryogenically cooled temperature, 426
Cryopump, 309
Crystal, 290
    defects, 138
    growth, 318, 322
Crystalline
    defects, 309
    scintillator, 269
    semiconductor, 368
CSD. *See* Chi-square distribution (CSD)
Cs spectrum, 388
CT. *See* Computed tomography (CT)
CTA. *See* Computed tomography
        angiography (CTA)
Cupping artifact, 217
Curie (Ci), 9
Current
    conduction, 394
    mode, 79, 90, 278
Curvilinear, 482
CVD. *See* Chemical vapor deposition (CVD)

Cyclotron, 45, 214, 455, 458, 459
Cyst, 485
Czochralski (CZ)
    growth, 296
    method, 293, 295, 297, 298, 337, 345,
        359, 418
    technique, 342, 347, 350
    technology, 418

## D

d-amyl-PPF, 256
Dark cond, 391
    current, 230, 373, 395, 402, 423, 431, 432
    electrical resistivity, 402
Daughter nuclei, 455
d-CHO-PPF2, 256
Dead layer, 143, 144
Dead time, 106, 177
Debye length, 144
Decay
    constant, 44, 338, 344, 347, 349, 369, 374,
        377, 412, 414
    time, 336, 374, 379, 421
Decaying source, 107
Deciduous trees, 71
Deep-level model, 136
Dees, 459
Defect
    center, 138
    formation by ionizing radiation, 277
Degenerately doped, 279
Degree of freedom, 124
De-ionization rate, 424
Delta function, 171
Dementia, 478, 490.
Dentistry, 228, 320
Density
    gradient, 299
    ionization, 275
Deoxyribonucleic acid (DNA), 3, 65, 66, 110,
    188, 453, 489
Depleted field effect transistor (DEPFET), 154
Detection efficiency, 90, 164, 215, 330, 358,
    401, 403
Detective quantum efficiency (DQE), 158, 159,
    167, 169, 231, 235, 241, 383
Detector
    counting efficiency, 265
    efficiency, 266
Diagnostic radiology, 464
Diagnostic test, 452
Diamond, 354, 355, 429–432

Diamond crystal, 307
Dielectric constant, 144, 323, 354, 358, 430, 438
Differential
    amplitude, 92
    amplitude increment, 243
    distribution, 93
    light output, 257
    mobility spectrometry-mass spectrometry (DMS-MS), 115
    quenching efficiency, 381
Diffused Li-layer, 435
Diffusion, 80, 143
Diffusivity, 300
Digital
    domain, 228
    image acquisition, 244
    imaging, 223
    imaging technology, 228
    infrared (IR) image, 192
    radiographic technique, 324
    radiography, 319–320
    subtraction angiography (DSA), 34
    X-ray imaging, 234
Dimercaptosuccinic acid (DMSA), 213
Direct conversion efficiency, 162
Direct detector, 157
Directional solidification, 318
Direct ionization, 273
Disintegration, 459
Distillation, 307
Divalent acceptor, 423
Divergence, 220
DNA chains, 191
Dopant, 263
Doped crystal, 375
Dose, 63, 109, 110, 396, 431, 462
    penalty, 223
    rate, 431
    reduction, 197
Dosimeter, 109–111
Dosimetry, 422, 429
Double C-arm cine fluoroscopy system, 202
Double-crucible, 298
Double escape peak, 98
Double sided HPGe, 441
Double sided silicon strip, 368
Double-sided strip detector (DSSD), 437
dpiX, 389

Dynamic
    method, 164
    range, 395
    renography, 213
Dynodes, 89

**E**
EBT. *See* External beam therapy (EBT)
ECG, 197
Echo signal, 484
EEG/MEG, 487
Effective
    atomic number, 421
    dose, 35
    mobility, 280
    radiation dose, 51
    Z, 425
Efficiency of a scintillator, 252
Efficiently quenched, 255
EGERT gamma ray, 60
Elapsed time, 265
Elastic cross section, 439
Elastic recoil spectrum, 99
Elastic scattering, 24, 28, 439
Electric field, 398
Electric field strengths, 135
Electroabsorption, 392
    signal, 393
    spectrum, 391
Electrode design, 135, 143
Electrodrift, 143
Electrodynamic gradient process, 332
Electroless deposition, 145
Electromagnetic radiation, 3, 20
Electromagnetic spectrum, 264
Electron
    life time, 136
    positron pair production, 52, 267, 269
    thermal velocity, 172
    trapping, 152
    vacancy, 274
Electron–hole (e–h)
    mobility, 135, 429
    pairs, 115, 272, 280, 369
Electronic
    noise, 136, 176
    recombination, 272
    transition in inorganic material, 275

Electro-optical modulator, 422
Elemental semiconductor, 354, 429, 434
Emission
  bands, 372, 383
  range, 370
  spectra, 371, 375, 418
  spectrometry, 95
  spectrum, 345, 410, 412
  wavelength, 319, 320, 350, 377
Endoscopy, 43, 191, 193
Energy
  fluence, 34
  gap, 273
  resolution, 95, 104, 136, 243, 320, 334, 350, 373, 386, 391, 405, 410, 418, 420–422, 439
  spectrum, 279
  X-ray-spectrometer, 103
ENt diagnosis, 230
Entrance exposure
  rate, 245
  value, 245
Epitaxially grown, 324
*Erigeron canadensis*, 69
Escape
  peak, 421, 428
  probability, 236
Exametazime, 452
Excitation luminescence, 254
Excited fluor, 260
Excited level, 274
Excitonic, 369
Exposure rate constant, 6
External beam therapy (EBT), 460, 461

**F**
False positive fraction (FPF), 242
Fano
  factor, 79, 244, 281
  noise, 387
Fast
  decay time, 404
  Fourier transform (FFT), 218, 383
  moving neutrons, 64
  neutron(s), 78
Fast neutron coincidence counter (FNCC), 344
Fast spin echo (FSE) technique, 207
Fermi level, 143
Ferromagnetic material, 481
FET amplifier, 173
F-H centers, 278
Field-effect transistors (FET), 173
Field focusing NMR (FONAR), 205

Field of view (FOV), 201, 208
Fifth generation CT, 212, 467
Film grains, 225
Film radiography, 225
Filtered back projection, 246
First generation CT, 211
Fission
  fragments, 45
  product, 455
Flashlight effect, 254
Flat panel detector (FPD), 72
Flat panel X-ray image detector, 324
Flicker noise, 173
Floating zone (FZ) growth, 293
Fluence, 7, 10, 428
Fluid dynamics, 291
Fluor, 259
Fluorescence, 188, 259
  quenching, 260
Fluorodeoxyglucose (FDG), 203
Fluoroscope, 465
Fluoroscopy, 34, 191, 198, 229, 244, 320, 324
  exposure, 390
  systems, 201
Flux, 300
Forbidden gap, 140
Forced convection, 293
Forensic dentistry, 30
Fourier transform (FT), 207, 240, 241, 400
Fourth generation CT, 211
Fractional standard deviation, 281
Frenkel defects, 278
Frequency
  distributions, 116
  domain, 217
Frisch collar, 428
Frisch grid, 146, 428
  design, 148
  geometry, 153
  radiation spectrometer, 147
Frisch-ring, 148
Full wave at half-wave maximum (FWHM), 90, 104, 142, 146, 162, 163, 239, 243, 387, 405, 406, 409, 411, 415, 428, 435–438
Functional magnetic resonance imaging (FMRI), 479
Fungi, 69
Fusion, 44
Fusion/hybrid technology, 485
FWHM. *See* Full wave at half-wave maximum (FWHM)
FZ growth, 295

## Index

**G**

Gadolinium silicate (GSO), 342–343, 417, 420, 422
Galatic, 60
Galatic cosmic rays (GCR), 71
Gamma ($\gamma$)
  and X-rays absorption coefficients, 371
  camera, 252, 464
  dosimetry, 431
  knife, 461–463
  knife surgery, 21
  photons, 471
  quanta, 13
  quantum, 254
    energy, 276
  radiation, 69, 86, 486
  ray(s), 2, 3, 6, 15, 20, 21, 60, 63, 74, 84, 91, 108, 126, 165, 197, 225, 245, 333, 378, 407, 412, 459
    absorption coefficient, 376
    camera, 215, 266
    detector, 422
    energy, 280
    imager, 14
    imaging, 43, 439
    photon, 18
    sensitivity, 78
    spectrometer, 427
    spectroscopy, 415
    spectrum, 393
  slope, 244
  sources, 166
Gantry, 212, 462
Gas proportional scintillation (PC), 82
Gastrointestinal, 201
Gaussian
  distribution, 122
  fit, 415
  noise, 169
Geiger–Muller (GM)
  counter, 75, 80, 86, 109, 118, 126
  discharge, 86
  mode, 89
  plateau, 86
  tubes, 78, 79, 86, 87, 112, 177
Generation–recombination (G-R) center, 138
Generator, 455
Geometrical unsharpness, 238
Geometrical weighting, 146
Geometric blurring, 41
Geophysical exploration, 316
Germanium (Ge), 71, 354, 359, 384, 437
GFR measurement, 452

Ghosting, 218, 399
Glomerular, 454
GM. *See* Geiger–Muller (GM)
Good linearity, 404
Gradient field strength, 207
Grain boundary, 252
Gray (Gy), 8, 10
  matter, 206
  scale, 214, 484
Grid
  frequency, 222
  ratio, 222
  strip density, 222
Ground state, 274
Growth-interface, 305, 309
Growth of single crystal
  of CZT, 353
  of GSO, 343
  of TlBr, 331
  of CWO, 347
Growth rate, 290
Guard ring, 435
Gunn effect, 422
Gynecological, 484
Gyromagnetic ratio, 481

**H**

Hadron(s), 278
  colliders, 288
  therapy, 245
Half-life(ves), 16, 45, 109, 265, 452, 456
Half-value layer (HVL), 22–23
Halide(s), 321
  materials, 272
  of heavy metals, 384
Hardening artifacts, 469
Hard ware induced artifacts, 473
Harmonic imaging, 481–482, 485
Hecht equation, 141
Helical scanning, 213, 467
*Helicobacter pylori* test, 452
Herbaceous plants, 71
He-spectrometer, 99
$HgI_2$ deposition, 330
High energy
  physics, 266
  resolution, 426
High flux isotope reactor (HFIR), 47
High frequency oscillator, 459
High pass and low pass filtering, 229
High pressure Bridgman method (HPBM), 175, 352

High-pressure/high temperature (HP/HT) method, 355
High pressure vertical Bridgman method (HPB), 302, 351
High-purity germanium (HPGe), 145, 321
  coaxial detector, 359
  detector, 440
High rate dose (HRD), 310
High resolution research tomography (HRRT), 417
High-vacuum pump, 309
Hole
  mobility, 391
  trapping, 152
Homogeneous Poisson point, 171
Homopolar, 277
Hot-wall epitaxy (HPE), 253, 381
Hounsfield units, 210
HPGe. *See* High-purity germanium (HPGe)
Huntington disease, 471
Hydrogenated silicon carbide, 430
Hydrothermal gradient, 295
Hydrothermal synthesis, 341
Hydrothermal treatment, 341
Hygroscopicity, 371

**I**

ICRP. *See* International Commission on Radiation Protection (ICRP)
Image
  intensifier, 198, 199
  lag, 399
  matrix, 216
  reconstruction, 245
  resolution, 141, 473
  transfer, 198
Image-guided cancer therapy, 464
Imaging
  modalities, 246, 471
  technology, 226
Impurity concentration, 163
IMRT. *See* Intensity-modulated radiation therapy (IMRT)
Index of refraction, 334
Indirect detector, 157
Indium tin oxide (ITO), 324
Individual quantum pulse mode, 279
Induced charge, 154, 279
Induction
  decay, 481
  effect, 339, 419
Inductively heating, 336

Inelastic scattering, 25, 26, 28
Infrared (IR), 3
  film, 193
  windows, 423
Ingot, 358
Inhaled radionuclides, 63
Inherent filtration, 36
Inorganic scintillating materials, 368
Inorganic scintillators, 96, 254, 272
Insects, 71
Insensitive time, 107
Integrated circuit (IC), 356
Integration time, 379
Intensity-modulated radiation therapy (IMRT), 460, 461
Interaction of neutron, 23
Interface morphology, 301
Interfacial layer, 144
Internal radionuclides, 63
International Commission on Radiation Protection (ICRP), 8
Inter pixel spacing, 136
Interstitial donor, 435
Interventional cardiology, 47
Intracoronary ultrasound (ICUS), 485
Intrinsic
  efficiency, 105
  resistivity, 438
  vacancy, 429
Inversion recovery, 481
Iodine, 7
Ion-chamber, 78, 79, 126
Ionic, 264
  compounds, 316
  conductivity, 402
  conductor, 404
  material, 262
Ionization
  current, 80
  energy, 269, 439
  mechanism, 274
  power, 271
Ionizing
  and non-ionizing radiation, 61
  energy, 67, 71
  radiation, 2–4, 15, 63, 65, 68, 70, 189, 207, 253, 482
IR. *See* Infrared (IR)
Irradiation, 397
Ischemic attack, 471
Ischemic heart disease, 478
Isotopes, 214
  produced in cyclotron, 460

# Index

Isotopic curve, 387
Isotropic, 207
ITO glass, 325, 346

## K

$K_\alpha$ photon, 30
Kapton, 150
Kerma, 10, 81, 222
Kidney, 108, 482
Kinetic Monte Carlo (KMC) model, 277
Klein-Nishina equation, 39
Krypton (Kr), 44, 454

## L

Lanthanum halide, 404, 405, 409
    La-bromide (LaBr3), 411
    La-chloride (LaCl3), 369, 407, 409
Lapse imaging, 479
Larmor frequency, 481
Latent heat, 302
Latent heat of fusion, 301
Latent image, 228
Lattice defects, 277
Lattice orientation, 252
Layered structure, 325
Lead iodide ($PbI_2$), 290, 369, 384, 387
Lead phantom, 399
Lead tungstate (PWO), 348, 420
Lead zirconate titanate (PZT), 482
Leakage current, 384, 387, 388, 435
Life time, 260, 418, 456
    of carriers, 153
Light collection, 265
Light emitting diode (LED), 122
Light escape cone, 371
Light output, 382
Light yield, 276, 277, 319, 320, 336, 347, 349,
    371, 379, 380, 408
    of inorganic scintillator, 275
Linear accelerator (LINC), 461, 464
Linear attenuation coefficients, 211, 268
Linear stopping power, 269
Line pair resolution, 399
Line-spread function (LSF), 156, 238, 387, 400
Lipids, 481
Lipophilic molecules, 452
Liquid crystal display (LCD), 229, 232
Liquid encapsulated CZ (LEC), 304
Liquid phase epitaxial deposition (LPD), 146
Liquid scintillator counting (LSC), 256, 257
Liquid–solid equilibrium conditions, 301
Lithium doped silicon (SiLi), 253
Lithium drifted silicon (Si-Li), 146

Liver, 108
Loadability, 33
Localized charge, 425
Logging, 420
Low-pressure Bridgman method, 351
LuAP, 350
Luminescence, 114, 254, 262, 265, 382, 393
    centers, 259, 262
    intensity, 420
    quenching, 259
Luminescent-physics, 316
Luminescent state, 379
Lung
    perfusion, 452
    ventilation, 452
Lutetium iodide (LuI):Ce, 335
Lutetium oxyorthosilicate (LSO), 263, 337,
    338, 340, 387, 417, 419
Lymphocyte depletion curve, 70

## M

Macroscopic cross section, 27
Magnetic fusion, 99
Magnetic resonance angiography (MRA), 465,
    467
Magnetic resonance imaging (MRI), 71, 175,
    188, 191, 195, 198, 205, 207, 208,
    214, 215, 229, 238, 241, 453, 467,
    478, 480, 487, 490
    or fMRI, 487
    or PET, 487
Magnetic resonance spectrometry, 95
Magnetization vector, 481
Magnetosphere, 60
Mammals, 71
Mammograms, 225
Mammography, 34, 220, 228, 310, 320, 479
Manganese (Mn), 7
Mass attenuation coefficient, 268, 269
Mass spectrometry, 95
Mass transfer surface kinetics, 291
MCA. See Multi channel analyzer (MCA)
Mean square voltage mode (MSV), 91
Medical imaging, 43, 177, 188, 195, 197, 289,
    326, 377, 388, 393, 464
    imager, 136
    system, 14
Medical photography, 191
Melanoma therapy, 48
Mercuric Iodide ($HgI_2$), 142, 326, 368, 369,
    384
Mesons, 278
Metabolic activity, 471
Metal oxide semiconductor (MOS), 71

Metal oxide silicon field effect transistor (MOSFET), 111
Metal solution growth (MSG), 305
Metastable state, 259
MIBG adrenals, 213
Micro-columnar structure, 321
Micro-electromechanical (MEMS), 348
Microgravity, 304
Microscopic cross section, 27
Microwaves, 3
Midgap level, 136
Milked, 455
Miniature type dosimeter, 112
Mitochondria, 62
Mobility, 136, 163, 392
Mobility-life time product, 139, 368, 414, 427, 429, 434
Modalities, 486
Modulation, 111
Modulation transfer function (MTF), 156, 159, 169, 231, 239–242, 383, 400, 401
Moire pattern, 223
Molecular biology, 188, 192
Molecular imaging, 43
Molten zone, 308
Monochromator, 345, 410, 412
Monoclonal antibody, 456
Monoenergetic spectrum, 100
Monomer, 257
Monte-Carlo
    method, 136
    technique, 108
Morbidity, 197
Motion artifacts, 469
MRI. *See* Magnetic resonance imaging (MRI)
MSV mode operation, 92
MTF. *See* Modulation transfer function (MTF)
Muffle furnace, 343
Multi channel analyzer (MCA), 73, 96, 413
Multiple sclerosis, 471
Multiple slice acquisition, 480
Multiplication, 82
Multizone vertical Bridgman method, 302
$\mu\tau$ product, 384
Mylar, 150
Myocardial, 490
    activity, 213
    perfusion imaging, 108
Myocardium, 452

**N**

NaI (Tl) detector, 96
NaI:Tl, 316, 370, 417
Nano-electromechanical (NEMS), 348

Neptunium (Np), 18
Neurosurgical, 462
Neutrino, 18
Neutron(s), 9, 23, 62–64, 76, 100, 278, 429, 455
    beam therapy, 461
    detector, 317
    scattering, 23
    shielding, 28
    source, 28
    spectra, 102
Nickel (Ni), 18
NMOS transistor, 72
Noise, 237, 239
Noise power spectrum (NPS), 158, 159, 167, 168
Non-invasive, 463
Non-paralyzable, 106
Non-paralyzable system, 177
Non-proportionality (nPR), 416, 417
Non-radiative decay, 272
Non-relativistic particle, 271
Nuclear
    energy, 125
    medicine, 2, 4, 391, 451, 489
    medicine imaging, 489
    radiation, 11, 61, 66, 70
    reactor, 455
    spectroscopy, 422
Nucleic acid, 62
Nuclides, 7, 11, 12
Number of projections in CT, 215
Number of quanta (NEQ), 241
Nyquist frequency, 235, 390

**O**

Oak pine forest, 69
OLED, 232
Oncology, 490
1,3-diphenyl-2 pyrazolines, 256
Optical
    coupling, 199
    coupling transmittance, 236
    density, 225, 227, 228
    excitation, 422
    fiber sensor, 421
    imaging, 188
Optically simulated luminescence, 73
Optically stimulated luminescent dosimeter (OSLD), 114
Optical transmission, 382
Organ counting, 108
Organic and Inorganic materials, 262
Organic scintillator, 254, 255, 257, 261

# Index

Orthogonal strip detector (OSD), 435, 436
Orthorhombic, 327
  β phase, 327
Orthovoltage X-ray machine, 461
Oscillator strength, 276
Over table (OT), 201
Oxidizing agent, 61

## P

Pair production, 21, 41, 96
Panel imager, 387
Parallel plate avalanche proportional
  counter, 82
Parallel plate avalanche counter, 85
Paralyzable, 106
  system, 177
Paranasal sinuses, 461
Parkin's disease, 471
Particle in binder (PIB) method, 389, 393,
  396, 397
Passivation layer, 229
$PbI_2$. See Lead iodide ($PbI_2$)
PECVD. See Plasma-enhanced chemical vapor
  deposition (PECVD)
Pen ionization dosimeter, 113
Percutaneous transluminal coronary
  angiography (PTCA), 48
Permittivity, 140, 164
PET. See Positron emission tomography (PET)
Phantom, 390
Phase
  encode, 207
  transitions, 264, 291
Phosphor, 76, 198, 236, 273
Phosphorescence, 259, 260, 374
Phosphor material, 198
Phosphorus implantation, 145
Photo absorption, 407
Photocathode, 87, 198
Photoconducting insulator, 386
Photoconductivity, 433
  photoconductor, 388, 390, 399, 430
  properties, 369
Photocurrent, 397, 398, 431, 432
Photodetector, 251, 322
Photodiode, 358
Photoelectric absorption, 40, 41, 96
Photoelectric coefficient, 402
Photoelectric effect, 20, 40, 41, 52, 96, 98, 267,
  269, 420
Photoelectron yield, 376
Photofraction, 376

Photomultiplier tube (PMT), 73, 75, 87–89, 96,
  115, 126, 162, 251, 256, 317, 345,
  369, 376, 378, 410, 412, 416, 418,
  419, 478
  current, 262
  photocathode yield, 344
Photon(s), 205, 224, 264, 372, 402, 409
  fluence, 34, 43
  pulse, 433
  recoil spectrometer, 99
  response, 377
  yield, 377
Photo peak, 98
  efficiency, 95, 104
  shift, 427
Photoresist, 394
Photo-sensors, 391
Physical vapor-deposited (PVD), 387
Physical vapor transport (PVT), 296, 300,
  327, 393
  and B–S processes, 299
  method, 301
Piezoelectric ceramics, 482
π-molecular orbitals, 255
Pincushion distortion, 201
PIN
  device, 253
  diode, 319
Pion, 429
Pitch, 399
Pixel, 210
Pixelated design, 153
Pixelated detector, 153, 156
Pixelated TlBr, 332
Planar design, 428
Planar detector, 216, 280
Planar geometry, 152
Planar images, 213
Planck constant, 267, 276
Plaque morphology, 485
Plasma, 310
Plasma-enhanced chemical vapor deposition
  (PECVD), 229, 233, 288, 310, 355
  grown diamond, 357
Plastic deformations, 291
Plastic scintillator, 73, 255
Platinum (Pt), 18
PLED, 233
Plutonium, 60
PM dynodes, 373
PMT. See Photomultiplier tube (PMT)
Pocket dosimeter, 113
Point spread function (PSF), 238

Poisson
  distribution, 118–122
  equation, 164
  process, 170
Polarization, 141, 142, 166, 387, 404, 424
  effects, 137, 143, 398
Polonium (Po), 12
Polychromatic, 217
Polycrystalline film, 317
Polycrystalline material, 322
Polyenergetic, 469
Polyethylene naphthalate (PEN), 233
Polymer, 73
Polymer binder, 331
Polymeric form, 327
Polymerization, 257
Polymerized, 310
Polypeptide receptor binding agent, 456
Poole–Frenkel effect (PFE), 391, 402
Porous graphite, 303
Positive sensitive proportional counter, 82, 84
Positron, 13
Positron emission tomography (PET), 13, 21,
      43, 71, 109, 177, 188, 189, 191, 195,
      197, 203, 204, 214, 215, 263, 266,
      336, 405, 420, 469–476, 486
  or CT, 474, 490
Positronium, 13
Power spectral density (PSD), 171
Pre-amplifier, 412
Preparation of radionuclides, 455
Pressure-temperature diagram, 305
Pressurized water reactor (PWR), 46
Primary decay time, 350
Primary transmission factor, 222
Principle decay time, 334
Pripyat, 68
Probability
  distribution laws, 116
  of success, 118
Probability density function (PDF), 124
Production of radionuclides, 44
Projection image, 226
Properties
  of CdTe and CZT, 354
  of C-S, 385
  of diamond, 358
Proportional counter (PC), 75, 78, 81, 82, 84,
      99, 126
Proteins, 62
Proton(s), 63, 278, 455, 481
  beam therapy, 460, 463
  density, 481

Pseudo binary phase diagram, 350
Pulse
  height, 280, 412
  height distribution, 94
  mode, 79, 91, 93
  processing, 136
  shape discrimination, 99, 376
PVT. *See* Physical vapor transport (PVT)

**Q**
Quality factor, 8, 386
Quanta, 20
Quantized conductance, 172
Quantum
  efficiency, 88, 89, 260, 261, 381
  energy, 37, 40
  mottle, 225
  noise, 176, 236
  radiation, 105
  yield, 260
Quartz fiber, 113
Quenching, 259, 260, 381
  constant, 260
  efficiency, 257
  parameter, 258
Q-value, 18, 78, 412

**R**
Rad, 8
Radiation, 292
  detectors (scintillators), 289, 316, 330, 354,
      369, 387, 388
  dose, 8, 469
  energy, 9, 109, 384
  hard sensors, 288
  length, 271, 371, 438
  oncology, 43
  therapy, 52, 460
Radiation biological effectiveness (RBE), 63
Radiation induced genomic instability
      (RIGI), 66
Radiative transition, 274
  C–V transitions, 276
Radioactive, 4, 11, 272
  decay, 11, 115, 121
  disintegration, 121
  lifetime, 109
  materials, 2, 50
  nucleus, 11
  tracer, 476
Radiograph, 112
Radiographic condition, 397

## Index

Radiographic grid, 220
Radiographic image, 238
Radiographic mode, 244
Radiography, 469
Radiography/fluoroscopic design, 201
Radioisotopes, 11, 13, 453, 454
   used in medicine, 457
Radiologic imaging, 201
Radiology, 15
Radioluminescence spectrum, 410
Radionuclide imaging, 188, 464
Radionuclides, 43, 108, 203, 455, 489
Radiopharmaceuticals, 452, 490
Radiophotoluminescence (RPL), 432
Radiotherapy, 15, 230, 431
Radiowaves, 3
Radium (Ra), 7, 12, 15, 63
Radon, 60
   page, 60
Raman scattering spectrometry, 95
Random process, 115
Rare earth (RE), 350
   ions, 273
Rate correction, 277
$RbCaF_3$, 274
$RbGd2Br_7$, 369
Reaction kinetics, 145
Reactive nitrogen oxide species (RNOS), 65
Reactive oxygen species (ROS), 64, 66
Reactor, 44, 46
Readout CCD, 158
Real time imaging, 157
Receiver operating characteristic (ROC) curve, 240–242
Receptor sensitivity, 237
Recoil
   interaction, 78
   ion energy, 102
Recombination, 138, 141, 434
   center, 423
   process, 163
Red forest, 68
Refractive index (RI), 252, 276, 319, 320, 344, 347, 350, 371, 377, 418
Relaxation of electrons, 275
Resistivity, 354, 384
Resolution, 237, 239, 382, 390, 400, 405
Respiratory motion (RM), 478
Restenosis therapy, 48
Reversible and irreversible reactions, 292
RGB
   Ce, 412
   crystal, 413
Richardson—Schottky effect, 402
Richardson's constant, 144

Richardson thermoionic emission, 144
Rigid body motion (RBM), 478
Rise time, 416
RNA, 188, 191
Rodent models, 188
Roentgen (R), 5, 29
Roentgen equivalent man (rem), 9, 63
Rose model, 240
Rubidium gadolinium bromide (RGB):Ce, 335


### S

Saddle point, 205
Saddle-shaped field distribution, 205
Saturation velocity, 358
Scalpel, 463
Scanning electron micrograph, 326
Scanning electron microscopy (SEM), 325, 395
Scan time, 467
Scattered radiation, 220, 225
Scattering, 23
   cross section, 38
Scatter to primary ratio (SPR), 220
Schief formula, 107
Schizophrenia, 471
Schottky
   barrier, 279
   CdTe diode, 423
   current, 165
Scintigraphy, 464
Scintillating fiber, 115
Scintillating material, 330
Scintillation detector, 251
Scintillation efficiency, 274
Scintographic biodistribution, 48
Screen film, 228
   radiography, 220, 469
   technology, 227
Screen mesh, 389
Screen printed, 390
   film, 324, 325, 397
   PIB, 399
Secondary electron escape. (SEE), 95
Secondary ionization, 83
Secondary X-rays, 272
Second generation CT, 211
Seed, 298, 308, 342
Segmented HPGE, 440
Segregation, 309
   coefficient, 351, 359
Selectivity, 222
Selenium (Se), 368
Self-absorption, 252, 370
Self propagating high-temperature synthesis (SHS), 346

514    Index

Self-trapped excitation, 278
Self-trapped exciton (STE), 406
Semiconductor
    crystals, 320
    detector, 278, 279, 424
    devices, 70
    diode, 251
    p-n diode, 252
    radiation detector, 253
    scintillators, 96
Sensitivity specks, 227
Sequenced array, 482
7-bis 9,9-dipropyl fluorine, 256
Shallow traps, 278
Shaping time, 403, 408
Shielding, 52
    materials, 16
Shifting of peak, 386
Shockley–Ramo theorem, 155
Shockley–Reed–Hall statistics, 136
Shot noise, 169
Si-APD, 406
Sievert (Sv), 10
Signal to noise ratio (SNR), 158, 167–169, 208, 214, 215, 235, 240–242, 468
Silicon (Si), 354, 356, 384, 435
    SiC Schottky diode, 102
    PIN diode, 252
    strip detector (SSD), 436
Single crystal
    of BGO, 343
    growth of LSO, 337
    growth of PWO, 349
    NaI, 372
Single-photon emission computed tomography (SPECT), 20, 43, 71, 109, 189, 195, 197, 205, 214, 215, 224, 229, 405, 452, 470, 476, 478, 485
    or CT, 245, 487, 490
    or CT scanner, 487
Si-PIN, 358
Slice, 480
    encode, 207
    thickness, 215, 468
Slow moving thermal neutrons, 64
Slow neutrons, 77
Slurry, 310, 325, 389
SNR. *See* Signal to noise ratio (SNR)
Sodium (Na), 7
Sodium iodide (NaI), 274, 318, 369
Soft filtration, 197
Solar particles, 60, 71
Sol–gel coating (SGC), 310

Solgel technique, 336, 348
Solid–liquid interface, 302
Solid-state photomultiplier (SSPM), 89–90
Solid state reaction, 338
Solid state scintillator, 254
Solid state synthesis, 349
Solubility, 359
Solute quenching, 260
Solution combustion synthesis (SCS), 343
Solution combustion technique, 346
Sonography, 481, 483
Source to image distance (SID), 34
Space
    charge, 143
    craft, 115
Spark chamber, 4
Spatial
    domain, 217
    filtering, 229
    frequency, 241
    resolution, 208, 224, 239, 253, 400
Specific activity, 11
Specificity, 240, 241, 393
Speckle filtering, 175
SPECT. *See* Single-photon emission computed tomography (SPECT)
Spectral peak broadening, 403
Spectrographic resolution, 150
Spin echo pulse, 481
Spleen, 482
Sputtered coated, 146
Squeeze, 310
Standard deviation, 104, 119
Statistical fluctuations, 90, 115, 282
Stereotactic radiosurgery (SRS), 460, 464
Stern–Volmer kinetic relation, 261
Stilbene, 76
Stochastic effect, 65
Stoichiometry, 330, 349
Stokes shift, 261
Stopping power, 161, 263, 321, 381, 401, 404, 420, 475
Strain engineering, 263, 264
Stroke, 471
Strontium (Sr), 44
Strontium iodide ($SrI_2$), 374
Supercooling, 341
Super saturation, 307
Surface
    channels, 144
    energy states, 139
    morphology, 325
    recombination, 139

# Index

recombination velocity, 139
state charge, 143
tension, 293
Swank factor, 159
Synchronizability, 103
Synchrotron
applications, 327
radiation, 434
Synthesized
Syn. LSO, 418, 419
Synthetic diamond, 433
crystal, 287, 288

## T

Tape automated bonding (TAB), 230
Targeted alpha therapy (TAT), 44
Tc 99, 453, 454
Technetium (Tc), 7, 44
Technology, screen printing, 330
method, 310
system, 330
Temperature gradient, 301, 320
Temporal resolution, 199
Terbium (Tb), 17
Terrestrial, 63
Tetraethyl orthosilicate (TEOS), 336–338
Tetragonal α-phase, 327
Tetragonal scheelite structure, 349
Tetrahedrally bonded, 326
TFT array, 230, 325, 330, 389, 395, 396, 399
Thallium (Tl), 44
Thallium activated NaI (Tl), 318
Thallium bromide (TlBr), 331, 369, 401
Theory
of crystal growth, 289
of nucleation, 289
Therapeutic examination, 452
Therapeutic radioisotope, 453
Therapy, 459
Thermal head
annealing, 310
conductivity, 301, 358
energy, 80
expansion, 377
gradient, 295
loadability, 33
neutron(s), 27, 76, 78, 412, 413
neutron detector, 391
noise, 173, 378
stress, 291

Thermodynamic equilibrium, 296
Thermogram, 192
Thermographic camera, 192
Thermography, 43, 191–193
Thermoionic emission, 144
Thermoionic noise, 88
Thermoluminescence, 264
Thermoluminescent dosimeter (TLD), 114
Thermoscopy, 192
Thick films, 310
Thin film transistor (TFT), 72, 174, 229, 232,
233, 322, 393, 401
Thin/thick film, 289
Third generation CT, 211
Third generation system, 215
Thorium (Th), 15
Three gradient magnet, 480
Thrombolytic therapy, 196
Thyroid, 103, 108
Thyroid metastases, 213
Thyrotoxicosis therapy, 452
Time of flight, 164, 266, 422
Time of flight PET (TOFPET), 476
Time resolution, 333, 413, 418
Tissue, 108
density, 206, 244
equivalent material, 431
Townsend
avalanche, 83
coefficient, 84
Transducer, 482
Transit charge transport (TCT), 164
Transmission probability, 172
Transmittance, 73
Transmitter, 484
Transparent dielectric media, 254
Transport kinetics, 386
Transport properties, 423
Trap(s), 419, 431
Trap density, 419
Trapezoid prism, 146
Trapping, 135, 136, 387
centers, 137, 138, 141, 424
and de-trapping centers, 135, 136, 138, 142,
280, 391, 399, 424
phenomena, 152
Traveling heater method (THM), 175, 304,
351, 427
Traveling magnetic field (TMF), 304
Triplet state, 255
Tritium (Tr), 44

Tritons, 101
Tumers, 452
Tunneling, 391
Turbo pump, 309
Two-zone controlled furnace, 300

**U**

Ultrasound (US), 71, 191, 229, 481, 482
   images, 175
Ultraviolet (UV)
   light, 255
   rays, 4
Under table (UT), 205
Uniform excitation of solid state detector, 280
Unpaired valence shell, 61
Uranium (U), 15

**V**

Vacuum deposited, 381, 391
   film, 323
Vacuum evaporation, 253, 310, 323
Vapor phase growth, 292
Vapor pressure controlled CZ (VCZ), 304
Vapor–solid equilibrium condition, 299
Vapor transport method, 323
Vascular, 195
Venn diagram, 188
Verneuil process, 296
Vertical and horizontal Bridgman method, 301, 351
Vertical gradient freeze (VGF), 304
Video camera, 199
Voltage mode, 91
Volume artifacts, 218, 469
Volume efficiency ratio, 322
Voxel to voxel, 155

**W**

Wavelength dispersive X-ray spectrometer (WDS), 103
Wavelength shifter, 257, 264
Weighting field, 150
Weighting potential, 146, 149, 154–155
Wet-gel, 310

White matter, 206
Wide bandgap, 289
Wiener–Khintchine theorem, 173
Wiener spectrum (WS), 167, 240

**X**

XDDI, 390, 399
*Xeroderma pigmentosum*, 66
X-ray(s), 2, 3, 6, 9, 15, 29, 61, 64, 74, 108, 125, 459
   and gamma ray, 142
   and gamma ray detectors, 322
   and gamma ray spectrometer, 387
   based CT, 189
   beam, 217
   camera, 266
   CT, 205, 344
   diffraction, 341
   diffractometer, 318, 324
   emission spectroscopy, 103
   exposure, 29
   fluoroscopy, 195, 197
   generator tube, 219
   imager, 396
   imaging system, 236
   photon, 396
   photo peak, 139
   spectroscopy, 84
   spectrum, 43
X-ray diffraction spectroscopy (XDS), 103
X-ray fluorescence spectroscopy (XRF), 103

**Y**

Yield, 265
Yttrium oxyorthosilicate (YSO), 417

**Z**

Z-factor, 263
Zinc (Zn), 7
   telluride (ZnTe), 387
   tungstate (ZnWO4), 380
Zone-melting technique, 308
Zone refining, 320, 327

Printed by Printforce, the Netherlands